MINERAL RESOURCE ESTIMATION CONFERENCE 2023

24–25 MAY 2023
PERTH, AUSTRALIA

The Australasian Institute of Mining and Metallurgy
Publication Series No 2/2023

AusIMM

Published by:
The Australasian Institute of Mining and Metallurgy
Ground Floor, 204 Lygon Street, Carlton Victoria 3053, Australia

ISBN 978-1-922395-19-1

ORGANISING COMMITTEE

René Sterk
FAusIMM(CP)
Conference Organising Committee Chair

Marat Abzalov
FAusIMM

Kathleen Body

Jeff Boisvert

Julian Ortiz Carbrera

Dhaniel Carvalho
MAusIMM

Isobel Clark

Jacqui Coombes

Scott Dunham
FAusIMM

Xavier Emery

Danny Kentwell

Ilnur Minniakhmetov
MAusIMM

Lynn Olssen

Oscar Rondon
FAusIMM(CP)

Leigh Slomp
FAusIMM

Mo Srivastava

Mike Stewart

Clint Ward
MAusIMM

AUSIMM

Julie Allen
Head of Events

Joelle Glenister
Manager, Events

Kathryn Laslett
Conference Program Manager

Samara Brown
Conference Program Manager

REVIEWERS

We would like to thank the following people for their contribution towards enhancing the quality of the papers included in this volume:

Kathleen Body

Dhaniel Carvalho

Isobel Clark

Jacqui Coombes

Scott Dunham

Xavier Emery

Ilnur Minniakhmetov

Lynn Olssen

Julian Ortiz

Oscar Rondon

Leigh Slomp

René Sterk

Mike Stewart

Clint Ward

FOREWORD

Welcome to the AusIMM's inaugural Mineral Resource Estimation Conference 2023, held in Perth, Australia. I'm incredibly proud of the conference organising committee's ability to pull together such a great event, and I hope it will make a long-lasting impact on you as a delegate.

When I floated the idea of this conference with some of the best practitioners I know who work in this space, I expected most to say they would be too busy. Only a week after sending out the emails, I found pretty much everyone jumping in with unbridled enthusiasm. The passion to bring this together is the reason I enjoy working in this industry so much, and the friendships, support, respect, and shared vision to make our sector a better place is truly inspiring.

Never before has there been one event where geologists, geostatisticians, software solution providers, and other resources professionals have come together to discuss what is undoubtedly one of the most intriguing parts of the mining value chain. Consider the often polar opposite personalities and skillsets required to master highly complicated mathematical concepts and deeply interpretive geological wizardry – for those of you who work in that very narrow space of overlapping Venn circles: you are truly special!

How is it that these different groups of people have never shared a room together to this extent and discuss how to push this discipline forward? I believe there is no other part of the industry where the words 'best practice' will invoke such vigorous debate and passionate reactions. Will we reach consensus on some points, or will we get lost in the woods that separate faulted geological domains from random variable theory? I can't wait to listen to the various paper presentations, and participate in debates to find out what is going on in our discipline, who is trying to solve which challenges, and who has come up with some new solutions for old or new problems.

You'll note that the format of this conference is a bit different to what you may be used to. The conference organising committee want to deliver an engaging and informative delegate experience via four dynamic discussion forums with high degrees of interaction between expert panels and attendees. But we want to balance this out with thought-provoking keynote presentations from esteemed industry leaders and robust technical presentations.

I hope you'll enjoy this format and that the conference is one of learning, meeting friends and colleagues and where we can collectively set the tone for future best practice.

Please find within this conference proceedings 22 papers written by global authors, who by sharing their experiences, learnings and expertise with you will help advance best practice within the mineral resource estimation discipline. I thank them for their time and contribution to this inaugural conference.

Yours faithfully,

René Sterk FAusIMM(CP)

Mineral Resource Estimation Conference Organising Committee Chair

SPONSORS

Major Conference Sponsor

BHP

Platinum Sponsor

rsc
MINING & MINERAL
EXPLORATION

Technology Sponsor

DATAMINE

The Parker Challenge Sponsor

RioTinto

Gold Sponsors

BMRC
Bedrock Mineral Resource Consulting

k2fly

RFD
Rock Flow Dynamics

srk consulting

Silver Sponsor

CSA

Premier Content Sponsor

SEEQUENT

Lunch Sponsor

CUBE
CONSULTING

Name Badge and Lanyard Sponsor

entech.

Welcome Reception Sponsor

VERACIO

Supporting Partners

**AUSTRALIAN
INSTITUTE OF
GEOSCIENTISTS**
Supporting Geoscientists

**BUSINESS
EVENTS
PERTH**

CONTENTS

Classification

Estimation methods

Failures and disasters

Geological modelling and estimation software – development, opportunities and limitations

Open to anything

Reconciliation of resource estimates

AI and machine learning

Introducing deep learning and interpreting the patterns – a mineral deposit perspective

D M First[1], I Sucholutsky[2], D Mogilny[3] and F Yusufali[4]

1. MAusIMM, Chief Geologist, StratumAI, Brisbane Qld 4075. Email: david@stratum.ai
2. VP Research, StratumAI, Richmond Hill ON L4C 0Y3, Canada. Email: ilia@stratum.ai
3. Co-founder, StratumAI, Richmond Hill ON L4C 0Y3, Canada. Email: daniel@stratum.ai
4. Co-founder, StratumAI, Richmond Hill ON L4C 0Y3, Canada. Email: farzi@stratum.ai

ABSTRACT

Machine learning is creating value in all facets of the mining industry, from exploration to production. The authors provide an accessible, high-level introduction to artificial intelligence (AI), machine learning (ML), and deep learning (DL); the latter being widely recognised as one of the most powerful forms of ML. In particular, the authors will introduce deep learning models known as convolutional neural networks (CNNs), how they are applied, and the economic considerations necessary for determining when DL may be the right solution to de-risking complex block modelling problems.

The authors present preliminary results from the mineral resource modelling study of the Jundee orogenic gold deposit, Yandal belt/gold province, Western Australia. The primary goal was to identify the direction and location of the narrow gold veins more accurately and demonstrate how non-linearly correlated elements are used as direct inputs to the resource model to assist with the target element's grade prediction. This demonstrates that: (1) existing techniques for finding correlations between assayed elements do not adequately reflect the complex geology of the asset, (2) non-linear correlations that are difficult to model as simple mathematical functions are representative of geological patterns in a deposit, and (3) non-linearly correlated assayed data, fed as inputs, increase the performance of the resource model as reconciled through blind tests.

To conclude, the authors hypothesise that the patterns represented by the DL block models may be revealing the results of overprinting geological processes that generated mineral deposits. For example, the primary hydrothermal processes that deposited metals created depositional patterns that become particularly complex as a result of being overprinted, in part or in whole, by secondary physicochemical processes. This may explain why non-linear geochemical relationships have the capacity to generate more accurate block models of mineral deposits.

INTRODUCTION

The Jundee gold mine is an orogenic gold deposit located at the northern end of the Yandal belt/gold province (Northern Goldfields region) of Western Australia, about 45 km north-east of Wiluna and approximately 520 km north of Kalgoorlie (Figure 1). Production commenced in the mid-1990s from several open pits and from the late 1990s from underground mines. Currently, ore is mined solely from the underground operations, primarily by uphole longhole open stoping sequencing. Ore processing is by conventional CIL, incorporating a gravity circuit after milling, with an average 92 per cent gold recovery. Jundee was acquired by Northern Star in 2014 (Smith, Haese and Grigson, 2017).

The deposit is an Archean orogenic lode gold deposit hosted within the northern Yandal greenstone belt, Yilgarn Craton (Vearncombe and Elias, 2017; Wyche and Wyche, 2017). The host rocks are dominated by a succession of diabase sills and tholeiitic basalt units. The mine sequence is composed of two basalt units separated by a sediment formation intruded by the sills. Locally the entire mine sequence is intruded by dacite, granodioritic porphyries and lamprophyres (Phillips, Vearncombe and Eshuys, 1998; Phillips, Vearncombe and Murphy, 1998; Smith, Haese and Grigson, 2017).

Gold is deposited and controlled by brittle-ductile shears within an array of transverse or oblique faults, primarily developed within the diabase and tholeiitic and to a lesser extent the felsic porphyries. The major oblique structures typically contain fault gouge with variable displacement up to 120 m with the granodioritic porphyries preferentially exploiting these faults. The faults and shear zones locally have variable orientation and dip (vertical to 20° in either direction) as conjugate subparallel shears, resulting in mineralised zones hundreds of metres wide.

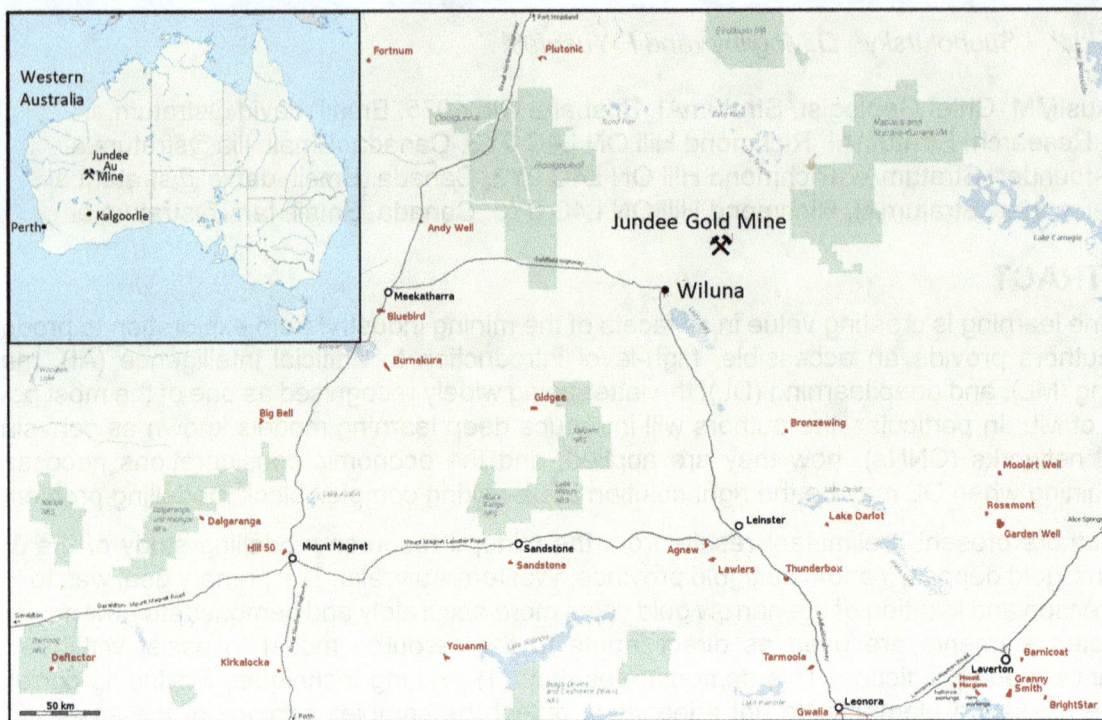

FIG 1 – Location map.

The mineralisation is hosted in numerous discrete high-grade tabular, narrow, and discontinuous fault-shear zones. High-grade mineralised zones typically range from less than 0.5 to 1.5 m and locally can be up to 5 m wide, with a vertical and lateral extent up to 500 m and 1000 m respectively, although within these zones, individual mineralised structures are highly discontinuous.

The host alteration is characterised by quartz–sericite–ankerite–sulfide (pyrite ± arsenopyrite ± chalcopyrite) assemblage. Gold mineralisation has a heterogenous distribution pattern, invariably restricted to the mineralised structure, with the local country rock barren. Course visible gold is a common occurrence, although the relative distribution of visible gold between areas is quite variable, resulting in a pronounced nuggety grade distribution pattern with grades ranging up to several thousand g/t Au.

At the Jundee mine the mineral resource team are in part tasked with identifying and predicting the direction of the relatively narrow high-grade veins for mine planning ore control, ore reserve estimation, which has proved challenging. It has been particularly challenging for the team to rely on the resource model to guide underground drilling or for mine planning in areas where underground drill spacing >20 m.

MACHINE LEARNING – BACKGROUND AND DATA SETS

As part of the review of potential methods to enhance the accuracy of the resource model at the Jundee mine, machine learning (ML) block modelling is being evaluated. The authors demonstrate that ML can produce more accurate models, with the result that they better target mineralisation than current drilling practices.

Recently, ML has emerged as a powerful tool for revealing complex patterns in data. At its core, ML algorithms learn from historical data to better forecast a future pattern or trend. The ML algorithms are written in Python, a programming language that distinguishes itself from other programming languages by its flexibility, simplicity, and large number of available open-source tools required to create modern software. Python helps the software engineers focus on solving logical problems rather than spending time on the basics of the programming language. This is one of the primary reasons that Python is the language of choice for machine learning and data science in general.

Pytorch is the ML library that houses the open-source tools and is used to construct neural network layers. These neural network layers are paired with CUDA (Compute Unified Device Architecture), a computing platform developed by NVIDIA to interact with the Graphics Processing Units (GPUs). NVIDIA is a technology company that designs and manufactures GPUs.

Deep learning (DL) is the term used for one of the most powerful ML algorithms, that uses multiple neural network layers of artificial neurons that composite into deep neural networks. The concept is loosely modelled on the way neuroscientists believe the brain functions, to identify patterns in very large data sets. DL has seen much success in the field of image recognition (eg medical imaging) as well as machine translation. This is the process of using artificial intelligence (AI) to automatically translate text from one language to another without human involvement, since 2011 (Goodfellow, Bengio and Courville, 2016).

For DL to be effective, large volumes of data together with high performance GPUs are required. GPUs of sufficient speed only became commercially accessible at scale over the past 6–8 years, in large part due to the development of the video game industry.

In general, to identify the best ML model parameters for an ore deposit employing DL, it requires training between 30 to 150 models. At Jundee each ML model was trained utilising 2 × A6000 GPUs for 90–150 hours and encompassed 150 iterations (epochs) of the entire data sets. As part of the data preprocessing stage, the models utilise the support of 96 vCPU (virtual Central Processing Unit) cores with 128 GB of RAM. After each model is trained, statistical inference must be performed on the output; ie predict the grade of every block in the block model of the deposit. Each model typically takes approximately 2.5 hours to predict the grade of an estimated 1 M blocks in a block model.

DL has seen limited use in mineral exploration, as most DL algorithms are designed to perform best when used with large amounts of highly dense data (eg imagery where pixel values are known), whereas geological and geochemical data is invariably sparse; eg large volumes of rock devoid of data between drill holes, rock-chip samples, etc. Only in the mining environment does the density of data from exploration drill holes, grade control and blastholes, provide sufficient data to take advantage of DL capabilities.

The primary goal of the ML study was to identify the direction and location of the narrow gold veins more accurately. This has proved to be a particularly challenging task, as the deposit's veins tend to have a width of 0.5–1 m, at grades >10 g/t Au.

The existing practice for the last ten years of production has been:

- Very tight spacing drilling in areas with potential to be added ore to the mine plan. There are over 500 000 m of underground drilling and 1 200 000 m of surface drilling.

- Create a kriging model based on highly constrained domains. This results in defining economic blocks (>2.2 g/t Au) based on whether they lie within the respective mineralisation domains, rather than producing an interpolation model based on the whole data set.

We evaluated the accuracy of the model through block level metrics of precision and recall:

- Precision is the percentage of blocks predicted as economic high-grade (HG) that are reconciled as HG in rock-chip data. It tracks the frequency of false occurrences; that is when a HG block or vein projected in the mine plan reconciles as waste.

- Recall is the percentage of reconciled HG that is predicted as HG. It tracks the frequency of false negative occurrences; that is veins that exist, but were missed by the resource model.

With any reconciliation conducted, the actual negative rate is likely higher than estimated because areas of missed mineralisation typically have limited production data; ie if ore is not predicted it is unlikely to be mined. Many true negatives that exist in the reconciliation data set are accidental discoveries due to infrastructural mining.

The kriging resource model has been used for mine operations, continuously being updated for about the last ten years. There are two data sets that were used to evaluate the efficacy of the January 2021 kriging resource model. The first being the underground drilling data set for 2021, and the second, the rock-chip data set for 2021. Both data sets are reasonably accurate proxies for ground truth; ie they determine whether a block is economic or not. The caveat being small block sizes are used, and rock-chipping is undertaken on the veins. It was determined that the assays are not representative of the average block, if the size of the blocks exceed 0.5 m × 0.5 m × 0.5 m.

The evaluation data sets use simplified ore control models created on 0.3 m blocks based on the following methodology: If there is only one sample inside the 0.3 m × 0.3 m × 0.3 m block, then the grade of the block is the grade of the sample. If there is more than one sample, then the grade of the block is the average grade of the sample assays. Each sample is capped at 30 g/t Au prior to the averaging process. Underground drilling is undertaken if drilling from surface has a drill spacing >20 m, while rock-chipping is undertaken where the drill spacing is <20 m and on average about 10 m.

The kriging model achieves low single digit precision and recall performance on the underground drill hole data set. For the rock-chip data set a performance of 26.5 per cent precision (~3/4 of blocks predicted economic grade were not) and 14.3 per cent recall (~6/7 of blocks reconciled as economic were not anticipated to be economic) was noted. The rock-chip block data set for 2021 (n=~10 000 blocks) was used for the remainder of the study.

These results indicate that the kriging model cannot be relied upon to guide underground drilling nor can it be reliably used for mine planning, particularly in areas where drill spacing >20 m. The rock-chip results are shown not to be too accurate; however, this is attributed to the challenge of modelling narrow veins and shears due in large part to biased sampling practises, a frequent issue experienced at many mines.

METHOD

Using the 2021 kriging model as a baseline, the authors design a ML alternative to the resource model that aims to address some of the challenges faced by the kriging model. The base structure of the ML models consists of convolutional neural networks (CNNs) (O'Shea and Nash, 2015) designed to estimate each block more accurately on a 0.3 × 0.3 × 0.3 m block size. The structure of a CNN can be seen in Figure 2. After training several CNNs, the authors composite the most promising models to create a more accurate 'ensemble' model (Dietterich, 2000).

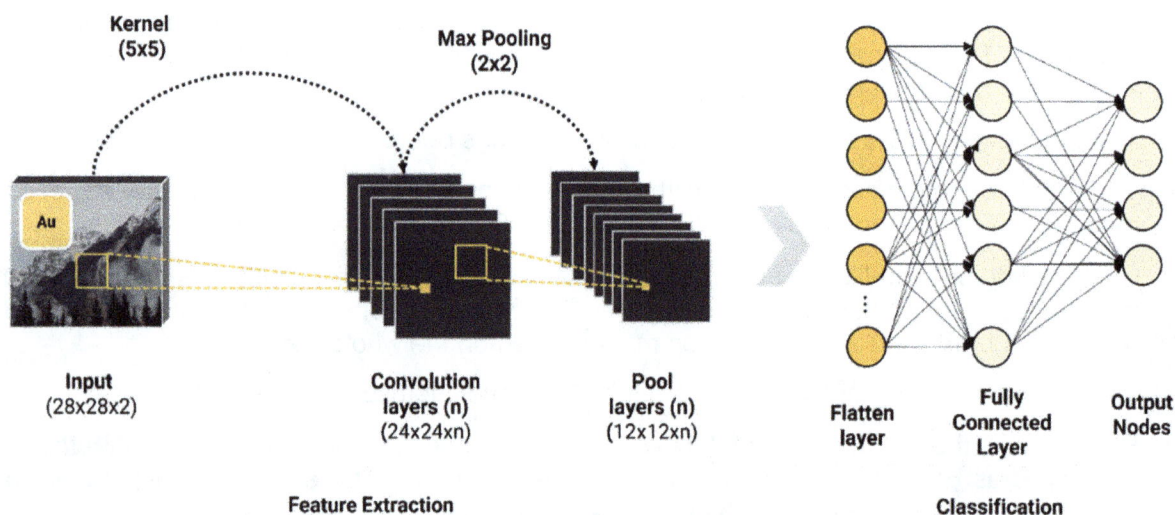

FIG 2 – Convolutional Neural Network (CNN).

This section outlines the general process of training and compositing ML models for resource estimation, as well as specifying how the authors trained and ensembled CNN models for the Jundee deposit. The objective and challenge for ML is to generate a final model that performs equally well on both the previously unseen (new) data and the original training data set, thereby enabling the model to be used to make predictions.

Training machine learning models

The purpose of ML models is to learn to map between input and output data. ML practitioners typically refer to the output data used during training as 'labels'. For example, when training a model to recognise objects in images, an image of a cat would be assigned the label 'cat' and the model would be trained to map from the image pixels to that label.

For resource estimation with ML data is included into the model as input data and as ground truth data. Input data is used as an input into the neural network with its use dependent on the CNN algorithm. The inputs can generally include almost any existing quantitative data, while the outputs ('labels') should be the grades of individual blocks. When using ML models to predict the grade of a particular block, the models are provided all available assay data within a certain range of that block as inputs.

Whereas ground truth data is used to correct the model's estimations for a specified block during the training process. For the Jundee gold model only gold assays were used as ground truth while any type of multivariate data can theoretically be used as model inputs. However, in this case study only geochemical assays were used and the rock-chip data set was restricted to input data for the model and excluded from the ground truth.

In order to train the model to learn to map the input assay data to the output block grades, the user must provide estimated block grades based on drill hole assays as the desired outputs for the training examples. That is the user must decide what the known assay values are for a selection of blocks. The larger the number of labelled blocks the more reliable the resulting model will be after training.

During the process of training ML models, the input data is typically split into a training set and a test set. The authors used approximately 80 per cent of the data for training and the remaining 20 per cent for testing. The model learns by predicting the grades of the output blocks in the training set based on the input blocks in the training set and then updates its estimation protocol to correct errors based on the discrepancies between its predictions and the actual values. Periodically, the model is evaluated on the hold-out test set to measure how well it performs on blocks it has not previously seen. To ensure that all samples can be included as part of both the training and test data sets, the 80:20 split is reshuffled periodically throughout the training process.

The authors find that, in general an orebody or deposit requires greater than 75 000 data points (assays) in the database to generate a productive ML model. With 25–75 000 data points it is debateable whether it is cost-effective to generate ML models. With less than 25 000 data points it is invariably more efficient/cost-effective to utilise traditional geostatistical tools, such as kriging.

In summary the critical data related parameters to iterate when training the most productive ML models for resource estimation are; the input range, the input data, the definition of ground truth (to use for output data), the dropout rate and the domain transform. More details on these and related parameters are provided in the following sections.

Model generalisation and overfitting

The ability to perform well on previously unseen (new) inputs is called generalisation. To improve generalisation, particularly with resilience to noisy or error prone data, another 35 to 70 per cent of input data is randomly removed within the range of each block so that the model learns to map even with noisy data. This popular data removal technique is known as 'dropout regularisation' and helps reduce overfitting and generalisation error (Srivastava et al, 2014; Goodfellow, Bengio and Courville, 2016).

Learning and generalising to each new case study can be a challenge. There are essentially three model outcomes:

1. Underfit Models: A model that fails to sufficiently learn the problem and performs poorly on a training data set and does not perform well on a test data set and generalises poorly.

2. Overfit Models: A model that learns the training data set too well, performing well on the training data set but does not perform well on the test data set and generalises poorly.

3. Good Fit Model: A model that suitably learns the training data set and generalises well to the test data set.

A model's fit can be considered in the context of the bias-variance compromise. An underfit model has high bias and low variance, regardless of the data in the training data set. Basically, it cannot learn to solve the problem.

An overfit model has low bias and high variance. The model learns the training data too well, trying to memorise not only the meaningful patterns but also the random noise. As a result, performance varies widely with new (unseen) data.

Underfitting can be addressed by increasing the capacity of the model. Capacity refers to the ability of a model to fit a variety of target functions; more capacity, means that a model can fit more types of functions for mapping inputs to outputs. Increasing the capacity of a model is achieved by changing the structure of the model, often by adding more layers and/or more nodes to layers. However, models with high capacity can overfit by memorising properties of the training data set that do not transfer well on the test data set (Goodfellow, Bengio and Courville, 2016).

The underlying objective of ML is to use historical data to approximate some unknown natural function that maps from the input to the output. For resource estimation, the relationship of a block's grade to the grades of neighbouring blocks depends on the geological processes that formed the deposit. As a result, a model is required that has sufficient capacity to approximate the complex patterns left by these processes. However, due to the limited amount of available historical data, an overly complex model may be able to simply memorise the training data instead of learning the underlying patterns. Therefore, evaluating the model on a hold-out test set is important for determining that it is not overfitting to the training data and that the patterns it learns generalise to previously unseen data (Reed and Marks, 1999).

An overfit model can be easily identified by monitoring the performance of the models during training by comparing the results from the training data set with the test data set. When a model's performance on the test data stop increasing (or even starts decreasing) while performance on the training data set continues increasing, this indicates that the model is no longer learning relevant patterns, but rather memorising noise in the training data.

Range

Like kriging models, ML models require a range parameter to identify the maximal range of data the model has access to when making predictions for a particular block. This parameter is typically set to 100–200 m with the best performance at the Jundee mine being 100 m. The reason why the ML parameter is often larger than the kriging range, is that ML is designed to capture non-linear patterns that can improve model estimation. These DL patterns can often extend over 200 m away from an orebody, though the reliability of predictions degrades as data density decreases.

Input data

ML models can access different data types. Priority is assigned to each based on error; for example, rock-chip samples are prone to bias due to sampling methodology, and QA/QC variability between mine site and certified independent off-site laboratories. From the perspective of the model, not all gold assays are equal, and the model handles each assay type separately. At Jundee, the diamond drill holes are assayed off-site, the RC drill holes and the rock-chip samples (highly biased due to rock-chipping only the veins and immediate wall rock) at the mine lab.

Understanding the differences is invaluable as it ensures the relatively low data quality from one form of data (ie rock-chip Au assays) does not significantly contaminate the interpretation of data when all assays are included in the model. For example, the model can determine that drill holes assayed at the off-site independent certified laboratory should be weighed more highly than RC chips assayed at on-site mine laboratory, and therefore being more representative of the ground truth.

The CNN can determine how representative the rock-chip assays are to ground truth. This is invaluable when handling lower quality, noisy and/or bias rock-chip data sets. The data sets should not be ignored as they still contain valuable special and assay information, but inherently there is too much uncertainty to be representative of a block's gold grade.

Normally, CNNs require dense, grid like input data (eg the pixels of an image). In order to make typically sparse mining data compatible with CNNs, the authors rasterise the input data into a grid of the same size as the desired block size.

Multivariate data and non-linear correlations

CNNs can leverage multivariate data sets (eg geochemical assays) by revealing the DL patterns generated by non-linear correlations, thereby corroborating and enhancing the primary models, particularly if there is a heterogenous (nuggety) grade distribution. This is unlike classical geostatistical interpolation techniques such as kriging, which are challenged when attempting to make use of multivariate data.

Only statistically significant non-linearly correlated elements are used as direct inputs into the resource model to assist with grade prediction for individual ore blocks, in order to reduce the probability of overfitting. However, it is recognised that the input elements are limited by their availability in the grade control data sets, as the mine site laboratories assay a restricted suite of elements (due to cost and analytical equipment) when compared to off-site certified independent laboratories.

Definition of ground truth

As discussed, to train an ML model, users need to provide labelled or known examples of 'true' grades for several blocks. However, it is important to consider the quality of data when defining the ground truth. For instance, some types of data, such as rock-chips, may provide valuable spatial and semiquantitative information, that is the assay may be geochemically reliable but does not represent the grade of a block. For example, a block with a chip sample of 20 g/t does not equate to the block having a grade of 20 g/t. It simply means there is a rock-chip sample of 20 g/t within the block.

The ground truth definition is crucial, as it determines which types of data should be used for error correction during the model's training process. By carefully selecting the ground truth data, the user can ensure that the ML model is learning from reliable sources and not incorporating potentially misleading information from less accurate data types.

Domain transform

For Jundee and many other gold mines, HG (above cut-off) samples make up less than 2–3 per cent of all gold assays. Modelling this type of gold grade curve is particularly difficult, as ML models will often aim to achieve 98 per cent accuracy by always predicting no economic gold! The risk is that if the model gets severely penalised for predicting an almost 0 g/t Au sample as 10 g/t Au, then it may exploit the grade curve and instead always predict below the cut-off grade. The easiest workaround is to convert the input and output data onto the log domain. The resultant domain would have the same distance between 0.1–1 g/t Au and 1–10 g/t Au allowing the model to be confident in predicting HG, while also learning that a 2.2 g/t Au sample (barely economic) is more similar (at least from an economic perspective) to 10 g/t Au than to 0 g/t Au.

Dropout regularisation

In ML, dropout regularisation refers to the practice of removing a percentage of the training data set at random during training. This regularisation approach helps prevent overfitting to the training data set by ensuring that blocks are not co-dependent on each other in a way that would otherwise allow the model to memorise noise in the data instead of learning real patterns (Srivastava et al, 2014).

Data is removed during the training stage with a frequency between 35 per cent to 70 per cent depending on the model parameters. The purpose of this is to improve generalisation. Conceptually, removing data in the training phase is a way to force the model to predict grade by recreating geological trends (patterns), rather than simply memorising the deposit. By removing data at random, prevents the deposit being memorised as the data being present to the model always looks a little different. If memorisation is not possible, the model is effectively forced to predict grade by learning to extrapolate geological patterns.

In a simplified sense, data removal stress tests the model so it learns to predict geological patterns; when large amounts of the data are randomly removed Thereby enabling the model to predict with greater accuracy the grade of each block during unstressed conditions; ie when the entire data sets are used to generate the block model.

Compositing machine learning models

After many iterations with the parameters listed above, a list of promising CNN models was produced and described below. For clarity, the notation DRC~DR is a simple way to notate the input data and the output data (estimated block grade based on drill hole assays). In this case, DRC refers to diamond drill holes (D), RC drill holes (R) and rock-chips (C) input data, while the definition of truth is based on diamond drill holes (D) and RC drill holes (R). Below is a diagram that summarises how data is used by the CNN. Where not explicitly stated, range is assumed to be 100 m and the data removal or dropout rate is 35 per cent. Figure 3 is a selection of models that achieved high block level accuracy.

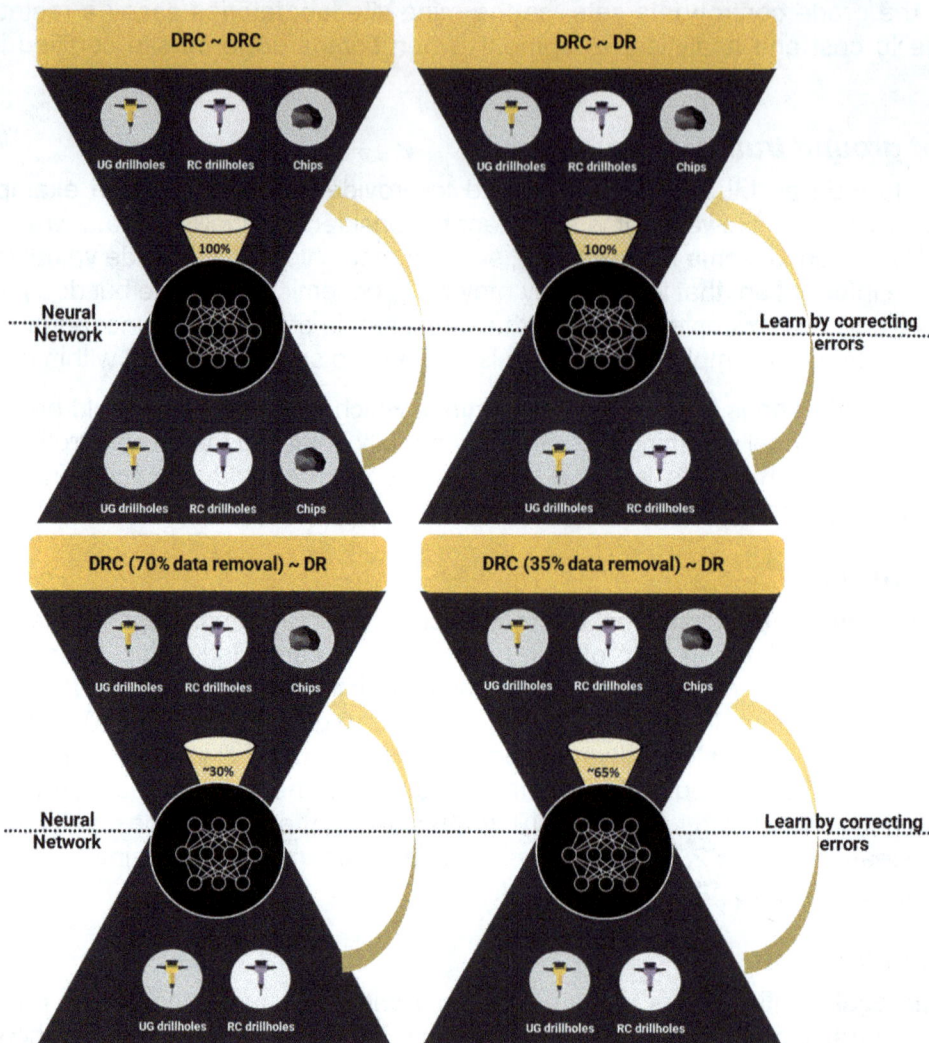

FIG 3 – Schematic diagrams for input ~ output models.

Ensembling

Once a set of promising ML models are produced, ensembling theory is applied. Ensemble learning is the process by which multiple models that are strategically generated are combined to solve a particular computational intelligence problem (Dietterich, 2000). Ensembling works because it averages out the errors of many models to make a more accurate 'ensemble' or composite model. In the case of mineral resource models, ensembling combines base models created with different data sets (eg different data types, variable dropout rates, and subsets of elements). If the models generate the same or similar results, there is higher confidence that the respective models are accurate. The kriging models can also be incorporated in the ensemble as they invariably provide higher quality predictions in low sample density areas.

The authors undertook a detailed evaluation of each base model developed in the previous section. If inclusion of a model in the ensemble improves the estimation of a particular metric during reconciliation and does not degrade or adversely affect performance, then that model is kept in. Models that do not improve the performance of the ensemble are systematically removed.

ML – kriging hybrid models

Interestingly, the use of ML or kriging for resource modelling is not a binary choice, but rather lies on a continuum. Kriging models can be included as part of the ensembling process and such hybrid ensembles are often more accurate than either individual kriging or ML models. The ensemble leverages the human pattern recognition of geologist created domains and the ability of ML to enhance and detect patterns that humans are incapable of identifying. For some studies (not included this study), it is possible to adjust the weights of models for the final ensemble based on each model's individual reconciliation performance or drill spacings. Thereby ensuring that the ensemble can optimally balance the respective strengths or weaknesses of geologist created domains and ML created estimations.

Representative examples of ML models generated by the CNNs are presented in Figure 4. The authors have used the results from an anonymised magmatic copper deposit, as the Jundee study results are confidential. Although it may appear unusual to use a magmatic copper deposit to compare model types, the heterogeneity of copper grade distribution is comparable to the nuggety gold grade distribution at Jundee. As a result, the models generated are not that dissimilar.

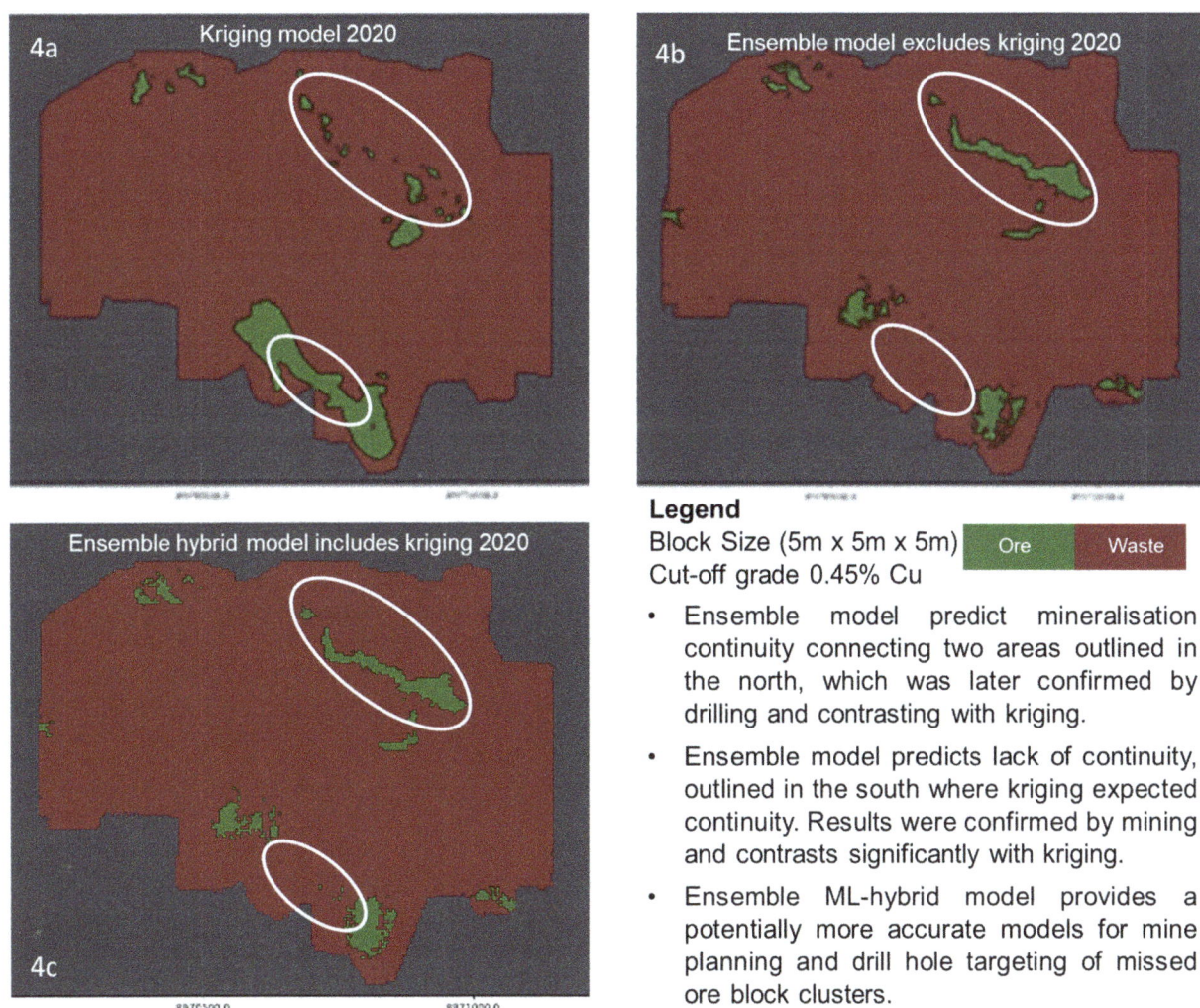

Legend
Block Size (5m x 5m x 5m) | Ore | Waste |
Cut-off grade 0.45% Cu

- Ensemble model predict mineralisation continuity connecting two areas outlined in the north, which was later confirmed by drilling and contrasting with kriging.

- Ensemble model predicts lack of continuity, outlined in the south where kriging expected continuity. Results were confirmed by mining and contrasts significantly with kriging.

- Ensemble ML-hybrid model provides a potentially more accurate models for mine planning and drill hole targeting of missed ore block clusters.

FIG 4 – Anonymised magmatic copper deposit demonstrates different model result: (a) Kriging model; (b) Ensemble model excluding kriging; (c) Ensemble hybrid model including kriging model.

RESULTS

The precision and recall results of individual models reveal the following trends. The exclusion of the rock-chip data set results in better recall models (ie reduced false negative rate). This demonstrates that while rock-chips can be useful for training, they are not as useful as arbitrators of ground truth.

A slight modification of the treatment of HG ore as inputs also yields higher recall where the model identifies the heterogeneity ('nuggetiness') within each block, thereby reconciling 65 per cent more HG ore than the kriging baseline model. However, this is offset by a decrease in precision (ie more false positives) as the model has 9 per cent lower chance of HG ore being predicted. Table 1 summarises the results of individual models described in Figure 3.

TABLE 1

Results from individual models; See Figure 3 for schematic input ~ output models.

Model name	Precision (%)	Recall (%)
Kriging	26.5	14.3
DRC ~ DRC	22.3	14.8
DRC ~ DR	24.3	23.7
DRC (70%dr) ~ DR	24.2	15.5
DRC (35%dr) ~ DR	24.2	7.9

D: diamond drill hole; R: RC drill hole; C: rock-chip samples; dr:% of data removal.

Therefore, optimising for one metric (eg precision) can cause the other metric (eg recall) to suffer; there is a negative correlation between the two metrics for this deposit, and as a result, a balance must be achieved in the model while optimising operations; eg reserve drilling, mine planning etc.

This results in the development of two ensemble configurations. One maximises precision (ie maximise HG prediction) and the other maximise recall (ie minimise missed mineralisation). The precision optimised ensemble resource model provides a high confidence estimate, of blocks that can be added to the mine plan. The recall optimised ensemble resource model serves as a guide for where additional resource drilling can be focused to de-risk the mine plan and increase *in situ* value.

As described, the ensemble resource models are created by compositing all the tested resource models together and removing those that do not improve the overall precision (in precision optimised model) or the overall recall (in recall optimised model) from the ensemble. The top precision optimised ensemble resource model yields a precision of 34.9 per cent, recall of 14.3 per cent while the recall optimised resource model yields 26.5 per cent precision, 26.2 per cent recall (Table 2).

TABLE 2

Results from kriging and ensemble models.

Model name	Precision (%)	Recall (%)
Kriging	26.5	14.3
Ensemble Precision Optimised	34.9	14.5
Ensemble Recall Optimised	26.5	26.2

In the precision optimised ensemble model, precision improves from kriging's 26.5 per cent to 34.9 per cent which is a 32.8 per cent relative precision improvement (32.8 per cent higher chance that block predicted as HG is HG) over kriging without sacrificing any recall (ie without missing more mineralisation). This is a substantial improvement that comes without having to trade-off missing more mineralisation. This can be used for substantially more accurate mine planning as every block to be HG will now have a 32.8 per cent higher chance of being HG when mined.

In the recall optimised ensemble model, recall improves from kriging's 14.3 per cent to 26.2 per cent, a relative 83 per cent improvement (83 per cent more blocks of mineralisation identified) without increasing false positive rate (ie without increasing the number of false HG predictions). This means 83 per cent more mineralisation (in MT) is identified as economic in reconciliation was originally predicted as economic.

The best precision optimised ensemble consists of six different models (five ML, one kriging). Each ML model has different features, yet most of them include rock-chip data as part of the output/ground truth. This is interesting as models without rock-chip data tend to be more accurate individually. Whereas ensembling has a 3 per cent relative precision improvement when the kriging model is included. Indicating that the best block model does not rely exclusively on DL patterns, but can take advantage of geologist drawn domains within a kriging model.

The best recall optimised model consists of five different models (four ML, one Kriging). The ML models with very high dropout rates tended to reduce the overall ensemble performance as those models did not have access to sufficient data to extract some of the more intricate patterns of the system that would have helped with identifying all possible HG gold.

DISCUSSION

As discussed, modelling nuggety narrow vein orogenic gold deposit is challenging using the existing geostatistical techniques (eg kriging). The ML modelling of multivariate data sets reveals complex DL patterns that kriging cannot easily generate, as correlations are complex. Likewise non-linear correlations are difficult to model by simple mathematical functions. Non-linearly correlated assays within the input data set provided a method to enhance the performance of the resource model during blind test reconciliation. As a result, kriging models are invariably ineffective at producing accurate geochemical models of complex nuggety ore deposits, such as Jundee.

The DL patterns provide a solution to modelling nuggety orogenic gold deposits more accurately. Although the essence of ML is to generate data patterns it remains an enigma as to what the pattern represent. It is hypothesised that the DL patterns revealed in block models may be revealing the overprinting geological processes that generated the deposit.

We know that hard rock ore deposits are crustal concentrations of useful elements or minerals that can be mined economically for profit. As with all near surface ore deposits, they consist of minerals and occasionally native metals, formed by overarching geological processes. There are four fundamental geological processes for the formation of an ore deposit (Figure 5a) (McQueen, 2005):

1. A source for the ore metals and minerals, both metals and their respective ligands.

2. A transportation mechanism that moves the components of the ore to the deposit site at a concentration to potentially allow an economic concentration or a transportation mechanism that removes non-ore components to allow a residual economic concentration.

3. A depositional mechanism or trap to fix the ore minerals or metals in place, together with their gangue.

4. A preservation process or geological setting that allows the ore deposit to be preserved.

After the initial formation of the mineral deposit, overprinting geological processes invariably occur that can both enrich and/or deplete the ore deposit (Figure 5b). For example, the primary hydrothermal processes that transported and deposited metals at Jundee created a depositional pattern that became more complex as a result of being overprinted, by secondary physicochemical and structural processes.

Although the distribution of the elements and ligands within an ore deposit may appear chaotic their deposition due to geological processes is not random. It is hypothesised that the geochemical DL patterns (ore grades) revealed within the block model reflect the primary and secondary geological and geochemical processes that produced the ore deposit. This might explain why the ML models predict grade more accurately than kriging (geostatistical interpolation) models.

These DL patterns may also result in more accurate modelling of ore types as well as geometallurgical and geotechnical parameter models.

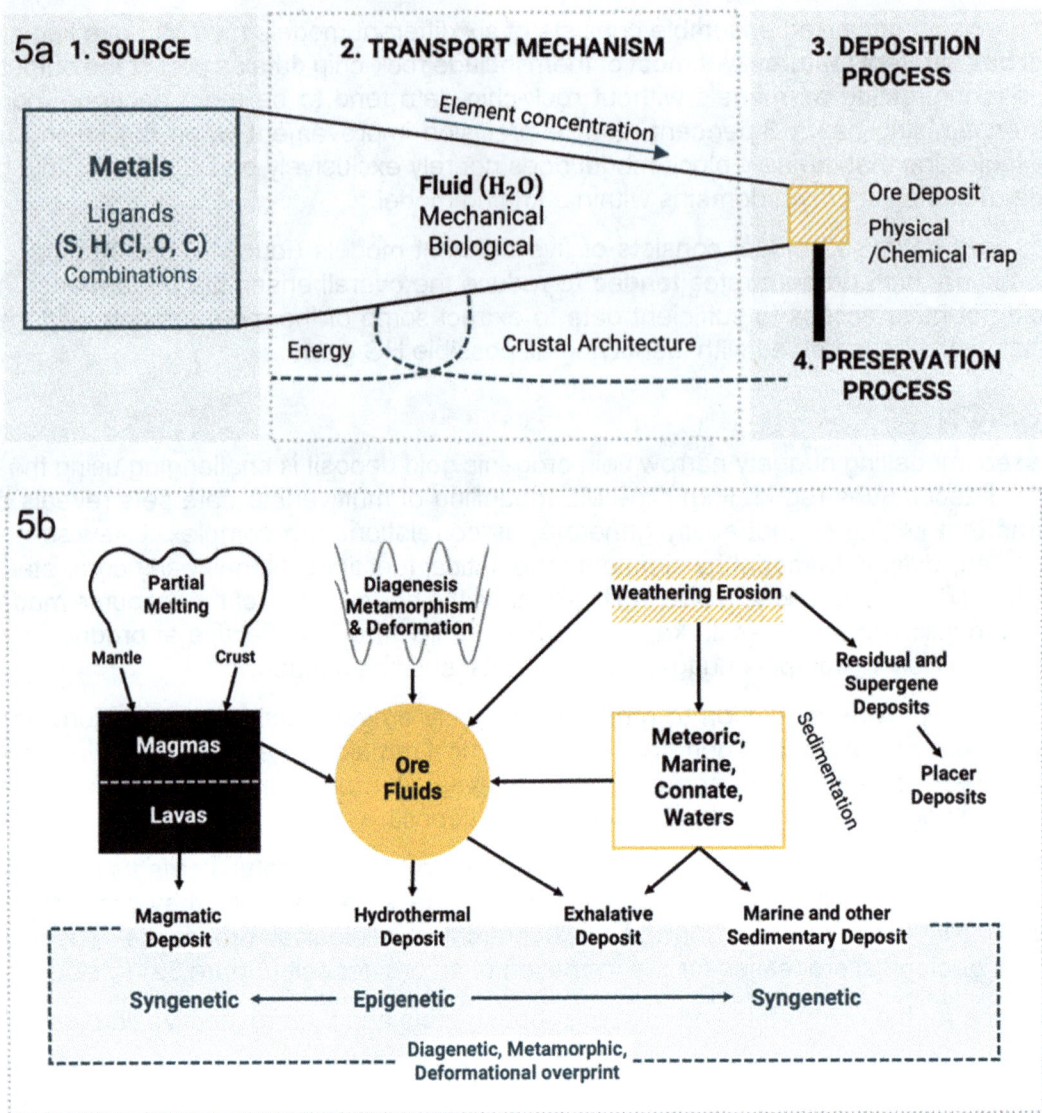

FIG 5 – (a) Geological processes for the formation of ore deposits; (b) simplified genetic classification outlining primary and secondary (overprinting) processes (after McQueen, 2005).

CONCLUSION

The study results demonstrate how ML can potentially improve mine planning and drill hole targeting at the Jundee mine, a complex orogenic narrow vein gold deposit. The ML approach was to apply CNNs that learn from historical production data and do not rely on pre-existing kriging domains. The CNNs revealed patterns in the data that relate directly to the primary, and overprinting secondary, geological processes that can be used to generate more accurate block values.

Rather than relying on a single ML model, the best results are achieved by designing different models and ensembling (averaging) the most promising models. This enables a better characterisation of ore block clusters for mine planning, targeted drilling as well as improved identification of unmined potentially economic ore block clusters that were poorly defined by the original kriging models.

Furthermore, it is demonstrated that it is not a binary choice between ML and kriging, but rather a continuum, as the most accurate models are achieved by averaging a composite of both ML and kriging predictions. The best recall and best precision models include both ML and kriging components. ML delivers the power of machine pattern recognition that can identify latent variables with complex relationships in the data, while kriging (particularly the geologist created domains) integrates geology intuition and structural interpretation in the modelling process particular in areas where data density is suboptimal.

The best precision optimised ML models have a similar recall to kriging, but a significantly improved precision (reduced false positive rate), enabling more accurate mine planning as predicted HG blocks will now have a 32.8 per cent higher chance of being HG when mined. This is a major improvement, as there is no compromise with a loss of ore mineralisation.

The best recall-optimised model has a similar precision as kriging but improves recall (reduced false negative rate) from 14.3 per cent to 26.2 per cent. This equates to an 83 per cent increase in identified ore tonnage, without a corresponding increase the false positive rate when reconciled.

It is demonstrated that it is possible to identify previously unrecognised mineralisation up to 50 m away from existing mine infrastructure. These ore cluster blocks that were previously missed could be incorporated as targets in a guided drilling program as part of routine mineral resource/ore reserve drilling.

ACKNOWLEDGEMENTS

The authors wish to acknowledge the anonymous reviewers who provide invaluable input to enhance the contents and structure of the paper. Northern Star Resources Ltd for permission to access the Jundee database and the interactive contributions by their employees; Heath Anderson, Leon Griesel, Patrick Moore and others, with respect to formulating the baseline and providing constructive feedback on the generated block models. A special thanks to Ms Ady Aguilar for drafting several of the figures.

REFERENCES

Dietterich, T G, 2000. Ensemble methods in machine learning, in *Multiple Classifier Systems: First International Workshop, MCS 2000,* Lecture Notes in Computer Science, 1857:1–15 (Springer: Berlin).

Goodfellow, I, Bengio, Y and Courville, A, 2016. *Deep Learning*, 801 p (The MIT Press: Cambridge).

McQueen, K G, 2005. Ore Deposit Types and their Primary Expressions, in *Regolith Expression of Australian Ore Systems: A Compilation of Exploration Case Histories with Conceptual Dispersion, Process and Exploration Models* (eds: S M Cornelius and I D M Robertson), pp 1–14 (Cooperative Research Centre for Landscape Environments and Mineral Exploration [CRC LEME]: Australia).

O'Shea, K and Nash, R, 2015. An introduction to convolutional neural networks, arXiv preprint *arXiv:1511.08458*. Available from: <https://arxiv.org/pdf/1511.08458.pdf> [Accessed: 14 April 2023].

Phillips, G N, Vearncombe, J R and Eshuys, E, 1998. Yandal greenstone belt, Western Australia: 12 million ounces of gold in the 1990s, *Mineralium Deposita*, 33:310–316.

Phillips, G N, Vearncombe, J R and Murphy, R, 1998. Jundee gold deposit, in *Geology of Australian and Papua New Guinean Mineral Deposits* (eds: D A Berkman and D H MacKennie), pp 97–104 (The Australasian Institute of Mining and Metallurgy: Melbourne).

Reed, R D and Marks, R J, 1999. *Neural Smithing: Supervised Learning in Feedforward Artificial Neural Networks*, 360 p (The MIT Press: Cambridge).

Smith, S, Haese R, and Grigson M W, 2017. Jundee gold deposit, in *Australian Ore Deposits* (ed: G N Phillips), pp 273–278 (The Australasian Institute of Mining and Metallurgy: Melbourne).

Srivastava, N, Hinton, G, Krizhevsky, A, Sutskever, I and Salakhutdinov, R, 2014. Dropout: a simple way to prevent neural networks from overfitting, *The Journal of Machine Learning Research*, 15(1):1929–1958.

Vearncombe, J R and Elias, M, 2017. Yilgarn Craton – mineral deposits and metallogeny, in *Australian Ore Deposits* (ed: G N Phillips), pp 95–106 (The Australasian Institute of Mining and Metallurgy: Melbourne).

Wyche, N L and Wyche, S, 2017. Yilgarn Craton geology, in *Australian Ore Deposits* (ed: G N Phillips), pp 89–94 (The Australasian Institute of Mining and Metallurgy: Melbourne).

ML and AI for resource estimation – what could possibly go wrong? Nothing! Everything!

M J Nimmo[1]

1. Principal Data Scientist, CSA Global (An ERM Group Company), Brisbane Qld 4000.
   Email: matthew.nimmo@casglobal.com

ABSTRACT

Machine Learning (ML), Artificial Intelligence (AI), and Swarm Intelligence (SI) techniques are extremely powerful tools for building predictive and generative models. ML can be used for building highly accurate regression and classification models. But without careful data science and statistical analysis, a regression or classification model could be highly biased and completely wrong – and we may not even realise. An example rock density estimation task will be used to illustrate the potential gains and possible pitfalls in using ML and AI in Mineral Resource estimation. What could possibly go wrong?

Nothing! For most simple regression or classification problems, very little. For density estimation, using the global mean may be all that is possible, but the resource classification would need to reflect the uncertainty in global estimation of tonnage.

Everything! From data collection and measurement, from selecting ML and AI techniques and designing the analysis workflow to our own cognitive bias. The things that could go wrong are vast and may include but not limited to asking the wrong question, not asking questions, data bias, measurement errors, insufficient data coverage, not collecting the right variables, not collecting enough observations, collecting clustered samples, unbalanced data, filtering out data without reason (outliers), clipping data (outliers), relying on automatic feature selection, including irrelevant variables, excluding relevant variables, geological interpretation, changing context, blindly following best practice, splitting small data sets, subsetting large data sets, using the wrong tool for the task, insufficient budget to complete the analysis, not allowing enough time for testing and experimentation (the science), focusing too much on building the model (the engineering), not learning from the data, our assumptions, our preconceptions, and our skill. This paper will explore the question of how training regression models to predict rock density are affected by gaps in the data (missing observations and missing variables). What could possibly go wrong?

INTRODUCTION

Estimation of global Mineral Resources typically focuses on the accurate spatial estimation of grade and the definition of the geological framework (interpretation of lithology, oxidation, alteration, and structure) in three dimensions. The geological interpretation allows estimation of volume. To estimate tonnage, a rock density estimate is required (Tonnage = Volume × Density). The tonnage estimates are then used for estimating contained metal and assessing the economic value of the mineral resource. Accurate prediction and spatial estimation of density is then a critical component of resource estimation along with estimation of grade and a good geological interpretation.

Estimation of rock density into the Mineral Resource block model can be accomplished by three methods. These are:

1. Spatial estimation (if there is sufficient data coverage in three-dimensions).

2. Regression analysis (if there are not enough observations).

3. Global mean value or mean by domain or factor (area, lithology, alteration, oxidation, mineralisation) – this is essentially a simple regression model.

Machine learning and artificial intelligence techniques can be used in regression analysis to build mathematical models that could be used to predict rock density. But they only exploit the correlations between variables, between independent variables and the target dependent variable, in this case rock density. Assuming adequate sampling, machine learning may be able to find excellent mathematical models that perform well not only on the training data (the data that was seen) but on out-of-sample data (the data not seen) if there are strong correlations between the independent and dependent variables. The weaker the correlations the worst the regression models will likely perform. For machine learning to be able to find a good model, the machine learning algorithm will need to identify the right variables to be included in the model, unless the structure of the model was not already defined manually and have enough samples to accurately estimate the coefficients of the model. Over-fitting and under-fitting of the model is a real danger which can result in a mathematical model not able to predict on unseen data or to extrapolate beyond the range of the training data. This will usually occur when there are too few data or too few variables – data gaps.

In ideal conditions where rock density observations cover the full multivariate space, that there are sufficient samples per discrete variable, and key predictor variables exist in the data, and that the data is valid, then potentially nothing could go wrong. The regression may be accurate enough for estimation of tonnage for Measured, Indicated and Inferred Mineral Resource estimation. In some instances, where rock density shows little variation then a global mean value for rock density could be used.

Regression analysis is typically performed using standard linear regression. But how does the model perform? How accurate is it under certain conditions such as lack of data coverage across the multivariate space (as opposed to 3D space) or missing variables? What problems might arise in building a predictive model for rock density that is material to the estimation of tonnage?

Rock density tends to be under-sampled and key factors that affect rock density are not all-ways measured or the key variables such as lithology and oxidation state are poorly logged or contain large number of errors. Rock density is directly controlled by the rock mineralogy and proportions of minerals. Mineralogy in-turn is influenced by lithology, alteration, oxidation state, and mineralisation. All the discrete variables lithology, alteration, oxidation state, and mineralisation directly reflect the geochemistry of the rock. In some instances, the correlations between assays can be purely the result of the factors and if these factors are accounted for then the correlations drop to near zero. These are confounding variables that must be included in any regression or classification analysis and modelling. However, as the rock geochemistry reflects the discrete variables lithology, alteration, oxidation state, and mineralisation and the rock mineralogy, then the rock geochemistry can be used as a proxy for the factors. Factor analysis and latent variable modelling could be used where geology logging is limited or erroneous to help identify an underlying rock factor that could represent the combination of the discrete variables lithology, alteration, oxidation state, and mineralisation. These latent variables could then be used in Machine Learning to help improve the accuracy of the mathematical model of density.

The things that can go wrong are large and range from errors in sampling, measurement, and storage of the rock density data, through to problems in machine learning from how it is applied to what techniques are used. Everything could go wrong, which could lead to poor estimates of rock density and wrong estimates of tonnage. This paper will explore the impact of gaps in the data including missing variables and missing samples, and selection of regression algorithm on the estimation of Mineral Resource tonnage.

METHODOLOGY

Experiments conducted for this paper are intended to illustrate the impact of various likely scenarios on building a predictive model of rock density for Mineral Resource estimation and what could possibly go wrong. The focus of the experiments is to test the effect of data coverage (missing samples) and available features (missing variables) on prediction of rock density.

The quality of the geology data including rock density is not assessed but assumed to have been thoroughly evaluated by the geologist responsible for building the Mineral Resource estimate. Issues with data can significantly and severely impact the resulting regression or classification model. Issues that are not assessed for this paper include distribution of missing values, presence and number of outliers, data imbalance (different frequencies of observations across discrete variable classes – important for classification), data collection and storage methods, method used for measuring rock density, and changes in context. Machine learning methods do not handle changes in context unless specifically designed to do so. Changes in context for estimation of density include the differences between the training data and the resource block model such as differences in how the discrete variables are encoded (typically grouped into domains for resource model and labelled differently) and differences in statistics of numeric variables (range and correlation).

Synthetic data

Synthetic data used for the experiments were randomly generated based on two real-world metalliferous deposits. The process of generating the synthetic data involved constructing a generative model followed by embedding discrete variables such as lithology, oxidation, and domain into a continuous numeric feature space, applying the dimension reduction algorithm for embedding discrete variables and for quantising the data into K disjoint data sets (Figure 1).

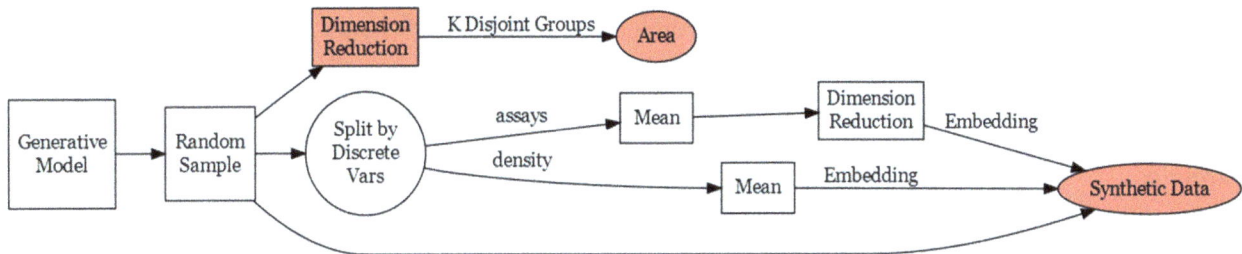

FIG 1 – Synthetic data preparation.

Two generative models were manually constructed using the causal models shown in the directed acyclic graphs in Figure 2. A generative model is a mathematical or statistical model that can generate data with certain statistical properties. A random sample with 10 000 observations were simulated from the generative models. Observations with values outside the expected range of 0–100 per cent were removed. The synthetic data contain no missing values which is a necessary condition for applying most statistical and Machine Learning techniques.

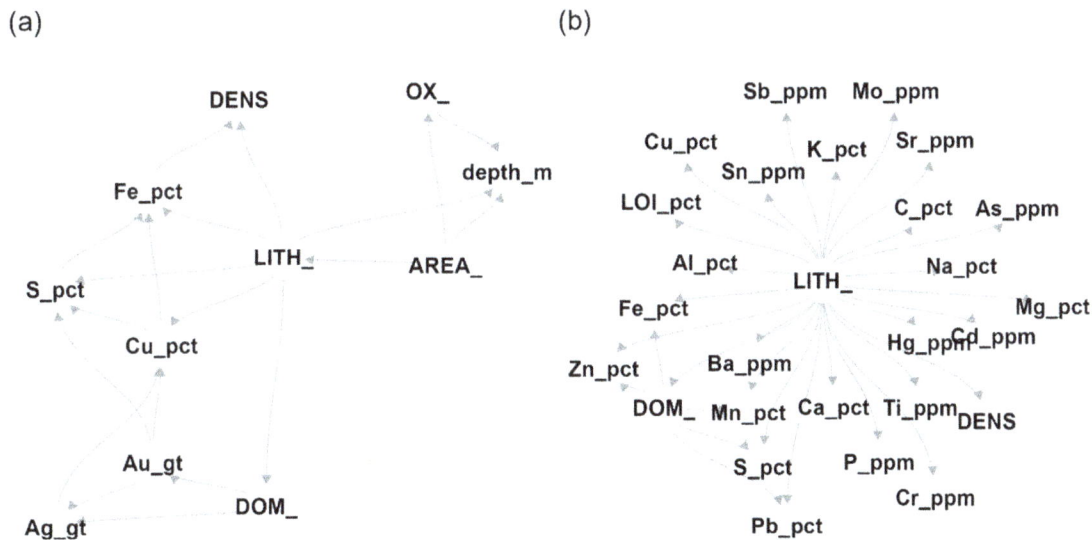

FIG 2 – Directed acyclic graphs showing the relationship between variables in the synthetic data sets: (a) Data set 1, (b) Data set 2.

Dealing with discrete variables

Majority of Artificial Intelligence and Machine Learning techniques can only handle numeric data. There are some machine learning techniques based on decision trees that can handle discrete variables directly. Ordinary least squares regression requires that discrete variables are encoded into numeric variables prior to model fitting, usually using one-hot encoding (binary dummy variables).

A problem that can arise when using discrete variables directly in machine learning is when different data sets contain different classes in a discrete variable. This can occur when splitting the training data into training, test, and validation data sets. It can also occur when there are multiple pits or deposits with different logging schemes or the logging scheme changes after the model has been trained. Or, when the logging codes in the drill hole data are simplified into geological domains, interpreted, and then encoded into a block model as different codes. Or when ore control data is collected with a logging schema that is different to that used for geological logging. Or when metallurgy samples are collected and tested. Discrete variables that describe area or domain are particularly vulnerable to this problem. The problem is that machine learning techniques cannot handle changing context even if the change is only cosmetic as in changes in logging codes but not meaning.

There are methods that can be used to convert discrete variables into numeric variables for use in machine learning. The conversion not only allows for all machine learning techniques to be used but also to alleviate the problems of changing context through changes in encoding of discrete variables by embedding the variable into a common feature space.

One common approach, usually used in linear regression, is one-hot encoding where each level in a discrete variable is encoded as a binary dummy variable. To ensure that the resulting dummy variables do not lead to a dummy variable trap, one level is usually dropped. The dummy variable trap is where two or more dummy variables are highly correlated making it difficult to interpret the regression results. A downside to one-hot encoding is the large increase in the number of terms that are added to a model which leads to a reduced degrees of freedom and potential over-fitting of the model. Other methods involve taking the mean value of the target variable and using that instead of the discrete variable. Other methods may involve deep neural networks to embed the discrete variables into a latent feature space.

To ensure consistency across all regression methods that were tested, two variants of embedding were used. The first embedding method used was the simple mean embedding where the discrete variable is converted into a numeric variable by replacing each class by the mean rock density for that class. The second method used a variant of mean embedding but used the mean values of the assays rather than density. The mean values were then projected into a low-dimensional feature space using dimension reduction method. Three dimensions were used for lithology (LITH_), two for domain (DOM_), and one dimension for the other discrete variables (OX_ and AREA_). The two embedding variables were then used in regression analysis instead of the discrete variables. This approach avoids the problems of having different classes across data subsets and allow for scaling to be easily applied to the data in a consistent manner.

Note that the embedding of the discrete variables into a n-dimensional feature space is dependent on the available numeric variables and how well those variables represent the discrete variable. It may be necessary to use mean embedding using only rock density to represent the discrete variable such as lithology.

Description of synthetic data

Data set 1 contains 11 variables and 7273 observations with no missing values. The variables include DENS, Au_gt, Ag_gt, Cu_pct, Fe_pct, S_pct, depth_m, AREA_, OX_, DOM_, and LITH_. The variables AREA_ (four classes), OX_ (three classes), DOM_ (14 classes), and LITH_ (22 classes) are discrete variables while the remaining variables are continuous numeric variables. Rock density measurements are recorded in the variable DENS as continuous numeric values ranging from 1.62 to 6.24 with a mean of 3 (n = 7273; sd = 0.47).

Data set 2 contains 27 variables and 10 000 observations with no missing values. The variables include DENS, Zn_pct, Pb_pct, Cu_pct, Al_pct, As_ppm, Ba_ppm, C_pct, Ca_pct, Cd_ppm, Cr_ppm, Fe_pct, Hg_ppm, K_pct, Mg_pct, Mn_pct, Mo_ppm, Na_pct, P_ppm, S_pct, Sb_ppm, Sn_ppm, Sr_ppm, Ti_ppm, LOI_pct, LITH_, and DOM_. The DOM_ (ten classes), and LITH_ (33 classes) are discrete variables while the remaining variables are continuous numeric variables. Rock density measurements are recorded in the variable DENS as continuous numeric values ranging from 2.04 to 4.16 with a mean of 2.92 (n = 10 000; sd = 0).

In both data sets, rock density (DENS) is directly affected by lithology (LITH_). For Data set 1 rock density is also directly affected by iron (Fe_pct) and indirectly affected by sulfur (S_pct) and copper (Cu_pct) For Data set 2, any correlations observed between assays and rock density are the result of lithology (LITH_) and minneralisation domain (DOM_). Figure 3 shows boxplots of rock density by lithology for both data sets and Figure 4 shows a series of scatterplots of rock density versus selected assays. The scatterplots for both data sets show strong positive correlation between the selected assay variables and rock density.

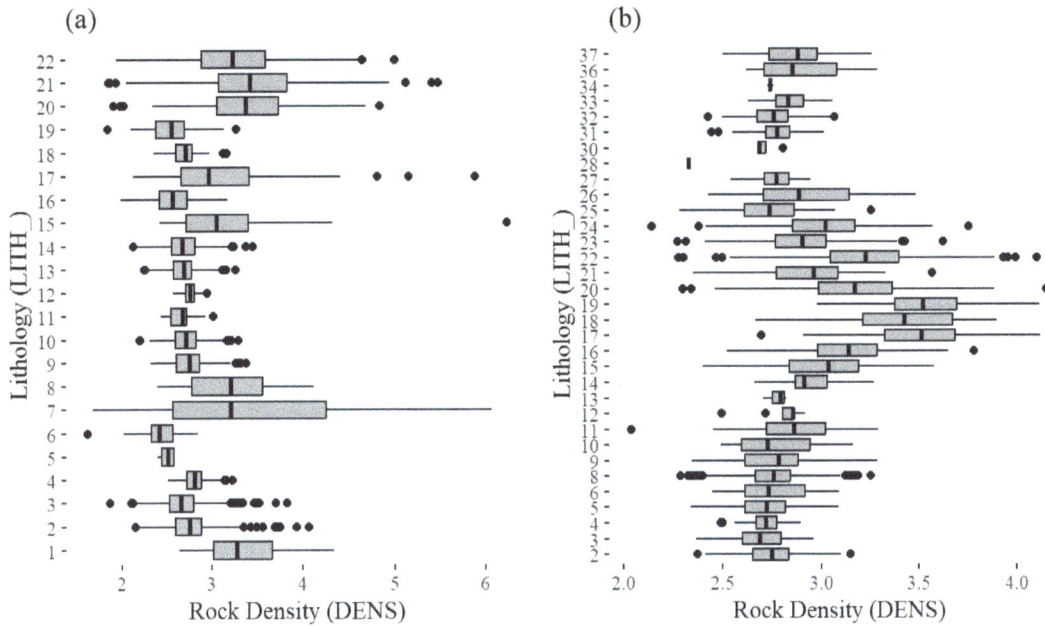

FIG 3 – Boxplots of rock density by lithology (LITH_): (a) Data set 1, (b) Data set 2.

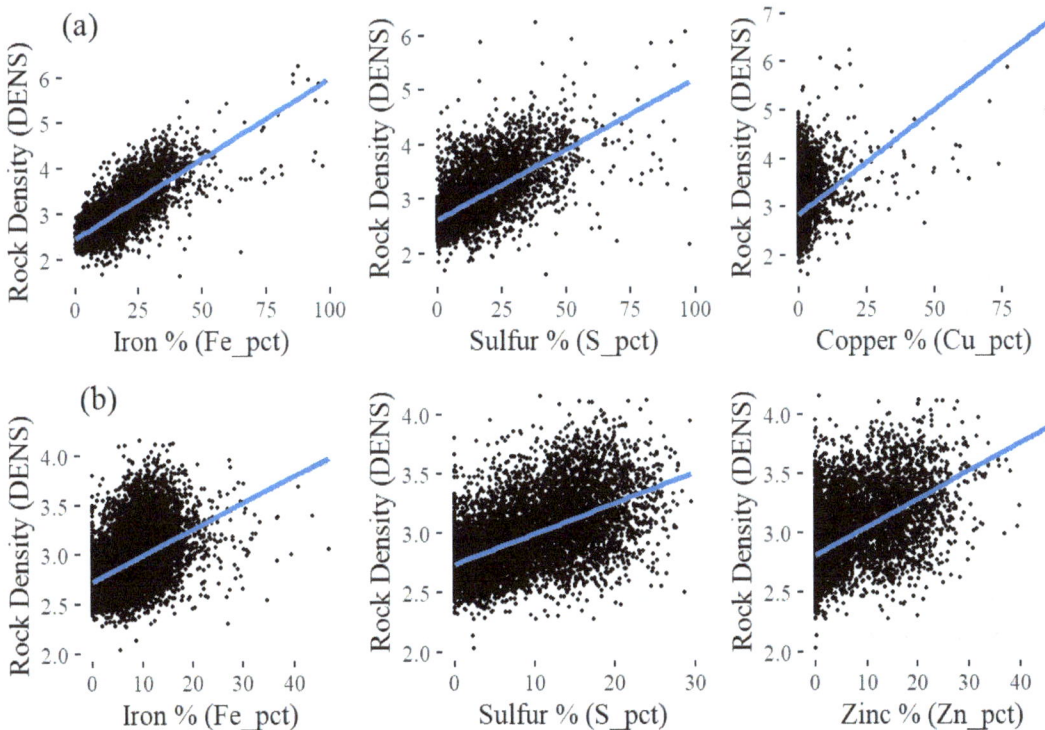

FIG 4 – Scatterplots of rock density versus selected assays: (a) Data set 1, (b) Data set 2.

Experiments

One of the biggest problems with trying to build a predictive model from data is the data itself. The quality of the data, the coverage of data in physical and multivariate space (including discrete variables such as lithology), the distribution of missing values and in the extreme, missing variables or variables that have a high missing value rate, and too few observations. How does gaps in the data affect the accuracy of a regression model trained on the data? Note that data gaps can be artificially introduced through inappropriate splitting of the data into training, test, and validation data sets (not tested for this paper).

Gaps in the data can lead to a common problem with machine learning where the model learns the data and the noise in the data and not the underlying relationships, referred to as model over-fitting. Model under-fitting is where the model is not complex enough to learn the relationships adequately enough. For model over-fitting the problem usually presents when there is insufficient number of observations given the number variables available to be used as covariates for regression modelling – the model is too complex. However, reducing the number of terms in a model to account for the number of observations can lead to not including key variables that directly affect the target variable resulting in poor model fit. Under-fitting can occur when key variables are not included in the modelling, usually because the variable is not measured, or the variable has too many missing values. In this case, many observations may exist but there are only a small number of variables that were measured such as only having assays for the metals of interest and not assaying for whole rock geochemistry or undertaking geological logging.

Gaps in the data can lead to miss-specifying the regression model and to miss-interpreting the results of the regression model. Regression analysis relies on correlations between the covariates and the target variable. Without strong correlations the resulting trained models will not be able to accurately predict the target variable, in this case rock density. However, having strong correlations between the independent and dependent variables does not guarantee that a good predictive model can be trained – one that accurately predicts the target variable for unseen (out-of-sample) data. Strong correlations between variables may be due to some other variable and if this variable is accounted for then the correlations may disappear or reverse. The missing variable that directly affects both the dependent and independent variables is called a confounder. Data set 2 is an example where the confounder variable, lithology (LITH_), is the only variable that is needed for predicting rock density – correlations between the assays and rock density are due entirely to lithology. Data set 1 also has lithology directly affecting rock density, but rock density is also affected by iron content as iron content is a direct indicator of massive sulfide which is identified also by lithology. Excluding confounders can result in an incorrect assessment of direct effects. An otherwise strong positive linear correlation between an independent variable and the dependent variable may reverse when accounting for the effects of the confounder – this is referred to as the Simpsons Paradox. Rather than just blindly using all available variables for building a predictive model it is better to develop a causal model that can then be tested. In the case of rock density estimation, the mineralogy of the rock directly affects rock density, mineralogy in turn reflects the broader lithology classes along with any changes the rock has undergone due to metamorphism, alteration, or oxidation and weathering. Experiments were conducted to test the impact of not including the most important variables that directly affect rock density in regression modelling.

A series of experiments were conducted using the two synthetic data sets to test the performance of machine learning for training regression models to predict rock density. The experiments do not test all possible scenarios that may impact on the quality of the predictive model and test what could possibly go wrong. Tests were focused on assessing the impact of data gaps on the accuracy and bias of trained regression models given gaps in the data. Gaps include missing observations (incomplete data coverage) and missing variables (not collecting values for key variables – the confounders). Impact of missing values (most geological data sets contain missing values which can be extremely problematic for machine learning), outliers (extreme values), and data errors (errors in geological logging, errors in rock density measurement, or errors in rock assays).

The experimental design (Figure 5) involved subsampling the synthetic data to build several training data sets, training the regression models to predict rock density, predicting rock density using the full synthetic data set, calculating the error metric, and then compiling the results.

Uniform Manifold Approximation and Projection (UMAP) was used to project the synthetic data onto a two-dimensional latent space representing the numeric multivariate space of the data. The discrete variables are not used in the projection. The UMAP latent space was used for selection of subsets of data for the experiments.

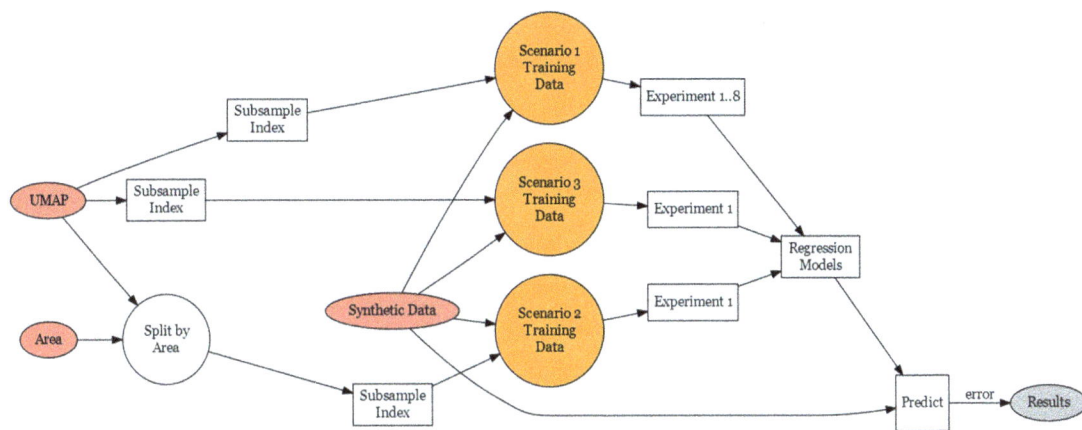

FIG 5 – Experiment design flow diagram.

K-means algorithm was used to partition each synthetic data set into six discrete subsets.

Latin Hypercube sampling were used to select samples globally and within each K-means partition. The global subset was used to assess impact of regression technique, effect of missing variables (especially confounder variables), and to illustrate that under ideal conditions (sufficient data coverage and all key variables observed and no missing values or data problems) Machine Learning can be an effective tool in building a predictive model for estimating rock density for Mineral Resource estimation.

Experiments were conducted across three sampling scenarios. The scenarios were designed to test sampling patterns that are commonly encountered for rock density sampling and measurement in Mineral Resource estimation.

Scenario 1 is where there are a many rock density measurements that have been collected that cover the full multivariate space (range of assays) and enough observations across all classes for all discrete variables such as different lithology. Observations were selected from the Truth data set at a rate of 1:20 observations.

Scenario 2 is where the sampling is limited to parts of the multivariate space. This scenario is designed to represent situations where rock density measurements are only collected from parts of a deposit such as only from the eastern side or western side of a deposit which have different lithology and assay values or from only one pit where there are multiple pits or from a single domain or from only one weathering zone (oxide or sulfide). As with Scenario 1, the number of observations selected from the Truth data set were at a rate of 1:20.

Scenario 3 is the like Scenario 1 but rather than using a fixed 1:20 subsampling rate, a series of subsets were created with 50, 100, 200, 300, 400, 500, 1000, 2000 and 3000 samples. This scenario was designed to test the effect of sample size on regression model error and look at model under-fitting (insufficient model complexity to accurately model the data) and model over-fitting (the model is too complex, too many parameters).

Eight experiments were conducted using the Scenario 1 data sets, Data set 1 and Data set 2 (Experiments 1 to 8). The experiments differ in what variables were included in the regression models, in what preprocessing was done, and in what error metric was used for training the regression models. Experiment 1 includes all available variables. Experiment 2 and 3 only include the metal assays. Experiment 4 and 5 only included the discrete variables which were encoded as numeric variables using mean embedding. One experiment was conducted using the Scenario 2 data (Experiment 9) and one experiment was performed using Scenario 3 data (Experiment 10).

Experiment 1

How good is a regression model for predicting rock density given the data has good multivariate coverage, has all key variables, sufficient observations for all classes for all discrete variables such as lithology, and there are enough observations given the number of terms (at least 15 observations for every model term)? Experiment 1 was designed to test the performance of different regression models and how choosing a certain regression model may adversely impact the accuracy of tonnage estimates for a Mineral Resource estimate. What could go wrong if we chose the wrong regression model to predict rock density?

Experiment 2

What if all we have are assays for the metals such as gold, silver, lead, zinc, or copper. Can we still get a reasonable estimate of rock density? In resource estimation it is very common to encounter the situation where a deposit only has assays for the metals of interest. In some situations, not even geological logging is available. Experiment 2 was designed to test the ability of metal assays to predict rock density. For Data set 1 only gold (Au_gt) and silver (Ag_gt) were included in the regression models. For Data set 2 only zinc (Zn_pct), lead (Pb_pct), and copper (Cu_pct) were used. What could go wrong if we do not measure key variables or include key variables in a regression model to predict rock density?

Experiment 3

What if we only have discrete variables such as lithology and domaining available for regression modelling? In this experiment only the discrete variables lithology and domain were used in the regression modelling. Lithology was embedded into a three-dimensional vector using UMAP applied to the assay variables (assuming we have assay variables to use) and rock density while domain was embedded into a two-dimensional vector using UMAP. What could go wrong if we only use lithology and domaining in regression modelling to predict rock density?

Experiment 4

What if all we have is lithology to predict rock density and we encode the variable as a three-dimensional numeric vector using mean embedding? Experiment 4 is like Experiment 3 but without the domain variable. Again, lithology was embedded into a three-dimensional vector using UMAP applied to the assay variables (assuming that the assay variables are available). What could go wrong if we only use lithology in regression modelling to predict rock density?

Experiment 5

What if, like Experiment 4, we only have lithology but encode the variable using only the mean values of rock density? This experiment differs from Experiment 4 in the way lithology is encoded as a numeric variable. Each class label was replaced with the mean rock density value for that class. This experiment is equivalent to using one-hot encoding of lithology typically used for linear regression. The benefit of using mean embedding is the significant reduction in the number of terms in the regression model which would be better for when there are limited number of observations. Discrete variables such as lithology may have many classes (n classes) and when combined with alteration which may also have many classes (m classes), the number of observations required to support the model to avoid over-fitting may be large (at least $15 * (n - 1) * (m - 1)$). What could go wrong if we only use lithology in regression modelling to predict rock density but use a different method for encoding the discrete variable as a numeric variable?

Experiment 6

What if we only use all assays and exclude all discrete variables? Like Experiment 2 but includes all assay variables.

Experiment 7

What if we assume that the correlations observed between numeric variables resulted from a hidden unobserved (latent) variable or variables. Would latent variable modelling be better than simply using the assays directly in regression analysis? Latent variable modelling techniques were used to convert the assay variables in Data set 1 and Data set 2 into a single hidden continuous numeric variable or a single discrete variable. Methods used include Principal Component Analysis (PCA), Independent Component Analysis (ICA), Multidimensional Scaling (MDS), ISOMAP, UMAP, t-SNE, Factor Analysis, and Gaussian Mixture Models (GMM). Would using latent variable modelling of the assay variables be better than just using the assay variables (Experiment 6)?

Experiment 8

What would happen if we used a different error metric, other than root mean squared error (RMSE), to train the regression model to predict rock density? Error metrics tested include mean absolute error (MAE), mean absolute percent error (MAPE), median absolute error (MDAE), mean squared error (MSE), mean squared log error (MSLE), R2, relative absolute error (RAE), relative squared error (RSE), root relative squared error (RRSE), root mean squared error (RMSE), standard error (SE), root mean squared log error (RMSLE), and symmetric mean absolute percentage error (SMAPE). Typically, the root mean squared error (RMSE) is usually used for training regression models, but can other error metrics perform better?

Experiment 9

What if we only have a small portion of the multivariate space covered by samples? How do we know if we have good coverage? Experiment 9 is like Experiment 1 but differs in only having samples from each K-Means group (Area). Scenario 2 subset data for Data set 1 and Data set 2 was used for the experiment. All available variables were included in the regression models. Is the subset data sufficiently different from the full data set to result in a regression model that does not predict rock density accurately?

Experiment 10

What if we only have a very small number of rock density measurements (only 50 observations or 100 observations)? How many observations do we need to build a regression model to predict rock density where the predictions are unbiased and have low prediction error? Experiment 10 is designed to test the effect of sample size on the quality of a trained regression model for predicting rock density. The Scenario 3 data was used for the experiment. All variables were included. Experiment 10 is like Experiment 1 but with different number of training samples.

Software, Techniques and Algorithms

The open-source R environment for statistical computing (R Core Team, 2022), R version 4.2.1 (2022–06–23 ucrt) Funny-Looking Kid distribution, was used. RStudio version 2021.9.0.351 Ghost Orchid was used as the integrated development environment along with the Rmarkdown technology to facilitate the automated compilation of the results and documentation and prepare an initial draft paper.

Machine learning techniques that were used for preparing the synthetic data included generative modelling (Bayesian Network), dimension reduction (UMAP), quantisation (K-Means), and sampling algorithms (LHS). The algorithms included:

- Bayesian network, R package bnlearn (Scutari, 2010)). Bayesian Networks are a statistical method that models the univariate distributions and correlations between variables – there are no input (predictor) variables or output (target) variables. They are a good technique to gain insight into the data and explore possible causal relationships and assist with selecting variables for classification or regression modelling. A Bayesian network can be trained on data or constructed manually or both.

- Uniform Manifold Approximation and Projection (UMAP), R package umap (Konopka, 2022). UMAP (McInnes, Healy and Melville, 2018) is a new machine learning technique for dimension reduction and visualising high-dimensional data. One benefit of using UMAP is the ability to add new data points to the coordinate system allowing separate data sets to be visualised in the same feature space.

- K-Means, base R (R Core Team, 2022). K-Means (Kaufman and Rousseeuw, 1990) is a machine learning algorithm for grouping data. It aims to partition the observations into a predefined number of groups or where each observation is assigned to the cluster with the nearest mean (the prototype vector for the cluster).

- Latin Hypercube Sampling (LHS), R package clhs (Roudier, 2011). LHS is a method for generating a near-random sample from a multidimensional distribution.

Machine learning algorithms used in the experiments included nine regression methods and eight dimension reduction methods. The dimension reduction methods were used for constructing latent (hidden) variables which were then used in regression modelling (Experiment 8). The R package caret (Kuhn, 2022) was used for training the regression models to predict rock density for all experiments.

The selected regression models included:

- Linear Regression (lm), base R (R Core Team, 2022). Linear regression is a statistical method for modelling the relationship between a dependent variable (response) and one or more explanatory variables (predictor). The linear regression model coefficients are optimised using Ordinary Least Squares (OLS). Coefficients are the weights that are applied to each of the explanatory variables. The regression is linear as the resulting model equation is a line in two-dimensions, a plane in three-dimensions and a hyperplane in higher dimensions. Non-linear relationships can be approximated using piecewise linear regression or Multivariate Adaptive Regression Splines (MARS). The standard linear regression model can be extended by including polynomial and functional terms such as x_2 or $\log(x)$ to handle non-linear relationships. Variants to the standard OLS regression include LASSO and RIDGE regression which add penalty terms to the loss function.

- Decision Tree (dt) or Classification and Regression Tree (CART), R package rpart (Therneau and Atkinson, 2022). CART algorithm produces a decision tree that classifies the data by recursive binary splitting based on several explanatory variables. The drawback to this method is that the resulting tree can be unstable to small changes to the data (including order) and can lead to over-fitting. The regression tree does not work well for continuous explanatory variables as they are discretised into a limited set of mean values (the means can be replaced with a regression function).

- Random Forest (rf), R package randomForest (Liaw and Wiener, 2002). Random Forest is an ensemble (multiple models) of decision trees that can be used to minimise over-fitting of the data but results in a complex model that may be difficult to interpret and deploy.

- Generalised Additive Model (GAM) with Spline (gamSpline), R package gam (Hastie, 2023). GAM (Hastie and Tibshirani, 1990) is a variation of linear regression where the target variable depends linearly on unknown smooth functions of some predictor variables.

- Cubist (cubist), R package (Kuhn and Quinlan, 2022). The Cubist algorithm is based on ideas from Quinlan (1992, 1993a, 1993b). It is a rule-based (if then outcome) regression model combining a decision tree with multivariate linear regression and nearest neighbour adjustments. The method also uses a boosting-like ensembling method (combining multiple decision tree-linear regression models).

- Multivariate Adaptive Spline (earth), R package earth (Milborrow, 2021). Multivariate Adaptive Splines (Friedman, 1991) are a variant of linear regression. The method automatically models non-linearity and interaction between variables and uses a hinge function. A hinge function takes the form of either $max(0, x - c)$ or $max(0, c - x)$ where c is a constant, called the knot.

- Least Angle Regression (lars), R package lars (Hastie and Efron, 2022). Least Angle Regression (Efron *et al*, 2004) is a method of fitting linear regression models to multivariate data.

- Weighted k-Nearest neighbour Classifier (kknn), R package kknn (Schliep and Hechenbichler, 2016). The K-Nearest Neighbour method (Hechenbichler and Schliep, 2004) uses the k-nearest observations (using Minkowski distance) from a training data set and a value is found via maximum of summed kernel densities. The standard rectangular kernel was used.

- Gaussian Process Regression with Linear Kernel (gaussprLinear), Radial Basis Function Kernel (gaussprRadial), and Polynomial Kernel (gaussprPoly), R package kernlab (Karatzoglou, Smola and Hornik, 2022). Gaussian process regression (Rasmussen and Williams, 2006) is essentially kriging but for regression. Gaussian Process Regression models are nonparametric kernel-based probabilistic models where the covariance function (kernel function) expresses the expectation that points with similar predictor values will have similar response values.

- Multi-layered Neural Network (mlp), R package RSNNS (Bergmeir and Benítez, 2012). MLP (Hastie, Tibshirani and Friedman, 2009) is a variant of feedforward Artificial Neural Network (ANN), a mathematical model that is loosely inspired by biological neural networks (the human brain). The neural network is a framework for many different machine learning algorithms. The ANN consists of a collection of connected processing units (neurons) typically formed into layers with different layers performing different transformations on their inputs. The neuron consists of a vector of continuous numeric inputs (typically scaled) which are multiplied by a set of weights then combined and fed into a nonlinear function. A feed-forward ANN with a single hidden layer containing a finite number of neurons can approximate continuous functions (universal approximation theory).

For Experiment 8, latent (hidden) variables were constructed using a variety of methods that included:

- Principal Component Analysis (PCA), base R (R Core Team, 2022). PCA is a statistical technique for dimension reduction by linearly transforming the data into a new coordinate system where most of the variation is in the first few components. PCA is commonly used as a data preprocessing step.

- Independent Component Analysis (ICA), R package ica (Helwig, 2022). ICA is a method of separating a multivariate signal into additive non-Gaussian subcomponents.

- Multidimensional Scaling (MDS), base R (R Core Team, 2022). Multidimensional Scaling is a machine learning algorithm for visualisation of high-dimensional data. It is a nonlinear dimensionality reduction technique. The algorithm aims to place each object in N-dimensional space such that the between-object distances are preserved as well as possible. Coordinates in each of the N-dimensions are then assigned to each object – typically two dimensions.

- ISOMAP, R package vegan (Oksanen et al, 2022). ISOMAP (Tenenbaum, vin de Silva and Langford, 2000) is a non-linear dimension reduction algorithm that extends MDS.

- Uniform Manifold Approximation and Projection (UMAP), R package umap (Konopka, 2022).

- t-Distributed Stochastic neighbour Embedding (t-SNE), R package Rtsne (Krijthe, 2015). T-Distributed Stochastic Neighbor Embedding (van der Maaten and Hinton, 2008) is a machine learning algorithm for visualisation of high-dimensional data. It is a nonlinear dimensionality reduction technique. The drawback to using this technique is that the algorithm is stochastic and uses a random number generator. This means that slightly different results can be obtained every time the algorithm is run. A random number seed can be used to ensure that the same results are obtained aiding reproducibility.

- Factor Analysis, R package psych (Revelle, 2022). Factor analysis is a statistical method that is used to describe the variability among variables in terms of a smaller number of unobserved (latent) variables called factors. The observed variables are modelled as linear combinations of the unobserved (hidden) variables and an error term.

- Gaussian Mixture Models (GMM), R package mclust (Scrucca et al, 2016). Gaussian Mixture Models are a statistical modelling technique that can be used for clustering. The approach is like the K-means clustering algorithm and can be considered a generalisation of K-means. Unlike K-means which gives discrete clusters, the Gaussian Mixture Model clustering assigns probabilities of being in a cluster to each data point. Rather than just using closeness to cluster centres, the algorithm finds a mixture of Gaussian distributions with mean vector and covariance and mixing proportion.

The R package Metrics (Hamner and Frasco, 2018) was used to calculate error variance for compiling the results of the regression models.

Metrics for evaluating regression models

For all experiments, the Root Mean Squared Error (RMSE) metric was used, except for Experiment 8 where several different error metrics were compared. The following error metrics were used to assess model error variance:

- Coefficient of Determination
$$R^2 = 1 - \frac{\sum_{i=1}^{n}(x_i - y_i)^2}{\sum_{i=1}^{n}(x_i - mean(x))^2}$$

- Mean Absolute Error
$$MAE = \frac{1}{n}\sum_{i=1}^{n}|x_i - y_i|$$

- Mean Squared Log Error
$$MSLE = \frac{1}{n}\sum_{i=1}^{n}\left(log(x_i + 1) - log(y_i + 1)\right)^2$$

- Mean Absolute Percentage Error
$$MAPE = \frac{1}{n}\sum_{i=1}^{n}\left|\frac{(x_i - y_i)}{x_i}\right|$$

- Mean Squared Error
$$MSE = \frac{1}{n}\sum_{i=1}^{n}(x_i - y_i)^2$$

- Median Absolute Error
$$MDAE = median(|x_i - y_i|)$$

- Relative Absolute Error
$$RAE = \frac{\sum_{i=1}^{n}|x_i - y_i|}{\sum_{i=1}^{n}|x_i - mean(x)|}$$

- Relative Squared Error
$$RSE = \frac{\sum_{i=1}^{n}(x_i - y_i)^2}{\sum_{i=1}^{n}(x_i - mean(x))^2}$$

- Root Relative Squared Error
$$RRSE = \sqrt{RSE}$$

- Root Mean Squared Error
$$RMSE = \sqrt{MSE}$$

- Root Mean Squared Log Error
$$RMSLE = \sqrt{MSLE}$$

- Symmetric Mean Absolute Percentage Error
$$SMAPE = \frac{100}{n}\sum_{i=1}^{n}\frac{|x_i - y_i|}{(|x_i| + |y_i|)/2}$$

The advantage of using the RMSE for training a regression model is that it is in the same units as the data. The disadvantage is that the metric is strongly biased by outlier values in the data. For this paper, the synthetic data is assumed to contain no outliers that could significantly impair model accuracy. The RMSE metric can also be difficult to interpret and can be difficult to compare results across different data sets with different variable data ranges.

Comparing experiment results, both the RMSE and the Mean Absolute Percent Error (MAPE) were used. The MAPE metric was selected as it presents the error in terms of a percentage which is useful for interpreting the results in the context of a Mineral Resource estimate. The MAPE value was used to assess the likely error, in percentage terms, of tonnage estimates for a Mineral Resource estimate.

RESULTS AND DISCUSSION

Before considering each of the experiments, a naive estimate of rock density was done for comparison. The naive model uses the global rock density value for each training data. The global rock density mean for Data set 1 was 2.9 (n = 7273; sd = 0.47) and for Data set 2 was 2.92 (n = 10 000; sd = 0.28). By coincidence both Data sets have very similar mean rock density values. The resulting global RMSE for Data set 1 is 0.47 and for Data set 2 is 0.28. In the naive model case, the RMSE is the same as the standard deviation, as the predicted values are the mean. Using mean rock density by lithology (LITH_), the RMSE drops to 0.3478007 for Data set 1 and to 0.1869227 for Data set 2.

The naive model is like using a Decision Tree for regression modelling or using linear regression where the independent variable is a discrete variable. However, both the Decision Tree and linear regression models can allow for inclusion of other variables for predicting the target variable, such as the domain variable (DOM_) in Data set 1 and Data set 2 and oxidation state (OX_) in Data set 1 and numeric variables such as assays.

Would it be better to use the naive model for estimating tonnage for a Mineral Resource estimate rather than building a regression model? For Data set 2, where lithology has a direct effect on rock density and where lithology also directly affects rock assays, accounting for all correlations between assays, perhaps. For Data set 1, maybe. This assumes that we have a good estimate of the mean rock density.

Under ideal conditions (Scenario 1), where the rock data contains all necessary variables that directly affect density (lithology, oxidation, other domains such as area or depth, and rock assays) and that the rock density data has good coverage spatially and in multivariate space, the best regression method (based on the lowest RMSE) for Experiment 1 Data set 1 was the Cubist algorithm (RMSE = 0.291; MAPE = 0.0699) and for Data set 2 was the random forest (RMSE = 0.199; MAPE = 0.0513) followed closely by the Cubist algorithm. Across all experiments and both data sets the overall best performing regression method was the Cubist algorithm followed by the random forest. Figure 6 shows the relative ranking of the regression methods across all experiments and both data sets, ranked by the error variance metric RMSE.

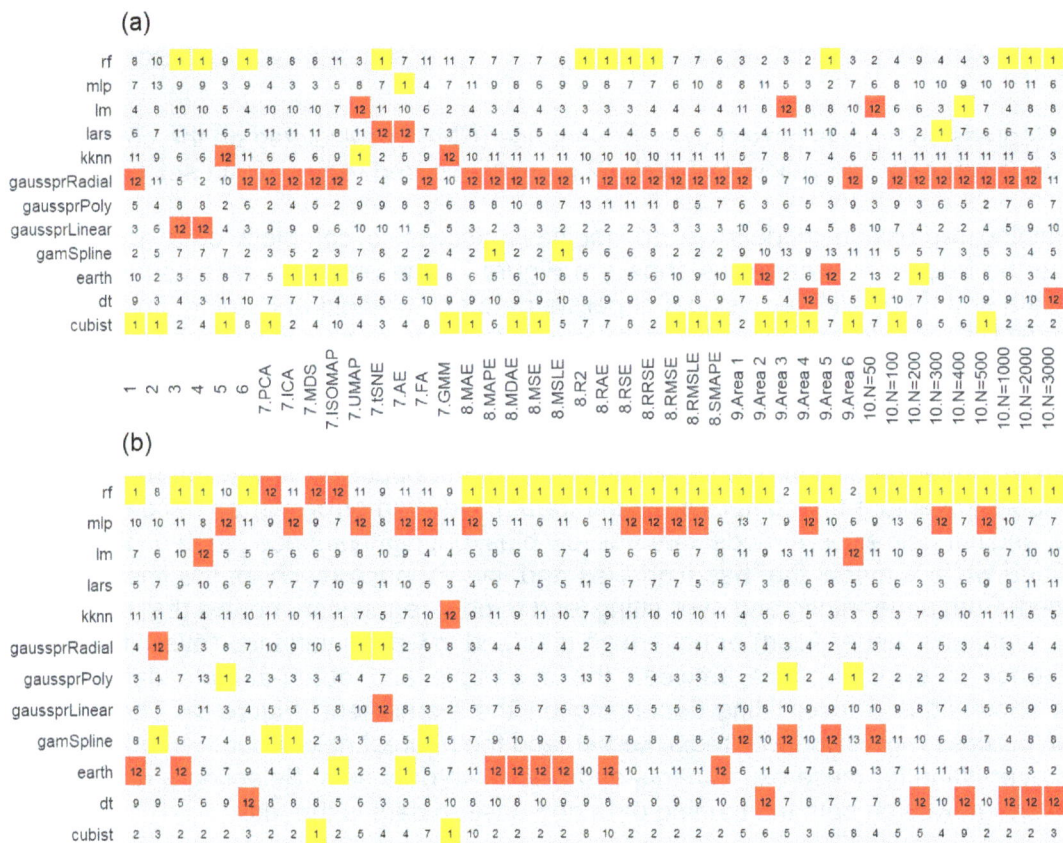

(a)

(b)

FIG 6 – Regression models ranked by error variance (RMSE) calculated on the complete data sets rank one (best) coloured yellow and rank 12 (worst) coloured red: (a) Data set 1, (b) Data set 2.

Under ideal conditions (Scenario 1) there is very little difference in regression model error variance (RMSE) for Experiment 1 (Figure 7). What could go wrong if we select a particular regression model? Nothing! for Experiment 1. The selection of regression method comes down to personal preference – the linear regression method being a good choice as it is simple to train and easy to interpret and use. However, when we do not have key variables (lithology) or not enough observations then the choice of regression method becomes more critical. The multivariate adaptive regression spline (earth) model has the largest range in RMSE, from 0.243 to 0.95 for Data set 1 and from 0.188 to 0.447 for Data set 2. However, for Data set 2, the multivariate adaptive regression spline performs poorly on one experiment but performs well on the rest. The Gaussian process regression models (gaussprRadial and gaussprLinear) and linear regression perform worse overall than the multivariate adaptive regression model. What could go wrong? Everything! for all experiments other than Experiment 1 the selection of regression method could result in an error up to three times worse and be worse than if the naive model was used.

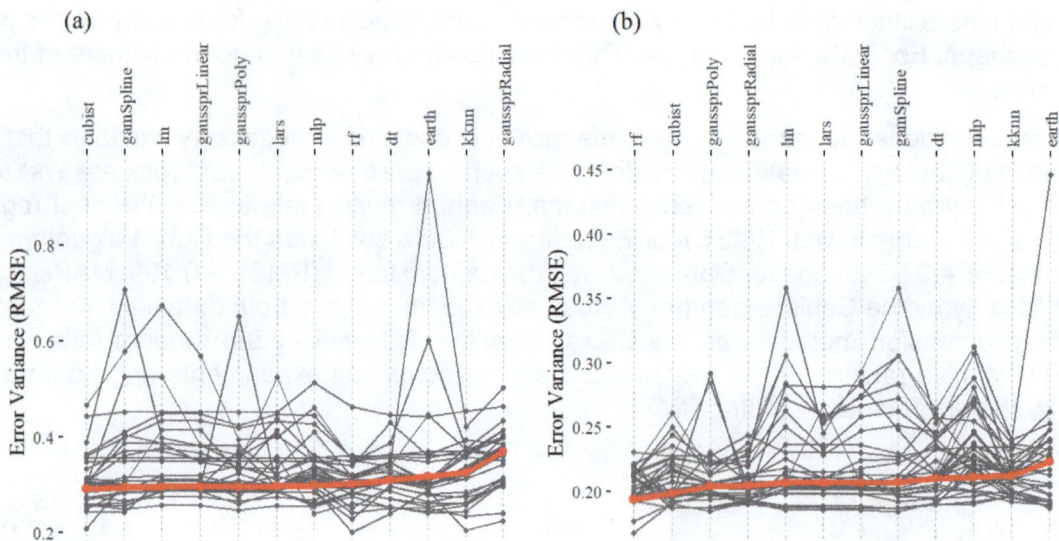

FIG 7 – Regression models ranked by error variance (RMSE) calculated on the complete data sets rank one (best) coloured yellow and rank 12 (worst) coloured red: (a) Data set 1, (b) Data set 2.

When training regression models we do not have access to complete data and are unable to determine the true error variance as shown in Figure 7. The error metric is usually applied to the training data and then validated using a separate data set withheld at training time. Using the error variance (RMSE) to select a regression method with the lowest value can result in selecting a model that performs poorly on unseen data. Training a regression model or any model using machine learning can lead to either over-fitting or under-fitting. Over-fitting is when the model is too complex, has too many parameters relative to the number of observations available (degrees of freedom) leading to an under-prediction of error. Under-fitting is the opposite, where the model is too simple and is unable to capture the complexity in the data. Prediction error can be decomposed into bias, variance, and intrinsic error. A model that has low bias, but high variance (such as the decision tree) is over-fitted while a model that has high bias and low variance (such as the linear regression) is under-fitted. Both under-fitting and over-fitting lead to poor regression models that have low bias but high error variance (under-fitted) or have high bias and low error variance (over-fitted). Comparing the difference in error variance between the in-sample and out-of-sample data all regression methods show signs of over-fitting depending on the experiment (Figure 8). The random forest, Gaussian process regression with radial kernel, and the nearest neighbour methods are particularly prone to over-fitting for these experiments and these two data sets. Selecting a regression method that will perform well on both the training data and unseen data is the trick.

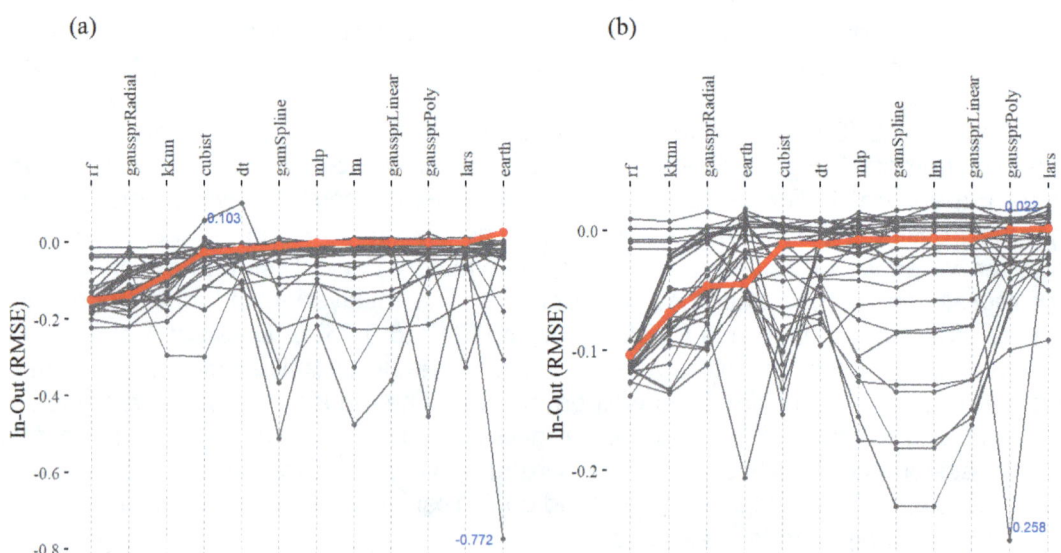

FIG 8 – Regression models ranked by error variance (RMSE) calculated on the complete data sets rank one (best) coloured yellow and rank 12 (worst) coloured red: (a) Data set 1, (b) Data set 2.

For prediction of density, based on the experiments performed on the two synthetic data sets, the Cubist method is a good method that will likely not be either under-fitted or over-fitted. In the authors experience, the Cubist algorithm tends to outperform other regression methods in most regression tasks done on geological and metallurgical data. It is the authors go-to method for most regression tasks due to the ability of the method to deal with non-linear relationships and disjointed subsets. These experiments showcase the utility and versatility of the regression method especially for prediction of rock density. The added benefit of the Cubist algorithm is that it can be easily interpreted and used in simple scripts or spreadsheets and requires less effort to train than does more complicated methods such as neural networks which also requires a lot of data.

The Cubist algorithm combines a linear regression model, decision tree, and nearest neighbour estimation into a single model. The decision tree splits the data into groups, the linear regression model is applied to each group to predict the target variable, and the nearest neighbour method is used to adjust the predictions. The method also takes advantage of ensembling, where multiple models are trained and combined to predict the target variable. Because the Cubist algorithm combines a decision tree with linear regression, the method tends to outperform both the decision tree and linear regression methods separately. The implementation of the Cubist algorithm used for these experiments also includes automatic feature selection where the linear regression model component can be complex containing all independent variables or as simple as a constant value (the mean of the data).

Interestingly Figures 7 and 8 show that some experiments perform better than Experiment 1.

The error variance (RMSE) for all regression models across all experiments for Data set 1 are compared in Figure 9 and in Figure 10 for Data set 2. The graphs highlight how well the Cubist algorithm performed across all experiments. The patterns in the graphs are similar between Data set 1 and Data set 2.

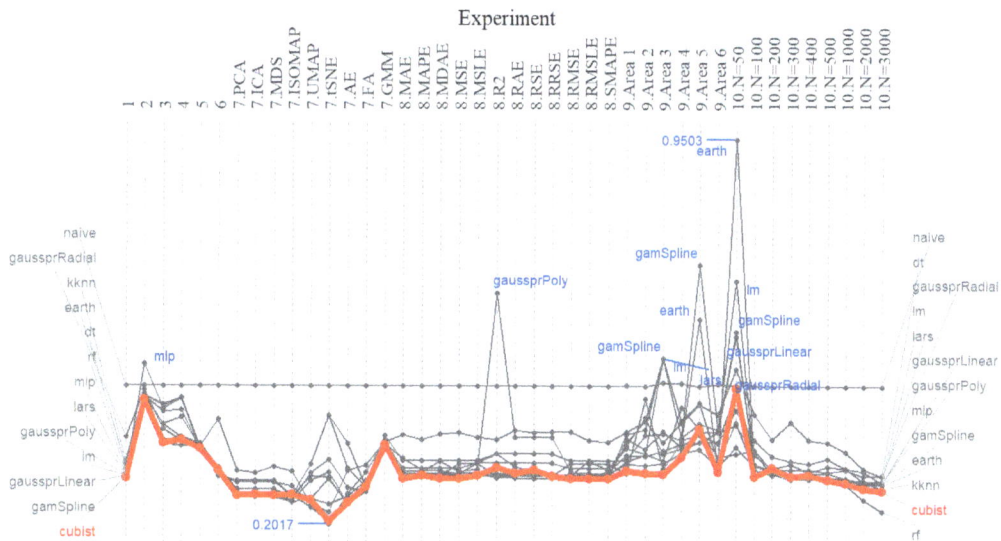

FIG 9 – Comparison of error variance (RMSE) calculated on the complete Data set 1 between the selected regression methods for across all experiments, with the Cubist algorithm highlighted in red.

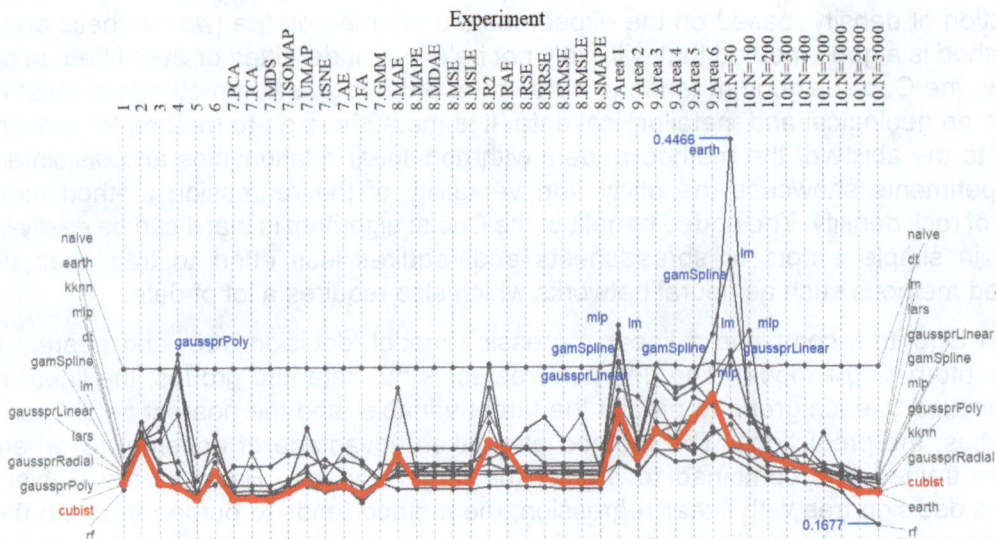

FIG 10 – Comparison of error variance (RMSE) calculated on the complete Data set 2 between the selected regression methods for across all experiments, with the Cubist algorithm highlighted in red.

Not including all key variables (lithology for Data set 1 and Data set 2 and iron or sulfur for Data set 1) resulted in higher RMSE error. Experiment 2 had consistently high RMSE error for all regression models while Experiments 3 and 4 had a wider range of RMSE error. Experiment 2 shows that not including the key variables can be costly in terms of error in predicting rock density. For Data set 1, no matter what regression method is used, the results for Experiment 2 are no better than using the naive model. In fact, using the naive model would be the best option if only metal assays are available given the simplicity of simply using the global mean. Experiments 3, 4 and 5 indicate that for Data set 1, only using lithology is better than only using metal assays (Experiment 2) but not as good as using all available assays which includes the key variables sulfur and iron.

Simply using the mean rock density value (Experiment 5) instead of the lithology class label proved to be better than using the mean assays and rock density and embedding the mean vector into a three-dimensional space using UMAP (Experiments 3 and 4). However, using the mean rock density as an embedding for discrete variables requires that there are sufficient observations to estimate the mean rock density values and that those observations are representative of the discrete classes used to subset the data. If the training data is incomplete, then the estimate of mean rock density values may not be accurate and using a different embedding method that does not use rock density would likely be a better choice.

Using latent variable modelling when only assays are available (Experiment 7) is better than using the assays directly (Experiment 6). This is likely because the data contains disjointed subsets split by lithology where relationships between independent variables are linear within subsets. The latent variable is replicating the role of the discrete variable lithology. For Data set 1, the best latent variable model is the t-SNE dimension reduction technique. However, using t-SNE as a preprocessing step does require careful selection of regression technique as the range between the lowest error (Cubist) and highest error (Gaussian process regression with radial kernel) is wider than other methods for Data set 1. Preprocessing Data set 1 using principal component analysis (PCA), independent component analysis (ICA), multidimensional scaling (MDS), ISOMAP, and UMAP resulted in similar error, and all have lower error than Experiment 1. For Data set 2, all preprocessing resulted in lower error than for Experiment 1.

For these experiments, training the regression model using a metric other than RMSE has little impact on the results. Except for using R2 for Data set 1 and MAE, R2, and RSE for Data set 2 where the resulting error is more variable depending on regression model.

Experiment 9, which looked at training regression models with only a small subset of the data that had limited coverage in multivariate space, had results ranging from an error like Experiment 1 to an error exceeding the naive model. This illustrates the danger in not having good data coverage in multivariate space. What could go wrong? Nothing, Everything! Resource estimation focuses on having good spatial data coverage, but data collection tends not to consider multivariate data coverage. When building regression models or using machine learning it is very important to consider data coverage in multivariate space. Questions to consider include: Does the data have enough observations across all discrete classes (lithology for instance)? Does the data range (minimum and maximums) match the expected target range?

Another aspect of using machine learning is the number of observations available in the training data. Small data sets tend to be problematic while having as much data as possible leads to better models. Figures 7 and 8 shows that as the number of observations increases from 50 to 3000 the error variance (RMSE) decreases. For these experiments having only 50 observations is not enough. In this instance it is better to use the naive model than perform regression modelling. Models trained from such a small number of observations are considered exploratory only.

Unlike Data set 1, the results for Data set 2 show that most models for most of the experiments beat the naive model. The exception being Experiments 4, 9 and 10. This is because lithology is the only variable required for building a good rock density predictive model. As with Data set 1, using latent variable modelling as a preprocessing step improves the accuracy of all regression models for predicting rock density. Increasing number of observations appears to result in a higher rate of error reduction compared with Data set 1. Again, only having 50 observations is insufficient for building a regression model, even for the simple linear regression model. Depending on the data (relationships between variables and direct effects on rock density) and the selected regression technique error ranges from slightly better than Experiment 1 to worse than the naive model.

Global error metrics such as RMSE or MAPE are useful for training and selecting regression models but hide the potential impact of regression modelling for predicting rock density when using varying cut-off grades for reporting tonnage for a Mineral Resource estimate. Figure 11 shows boxplots of mean absolute percent error (MAPE) by cut-off grade. The MAPE metric was used as it is easier to interpret the resulting error values with respect to impact on tonnage estimation. The graphs show that as the cut-off grade increases, the error (MAPE) also increases. For Data set 1, using a small cut-off grade (>0.5) results in an error of over 10 per cent. A cut-off of 2.5 or higher means that the estimates start to exceed 20 per cent error. For Data set 2, a very high cut-off grade (>17.5) is required before error starts to exceed 10 per cent. Depending on the data, applying a cut-off grade to a resource model and reporting tonnage derived from rock density values predicted using a regression model may result in high uncertainty in the estimated tonnage. So much so, that the assigned resource classification may need to be downgraded. For Data set 2, the estimates of tonnage are likely to be suitable for a classification up to Measured depending on other factors while for Data set 1 the tonnage estimates may only be suitable for an Inferred classification. Not tested for this paper, but one possible approach to limiting the error at higher cut-off grades is to build models across different cut-off grades. This will increase the workload and number of regression models but may help to reduce the error at high cut-off grades. However, the problem of increasing error at increasing cut-off grade may not be eliminated meaning using the rock density estimates for anything other than a global estimate of tonnage is risky. Any mine planning or financial analysis using higher cut-off grades will need to factor in the increased risk in tonnage estimation.

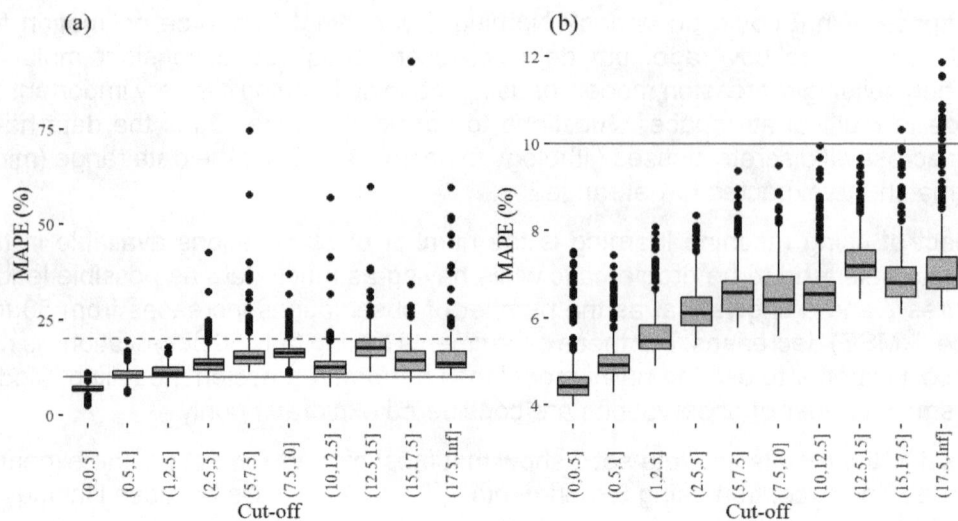

FIG 11 – Boxplots of mean absolute percent error (MAPE) by cut-off grade: (a) Data set 1, (b) Data set 2.

CONCLUSIONS

Machine Learning techniques are a powerful tool for use in Mineral Resource estimation and can be used for building regression models to predict rock density.

In the case of estimating rock density under ideal conditions, where there is good data coverage across the multivariate space (and possibly physical 3D space) and key variables are available, then building a regression model to predict rock density is viable. If key variables are not available, then a good regression model will depend on the presence of correlations with other proxy variables, but this may lead to an increased risk of model over-fitting or under-fitting resulting in higher out-of-sample error.

The standard linear regression model, commonly used for regression tasks, is neither the worst nor best regression method for predicting rock density. There are better regression techniques, such as the ubiquitous Random Forest method, that can be used for regression analysis to predict rock density for Mineral Resource estimation, or even better, the Cubist method which incorporates a decision tree with linear regression in an ensemble of models.

The choice of how discrete variables are encoded as numeric variables affects the accuracy of a regression model. Simple mean embedding looks to be a viable option rather than using one-hot encoding which can result in many model terms requiring a larger number of observations. A good choice for embedding discrete variables into a numeric variable is through dimension reduction techniques applied to a multivariate mean vector.

Using latent variable modelling should be strongly considered where there are strong relationships between independent variables and where discrete variables such as lithology, oxidation state, or mineralisation domain are not available. In the case of building regression models to predict rock density, using latent variable modelling techniques as a data preprocessing step appears to result in better regression models than using the data directly. Latent variable modelling can be done using any of several dimension reduction techniques such as PCA, ICR, MDS, UMAP, or t-SNE or a more statistical approach could be taken by using Factor Analysis. Factor Analysis is a good option whereas it allows gaining insight into the data from a causal perspective.

Unfortunately, the ideal conditions for good prediction of density are usually not likely to be present due to constraints on budget, ability to measure rock density such as poor rock competency or high porosity or focus on only the primary metal grades. It is common to have biased sampling and limited data coverage of rock density and key rock density predictors not estimated in the block model and under assayed in the drill hole data. Under these conditions the error in estimating rock density can be high leading to incorrect estimation of tonnage which will in turn affect the assigned Mineral Resource classification. Density should be treated equally as importantly as the primary metal or element of economic interest.

What could go wrong?

Nothing! If we have good data coverage, have all the key variables that affect rock density, have enough observations (at least 50 observations per lithology or domain class), select the right regression method, and report tonnage globally, then nothing. This will also depend on other factors also aligning such as our own cognitive bias that affect our decisions. Predictions of rock density from geological features such as assays, lithology, oxidation state, and domain using a regression model will likely be sufficient to support a Mineral Resource estimate. The confidence in the estimate will depend on factors such as data quality, data coverage, regression modelling, geological interpretation and volume modelling, and spatial estimation.

Everything! Limited data, data collected in only one domain or one area, not collecting the right variables (assays, lithology, oxidation state), could lead to inaccurate rock density predictions that are not suitable for estimation of tonnage for a Mineral Resource for any level of confidence other than Inferred. The error in rock density estimates increases with increasing cut-off grades which could lead to disastrous consequences if the tonnage estimates are used in mine planning or financial analysis.

REFERENCES

Bergmeir, C and Benítez, J M, 2012. Neural Networks in R Using the Stuttgart Neural Network Simulator: RSNNS, *Journal of Statistical Software*, 46(7):1–26. Available from: <https://www.jstatsoft.org/v46/i07/>

Efron, B, Hastie, T, Johnstone, I and Tibshirani, T, 2004. Least Angle Regression, *The Annals of Statistics*, vol 32.

Friedman, J H, 1991. Multivariate Adaptive Regression Splines, *The Annals of Statistics*, vol 19.

Hamner, B and Frasco, M, 2018. Metrics: Evaluation Metrics for Machine Learning. Available from: <https://CRAN.R-project.org/package=Metrics>

Hastie, T and Efron, B, 2022. Lars: Least Angle Regression, Lasso and Forward Stagewise. Available from: <https://CRAN.R-project.org/package=lars>

Hastie, T J and Tibshirani, R J, 1990. *Generalized Additive Models* (Chapman & Hall/CRC).

Hastie, T, 2023. Gam: Generalized Additive Models. Available from: <https://CRAN.R-project.org/package=gam>

Hastie, T, Tibshirani, R and Friedman, J, 2009. *The Elements of Statistical Learning: Data Mining, Inference, and Prediction* (Springer: New York).

Hechenbichler, K and Schliep, K P, 2004. Weighted k-Nearest-Neighbor Techniques and Ordinal Classification, Discussion Paper 399, SFB 386, Ludwig-Maximilians University Munich. Available from: <http://www.stat.uni-muenchen.de/sfb386/papers/dsp/paper399.ps>

Helwig, N E, 2022. Ica: Independent Component Analysis. Available from: <https://CRAN.R-project.org/package=ica>

Karatzoglou, A, Smola, A and Hornik, K, 2022. Kernlab: Kernel-Based Machine Learning Lab. Available from: <https://CRAN.R-project.org/package=kernlab>

Kaufman, L and Rousseeuw, P J, 1990. *Finding Groups in Data: An Introduction to Cluster Analysis* (Wiley: New York).

Konopka, T, 2022. Umap: Uniform Manifold Approximation and Projection. Available from: <https://CRAN.R-project.org/package=umap>

Krijthe, J H, 2015. Rtsne: T-Distributed Stochastic Neighbor Embedding Using a Barnes-Hut Implementation. Available from: <https://github.com/jkrijthe/Rtsne>

Kuhn, M and Quinlan, R, 2022. Cubist: Rule – and Instance-Based Regression Modeling. Available from: <https://CRAN.R-project.org/package=Cubist>

Kuhn, M, 2022. Caret: Classification and Regression Training. Available from: <https://CRAN.R-project.org/package=caret>

Liaw, A and Wiener, M, 2002. Classification and Regression by randomForest, *R News*, 2(3):18–22. Available from: <https://CRAN.R-project.org/doc/Rnews/>

McInnes, L, Healy, J and Melville, J, 2018. UMAP: Uniform Manifold Approximation and Projection for Dimension Reduction. Available from: <https://arxiv.org/pdf/1802.03426.pdf>

Milborrow, S, 2021. Earth: Multivariate Adaptive Regression Splines. Available from: <https://CRAN.R-project.org/package=earth>

Oksanen, J, Simpson, G L, Blanchet, F G, Kindt, R, Legendre, P, Minchin, P R, O'Hara, R B, Solymos, P, … Weedon, J, 2022. Vegan: Community Ecology Package. Available from: <https://CRAN.R-project.org/package=vegan>

Quinlan, J R, 1992. Learning with Continuous Classes, in *Proceedings of Australian Joint Conference on Artificial Intelligence*, pp 343–348.

Quinlan, J R, 1993a. *C4.5: Programs for Machine Learning*, 302 p (Morgan Kaufmann Publishers Inc.: San Francisco).

Quinlan, J R, 1993b. Combining Instance-Based and Model-Based Learning, in *Proceedings of the Tenth International Conference on Machine Learning, ICML'93*, pp 236–243 (Morgan Kaufmann Publishers Inc.: San Francisco).

R Core Team, 2022. R: A Language and Environment for Statistical Computing, R Foundation for Statistical Computing. Available from: <https://www.R-project.org/>

Rasmussen, C E and Williams, C K I, 2006. *Gaussian Processes for Machine Learning* (MIT Press).

Revelle, W, 2022. Psych: Procedures for Psychological, Psychometric, and Personality Research. Available from: <https://CRAN.R-project.org/package=psych>

Roudier, P, 2011. Clhs: A r Package for Conditioned Latin Hypercube Sampling.

Schliep, K and Hechenbichler, K, 2016. Kknn: Weighted k-Nearest Neighbors. Available from: <https://CRAN.R-project.org/package=kknn>

Scrucca, L, Fop, M, Murphy, T B and Raftery, A E, 2016. mclust 5: Clustering, Classification and Density Estimation Using Gaussian Finite Mixture Models, *The R Journal*, 8(1):289–317. Available from: <https://doi.org/10.32614/RJ-2016-021>

Scutari, M, 2010. Learning Bayesian Networks with the Bnlearn r Package, *Journal of Statistical Software*, vol 35. Available from: <http://www.jstatsoft.org/v35/i03/>

Tenenbaum, J, vin de Silva, B and Langford, J C, 2000. A Global Geometric Framework for Nonlinear Dimensionality Reduction, *Science*, vol 290.

Therneau, T and Atkinson, B, 2022. Rpart: Recursive Partitioning and Regression Trees, Available from: <https://CRAN.R-project.org/package=rpart>

van der Maaten, L J P and Hinton, G E, 2008. Visualizing Data Using t-SNE, *Journal of Machine Learning Research*, vol 9. Available from: <http://jmlr.org/papers/volume9/vandermaaten08a/vandermaaten08a.pdf>

Best practice

A guide to reporting Mineral Resource exclusive of Mineral Reserve

T Rowland[1], H Arvidson[2], M Noppé[3], M Mattera[4], B Parsons[5], V Chamberlain[6] and R Marinho[7]

1. Industry Consultant, Tim Rowland Consulting, Rant-en-dal, 1739 Krugersdorp, Gauteng, South Africa. Email: tim.william.rowland@gmail.com
2. MAusIMM, Chief Geoscientist, K2fly Pty Ltd, Perth WA 6000. Email: heath.a@k2fly.com
3. FAusIMM(CP), Director, WH Bryan Mining Geology Centre, The University of Queensland, Brisbane Qld 4072. Email: m.noppe@uq.edu.au
4. MAusIMM, Mining Industry Process Consultant, Dassault Systemes, Perth WA 6000. Email: michael.mattera@3ds.com
5. MAusIMM(CP), Practice Leader/Principal Consultant (Resource Geology), SRK Consulting, Denver USA 3461. Email: bparsons@srk.com
6. FAusIMM, Senior Vice President Strategic Planning, AngloGold Ashanti, Johannesburg Gauteng 2196. Email: vchamberlain@anglogoldashanti.com
7. Technical Director Reserve Evaluation, Teck Resources, Vancouver BC Canada V6C 0B3. Email: rodrigo.marinho@teck.com

ABSTRACT

With the introduction of CRIRSCO (2019) based mineral asset disclosure, Regulation S-K part 1300, 2019 (S-K1300) by the United States (US) Securities and Exchange Commission (SEC), applicable to companies listed on the New York Stock Exchange from 2021, Mineral Resource is now *required* to be reported Exclusive of Mineral Reserve, referred to here as 'Exclusive Mineral Resource' (EMR). EMR is equivalent to reporting Mineral Resource additional to Mineral Reserve, which has been permitted in jurisdictions outside the US for many years. Despite this, there are no practical guidelines in the literature or relevant reporting codes and rules for Competent/Qualified Persons to refer to, which creates a risk that disclosure of EMR estimates across the industry will be inconsistent and potentially ambiguous for analysts, investors, potential investors, and other stakeholders.

INTRODUCTION

S-K1300, like other disclosure guidelines and reporting standards, does not discuss *how* the reporting of EMR should be done from a technical point of view. In this paper we propose technical definitions and discuss the merits of various approaches and issues that Competent Persons should consider when reporting.

In simple terms we define EMR as the Inclusive (total) Mineral Resource (IMR) less the portion converted to Mineral Reserve before dilution and other modifying factors are applied. Importantly, consideration of the spatial conditions is critical so that estimating EMR is not a straightforward 'accounting' approach.

We discuss and recommend an approach for addressing the fundamental issue of determining Reasonable Prospects for Eventual Economic Extraction (RPEEE) of the resultant EMR. The term RPEEE is used in CRIRSCO, the JORC Code (2012) and SAMREC Code (2016), whereas S-K1300 uses a subtly different version – Reasonable Prospects for Economic Extraction (RPEE). There are important nuances between RPEEE and RPEE that may need careful consideration when companies are reporting in multiple regulatory jurisdictions.

Application of modifying factors such as mining dilution and mining recovery to three-dimensional block models means that Mineral Reserve comprises a different *spatial* definition compared to Mineral Resource (this may not be the case when considering two-dimensional modelling techniques such as polygons or grids for example). This difference in spatial definition is why geoscientists and mining engineers often create separate models in support of Mineral Resource and Mineral Reserve. We discuss the nuances that need to be considered for different modelling and mining methods and EMR delineation and reporting.

Future revisions to reporting codes in all jurisdictions are encouraged to incorporate appropriate guidance for reporting EMR in public disclosures.

BACKGROUND

The term 'Exclusive' does not appear in the JORC Code, which instead uses the phrase 'Mineral Resource *additional* to Ore Reserve'. SAMREC takes a similar approach to JORC, whereas S-K1300, and CRIRSCO use the term 'Exclusive' when clarifying the basis of reporting. Transparency, materiality, and competency are the common principles across all codes, and we take the terms 'additional to Mineral Reserve' and 'Exclusive of Mineral Reserve' to have the same intent.

In this paper the term 'Mineral Reserve' is used because the motivation for the paper came from the introduction of S-K1300 and that is the term defined in that rule. Mineral Reserve is synonymous with the preferred term 'Ore Reserve' in JORC.

DEFINITIONS

The EMR is derived from the IMR less the material used to generate the Mineral Reserve *before dilution and other modifying factors are applied*. To preserve the link between the EMR and the Mineral Reserve, all entities should be reported from the same date-stamped Mineral Resource model. Both IMR and EMR are underpinned by the same input parameters and assumptions (ie no additional input parameters or modifying factors are required for EMR).

EMR is determined through the assessment of five principal components, described below as Portions, which are defined in Table 1.

TABLE 1

Definitions of portions.

Portion	Open Pit	Underground
1	The Mineral Resource between the Mineral Reserve pit design and the Mineral Resource pit shell defined by the Mineral Resource criteria.	The Mineral Resource that lies outside an underground mine design used to generate the Mineral Reserve but within conceptual mineable stope shapes (MSS) (eg using a Mineable Stope Optimiser (MSO) software) defined by the Mineral Resource criteria.
2	Measured and Indicated Resource inside the Mineral Reserve pit design that do not meet the Mineral Reserve criteria and have not been incorporated as dilution within the Mineral Reserve.	There is no underground equivalent for Portion 2. Any Measured and Indicated Resource inside the MSS will be reported as Mineral Reserve because they typically cannot be mined separately.
3	Inferred Resource inside the Mineral Reserve pit design that has not been incorporated as dilution within the Mineral Reserve.	Inferred Resource inside the Mineral Reserve mine designs where this material has not been incorporated as dilution within the Mineral Reserve.
4	The Mineral Resource where technical studies to generate Mineral Reserve have not yet been completed or where economics (RPEEE or RPEE) support Mineral Resource but do not meet Mineral Reserve criteria.	
5	Stockpiles or tailings storage facilities which contain mineralised material that qualifies as Mineral Resource but not as Mineral Reserve. Mineral Resource criteria and RPEEE or RPEE principles will need to have been applied as appropriate.	

For illustration, Figure 1 (open pit) and Figure 2 (underground) show the various Portions and represent common approaches for constraining the Mineral Resource by using optimisation software to create volumes (pit designs, shells or stope optimisation shapes) above different cut-off criteria (noting that grade is often not the sole criterion). Mining method and selection of cut-off criteria are important constraints used to define Mineral Resource (amongst other criteria). Most commonly companies report Mineral Resource at a lower cut-off grade, typically reflecting a higher commodity price, compared to Mineral Reserve, however, using the same cut-off grade is sometimes done –

the impact on EMR is that when the same cut-off grade is used there will be less EMR. The methods of constraining Mineral Resource shown here are examples only and can be done in other ways at the discretion of the Competent Person. Whatever method is adopted will require supplementary notes in disclosures to advise the reader on how to interpret the EMR estimates.

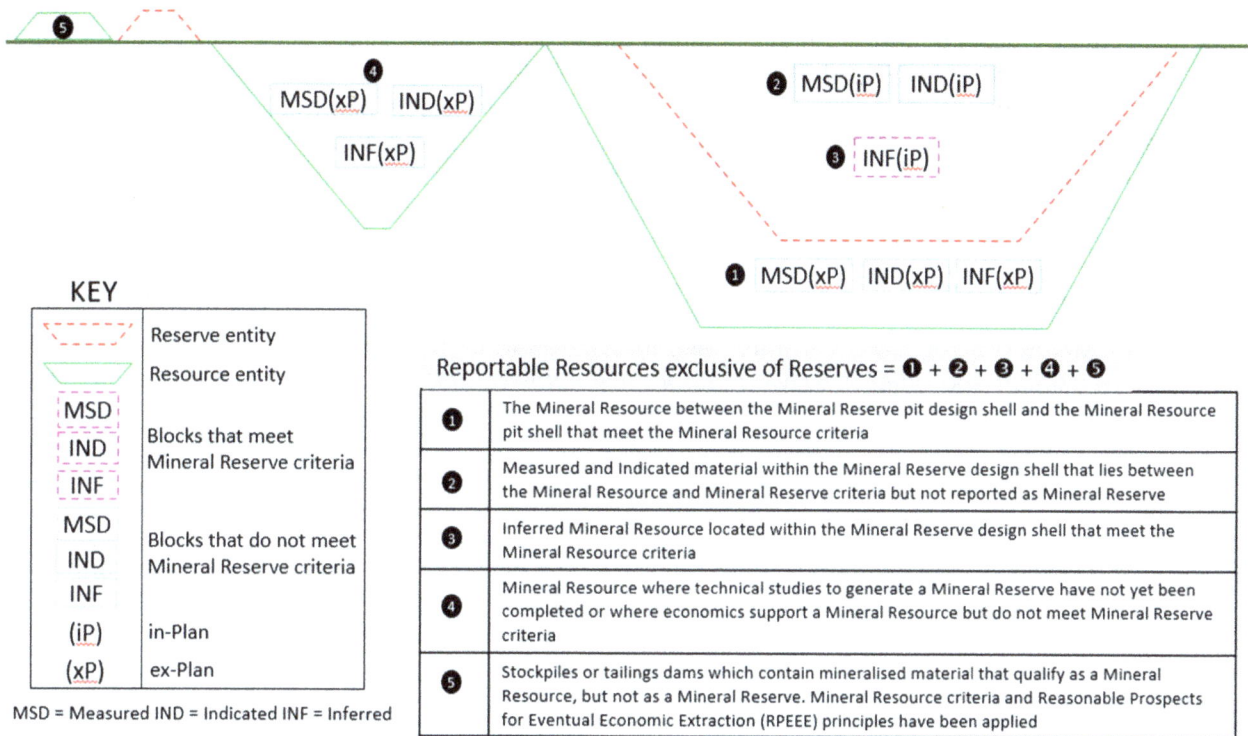

FIG 1 – Schematic representation of Mineral Resource and Mineral Reserve entities and Portions for open pit mines.

Reportable Resources exclusive of Reserves = ❶ + ❷ + ❸ + ❹ + ❺

❶	The Mineral Resource between the Mineral Reserve pit design shell and the Mineral Resource pit shell that meet the Mineral Resource criteria
❷	Measured and Indicated material within the Mineral Reserve design shell that lies between the Mineral Resource and Mineral Reserve criteria but not reported as Mineral Reserve
❸	Inferred Mineral Resource located within the Mineral Reserve design shell that meet the Mineral Resource criteria
❹	Mineral Resource where technical studies to generate a Mineral Reserve have not yet been completed or where economics support a Mineral Resource but do not meet Mineral Reserve criteria
❺	Stockpiles or tailings dams which contain mineralised material that qualify as a Mineral Resource, but not as a Mineral Reserve. Mineral Resource criteria and Reasonable Prospects for Eventual Economic Extraction (RPEEE) principles have been applied

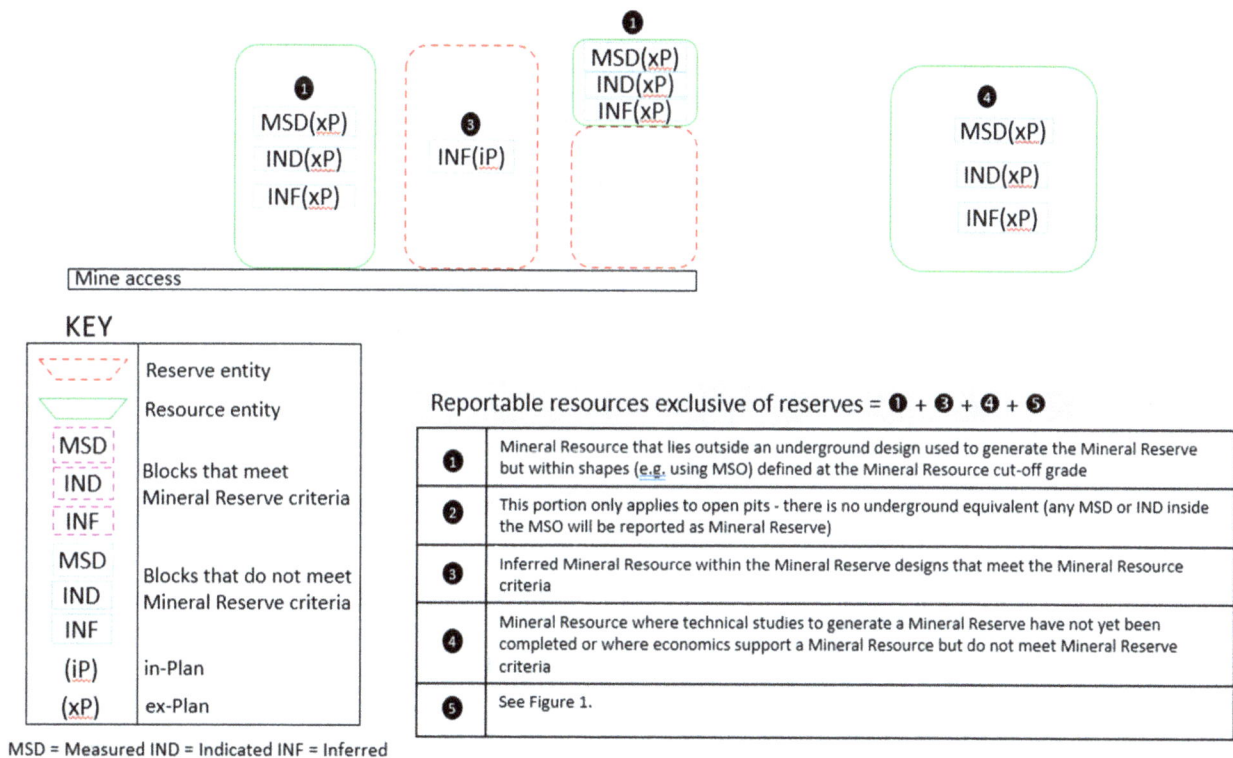

FIG 2 – Schematic representation of Mineral Resource and Mineral Reserve entities and Portions for underground mines (MSO = Mineable Stope Optimiser).

Reportable resources exclusive of reserves = ❶ + ❸ + ❹ + ❺

❶	Mineral Resource that lies outside an underground design used to generate the Mineral Reserve but within shapes (e.g. using MSO) defined at the Mineral Resource cut-off grade
❷	This portion only applies to open pits - there is no underground equivalent (any MSD or IND inside the MSO will be reported as Mineral Reserve)
❸	Inferred Mineral Resource within the Mineral Reserve designs that meet the Mineral Resource criteria
❹	Mineral Resource where technical studies to generate a Mineral Reserve have not yet been completed or where economics support a Mineral Resource but do not meet Mineral Reserve criteria
❺	See Figure 1.

Accumulation of the Portions to estimate EMR

Derivation of the estimate is based on the sum of the Portions to determine the EMR. That is, EMR = Portion 1 + Portion 2 + Portion 3 + Portion 4 + Portion 5.

Important notes on the EMR estimation process:

- The RPEEE assessment is only completed once for the Mineral Resource. This demonstrates viability for both the IMR and EMR. There is no RPEEE test on the remaining standalone EMR without the Mineral Reserve. The EMR component, by itself, does not warrant a discrete or abstract evaluation for mining, as it will always be considered a subpart of the greater Mineral Resource footprint. Also, the expectation is the EMR volume and spatial conditions are appropriate for meeting the RPEEE or RPEE assessment as determined when generating the IMR. This, however, may warrant a reasonableness check at the time of reporting.

- Both EMR and IMR should be underpinned by the same input parameters and assumptions.

- The EMR is derived from the IMR, less the material used to generate the Mineral Reserve *before dilution and other modifying factors are applied*. The expectation is that in most cases there will be approximate equivalency in total metal reported under IMR and EMR (in relation to the material used to generate the Mineral Reserve). Meaning that under the EMR process it is not anticipated that there will be a material or apparent 'loss' of metal compared to the IMR.

In terms of a decision process flow, Figure 3 shows how an IMR is converted to EMR, via the path through application of modifying factors in generating the mine plan, then consideration of material that does not qualify as Mineral Reserve and is therefore additional to the Mineral Reserve.

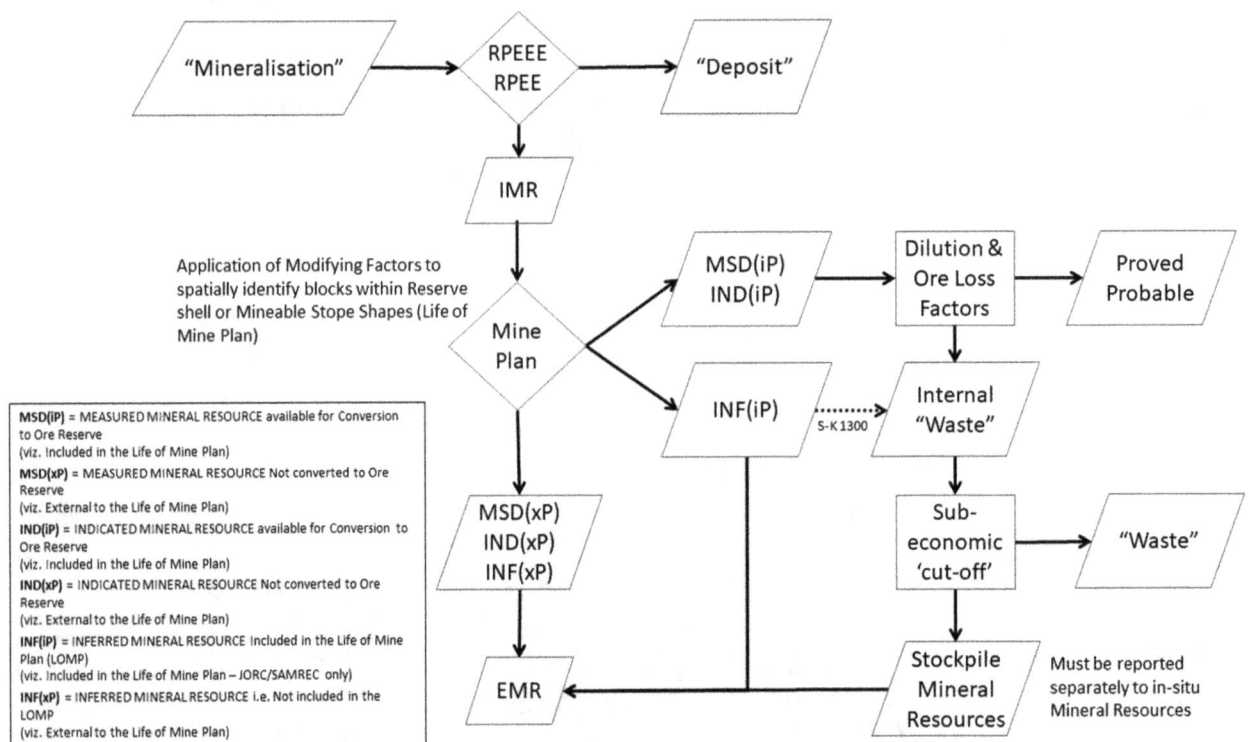

FIG 3 – Decision process flow diagram from IMR to EMR.

TREATMENT OF 'REASONABLE PROSPECTS' IN VARIOUS CODES

The definition and description of requirements for assessing reasonable prospects 'for economic extraction' or 'eventual economic extraction' vary among the different disclosure codes and regulations. It is important for Competent Persons to understand the nuances of these as they may need careful consideration when companies are reporting in multiple regulatory jurisdictions. All reporting codes aligned with the CRIRSCO reporting template require an assessment of reasonable prospects and the basis of the assumptions to be disclosed.

Examples of such differences include requirements and specific reporting for mining remnants, pillars, or low-grade mineralisation under the JORC and SAMREC Codes, which are not defined in the Canadian Institute of Mining standards (CIM, 2014) or S-K1300.

The US SEC requires the assessment of the Mineral Resource to be demonstrated at the time of reporting and not at an undefined (eventual) date in the future. This could potentially impact on several input criteria and key assumptions such as potential markets, commodity or planning price decks or processing recovery technologies. It should be noted that the SEC has been specific in stating that this does not mean that extraction is assumed to occur immediately, but rather it is expected that extraction will occur over a temporal period, which may vary depending on the commodity, and this assumption of the period for extraction should be disclosed within any Technical Report Summary.

One consistency across reporting codes is a consensus that the Mineral Resource is not simply an inventory of all mineralisation drilled and sampled. Under S-K1300 the minimum requirement for reporting Mineral Resource is that an initial assessment must be completed by a Qualified Person(s) and include a qualitative assessment of the assumed technical and economic factors, together with operational factors. As with all reporting codes this analysis must include and disclose the basis for the cut-off grade including assumed mining costs, mining methods, metal prices, and processing recoveries, as appropriate. It is this context that should be considered as the basis purely for reporting of Mineral Resource, which is consistent with the guidelines of other CRIRSCO based codes. The process should not imply a more detailed level of engineering study has been completed to misconstrue the Mineral Resource as Mineral Reserve.

MODELLING AND MINING METHODS

Application of modifying factors such as mining dilution and mining recovery to three-dimensional block models, which are dependent on the mining method, mean that Mineral Reserve may comprise a different *spatial* definition compared to Mineral Resource (see Figure 4). This difference in spatial definition is why geoscientists and mining engineers often create separate models of Mineral Resource and Mineral Reserve.

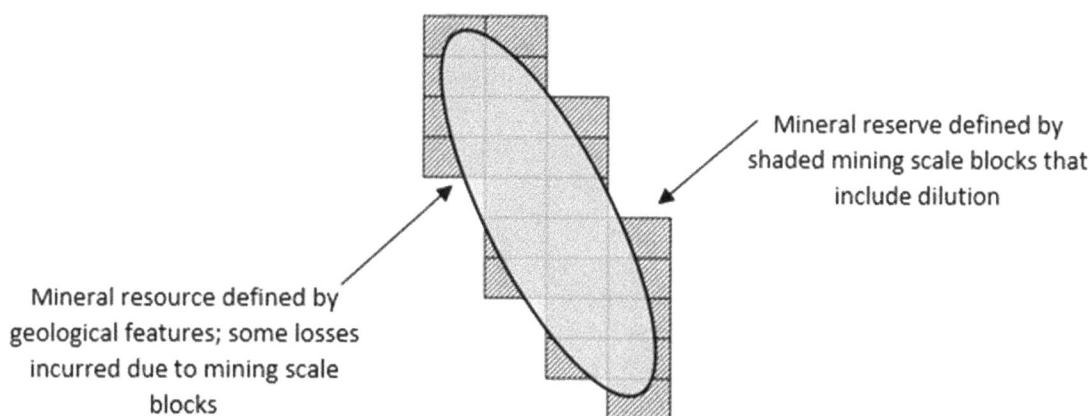

FIG 4 – Schematic example of the spatial relationship between the Mineral Resource and Mineral Reserve.

One method to exclude the Mineral Reserve from the Mineral Resource is to apply spatial 'cookie-cutting' to the block model (cookie-cutting colloquially describes the process of selecting a specific volume in the block model using a three-dimensional wireframe model of that volume). However, the mine design will incorporate mining modifying factors and may leave remnants behind or include subeconomic or non-mineralised material.

When updating Mineral Resource estimates, caution should be taken if this is undertaken by subtracting yearly production (ie the accounting method) without spatial analysis as this can lead to progressive deviation from actual spatial extraction if applied over several reporting cycles. When the actual mined-out shapes are applied to the Mineral Resource and the EMR is re-evaluated based on spatial analysis, the difference could be material due to the actual density of the material extracted being different to estimates and the shape of the extracted areas impacting the ability to mine portions of the EMR. Furthermore, the accounting approach ignores the fact that economic criteria can change over time, which requires re-evaluation of RPEEE/RPEE and modifying factors. This issue of currency is highlighted in requirements in S-K1300, for example. If depletion of an existing Mineral Reserve or EMR is undertaken for more than three consecutive reporting cycles, it is recommended that a comment as to the continued applicability of the underlying assumptions be provided by the Competent Person.

After cookie-cutting with the mine design geometry it is then questionable whether remnants should continue to be reported as Mineral Resource. If extracting the Mineral Reserve leaves insufficient width of mineralisation or critical mass to support mining, one could argue that the remnant Mineral Resource has been sterilised and should not be reported. If so, the technical mining professional must now create a sterilisation model and apply that to the Mineral Resource model.

The topic of 'remnant resource' is particularly important in some situations, for example mature caves, where the remnants may have been 'carried forward' by depletion but, as the cave nears the end of its operation, those remnants may not be recoverable and/or overstate the potentially remaining metal (this may be relevant when calculating impairments). In this situation these previously quoted remnants are likely to be written-off since they no longer meet the criteria that Mineral Resource must have RPEEE, (ie 'more likely than not') and would not meet the S-K1300 requirement for RPEE.

The JORC Code (clause 41) specifically refers to mineralised fill, remnants, pillars, low-grade mineralisation etc. It states that this material is subject to the same reporting criteria as in situ Mineral Resource where they satisfy RPEEE, and where extraction is reasonably justifiable in the case of Mineral Reserve. The guidance in the JORC Code goes on to say that 'judgements about the mineability of such mineralised material should be made by professionals with relevant experience'. Clearly, if the RPEEE assessment is negative for all or part of the mineralised material then this material cannot be classified as either Mineral Resource or Mineral Reserve. If some portions are subeconomic, but there is a reasonable expectation that it will become economic, then this material may be classified as Mineral Resource. It is recommended that remnants should be written off as soon as possible in the reporting cycle if the relevant experts, using appropriate methodology and long-term assumptions, determine the remnant tonnage and grade to be 'uneconomic'.

The JORC Code guidance recommends that the tonnage and grade estimates for 'marginal grade material', which may be intended for treatment towards the end of mine life, be itemised separately in Public Reports, although they may be aggregated with total Mineral Resource and Mineral Reserve figures.

In the case of a block cave, once the footprint has been defined, there is little flexibility to change it (apart from stopping extraction from a given drawpoint), so the remnant Mineral Resource and Mineral Reserve must be defined within the confines of the existing block cave geometry. In this case the Mineral Resource will be the same as the Mineral Reserve (this mining method comes with a very high level of dilution that will have to be applied to the Mineral Resource). This is also the case for sublevel caving, although there is a little more flexibility due to the larger amount of drilling and blasting information and the reduced distance between sublevels.

Often, in open pit mines, the mining methods will be applied at different scales. For example, a truck and shovel fleet can mine at a range of bench heights and this difference in scales will change the estimated Mineral Reserve because it changes the amount of dilution and recovery, but it does not necessarily change the Mineral Resource (only the level of dilution and ore loss changes for the same mining volume).

An operation may adopt different selective mining methodologies in different parts of a deposit due to the diverse geological features (eg thickness, shape, continuity, and content) or at different times due to economic conditions. For example, if under certain circumstances a bulk commodity is more valuable when it contains less contaminants, then an operator may decide to mine more selectively to reduce mining dilution (contamination), which means the Mineral Reserve is changed.

Presenting the Mineral Resource as stable relative to the Mineral Reserve, which is an advantage of reporting IMR, allows for a degree of change to mining methods and mine designs whilst ensuring that an investor retains a clear view of the future potential of the property.

Faced with the requirement to report EMR it will be tempting for some registrants to avoid the additional work and simplify the problem using an accounting approach (ie IMR – Mineral Reserves = EMR), then argue that the difference is immaterial. If such an approach is adopted it should be made transparent to the investor with supporting rationale, but even if the result is immaterial the accounting approach is not considered good practice or proper diligence because it ignores the spatial component, and it is not endorsed here. The authors consider it good practice that when EMR estimates are reported the IMR is also reported for completeness and full transparency; some companies already disclose this way but not all.

OTHER CONSIDERATIONS

The impact of applying the modifying factors (and other changes to the underlying model of Mineral Resource) are clearer in the conversion of Mineral Resource to Mineral Reserve when EMR is used for reporting compared to when IMR is used, as IMR could mask these changes. Presenting the changes in EMR versus Mineral Reserve using a waterfall chart can make the conversion process more transparent because it highlights the inverse relationship. Changes to the estimate of either the EMR or Mineral Reserve should be reflected as a negative change in one with an approximately equivalent positive change in the other (or vice versa) as illustrated in Figure 5. The waterfall chart example reflects unit changes allocated to five key criteria impacting the difference (reconciliation) between annual opening and closing balances.

EMR is more transparent for investors as the portion of Mineral Resource not converted to Mineral Reserve is clear. In the authors experience, people not familiar with the subtle distinction between Mineral Resource and Mineral Reserve often add the two, seeking to understand the total potential of the project or operation. Reporting EMR minimises the chances of inappropriate addition leading to incorrect conclusions of the total and potentially misleading reporting.

Another relevant topic that may be important to consider in future guidelines for EMR estimation includes the reporting of grade-tonnage curves, where an EMR grade-tonnage curve could potentially be misleading about the nature of the IMR due to the spatially different volumes being reported. If presenting grade-tonnage curves, then both EMR and IMR grade tonnage curves should be presented for the sake of transparency and clarity.

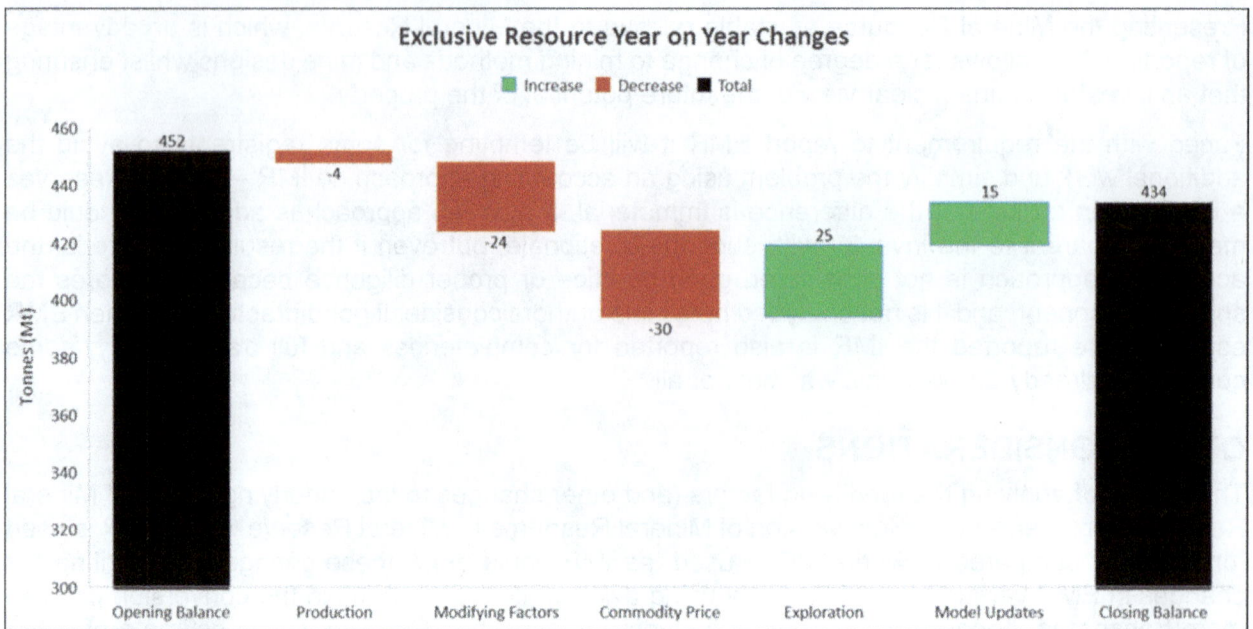

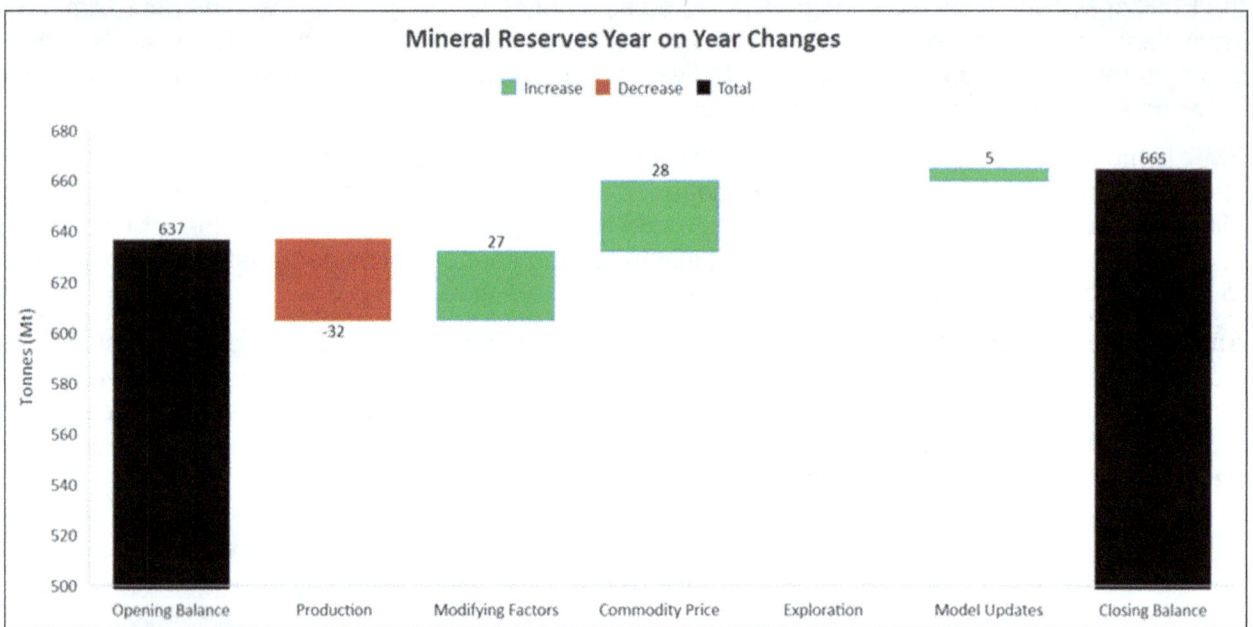

FIG 5 – Waterfall charts of changes to EMR (above) and Mineral Reserve (below).

CONCLUSIONS

Presented here is a definition based on five Portions describing mineral entities for underground and open pit configurations, that sum to provide the reportable EMR estimate. The EMR is derived from the IMR less the material used to generate the Mineral Reserve *before dilution and other modifying factors are applied*. However, the process is not a simple accounting procedure, and the authors discourage that approach because it disregards important spatial conditions and considerations.

Considerations of related issues such as RPEEE, RPEE, spatial modelling, mining methods, cut-off grades, grade-tonnage curves, year-on-year changes and the discretion and judgement of Competent Persons in relation to these matters have been discussed.

One of the more significant issues discussed was the treatment of RPEEE or RPEE when considering EMR; the recommendation is that RPEEE or RPEE is assessed for an entire deposit (IMR) rather than for EMR separately. The nuances between S-K1300 and other codes in relation to RPEEE and RPEE are important to understand for companies reporting in the US, and especially for multi-listed companies who face the challenge of satisfying the various codes with consistency.

Reporting both IMR and EMR offers investors full disclosure and pre-empts the potential for confusion or misunderstanding. Each approach has positive and negative aspects: IMR potentially remains more stable over time and shows the total Mineral Resource without mixing Mineral Resource and Mineral Reserve estimates. EMR presents only that part of the Mineral Resource not converted to Mineral Reserve, which allows clearer understanding of changes year-on-year and mitigates the risk that investors inappropriately sum IMR and Mineral Reserve to understand the total potential of a project or operation.

Future updates or revisions to the various reporting codes and rules going forward are recommended to incorporate meaningful guidance on EMR estimation to converge the industry's understanding and interpretation of this important component of Mineral Resource and Mineral Reserve reporting. Driving alignment on EMR disclosure will avoid ambiguity and potential confusion amongst analysts, investors, and stakeholders. This will support a level playing field, help facilitate improved equivalency in reporting and assist comparative industry analysis.

For a variety of reasons, the mining industry is coming under increasingly intense scrutiny. Transparent and consistent reporting of Mineral Resource and Mineral Reserve is just one way of ensuring regulatory and investor confidence.

ACKNOWLEDGEMENTS

The authors wish to acknowledge discussions and peer review of this paper by Andre Badenhorst, Dr Winfred Assibey-Bonsu and Malcolm Thomas of Goldfields Limited and Andrew Weeks of 2020Resources.

REFERENCES

CIM, 2014. Canadian Institute of Mining, Metallurgy and Petroleum (CIM) Definition Standards for Mineral Resources and Mineral Reserves. CIM Standing Committee on Reserve Definitions.

CRIRSCO, 2019. International Reporting Template for the public reporting of Exploration Targets, Exploration Results, Mineral Resources and Mineral Reserves. Committee for Mineral Reserves International Reporting Standards.

JORC, 2012. Australasian Code for Reporting of Exploration Results, Mineral Resources and Ore Reserves (The JORC Code) [online]. Available from: <http://www.jorc.org> (The Joint Ore Reserves Committee of The Australasian Institute of Mining and Metallurgy, Australian Institute of Geoscientists and Minerals Council of Australia).

SAMREC, 2016. The South African Code for the Reporting of Exploration Results, Mineral Resources and Mineral Reserves. The South African Mineral Resource Committee.

S-K 1300, 2019. United States Securities and Exchange Commission Regulation S-K part 1300.

Environmental, social and governance considerations in public mineral reporting

T Rowland[1], J Joughin[2], F Cessford[3], V Chamberlain[4], T Flitton[5], H Arvidson[6] and N Pollock[7]

1. Industry Consultant, Tim Rowland Consulting, Krugersdorp 1739 Gauteng, South Africa. Email: tim.william.rowland@gmail.com
2. Corporate Consultant (Sustainability), SRK Consulting, Cardiff CF10 2HH, UK. Email: jjoughin@srk.co.uk
3. Corporate Consultant ESG, SRK Consulting UK Ltd, Cardiff CF10 2HH, UK. Email: fcessford@srk.co.uk
4. FAusIMM, Senior Vice President Strategic Planning, AngloGold Ashanti, Johannesburg 2196 Gauteng, South Africa. Email: vchamberlain@anglogoldashanti.com
5. VP Resource and Reserve, AngloGold Ashanti, Johannesburg 2196 Gauteng, South Africa. Email: tflitton@anglogoldashanti.com
6. MAusIMM, Chief Geoscientist, K2fly Pty Ltd, Perth WA 6000. Email: heath.a@k2fly.com
7. AAusIMM, Chief Executive Officer, K2fly Pty Ltd, Perth WA 6000. Email: nic.p@k2fly.com

ABSTRACT

Environmental, social and governance issues (ESG) have become a defining feature in the marketplace to differentiate preferred investments. With the sustainability commitment and reporting landscape anticipated to evolve exponentially, it is now imperative to rapidly develop the approach taken by mineral companies to adequately identify and address ESG risks and opportunities in the reporting of Mineral Resource and Mineral Reserve (MRMR). MRMR cannot be effectively reported as technically and financially assured and realistically executable, if ESG is not integrated into the strategy, business model, input assumptions and process that defines and drives robust MRMR public reporting.

Regulatory reporting codes across the globe are presently in various stages of being updated and modified to ensure ESG factors are appropriately covered as a key part of diligent and transparent disclosure. Security exchanges are making it clear that companies must not omit ESG risks from reports to investors.

This paper discusses associated challenges for MRMR reporting and suggests avenues for improvement. Key themes include: the importance of procedure rather than prescription and checklists; how ESG criteria impact Modifying Factors and determination of Reasonable Prospects for Economic Extraction (RPEE); the need for financial models to account for material ESG commitments; the uniqueness of each mineral asset; the need for education, upskilling and accreditation of both ESG experts about MRMR reporting and equally Competent Persons in ESG matters; and the need for clear and balanced reports that meet the needs of end users. (RPEE is adopted here for simplicity and consistency, recognising that some reporting codes still include 'Eventual' in the term (ie RPEEE). Whilst RPEE and RPEEE are subtly different the intent is to point to the process of assessing ESG factors in determining reasonable prospects.)

The principles of transparency, materiality and competence remain paramount when considering the impact of ESG matters on MRMR reporting, and this necessitates an inclusive, cohesive, and integrated approach to assessing ESG issues to ensure the necessary expertise and diligence is applied.

INTRODUCTION

Global demand for minerals to facilitate a transfer to more sustainable energy supply is growing exponentially and this requires more mining (Hund *et al*, 2020). While there is increasing stakeholder interest in mining and the ESG credentials of miners and their products, the industry suffers from a lack of trust about ESG performance and is being pressurised by multiple stakeholders to increase transparency. Coupled with this, there has been an explosion of ESG disclosure and performance standards and emergence of new schemes for validation of performance. Rating and benchmarking of mining companies by rating agencies and specialist data providers for investors intensifies pressures on miners.

ESG considerations are creating risks and opportunities for financial institutions and mining companies, adding complexity to the comparative valuation of assets (Mitchell, 2022; Vagenas, 2021). Investor concerns and expectations about operational and financial risks are being compounded by global trends of declining resources. The observed exploitation of the 'Green Investor Market' by some companies with little or no foundation is disingenuous and formal guidance on ESG integration needs to address this to protect the investor.

Currently, interpretations of 'sustainable development' by the mining industry, reporting code committees and security exchange regulators vary, despite and perhaps because of the abundance of standards. Now is the ideal time to future proof and converge global mineral reporting codes and related securities exchange regulations to frame current expectations of investors and be flexible enough to accommodate future requirements. However, there can be no dispute that ESG is integral to responsible business and long-term value and as such must be integral to the estimation of a mineral company's fundamental asset base, its Mineral Resource and Mineral Reserve.

Importantly, ESG is already flagged in the numerous reporting codes, albeit with varying levels of clarification on its assessment and application. Along with mining, metallurgical, processing, infrastructural, economic, marketing and legal criteria, ESG is already part of the spectrum of specified Modifying Factors to be considered in the relationship between exploration, Mineral Resource and Mineral Reserve. It is therefore not new and does not warrant a special niche in mineral reporting but as emphasised in this paper, ESG needs to be fully integrated into company strategy, business and operating processes with a clear 'golden thread' linking to mineral reporting.

The process of integrating ESG into global reporting codes is a rare opportunity to steer and enable continuous improvement in mineral reporting, emphasising the critical importance of ESG factors in underpinning realistic MRMR disclosure. A company's mineral reporting needs to integrate ESG commitments with its core business processes and value chain, without which MRMR cannot be declared as realistic, assured and executable.

This paper sets out key themes, emerging challenges, contentious topics and core expectations associated with ESG integration with mineral reporting. Options are proposed to improve outcomes for the industry. Importantly, this paper provides a discussion document to assist with continuous improvement of the various global reporting codes when making revisions and modifications.

The burning platform is not founded on the need to facilitate change but rather on the consequences of not expediting it in a manner that is both practical and reasonable to enact for small, medium and large mining companies, and delivers true value creation to all stakeholders.

In this paper the term 'mineral property' is used to capture both projects and operations. Following discussion of standards and scope, industry trends and expectations, the paper leads into observations of mineral reporting codes and how key aspects of those codes can be considered through the ESG lens. A discussion of relevant roles and ways of working to improve outcomes is followed by a framing for minimum disclosure. Guiding principles for improvement are distilled from all the above prior to concluding that whilst ESG integration into mineral reporting is challenging, there are pragmatic options to consider and optimism that substantial benefits will accrue.

STANDARDS AND FRAMING THE SCOPE OF ESG IN MINING

There is no universally accepted definition of ESG, a web-search will not provide a definition but will list varying issues covered by the term. The term has direct links with responsible investment, having been coined by 20 financial institutions in a report called 'Who Cares Wins' in 2004 (United Nations Global Compact, 2004). This report was the springboard for the formation of the United Nations Principles for Responsible Investment (UN PRI). This investor ESG initiative reported having 3826 signatories to the PRI, comprising 3404 investors and 422 service providers, and representing collective assets under management of US$121 trillion as of 31 March 2021 (UN PRI, 2021).

The PRI commit signatories to incorporate ESG issues into investment analysis and decision-making processes, to active ownership with ESG incorporated into ownership policies and practices, and to seeking of ESG disclosure from investees. While ESG is central to the PRI, the PRI asserts that a definitive list of ESG issues does not exist, stating:

> *It would not be possible or desirable to produce a list, or a set of definitions, that claimed to be exhaustive or definitive. Any such list would inevitably be incomplete and would soon be out of date.* (UN PRI, 2018)

Framing the scope of ESG for miners

In the mining industry, there is wide acceptance that the term 'ESG' embraces all aspects of sustainability (and is often used inter-changeably with sustainability) and clearly links directly to corporate responsibility as first defined in the UN Global Compact. This states its aim to align [business] strategies and operations with universal principals on human rights, labour, environment and anti-corruption, and take actions that advance societal goals.

Quality ESG reporting in the context of mineral reporting is likely to help differentiate companies as preferred investments; in fact this is a rapidly accelerating trend observed in the last two years as investors allocate increased weight to sustainability in the mining sector.

The protection against an excessive reporting burden and potential duplication with other sustainability reporting is important. If not checked, mushrooming reporting burdens can easily become too onerous and impractical to embrace and administer, given that reporting capabilities and capacities are not unlimited across industry.

Potentially serious disconnects can occur when material commitments are made in the public domain in separate sustainability and annual reports, mineral reports and climate action reports and other communications with investors. These are effectively constructive obligations and need as much attention as material commitments made in contracts with the government and conditions of approvals – lip service and greenwashing must be avoided. It is important to recognise that project and life-of-mine cash flows and financial models should integrate and reflect all input costs, including ESG-related costs, necessary to execute the plan, as appropriate to the life cycle stage of the asset.

Consistency between company ESG reporting, annual reports and financial statements, including life-of-mine provisioning, eg decarbonisation, water management, end land use etc is often lacking and warrants close attention – inconsistencies have been noted with some companies where life-of-mine cash flow models have not appropriately integrated these costs. Residual and latent liabilities will also need disclosure but the challenge, though, is how to quantify this upfront?

Figure 1 summarises the key ESG topics covered in responsible mining standards – some of which are identified in the next section. Recognising that the Pan-European Reserves and Resources Reporting Committee (PERC, 2021) and other codes include or are likely to include ESG as a specific Modifying Factor, the Committee for Mineral Reserves International Reporting Standards (CRIRSCO) has acknowledged a need to derive an ESG definition that is directly applicable to MRMR reporting and has therefore requested its ESG subcommittee to develop one.

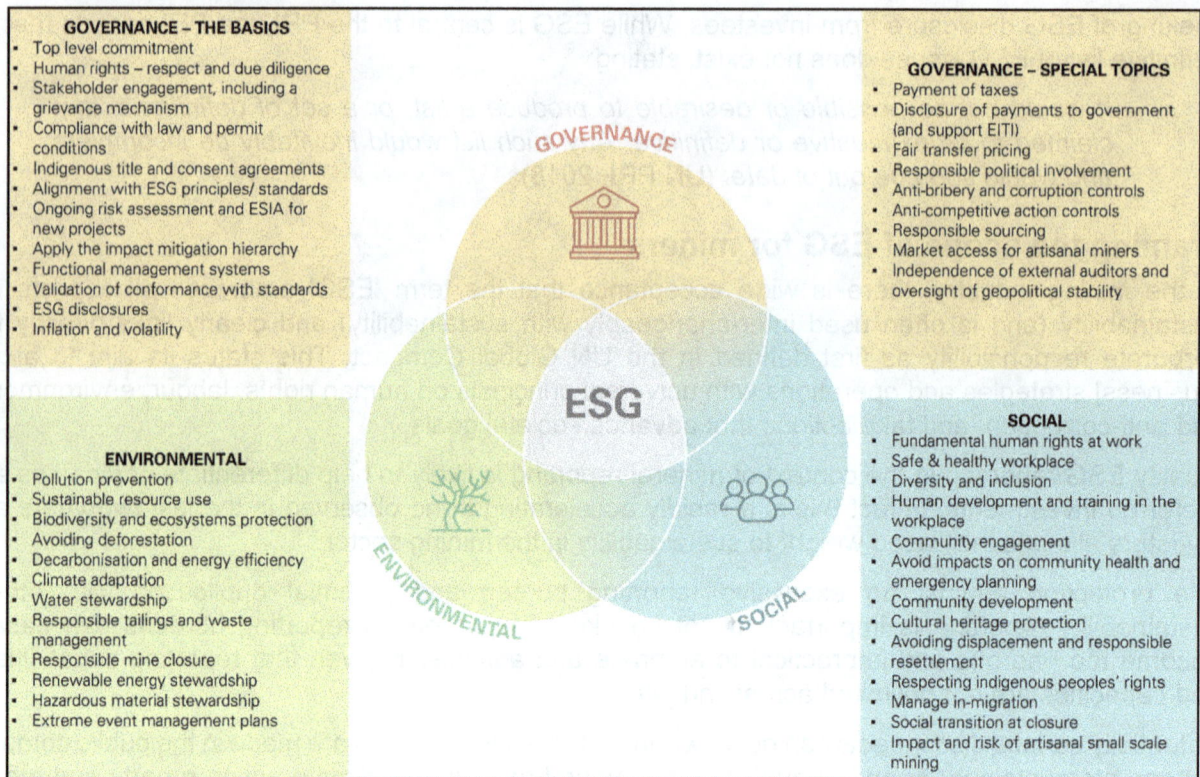

FIG 1 – ESG topics relevant to mineral exploration and mining.

Members of the mining industry are recognising that the separation of environment, social and governance is artificial, as the subjects shown in Figure 1 are typically cross-cutting and often closely inter-related. For example, the topic of responsible mine waste management defined in the Global Industry Standard on Tailings Management (GISTM; Global Tailings Review, 2020) covers most of the governance topics on the left side of the image and many of the listed environmental and social topics. The same is true for the subject of water stewardship. Standards on climate disclosure also place importance on governance. Human rights are often placed under the social topic but respect for human rights and human rights due diligence are fundamental to governance. Respect for human rights underpins many of the other listed ESG topics.

The environmental and social components to be considered in MRMR reporting may be constrained to a 'zone of influence' of the mineral property. This requires at least a preliminary understanding of the bio-physical and social context, which extends beyond the direct area of the concession and includes wider catchment (air, water, labour etc) considerations. Material environmental and social factors that have the potential to become Modifying Factors for MRMR reporting can then be identified through a process of:

- Analysing and evaluating this context.

- Understanding the issues and concerns of potentially affected stakeholders.

- Predicting potential mineral property-induced impacts that may arise because of interaction between this context and the proposed mineral property activities.

- From these, identifying risks (threats and opportunities) that could influence the ability of the reporting entity to sustainably explore for, develop, operate and close the mineral property.

Of the E, S and G, governance may be most difficult to define as there is active debate about what should be included and excluded in MRMR reporting. Is it external-governmental and/or legal and regulatory (resource) governance and/or internal-corporate governance and/or in-house mineral property governance? The balance of opinion appears to be tipping toward restricting the definition of G for MRMR reporting to mineral property governance, as corporate governance is better covered under the requirements for annual reporting. The view that governmental and regulatory governance factors are integral to the concept of ESG is endorsed.

Governance in MRMR reporting should aim to avoid repeating other company governance disclosures with a focus on governance 'in action' and what oversight is in place to provide the company executive and Board of Directors with the necessary comfort on the due diligence and assurance underpinning MRMR statements. Confirmation of the economic viability and impairment status of assets is necessary to support the veracity of MRMR disclosure. Financial and tax transparency, including verification that operating and capital expenditures incorporate all costs required to execute the plan, including delivery on ESG commitments, cannot be overstated.

Table 1 presents governance concepts relevant to project planning and MRMR estimation based on principles in ESG standards and what it means for Competent Persons and ESG subject matter experts (SMEs). There is a need for awareness, training and multi-disciplinary mineral property teams to understand and meet these responsibilities.

TABLE 1

Governance concepts in mineral property planning and MRMR estimation.

Topic	Governance concepts applicable to mineral property planning and MRMR estimation
Leadership/top level commitment to ESG	• The mineral property leaders and Competent Persons are responsible for ESG integration into decision-making.
Human rights	• Respect for human rights needs to be applied in all decision-making and considered in MRMR estimation and reporting.
Compliance with law and regulatory frameworks	• Understanding of the law and compliance obligations is fundamental. • Appreciation that the bulk of the compliance obligations are ESG obligations, which do not relate only to protection of people and their rights and the environment, but also to value creation for host countries and host communities.
Stakeholder engagement	• Address the interests and expectations of stakeholders (host country, communities, business and other stakeholders) in mineral property planning and impact and risk assessments.
Impact and risk assessment and management	• Affected people, biodiversity and ecosystems and the resources they depend on need to be considered in impact and risk assessment; adequate contextual information is needed to inform the assessments commensurate with mineral property planning. • Identified impacts and risks should be designed out where possible or addressed in management plans (applying the mitigation hierarchy of avoid, minimise, compensate/offset). • Water management, waste management, dust or air quality management, climate resilience and closure vision and provision are fundamental to mine planning, engineering and operation, and should not be seen as externalities. • Mineral property management systems to address ongoing impact and risk management.
Transparent and material disclosure	• Disclosure of material ESG risks and mitigating actions whilst avoiding perceptions around greenwashing.
Diligence	• Review of the adequacy of data for decision-making and estimation.

Standards

The mining industry faces a plethora of standards, which on first appearance can be daunting. However, there are common themes running throughout, most of them align with the ESG concepts outlined above. In simple terms, standards can be grouped as summarised below:

- Overarching standards reflecting widely accepted global norms: This includes things like the UN Global Compact mentioned above, the UN Sustainable Development Goals (United Nations, 2023), the UN Guiding Principles on Business and Human Rights (United Nations Global Compact, 2011), the International Labour Organisation (<https://www.ilo.org/global/lang--en/index.htm>) Declaration on the Fundamental Principles and Rights at Work and International Organization Standards (ISO). These may not be directly relevant to MRMR reporting but they directly influence business strategies and practices, that in turn are reflected in how mineral properties are developed.

- Disclosure standards (or laws, in some cases): These are driven by investors demand for transparency on both ESG risks to a company, as well as the impacts arising from a company (concept of double materiality). Examples include: the Global Reporting Initiative (<https://www.globalreporting.org/>); the European Parliament (2022); Extractive Industry Transparency Initiative (EITI, <https://eiti.org>) and the new disclosure standards being drafted by the International Sustainability Standards Board (<https://www.ifrs.org/groups/international-sustainability-standards-board/>) of the International Financial Reporting Standards (IFRS). Companies are rated and benchmarked by data providers for investors using the ESG disclosure information. MRMR codes are also disclosure standards, so could be placed in this group.

- Lender standards associated with debt or corporate financing: The most widely recognised of these are the Equator Principles EP4 (<https://equator-principles.com>), and the International Financial Corporation (IFC, 2012) Performance Standards, though some banks have their own sets of standards. They generally set performance expectations associated with human rights, pollution control, natural resource efficiency, and protecting and conserving biodiversity, ecosystem services and cultural heritage. They are applicable to MRMR reporting in that they set out the environmental and social processes by which companies engage with stakeholders, establish the bio-physical and social context, assess impacts and derive appropriate design and operational controls iteratively throughout mineral property development.

- Responsible mining or sourcing requirements: These are linked to supply chain demands for responsible minerals and are also performance based. Some are industry led, such as the International Council on Mining and Metals (<https://www.icmm.com>) Principles and Performance Expectations (and extensive associated guidance). Others are supply-chain led and are often sector specific such as The Copper Mark (2020) or World Gold Council (2019). Probably the most stringent set of performance expectations are set out by the Initiative for Responsible Mining Assurance (IRMA, 2018) based on a broad multi-stakeholder process.

- Issue specific standards: These are generally more technically orientated establishing standards or performance expectations around a specific topic and thus each MRMR mineral property would need to establish which of these would be applicable (in association with their business and regulatory stakeholders). Examples include the International Cyanide Management Code (International Cyanide Management Institute, 2021), Voluntary Principles on Security and Human Rights (2023) and World Health Organisation (2023) guidelines.

- Regional standards: These are standards or laws specific to a particular region of the world such as Europe (EU Directives) or Economic Community of West African States. As with global standards they may not be directly applicable to a mineral property but will set out wider ESG concepts that could influence regulators and investors.

Two subject-specific standards that have effected great change in the mining industry in the last couple of years are the Task Force for Climate-Related Financial Disclosure (TCFD, <https://www.fsb-tcfd.org>) and GISTM. For MRMR estimation and reporting, there is a need to pay more attention to decarbonisation and climate resilience and to mine waste management.

GROWING INDUSTRY EXPECTATIONS AND TRENDS

The scope of ESG applicable to mining is vast. However, selected expectations emanating from global audit companies can be interpreted and are summarised here. Capital allocation to support delivery on ESG strategies and public commitments is vital, as without appropriate funding those strategies are not realistic and deliverable, especially under challenging metal price and carbon price regimes. Once ESG uncertainty and opportunity is quantified by a company, the level of risk associated with social or environmental criteria or potential failures and opportunities concerning mineral development and reporting can be profiled. It then becomes easier to make the case for capital allocation into ESG driven initiatives, organisational restructures, resource allocation and the creation of new roles.

Most mining companies have strategies encompassing ESG visions and missions, but not all have looked beyond these to a purpose that resonates with business, regulators and investors regarding how they are 'wired to mineral reporting' in a meaningful and understandable manner. For example, incorporating ESG elements in company capital investment frameworks will enable a culture of excellence based on leading-practice asset and mineral property management. This equates to a better informed investment decision, minimising and controlling the business risk and maximising return. If this can be addressed, it should, in time, bring new investors into the market, particularly for those companies with strong, sustainable track records and transparent reporting.

Where ESG initiatives are typically led by the sustainability function in a company, it is important this function is elevated to have sufficient corporate representation and involvement with the MRMR estimation and reporting process. Good practice dictates that all disciplines should embed ESG into their work. Collaboration across disciplines and integrated planning is vital to addressing ESG risks and wider sustainability challenges, and realising new opportunities created by the energy transition. It is suggested that embedding ESG with the support of subject matter experts (SMEs) into the MRMR reporting framework is essential and their participation in company Resource and Reserve Committees is vital. Many Competent Persons and executive's balanced scorecards will include emphasis on technical assurance and risk management in reporting, but ESG also needs to be effectively captured in company standards and performance management tools to help drive integration.

Trending topics flagged by recent industry reviews include, but are not limited to, the high cost of decarbonisation commitments, preparing operations for climate change and the cost of implementing new technology. In addition, re-shaping traditional mining value chains, building flexibility and resilience in the face of regulatory uncertainty, embedding ESG into organisations and establishing a new paradigm for relations with indigenous people are also frequently raised.

The ESG linkage to and associated impacts on mining cut-off grades, mine planning, cash flow models and free cash flow margin, All-in-Sustaining Cost and Net Present Value metrics also need to be better mapped out and manifested in business models. It is important to create operating models to support the implementation of ESG commitments. This goes for the business and life-of-mine planning processes as a whole, as well as integrating ESG actions, funding and resourcing provisions into the mineral reporting process.

The creation of senior-level accountability in a company's organisational structure is a key enabler along with embedding ESG into core roles and incentives. Adoption of an ESG lens for smart capital allocation has the potential to assist with aligning capital provisioning to support ESG strategies. The creation of an advantaged sustainable portfolio involving a range of initiatives spanning four broad categories of investments can be considered:

- Investments that help create a strategically sound portfolio that is competitively positioned, has the right balance of innovation and leverages synergies within the portfolio.

- Investments that create value through maximising intrinsic value, address any gap with respect to market value and establish whether the company is in fact the right owner for an asset in the long-term.

- Investments that make the company more resilient by balancing feasibility and risk, building optionality and ensuring the organisation's survival through different scenarios.

- Investments that make the company more sustainable through creating social, environmental and economic value.

MINERAL REPORTING CODES

Many Mineral Resource statements have used non-specific wording, similar to the following, to avoid expanding on ESG related topics: *We are not aware of any factors (environmental, permitting, legal, title, taxation, socio-economic, marketing, political, or other relevant factors) that have materially affected the mineral resource estimate.*

Most of the reporting codes have had some expectation of permitting, environmental and social disclosure for years, particularly for reporting MRMR that fall in protected areas. There are differences in the detailed requirements between the various codes, but this basic concept appears in all of them to a lesser or greater degree. The challenge the reporting codes face is the current ESG requirements are either too generic or alternatively focus only on very specific risks, such as those related to tailings and waste rock management, without fully considering the site-specific circumstances of each asset.

Interpretation of these requirements varies widely by practitioners, and adherence to disclosure requirements is often not picked up or is poorly enforced by the security exchanges. With historical lack of action by investors on these types of issue, legal precedent on how the clauses should be interpreted has not been established. With increasing global recognition of the importance of ESG, such statements may no longer be considered sufficient by the securities exchanges and company investors.

Recognising this, a 2021 roundtable of the CRIRSCO members and advisors concluded that historical disclosure of ESG factors has been weak even though requirements for disclosure of these types of topics has been around for years. The roundtable concluded that even if major changes to the wording of the codes is not considered necessary, guidance on stricter application of ESG requirements is expected in the near term.

A review of CRIRSCO's (2019) Template Table 1 highlights that elements of ESG are both implicit and explicit throughout the requirements and not restricted to one line of the table. This reflects the changing attitudes of businesses who are now embedding ESG into every aspect of their decision-making. The paradigm is changing such that ESG is no longer confined to a chapter summarising the environmental and social impact assessment process, but rather is picked up throughout study reports to reflect how each step of the development process and each decision has considered ESG factors.

Legal and permitting

In respect of Mineral Reserves, it is understood that appropriate steps are being taken by the company to secure the necessary permits and that the timing of the permits will not jeopardise the life-of-mine plan.

Competent Persons are not lawyers or permitting experts and so there are often clauses like the above used in MRMR reporting when it comes to providing comment on how legal and permitting issues may influence RPEE and/or Modifying Factors for reporting of Mineral Reserve. Although some codes may allow for the Competent Persons to rely on legal information provided by the company or third-party experts, such general statements are not likely to meet the expectations of shareholders.

Ontario Securities Commission (OSC, 2005) NI 43–101 (Item 3, Form F1) states the Qualified Person (QP – functionally equivalent to Competent Persons) may include a disclaimer of responsibility for legal, political, environmental, or tax matters relevant to the technical report, on the condition they identify:

- The source of the information relied upon (date, title, author).
- The extent of reliance.
- The portions of the technical report to which the disclaimer applies.

United States Securities and Exchange Commission (US SEC, 2020) Rule S-K 1300 (Paragraph 15) says the QP may rely on information provided by the company for, among other things: legal matters, environmental matters and governmental factors, however, such reliance must be clearly stated and the justification for the reliance must be given.

There is an assumption the Competent Person has, as a minimum, applied some level of due diligence in terms of doing some homework and/or referring to specific advice or opinion from a permitting/legal expert, rather than just relying on the company's word. It may therefore be acceptable for the Competent Person to make a generic statement such as the one above, if there is clear evidence either from the company itself or from a relevant expert, that can be cited in the report to provide the necessary comfort (whilst also recognising the nuances between the requirements in the different codes).

A substantial portion of the conditions of permits and agreements relate to ESG. Multi-disciplinary technical competence is often required instead of, or alongside, legal input to comprehend materiality and check compliance with material conditions of approval.

The various codes often use terms like legal, permitting, regulatory and governance without necessarily accepting that these are integrally linked. There is also often a separation between disclosure of mineral licenses (security of tenure), environmental permitting and social agreements, however there is usually a close relationship between these approvals. Environmental, social, labour and human rights protections are often embedded in mineral law, mining concession agreements and/or rights to operate.

There also needs to be recognition that not all 'agreements/approvals' may be legally controlled. Current report code clauses often limit disclosure to legal agreements. This may result in traditional rights (of, for example, indigenous people) not being handled sensitively and this in turn could lead to major risks for mineral properties in the future if communities feel their interests are not appropriately being considered.

In addition, and as noted above, commitments made by a company in the public domain (eg website, public meetings and press releases) may constitute constructive obligations that have the potential to significantly affect timelines and costs of the approval process if enforced. For example, statements around decarbonisation targets imply the choice of equipment, power supply and processing methodologies presented in mineral property studies and mine plans reflect the company's ability to meet the stated decarbonisation targets in the time frames specified and within the estimated capital and operating expenditures. Competent Persons will be responsible for confirming and satisfying themselves that this is the case.

Transparency schemes have been established to raise public awareness of conditions of approvals and agreements and thereby promote compliance with these. Freedom of information laws in many countries and the EITI are examples. Increasing transparency increases risks associated with non-compliance with law.

Materiality and risk

The materiality principle remains fundamental in weighing and disclosing ESG factors impacting MRMR reporting. When used to qualify a requirement for providing information as to any subject, the term 'material' limits the information required to those matters to which there is a substantial likelihood that a reasonable investor would attach importance in determining the quality and value of the mineral asset.

Full consideration and assessment of ESG factors that could have a material effect on the RPEE and outcome of a mineral property must be undertaken. The determination of materiality depends on many criteria, including but not limited to the mineral property type, location (context), mineralisation or orebody style, jurisdiction, legal requirements, financing, investors and stakeholders.

Typical aspects for the Competent Person to consider concerning 'ESG materiality' include (but are not limited to):

- What could potentially stop the project or mining operation?

- What could materially impact the proposed mine schedule, costing and/or value proposition?
- What could affect the company's reputation?
- What could have a material impact on the risk profile and ability to execute the project or life-of-mine plan?
- Could sales of products be impacted if an embargo exists?
- How can the mineral property be set-up for success or sustain the operation successfully in respect of proactively managing ESG risks?

In preference to a generic checklist methodology, which encourages reliance on a tick box approach, a procedural methodology to identifying and considering ESG factors is preferred. It should be administered by ESG, technical and financial SMEs in support of the Competent Person. This will enable an inclusive process that delivers balanced and integrated assessment resulting in meaningful output through consensus decision-making.

A scorecard approach to assessing 'ESG Materiality' is often favoured by industry and can assist with ranking and prioritising factors to support transparency and appropriate disclosure. Referencing company strategic and operational risk registers will also assist with capturing significant factors.

MRMR risk assessments conducted by a multi-disciplinary team with relevant expertise is suggested as the primary procedural approach to ESG in mineral reporting. Identifying and clearly describing the ESG risks associated with a mineral property is fundamental to sustainability and assured mineral reporting.

ESG risks can be difficult to assess using quantitative scientific or knowledge-based considerations since they are dependent on the perception of companies, different stakeholders and regulators. However, the Competent Person is expected to provide transparency and clarity on these risks, how they are managed and responsibly financed.

For improved transparency, materiality needs to be defined and clarified for each public MRMR report. Circumstances are considered material if omission or misstatement of the associated factor or information could influence the economic decisions of users. ESG risks relating to RPEE and Modifying Factors should be rationalised together with all technical and financial risks and other Modifying Factors to ensure equivalency and a balanced approach.

The Competent Person will need to consider both *quantitative* and *qualitative* factors when determining materiality. Examples of typical *quantitative* measures include:

- Impact on expected progressive reclamation, mine closure costs and environmental liability.
- Impact on the estimated Mineral Resource of the mineral property.
- Impact on the estimated Mineral Reserve of the mineral property.
- Impact of public ESG commitments and obligations on costs, eg concerning decarbonisation, water preservation and sustainable energy.
- Impact on the forecast project, operating and capital expenditures.
- Impact on revenue, free cash flow, All in Cost (AIC/oz), EBITDA and tax regime.

Beyond the quantitative measures, the Competent Person will need to consider more *qualitative* measures in the materiality analysis. These include topics such as:

- Importance of delivering on long-term strategic goals and ESG commitments, eg vehicle electrification and decarbonisation.
- Social/environmental materiality.
- The adequacy of the knowledge base to assess social/environmental materiality.
- The adequacy of management plans and systems to address material social/environmental matters.

- The materiality of technical factors: for example, an operation with a tailings storage facility viewed as a material risk to the overall company due to the high risk and consequences of potential failure.

- Technological innovation, where that innovation may be considered material to the future of the business.

RPEE

The assessment of the RPEE of Mineral Resources, and that extraction could reasonably be justified for Mineral Reserve at the time of public disclosure, is a critical aspect of mineral reporting. RPEE assessments will need to include full consideration of the direct and indirect environmental and social impact and cost of extraction, processing and end use in terms of environmental degradation, biodiversity, cultural destruction, or climate change. Where communities' human rights are potentially affected by proposed mineral properties, measures to enable their engagement and consent will warrant emphasising as important considerations.

Early-stage exploration projects are, by their nature, often unable to define ESG factors and costs comprehensively and in any definitive detail. In such cases, it is expected the Competent Person will require a view on at least framing the identified and potential ESG threats and opportunities, to inform the investor of prevailing or possible future ESG issues and how development going forward will address these.

The Competent Person will be required to motivate whether there is a reasonable basis to believe that all ESG permits, authorisations and licences can be obtained within the time frame envisaged for the project. To achieve this the Competent Person will typically need to seek further guidance from SMEs on how ESG factors should be incorporated into MRMR disclosure and the valuation of mineral assets.

Achievable mine closure and progressive rehabilitation is another important criterion that warrants consideration in this context and is a crucial input to an operation's overall strategy, scheduling, financial modelling, valuation and economic viability. If this aspect is excluded from the RPEE assessment, planning, estimation and valuation of MRMR, it will compromise transparency and materiality, resulting in misleading investors and stakeholders concerning overall costs, economic viability and the valuation of assets.

Role of the Competent Person and Subject Matter Expert

Most codes and guidance indicate the information disclosed in public reports is based on work that is the responsibility of suitably qualified and experienced persons who are subject to an enforceable Professional Code of Ethics. In many cases, the Competent Person will review ESG information that has been assessed and synthesised by SMEs in the ESG disciplines. It remains incumbent on the SMEs to have appropriate levels of knowledge, relevant expertise and experience for the type of mineral property under consideration and in the activity which the SME is undertaking. However, the Competent Person must be evidentially satisfied that they can face their peers and demonstrate appropriate competence has been applied to the ESG information disclosed and the reliance placed upon it.

The Competent Person is expected to substantiate and validate conformance and compliance statements, stating the applied reliance included in disclosure, especially when linked to the standards described above. The use of disclaimers is generally not endorsed by regulators unless in specific situations clearly outside of the Competent Persons reach and jurisdiction. Importantly, there should be no dichotomy between the Competent Persons expected company role and duties versus the mandate for regulatory compliance. Company recognition and support for the growing accountability and potential liability placed on Competent Persons is a subject deserving more industry focus.

Collaborative and multi-disciplinary approach

It is deemed good practice that when SMEs are relied upon by Competent Persons (where Competent Persons do not have the required specific area of expertise), they should prepare supporting materials and be consulted when interpreting ESG information. Subject to the stage of development, a wide breadth of disciplines and local experts may be needed to fully understand the environmental and social conditions, legal frameworks, context, materiality and risks for a mineral property. It is recommended to include multiple disciplines and areas of expertise in project management, mine planning and operational implementation to fully address ESG aspects of a property. It is also viewed as necessary to have subject leads and experts work in close collaboration as a multi-disciplinary team, even in the early phases. Early assessment and involvement of ESG criteria and factors in the MRMR estimation process is vital and it should not be an area that receives late attention.

The Competent Person must confirm that capital allocation to support the delivery of key ESG strategies and commitments is in place and provided for in the financial models. If not in place, those strategies and constructive obligations in the public domain cannot be seen as realistic and executable, putting the delivery of anticipated value from mineral properties at risk.

Joint venture (JV) scenarios can present challenges for the Competent Person where the JV partner manages the asset; it remains incumbent on the Competent Person to consider the relevant ESG criteria to ensure the mineral reporting is aligned to the respective prevailing code. Where the JV partner is unable to demonstrate due diligence has been applied to necessary ESG criteria or Modifying Factors and where material risk exists concerning the execution of the project or life-of-mine plan, the Competent Person should disclose the inadequacies in alignment with reporting transparency, materiality and competency.

Framing minimum disclosure

Achieving minimum disclosure relies on a sharp focus on adding value to the users of the information and highlighting matters based on materiality, risk and transparency. Mineral reporting formats such as Annual MRMR statements, Supplements to Integrated Annual Reports and Competent Person Reports are not the appropriate settings for exhaustive and in-depth 'cover-all' ESG and corporate sustainability disclosure.

The minimum requirement is seen as transparently disclosing, on a robust and defendable basis, relevant ESG risks, uncertainties, commitments, permitting requirements and other information that could materially influence the economic value and likelihood of economic mining, processing and responsible closure of an asset. ESG information and statements in disclosures must be truthful, accurate, clear, unambiguous, fair and meaningful. Important information should not be hidden or omitted and greenwashing and boilerplate statements avoided in preference for granularity in reporting as and when deemed important to the investor. MRMR reports must provide stakeholders with ESG information that equips them to make informed decisions.

Taking a procedural approach to identifying material matters

A procedural approach to ESG in MRMR involves application of the basics of risk assessment, which are having a good understanding of the context and paying attention to the expectations of stakeholders when identifying material matters. Recognising each minerals property has an ESG context that is truly unique and is often complex, examples of topics and questions that should be considered when applying this approach are given in Table 2. Table 2 assumes mine waste and water management, emission control and mine closure are viewed as integral to the planning of the mineral property. Hence, it does not ask: 'Have these fundamental components of any mine been considered in planning and estimation?'

The relevance and materiality of these matters needs to be considered in relation to the phase of the mineral property and the disclosure detail commensurate with the phase. Reporting disclosures should be considered as additive as the asset moves from exploration to studies, operation and closure. The concept of double materiality should be applied: where potentially material matters may arise as a result of a property's ESG setting in the form of impacts from the mineral property or impacts to the mineral property.

TABLE 2
Topics and questions for assessing materiality.

Topic	Some questions pointing to potentially material factors
Environmental and social context Information on context can be obtained from site visits, satellite imagery, public domain websites and, if available, from ESIA and specialist reports, management plans and monitoring programmes	
People: Who could be affected? (Respecting the rights of host communities is core to ESG and failure to respect these rights can result in salient human rights transgressions that are highly material to investors.)	• Are people and/or communities using the land, water, and other resources on, around and downstream of the deposit and mineral property infrastructure? • Who might be physically or economically displaced by the development? • Are there any vulnerable people who could be disproportionately affected? • Are there indigenous peoples who could be affected? • Will natural resources that people depend on be affected? • Will cultural heritage be affected? • Will public infrastructure be affected? • How else could people be affected (hazards, nuisance, air pollution, in migration...)? • Are there notable social and political contextual issues that must be considered? (See stakeholder engagement below)
Biodiversity and ecosystem services	• What habitats are there on, around and downstream of the site and are these of notable conservation importance? • Are there species of conservation importance within the area? • Could deforestation be an issue? • How do host communities use these biological resources? • What ecosystem services does the area provide? • Are there requirements to achieve no net loss or a net gain in biodiversity specific values?
Water resources (These are frequently in the frame of potential material risks.)	• Which drainage basin and downstream water bodies could be affected? • Who are the water users in these catchments that could be impacted by or impact upon the project? • Are there any issues relating to water security, site water cycle and treatment requirements, including the status of water supplies to host communities and catchment management? • Are there any savings in water use or reduced consumption impacting the economics of the mineral property?
Carbon budget/ limit/target and carbon taxes	• Are there any national or local carbon budgets, mineral property-specific government limits and/or corporate commitments/targets that must be considered? • What carbon taxes are applicable?
Existing liabilities, closure liabilities, hazards, and hazardous substances	• Are there material issues related to the following: • Pre-existing liabilities, such as existing contamination/pollution/damage • Retrenchment costs • Financial provisions needed for closure • Hazardous substances

Topic	Some questions pointing to potentially material factors
	• Mine or process wastes classified as hazardous
	• Climate hazards and climate change hazards
	• Water hazards – floods, inrushes to workings
	• Seismic and geotechnical hazards
	• Transportation hazards and traffic
	• Artisanal and small-scale mining encroachments
	• In-migration and settlement in hazardous areas
	• War/conflict/unrest and associated security and safety risks, as well as human rights risks
	• Bribery and corruption
Protected areas	• Are there any protected areas that could be affected (from a biodiversity, cultural, or other perspective)? Have these been mapped?

Legal and regulatory context

Information needed includes an understanding of the relevant legislation, exploration and mining permits, minerals agreements with the host government, agreements with host communities, primary environmental approvals, binding management plans and legal insight on surface rights – including customary land rights and rights of indigenous people who may be affected by the development, government inspection reports, legal audit reports and enforcement notices.

Rights and agreements linked to minerals and surface tenure and approvals to operate	• Is the deposit/resource/reserve within a valid concession?
	• What is the nature and duration of company's rights of exploration and extraction?
	• What are the primary environmental permits and land use consents required?
	• How will surface rights be obtained for all areas needed for the development? Do customary land rights apply? Do rights of way need to be obtained?
	• Do water rights need to be obtained?
	• Are agreements with the host government needed?
	• Are agreements with host communities needed or expected?
	• Do any environmental management, resettlement action, labour, procurement and/or community development plans need to be formally approved?
	• Are the needed approvals and agreements in place? If not, how will these be obtained, what process must be followed and how long will it take?
	• What are the material conditions in the approvals and agreements?
	• What are the material commitments in binding plans?
	• Is there compliance with the material conditions? Does the company have systems in place for recording compliance obligations and tracking compliance?
	• Have enforcement notices been issued? If so, what are the implications?
	• Is the asset subject to any claims or disputes?

Stakeholder expectations and commitments made to stakeholders

Information needed to understand local stakeholder expectations includes records of engagement with host government authorities and host communities – through permitting

Topic	Some questions pointing to potentially material factors
	processes and agreements these expectations are often translated into conditions of approvals and agreements. Commitments to business stakeholders may be given in press releases, annual reports, sustainability reports and climate disclosure reports. Agreements with business stakeholders may include off take agreements, insurance agreements, loan agreements and agreements with independent power producers and leasing agreements.
Government and host community expectations	• Are there significant expectations the mines will create value for the host nation and host communities? • Are there unmet expectations or objections that could lead to material hindrances such as protracted permitting processes or disruptive blockades along transport routes important to operational mines? • Is there a need to achieve free, prior and informed consent (FPIC) where indigenous people could be affected by a development?
Business stakeholder expectations	• Has the company made any material commitments to business stakeholders? For ESG consider commitments to align with specific standards or targets like decarbonisation targets. • Are there material commitments in agreements with stakeholders?
Mining options and opportunities	
Mining options	• What innovative project and mining initiatives, options and alternatives have been analysed to improve the sustainability outcomes for the mine and mineral products? • What planning and design opportunities exist to avoid/minimise negative impacts on air, water, people, biodiversity and ecosystems, and to enhance positive impacts?

GUIDING PRINCIPLES FOR IMPROVEMENT

Having considered standards, scope, MRMR reporting codes, industry trends and expectations, the following guiding principles are recommended for improving the ESG performance of the mining industry and the quality of MRMR reporting (some of these principles are included or implicit in current reporting codes but are included here for completeness):

1. Recognise that mine waste, water and energy management and closure planning are integral to mineral property planning and not ancillary factors.

2. Elevate attention to energy and water demand, supply and efficiency, and climate resilience in project planning.

3. Update the language in MRMR codes and standards to align with ESG terminology, for example referring to biodiversity and ecosystem services rather than flora and fauna.

4. Promote reporting to be clear, concise, simplistic, relevant and avoid onerous and voluminous expansion of reports.

5. Differentiate broader sustainability reporting associated with annual reporting from that specific to mineral reporting.

6. Encourage the use of links and/or references to other publicly available sources of information to assist streamlining reports and avoid replication.

7. Drive convergence in company ESG reporting to streamline reports and avoid duplication and over-reporting through a company's various public communication channels.

8. Elevate but do not over-amplify consideration of ESG matters as a key component of balanced material, transparent and competent mineral reporting.

9. Provide guidance that can be realistically and reasonably enacted by the spectrum of listed companies, from small local juniors to large international corporations.

10. Promote consistency in how ESG is integrated with mineral reporting to drive equivalency and enable comparative analysis between companies.

11. Focus on satisfying the needs of the end user of the information to equip them with the context and insight to assess ESG factors, risk, compliance and leading practices.

12. Develop guidance on minimum and leading practice reporting that traverses all life cycle phases including exploration, project studies, construction and commissioning, mining operations, mine closure and post closure.

13. Proactively engage with security exchanges and regulators in different jurisdictions regarding future and anticipated ESG disclosure requirements to aid alignment with the respective listing rules.

14. Clarify the role of ESG in the different stages of reporting of mineral asset valuations.

15. Provide advice on how the materiality of ESG factors can be determined.

16. Provide specific, practical direction on how ESG matters should be incorporated into assessments of RPEE and Modifying Factors for conversion of Mineral Resource into Mineral Reserve.

17. Provide guidance regarding ESG competence and Competent Person and SME accreditation and continuous professional development.

18. Clearly explain governance including how ESG components impact on risk, reputation and liability.

19. Support a procedural approach for ESG integration as opposed to reliance on a checklist or tick box solution.

20. Encourage ESG assessments to be undertaken by SMEs in support of Competent Persons, to ensure a balanced approach with the requisite expertise and insight applied.

21. Where company executive compensation plans include ESG metrics, wire these into MRMR strategy and reporting to ensure the 'golden thread' is intact and there is no disconnect between MRMR statements and company strategy and execution.

CONCLUSION

Many large institutional shareholders are asking companies to focus more, do more and disclose more about their ESG efforts. In fact, ESG is now the topic most often covered during shareholder engagements that include company directors.

Focus must be on ensuring the necessary diligence and assurance is incorporated to provide business, investors, mining analysts, regulatory bodies, stock exchanges and other stakeholders with appropriate information to assess and compare mining companies concerning their coverage of ESG factors as entrenched in MRMR disclosure.

Consideration of ESG strategy, commitment, implementation, risk management and determination of associated Modifying Factors is a critical component of mineral reporting. A demonstrable linkage between ESG and company strategy, project development, business and life-of-mine planning and operational delivery is necessary for the veracity of MRMR disclosures. The continuous assessment of a mineral property as it progresses is critical because changes in ESG aspects over time may contribute to or become material aspects and risks that significantly affect Modifying Factors, costs, delivery timelines and value.

Emphasis needs to be retained on the various stakeholders involved, and that their respective expectations and needs are kept front and centre, and most importantly, up to date. Amplifying sustainability topics above and beyond the other criteria and Modifying Factors that require consideration, such as mineral estimation, geotechnical, mining, engineering, metallurgy and processing must be avoided to maintain overall balance. Integration of all of these is key.

As ESG integration into mineral reporting codes progresses across global reporting jurisdictions, it will be important to drive convergence as opposed to codes evolving independently. Deviation would only serve to exacerbate complexity and workload, especially for dual listed companies. The CRIRSCO mandate is pivotal to achieving convergence across the global industry.

The challenge of integrating ESG with mineral reporting can be intimidating, but if the current opportunity for transforming industry reporting codes retains a sharp focus on the end-users of the information and how their assessment of MRMR will benefit from the increased disclosure, the outcome will be successful and value accretive.

Transparency and materiality are critical criteria implying the Competent Person and SME need to be prepared, empowered and equipped to transparently disclose, on a robust and defendable (competent) basis, material ESG risks and uncertainties that have the potential to impact the value outcome of the mineral property. As such, the role of the Competent Person and the associated accreditation and accountability to regulators and recognised professional organisations (RPOs with enforceable disciplinary procedures) deserves recognition and support within organisations. Competent Persons need to be empowered to deliver results not only in compliance with codes and regulations but also deliver outcomes consistent with stakeholder expectations, for the advancement of mining and humanity in general.

REFERENCES

Committee for Mineral Reserves International Reporting Standards (CRIRSCO), 2019. International Reporting Template for the public reporting of Exploration Targets, Exploration Results, Mineral Resources and Mineral Reserves.

Copper Mark, The, 2020. Copper Mark Criteria for Responsible Production – Criteria Guide, February 2020. Available from: <https://coppermark.org/standards/criteria/>

European Parliament, 2022. Corporate Sustainability Reporting – Directive (EU) 2022/2464, 14 December 2022. Available from: <https://eur-lex.europa.eu/legal-content/EN/TXT/?uri=CELEX:32022L2464>

Global Tailings Review, 2020. Global Industry Standard on Tailings Management. Available from: <https://globaltailings review.org/global-industry-standard/>

Hund, K, La Porta, D, Fabregas, T P, Laing, T and Drexhage, J, 2020. Minerals for Climate Action: The Mineral Intensity of the Clean Energy Transition, International Bank for Reconstruction and Development/The World Bank. Available from: <https://pubdocs.worldbank.org/en/961711588875536384/Minerals-for-Climate-Action-The-Mineral-Intensity-of-the-Clean-Energy-Transition.pdf>

Initiative for Responsible Mining Assurance (IRMA), 2018. Standard for Responsible Mining IRMA-STD-001. Available from: <https://responsiblemining.net/resources/#resources-standard>

International Cyanide Management Institute, 2021. The International Cyanide Management Code For the Manufacture, Transport, and Use of Cyanide In the Production of Gold. Available from: <https://cyanidecode.org>

International Financial Corporation (IFC), 2012. Performance Standards. Available from: <https://www.ifc.org/wps/wcm/connect/Topics_Ext_Content/IFC_External_Corporate_Site/Sustainability-At-IFC/Policies-Standards/Performance-Standards>

Mitchell, P, 2022. Top 10 business risks and opportunities for mining and metals in 2023, Ernst and Young, September 2022. Available from: <https://www.ey.com/en_gl/mining-metals/risks-opportunities>

Ontario Securities Commission (OSC), 2005. National Instrument (NI) 43–101: Standards of Disclosure for Mineral Projects. Available from: <https://www.osc.ca/en/securities-law/instruments-rules-policies/4/43–101/national-instrument-ni-43–101-standards-disclosure-mineral-projects-0>

Pan-European Reserves and Resources Reporting Committee (PERC), 2021. The PERC Standard for Reporting of Exploration Results, Mineral Resources and Mineral Reserves. Available from: <https://percstandard.org/perc-standard/>

United Nations Global Compact, 2004. Who Cares Wins – Connecting Financial Markets to a Changing World. Available from: <https://www.ifc.org/wps/wcm/connect/de954acc-504f-4140–91dc-d46cf063b1ec/WhoCaresWins_2004.pdf?MOD=AJPERES&CVID=jqeE.mD>

United Nations Global Compact, 2011. Guiding Principles on Business and Human Rights. Available from: <https://unglobalcompact.org/library/2>

United Nations Principles for Responsible Investment (UN PRI), 2018. PRI Reporting Framework – Main definitions. Available from: <https://www.unpri.org/Uploads/i/m/n/maindefinitionstoprireportingframework_127272_949397.pdf>

United Nations Principles for Responsible Investment (UN PRI), 2021. Annual Report 2021. Available from: <https://www.unpri.org/annual-report-2021/how-we-work/building-our-effectiveness/enhance-our-global-footprint>

United Nations, 2023. Sustainable Development Goals. Available from: <https://www.un.org/sustainabledevelopment/>

United States Securities and Exchange Commission (US SEC), 2020. Regulation S-K, Section 155, subpart 1300, S-K1300. Available from: <https://www.sec.gov/divisions/corpfin/guidance/regs-kinterp>

Vagenas, J, 2021. 7 reasons why ESG issues present the biggest risk for the mining sector today, *AusIMM Bulletin*, December 2021. Available from: <https://www.ausimm.com/bulletin/bulletin-articles/7-reasons-why-esg-issues-present-the-biggest-risk-for-the-mining-sector-today/>

Voluntary Principles on Security and Human Rights, 2023. Security and Human Rights Initiative. Available from: <https://www.voluntaryprinciples.org/>

World Gold Council, 2019. Responsible Gold Mining Principles (RGMPs). Available from: <https://www.gold.org/industry-standards/responsible-gold-mining>

World Health Organisation, 2023. Home page. Available from: <https://www.who.int/>

Application of three lines model in resource and reserve estimation and reporting

D K Mukhopadhyay[1], D Hope[2] and J D Harvey[3]

1. MAusIMM(CP), Senior Manager, Resource and Reserve Governance, South32, Perth WA 6000. Email: dk.mukhopadhyay@south32.net
2. Senior Manager, Tenure, Mining Rights and Resource Geology, South32, Perth WA 6000. Email: david.hope@south32.net
3. Lead Resource Geology, South32, Perth WA 6000. Email: joshua.d.harvey@south32.net

ABSTRACT

South32 (S32), a globally diversified mining and metals company, is listed on securities exchanges around the world (ASX, JSE, LSE), with its primary listing being on the Australian Securities Exchange (ASX).

Chapter 5 of the ASX Listing Rules (Australian Securities Exchange (ASX), 2019b) requires a mining entity to provide a summary of governance arrangements and internal controls with respect to its estimation processes, estimates and reporting of Mineral Resources and Ore Reserves annually. As a global company, it is also important to design and implement all governance processes and internal controls which can be used consistently across the organisation.

The three lines model (previously known as the 'three lines of defense' model) is employed across South32 to manage risk where the first line (Practitioners) designs, implements and executes processes and controls to manage risk; the second line (Stewardship) identifies material gaps and improvement opportunities in the first line by monitoring conformance with the group requirements, legislative and regulatory obligations. Second line also assists with the development and improvement of core fundamentals based on industry good practise. The third line (Group Assurance) provides independent and objective assurance over the Group's system of risk management and control.

The estimation of resources and reserves for most of the operations within South32 is managed by the Planning function. The estimation process is part of the annual Life of Operation Plan (LoOP). The LoOP identifies and assesses the strategic alternatives for the development of an operation and recommends the preferred pathway based on value and risk. The LoOP starts with opportunity framing and the updated resources as an input and ends with the short-term plan and annual budget (the outputs). Ore Reserves estimates are one of the outputs from the LoOP. All the inputs and outputs follow internal standards and procedures which includes peer review and performance reconciliation. Stewardship (Second line) which is a centralised function has standards in place to ensure compliance to internal controls and external regulatory requirements. This includes assessment of the competence of Competent Persons, technical reviews, independent audits and training in the relevant chapters of the ASX Listing Rules, Guidance Notes and Appendices. Group Assurance (Third line) which reports to the Risk and Audit Committee of the Board of Directors ensures the design and operating effectiveness of the processes, standards and procedures associated with the estimation and reporting of resources and reserves. While the level of scrutiny has a different intent for each line in the three lines model, the overall purpose is to achieve a consistent outcome across the organisation.

INTRODUCTION

South32 (S32), a globally diversified mining and metals company, is listed on securities exchanges around the world (ASX, JSE, LSE), with its primary listing being on the Australian Securities Exchange (ASX). The company's purpose is to make a difference by developing natural resources, improving people's lives now and for generations to come. South32 is trusted by its owners and partners to realise the potential of their resources. The company produces commodities including bauxite, alumina, aluminium, copper, silver, lead, zinc, nickel, metallurgical coal and manganese from its operations in Australia, Southern Africa and South America. With a focus on growing its base metals exposure, South32 also has two development options in North America and several partnerships with junior explorers around the world. South32 strives to create enduring value through strong governance and management of our performance.

Chapter 3 of the ASX Listing Rules (ASX, 2019a) sets out continuous disclosure requirements that an entity must satisfy. Chapter 4 (ASX, 2013a) sets out relevant periodic disclosure requirements in relation to each quarter, half year and end of year. For a mining company listed on the ASX, Chapter 5 (ASX, 2019b) sets out additional reporting and disclosure requirements on its exploration and mining activities. Additional context on Chapter 5 is available in Guidance note 31 (ASX, 2013b). While the requirement for reporting of Exploration Targets, Exploration Results, Mineral Resource and Ore Reserve is outlined in the JORC Code (2012), the requirement for reporting a production target is provided exclusively in Chapter 5 (ASX, 2019b).

For a globally diversified organisation, consistent implementation of listing rules and complying with all regulatory requirements is complex. All the processes, procedures and standards are developed considering industry best practice; statutory requirements of each state region and country where the company operates; and requirements outlined in the ASX Listing Rules. The approach to risk management is usually governed by the company's risk management framework. The minimum mandatory requirements for the management of risks that can materially impact the ability to achieve purpose, strategy and business plans are defined in internal risk management standards.

The three lines model helps organisations to identify structures and processes that best assist the achievement of objectives and facilitate strong governance and risk management. The model is optimised by adopting a principle-based approach focusing on contribution risk management makes in achieving objectives and creating value with a clear understanding of roles and responsibilities and implementing measures to ensure activities and objectives are aligned with the prioritised interests of stakeholders. The model can be designed such that it is fit for purpose based on the size of the organisation and can be implemented in all areas.

The estimation of resources and reserves (including the frequency of updates) varies from company to company. In most companies, this activity is embedded in the annual planning process. The quantum of work required for each operation is reviewed based on additional data generated during the year, feedback from operational performance, change in external factors and assessment of business strategy. For development projects, routines are based on the status of the project and level of study the project is planning to undertake or are completing. One of the material risks for the business relevant to all investors is 'Incomplete and/or inaccurate resources and reserves reporting'. It becomes important, as a result, that there are embedded governance processes to review each of these steps to provide the investor with the appropriate outcome for the year in our reported resources and reserves estimate in our annual report (South32, 2022) or when there is 'material change' to a 'material mining project' (as defined in Chapter 19 of the ASX Listing Rules (ASX, 2016)).

Through this paper, we are providing an overview of our application of the risk management framework in resources and reserves estimation processes, the three lines model and our implementation of the three lines model in the estimation and reporting of resources and reserves.

RISK MANAGEMENT FRAMEWORK

Risk management is governed by a risk management framework (the author would like to point out that the commentary included in this paper is an excerpt from the annual report relevant to estimation and reporting of Resources and Reserves. The full section is available in the Annual Report (South32, 2022)). The minimum mandatory requirements for the management of risks that can materially impact the ability to achieve an organisation's purpose, strategy and business plans are typically articulated through business standards and procedures aligned with relevant international standards (eg the International Standard for Risk Management AS/NZS ISO 31000:2018). The risks are regularly assessed and managed at both a company-wide strategic level and at a tactical level for operations and projects.

Risk appetite

Risk appetite is the level of residual risk that a company is willing to accept in pursuit of its strategic objectives, which is determined based on the operating environment. A Risk Appetite Statement outlines the extent to which a company is willing or not willing to engage with higher levels of risk (both threats and opportunities) to realise greater benefit in the pursuit of its purpose and strategy. Key risk indicators (KRIs) are set by business and used to monitor performance against the risk appetite. Understanding the risk appetite across the strategic risks assists in decision-making across a company.

Material risks

The system of risk management can be based on the three lines model which describes how key organisational roles work together to facilitate strong risk management and assurance. This approach is used to manage the material risks and is used to:

- Provide stable and consistent processes, tools, and routines to identify and regularly assess the most impactful threats and opportunities.

- Deliver predictable outcomes and prevent unforeseen events with material impacts.

- Understand our risks and manage these at all levels of the organisation.

- Reduce the impact or eliminate risks where appropriate or improve our processes using a risk-based approach.

The effective management of material risks is routinely assessed and reviewed which assist the Board to carry out its role of overseeing the risk management and assurance practices. Reliable data on material risks contributes towards the monitoring and management of strategic risks. This provides insight into trends and emerging themes that can trigger a review of business plans or inform a change in strategic direction.

As indicated previously, the material risk related to resources and reserves reporting from an investor's view can be 'incomplete and/or inaccurate resources and reserves reporting'. A Risk bow-tie method can be used to assess the risk. The bow-tie can be defined based on a causal pathway map. Proactive and reactive controls can be formulated based on the risk bow-tie. Some of the controls, should be identified as critical controls. The critical controls can form the framework for second line assurance process in 'the three lines model' which is detailed in the 'Implementation' section.

In addition to the critical controls, there can be underlying risks which have operational and functional level controls to manage the above-mentioned material risk. These risks should be managed by the first line of the three lines model. Examples of such risks from a South32 context are included in Table 1.

TABLE 1

Risks supporting the Incomplete and/or inaccurate resource and reserve reporting.

Area	Risk
Tenement Management	Failure to maintain prospecting or mineral title/rights
Exploration	Failure to maintain, realise or enhance the Mineral Inventory through brownfields exploration
Mineral Resource	Failure to deliver a reliable resource estimate or mineral inventory
Life of Operation Plan	Failure to maximise value/select optimal pathway through the Life of Operations Plans
Ore Reserve	Significant Inaccuracy in Reserve Estimation and Reporting
Closure	Failure to identify and undertake progressive rehab; early closure; or relinquishment opportunities resulting in loss of business value

Strategic risks

Strategic risks are risks which can affect our ability to achieve strategic objectives. They have the capacity to affect the whole, or a significant part, of the organisation and therefore tend to have significant impacts, both negative and positive.

The strategic risk related to Mineral Resources and Ore Reserves estimate is outlined in South32's Annual Report (2022) as below.

Maintain, realise or enhance the value of our Mineral Resources and Ore Reserves

We intend to realise the potential of the resources and reserves we are entrusted to develop. We work to continually optimise our operations through sound technical and economic understanding of our resources and reserves. Our most recent assessment of the strategic risk is provided:

- Opportunities: We continue to enhance our understanding of our resources and reserves. We leverage this enhanced understanding through the annual planning cycle to define and assess additional opportunities to add value to our business.

- Threats: If we fail to continually optimise our operations and projects, it will have a significant impact on shareholder returns, the benefits our stakeholders receive and ultimately, the sustainability of the company.

- Risk appetite: We are not willing to take risks that inhibit our ability to realise the potential of the resources and reserves we are entrusted to develop.

THREE LINES MODEL

The 'three lines of defense' combined assurance model was developed for HSBC by KPMG in the United Kingdom in the 1990s. It was later adopted by the Basel committee on Banking supervision. It was designed to ensure that all assurance activities are visible to the Audit Committee and Senior Management. The first line was developed to review governance and compliance arrangements to demonstrate that all the 'checks and balances' are working effectively. The role of the second line was to confirm the effectiveness of governance and compliance and to identify and action improvements. The third line independently assured the effectiveness of the first and second line. The framework initially developed is provided in Figure 1.

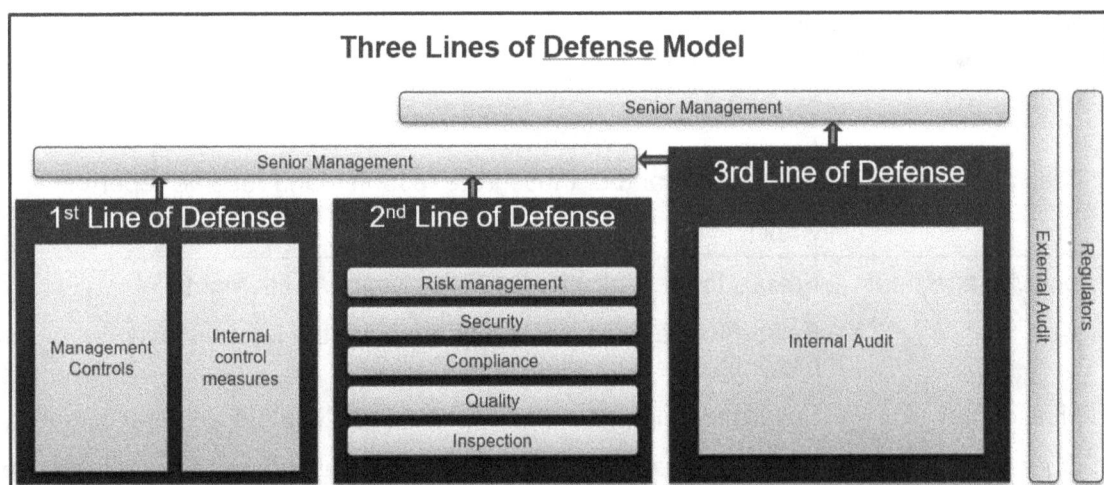

FIG 1 – 'Three lines of defense' framework (Source: Telem (2017)).

The Institute of Internal Auditors (IIA) adopted the 'three lines of defense' model, updated and redefined as the 'Three Lines Model' (IIA, 2020). According to IIA in their website, *'The Three Lines Model is a fresh look at the familiar 'Three Lines of Defense', clarifying and strengthening the underpinning principles, broadening the scope, and explaining how key organisational roles work together to facilitate strong governance and risk management'*. The revised framework is provided in Figure 2.

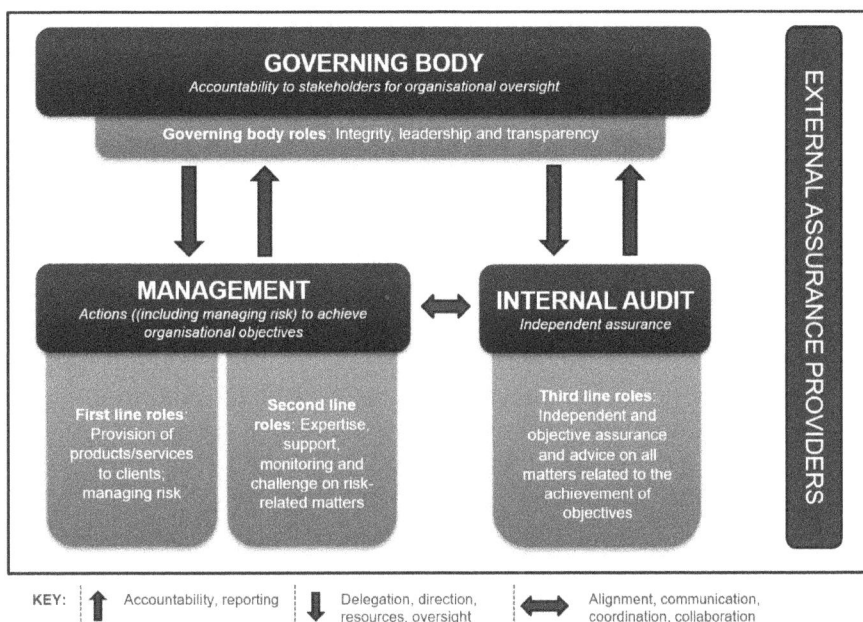

FIG 2 – Three Lines Model framework (Source: IIA, 2020).

The IIA's three lines model is based on six principles. Salient features of each of the principles is shown in Table 2.

TABLE 2

Three lines principles.

Principles	Title	Description
Principle 1	Governance	Governance of an organisation requires appropriate structures and processes that enable accountability, actions, assurance, and advice
Principle 2	Governing body roles	The governing body ensures appropriate structure and processes are in place for effective governance
Principle 3	Management and first and second line roles	Management's responsibility is to achieve organisational objectives comprising both first and second line roles
Principle 4	Third line roles	Internal audit provides independent and objective assurance and advice on the adequacy and effectiveness of governance and risk management
Principle 5	Third line independence	Internal audit's independence from the responsibilities of management is critical to its objectivity, authority, and credibility
Principle 6	Creating and protecting value	All roles working together collectively contribute to the creation and protection of value when they are aligned with each other and with the prioritised interests of stakeholders

The roles and responsibility of the three lines *viz*, first, second and third, as outlined in the IIA's three lines model is included below.

First Line: The focus of the first line is to achieve the business objective of the organisation. The primary role of first line is to maintain structures and processes in place for management of operations and risk. It ensures compliance with legal, regulatory and ethical expectations.

Second Line: Second line provides complimentary expertise, support, monitoring and challenge related to the management of the risk. This is achieved by conducting analysis on adequacy and effectiveness of risk management.

Third Line: The third line maintains its primary accountability to the governing body and provides independent and objective assurance and advice to the management and the governing body on the adequacy and effectiveness of governance and risk management.

Like any other metric, the three lines model will work efficiently if the communication between the three lines is open and transparent, and the relevance and effectiveness of the risk is constantly reviewed through the cycle of PDCA (Plan-Do-Check-Act) (Deming, 1950). When all the processes are mapped in the framework of SIPOC (**S**uppliers, **I**nput, **P**rocess, **O**utput and **C**ustomers), any change in relation to these elements for a particular process due to internal or external factors can be identified promptly. Additional new or changed controls can be implemented to enable dynamic risk management. In the following section, we will elaborate how the three lines model works effectively in resources and reserves estimation and reporting.

THREE LINES MODEL IMPLEMENTATION AT SOUTH32

Post demerger in May 2015, all the material risks that could impact the achievement of the business plans were analysed and assessed across the business. Controls were implemented and verified on an ongoing basis, ensuring that the level of risk was monitored and managed.

Resource and reserve estimation and reporting was centralised as a function from day one. Geologists, Mining Engineers, Asset Planners and the valuation team (with support from other professionals) worked on the strategic plan for each operation in an annual strategic planning cycle (LoOP).

The Resource and Reserve Governance team was set-up to oversee the estimation and reporting process to ensure legal and regulatory compliance. The group was tasked to implement processes to independently review the work done annually and/or to conduct independent audits when there was a 'material change'. This group was also entrusted to review and approve the nomination of Competent Person by verifying competence, and to interact with other outward focused teams within the company (Investor Relations and Company Secretariat) to review public reports containing information on exploration targets/results, resources, reserves and production targets in order to mitigate the risk of 'Incomplete and/or inaccurate resources and reserves reporting'.

Group Risk and Assurance ensured the efficiency and effectiveness of the Resource and Reserve Governance team (R&RG) in managing the processes and procedures of risk mitigation related to resources and reserves. Thus, the 'three lines of defense' model was born within our company without naming it as such.

We continued to work on our risk management approach and identified new tools for the effective management of risk. In 2019, we formally announced the implementation of the 'three lines of defense' model across the company. Since then, we have been working towards optimising the framework to make it a fit-for-purpose process for the company. Significant efforts have been invested to strengthen the processes in the first line for effective management of risk which included the definition of additional resources and reserves estimation related risks. More clarity was introduced to the role of the second line to ensure that it carries the added responsibility of stewardship. The most recent framework with roles and responsibility for each line as defined within South32 is provided in Figure 3.

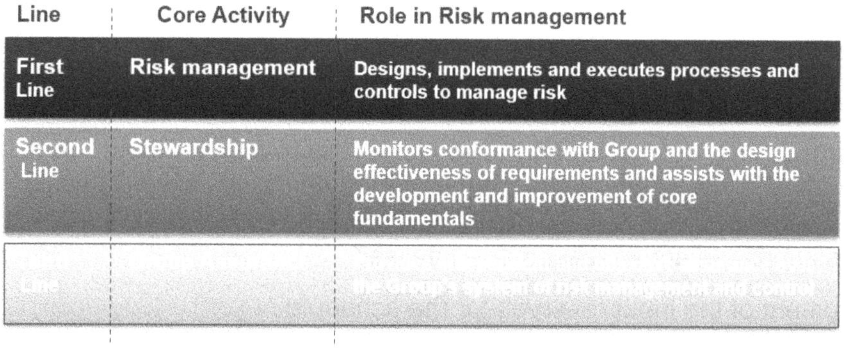

Line	Core Activity	Role in Risk management
First Line	Risk management	Designs, implements and executes processes and controls to manage risk
Second Line	Stewardship	Monitors conformance with Group and the design effectiveness of requirements and assists with the development and improvement of core fundamentals
Line		the Group's system of risk management and control

FIG 3 – Risk management roles.

First line

South32's first line framework consists of multiple activities conducted in conjunction with the LoOP and our annual resources and reserves updates. These measures were designed to align with our risk management framework. The framework forms the basic set of tools used to support the control management activities to manage the overall risks of maintaining tenure, and the development and reporting of resources and reserves, as well as the development of the LoOP.

Underpinning the first line are our Level 1 global procedures for Tenement Management, Resource and Reserve Estimation. These procedures have been developed incorporating best practices in all areas of data collection and interpretation.

Exploration is the primary source of data acquisition and is managed and executed at the operations. To ensure the quality and integrity of the input data, 'Competent Person site checklists' are used. The checklists are designed to verify that operation specific procedures (including competency requirements of personnel responsible for carrying out the work) meet the requirements of the global guidance framework. The checklist is completed on a quarterly basis by the Competent Person and includes verification of all activities carried out at the operations and external laboratories (where relevant). Table 3 summarises the key areas addressed by the resource geology checklists.

TABLE 3

Resource competent person checklist.

Section	Key observation areas
Mapping	• Personnel training and competency
Drilling and Sampling	• Appropriateness of the sampling procedure and quality and representivity of the sample collection process
Logging	• Appropriateness of the logging criteria • Training and competency of personnel
Laboratory (internal or external)	• Procedural compliance • Quality and assurance measures used • Training and competency of personnel

Verifications are designed to meet our internal requirements and to ensure that the criteria outlined in Table 1 of the JORC Code can be addressed on an 'if not, why not' basis. This is done through a continuous peer review process which is initiated for each resource and reserve model update. A continuous peer review is conducted concurrently with the update of the models on a one-to-one basis, or in a group environment. The peer reviews are conducted by Competent Persons or practitioners responsible for other deposits within the company and are designed to test both the application and quality of the estimates. Through the peer review process, practitioners are challenged to explain, justify and defend their interpretations, the application of their assumptions, and the selection of methods while preparing for and undertaking a resource or reserve estimate. A tollgate approach ensures that any outcomes of the review, recommendations or issues can be addressed prior to the initiation of the next stage of work. Tollgates are required to be signed off by the peer reviewer and occur after:

- data extraction from the database, validation and QAQC assessment
- geological modelling, estimation and validation of the resource model
- classification and reporting of the Mineral Resource
- planning assumptions are assessed and finalised
- the development of the input reserves for the schedule
- the application of economic factors
- the classification and reporting of the Ore Reserve.

A more detailed outline of the peer review is provided in Table 4 which can be linked to the relevant resource and reserve critical elements.

TABLE 4

Peer review checklist.

Critical element	Section	Key observation areas
Database management/Drilling and Sampling/Geological modelling	Scope and Tenement Status	• Model objective and scope in the context of the geological setting • Drilling and assay reports, and laboratory audits • Tenement status and standing
	Data extraction and Manipulation	• The official database compared to application data • Measures taken to prepare and validate data, and the treatment of invalid data and supporting documentation • Database structure, integrity and security measures used to assess the quality of the data
Geological modelling	Topography and Geological Modelling	• Topography, survey data and collar location accuracy • Geological surfaces versus input data • Geological model representivity and geological domaining • Validation of input data to the model
Resource estimation	EDA	• Descriptive statistics and sample treatment (eg outliers and top-cutting)
	Variogram Analysis	• Spatial continuity models • Appropriateness and application of the estimation approach • Verification of the parameters and documentation of the inputs and results
	Block model Set-up	• Verification of scripts/macros • Block size selection and orientation • Search and estimation criteria; constraints and the basis of application

Critical element	Section	Key observation areas
		• Appropriateness of dynamic search options
	Estimation	• Estimation method, parameters and results
		• Estimation and post-estimation categorisation assignment
		• Review supporting documentation
	Model Validation	• Assessment of estimation quality (including visual and numeric validations, swath plots, scatter plots, estimation quality statistics, smoothing assessments, change of support, local and global reconciliations, grade tonnage curves and changes to previous models)
	Classification and Reporting	• Documentation and justification supporting the classification of the Mineral Resource
		• Verification of the application of the classification criteria
		• Reporting scripts and macros
		• Review of JORC Table 1 and technical reports against the South32 template and industry best practice
	Handover	• Model storage and version control verification
		• Resource model handover and signoff documentation
Determination and verification of modifying factors	Modifying Factors	• Basis and versions of key assumptions with associated signoff
Generation of a reserve model based on existing assumptions and modifying factors	Cut-off Grades	• Basis and application of cut-off criteria
	Ore Reserve Modelling	• Modifying factor application and underlying models and assumptions
Determination of economic limits and development of the mine design	Economic Mining Limits	• Documentation and endorsement of key criteria
Scheduling the Ore Reserve using approved assumptions and parameters	Mine Design	• Verification of the economic, and other relevant constraints
	Ore Reserves Mine Schedule	• Application of relevant metrics and assumptions used in scheduling
	Economics/Valuation	• Validation of the source and appropriateness of inputs into the valuation
		• Sensitivity analysis
	Risk to mine plan	• Risk assessment related to the current mine plan and the management of the risks within the risk management framework
Generation of a reserve model	Reconciliation	• Spatial and numerical reconciliations against our internal standards

Critical element	Section	Key observation areas
based on existing assumptions and modifying factors	Market Assessment	• Verification of the level and appropriateness specific to the commodity
	Environmental and Social	• ESG considerations and levels of associated study to support the assumptions
Determination of economic limits and development of the mine design	Closure Planning	• Verification of the level of closure planning conducted
	Infrastructure	• Relevant elements to support the mine plan (eg power, utilities etc)
Classification and reporting	Classification	• Documentation and justification for classification
	Audits/Peer Reviews	• Review of audits/reviews and outstanding actions

Both the Mineral Resources and Ore Reserves estimates are subject to a Risk Review (second line activity) where outcomes of the first line activities are presented to a collective of our Competent Persons and practitioners, external subject matter experts, and the second line team (R&RG). Key risks and issues that were identified during the first line activities are reviewed, and actions taken throughout the year, to mitigate or resolve, are presented to the group. The outcomes are reported and assessed in line with our second line processes and are described in more detail below.

In addition to meeting the requirements of our risk management framework, these activities support the Competent Person in their assessment of the integrity of the information and approach used to develop the model, and their ability to defend their work amongst their peers. Additionally, this approach also creates an environment of sharing and learning.

Second line

Five core fundamentals have been designed to establish the second line as a stewardship and assurance activity. These fundamentals (Critical processes, Group Standard, Parent Controls, first line implementation and verification) help with the effective management of risk. The role of the second line within the company is designed to manage the material risk of 'incomplete and/or inaccurate resources and reserves reporting' and to conduct periodic assessment of the strategic risk to 'maintain, realise or enhance the value of our Mineral Resources and Ore Reserves' for each operation and project from a legal and a regulatory compliance perspective.

Most of the second line activities are designed to follow the annual cycle of the LoOP to ensure that once an activity is completed, it is reviewed immediately not only from a technical perspective but also from an internal compliance or a regulatory (Chapter 5 (ASX, 2019b)) perspective. Most of the additional assessments branch out of this primary activity and are critical. The annual review process along with the associated activities are outlined in Figure 4.

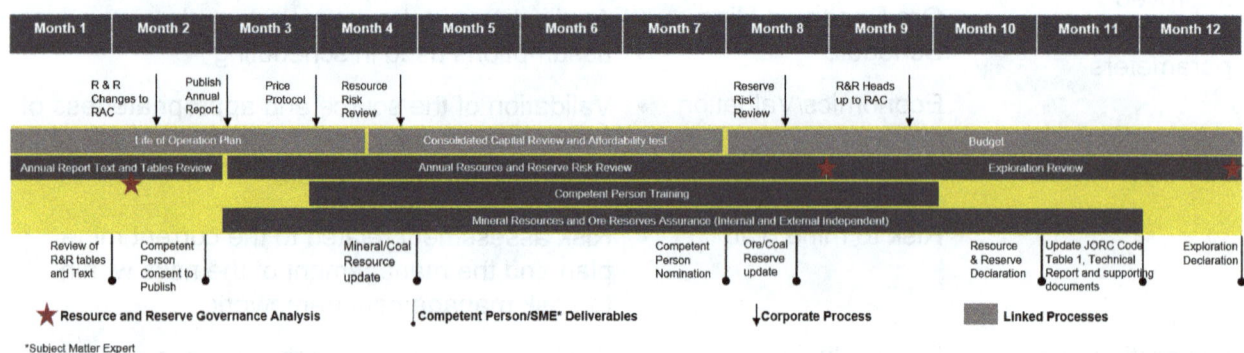

FIG 4 – Integrated Planning and R&R Governance calendar.

Mineral Resource Estimation Conference 2023 | Perth, Australia | 24–25 May 2023

The main activities that help us to manage second line are provided in Table 5.

TABLE 5

Second line activities.

Sl. No.	Process	Frequency	Context
1	Materiality Assessment	Annual	As a mining entity, we have additional disclosure requirements defined by the ASX when there is a 'Material Change' to a 'Material Project'. Guidance from the ASX is to use Accounting Standard AASB 1031. Under paragraph 15 of the Accounting Standard, Materiality is defined as an amount which is equal to or greater than 10% of the applicable base amount, generally presumed to be material and an amount which is equal to or less than 5% of the applicable base amount, generally presumed not to be material unless, in either case, there is evidence or convincing argument to the contrary.
2	Assessment and approval of New Competent Person	Annual or in case of maiden disclosure or when there is material change	We verify the Competence of the Competent Persons using the guidance stipulated in the JORC Code. This is done using a defined checklist which allows the competent persons to provide information on their qualification, professional membership and relevant experience as required for individuals to act as Competent Person. This is reviewed by a committee and the outcome is communicated.
3	Competent Person training and Governance training	Once in three year or earlier	All the Competent Persons are required to undertake this training once every three years as a refresher or when appointed for the first time or when there is significant change in internal requirement or external regulatory framework (eg update to JORC Code). The training covers any updates to the JORC Code, relevant chapters of ASX Listing Rules, Australian Securities and Investments Commission (ASIC) guidance and internal South32 requirements. The training is delivered by a team of experts from within and outside the company. Case studies are discussed, and an assessment is completed. Training is also provided to key personnel responsible for public reporting.
4	Mineral Resources and Ore Reserves reporting risk review	Annual	This process identifies and manages Mineral Resources and Ore Reserves reporting risks, established controls in the form of governance processes and the activities to meet public reporting regulatory requirements. This process is completed in a workshop format where the Competent Persons present to their peers and discuss the risks and mitigating measures implemented to control the risk.
5	Mineral Resource and Ore Reserve Stewardship	Once in three year or earlier based on outcome from Sl. No.4.	We confirm that all controls are in place for estimation and reporting of resources and reserves at each of our operations and projects. This is based on a framework to confirm the design and operating effectiveness of controls.
6	Independent audit	Need basis, mainly driven by material change or significant change in assumptions or processes	Independent audits (using consultants external to South32) of exploration results, resource and reserve estimates are conducted when there is 'material change' to outcome, or in the case of a maiden disclosure or significant change in processes.

All these activities are designed considering the internal South32 requirement, JORC Code, and relevant ASX Listing Rules (Chapters 3 and 5 (ASX, 2019a, 2019b)) and associated guidance note and relevant ASIC guidance. Appropriate actions are defined to close the gaps identified from the above processes and lodged in our risk management application for timely completion by relevant subject matter experts.

We have a centralised database to store resource and reserve information and supporting technical documents. For consistency across all operations and projects, we have nine procedures for practitioners to analyse and provide 'health check'.

The second line reports to South32's Risk and Audit Committee twice a year to update the committee on the health of resources and reserves by providing the summary of the outcome from all of the above activities.

Third line

Internal control helps the Group achieve its objectives and sustain and improve performance. We have utilised the Committee of Sponsoring Organizations of the Treadway Commission (COSO, 2017) Internal Control—Integrated framework to classify internal audit findings. An outline of the framework and the selected elements is included in Figure 5.

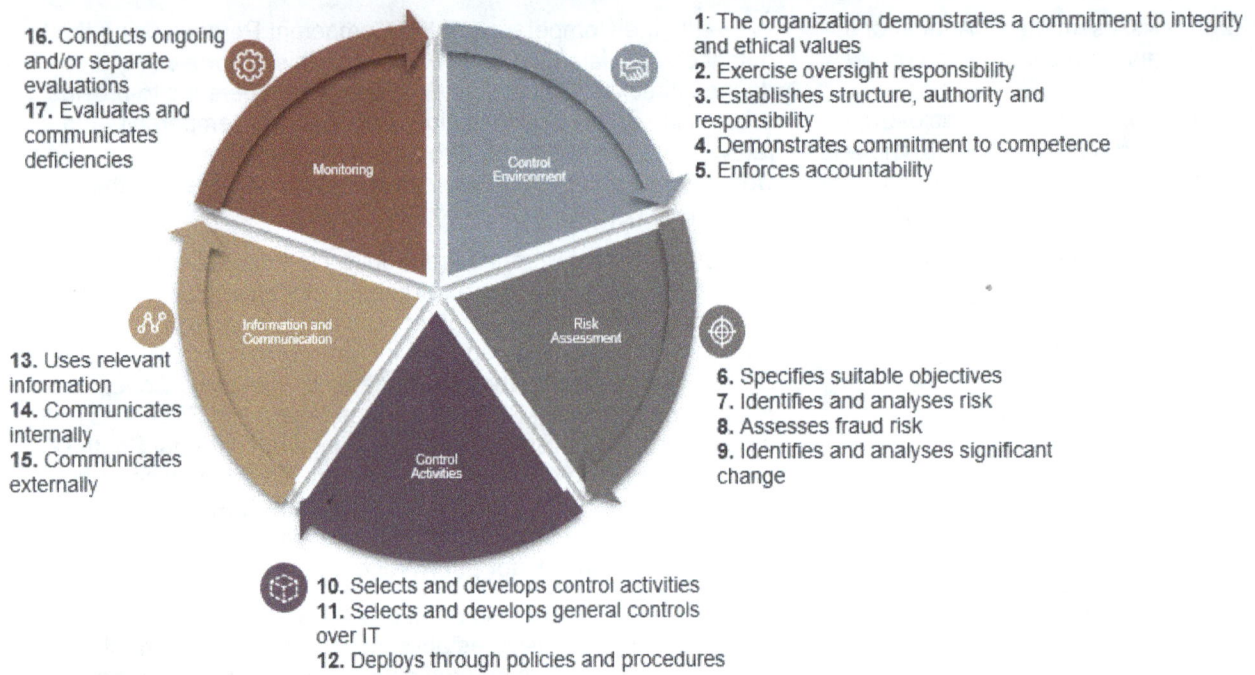

16. Conducts ongoing and/or separate evaluations
17. Evaluates and communicates deficiencies

Monitoring

Control Environment

1: The organization demonstrates a commitment to integrity and ethical values
2. Exercise oversight responsibility
3. Establishes structure, authority and responsibility
4. Demonstrates commitment to competence
5. Enforces accountability

13. Uses relevant information
14. Communicates internally
15. Communicates externally

Information and Communication

Risk Assessment

6. Specifies suitable objectives
7. Identifies and analyses risk
8. Assesses fraud risk
9. Identifies and analyses significant change

Control Activities

10. Selects and develops control activities
11. Selects and develops general controls over IT
12. Deploys through policies and procedures

FIG 5 – COSO framework (COSO, 2017).

Group Assurance manages the third line and owns the Assurance Standard. Group Risk own the risk management framework and the Risk Management Standard. The second line stewardship standard has been developed to ensure consistent implementation in all areas of business.

The Assurance standard sets out planning, execution and reporting of internal audit. Group Assurance conducts internal audits to evaluate and identify areas where management should improve the effectiveness of the Group's risk management, control, compliance and governance processes.

BENEFITS

The three lines model when appropriately implemented creates significant benefit. The implementation allows us to appropriately define processes for each of the activities that we undertake and implement it across the organisation consistently. This is a significant advantage for a globally diversified metals and mining company. Each of the Competent Persons are offered the opportunity to participate in first (peer review) and second line (Mineral Resource and Ore Reserve assurance and risk reviews) verification steps which allows Competent Persons to gain knowledge in different commodities. The Competent Persons not only learn from each other through the verification process, but also get the opportunity to gain a detailed understanding of other operations and projects within the organisation. They can identify and implement improvement in their own operation as well.

From an Organisational perspective, this model allows integration of verification processes within the work program with no requirement of additional resources. In addition, the risk of knowledge loss created due to employee turnover is significantly reduced. From a risk management and governance perspective, maximum assessment and verification is done where the activity is owned (first line) and this results in significant risk reduction at the source.

It is difficult to identify limitations of this process. However, one major consideration is to ensure sufficient review time is incorporated into the first line process which requires a disciplined approach.

CONCLUSIONS

South32 has significantly benefited from implementation of the three lines model into all its activities in the development and reporting of our resource and reserve estimates. Since its implementation, we are constantly working on optimising our standards, processes and procedures to realise maximum benefit. As a direct result, our technical teams have benefited positively due to the collaborative nature of the process. Overall, the model supports transparency, materiality and competence, the guiding principle of the JORC Code.

ACKNOWLEDGEMENTS

The authors would like to thank South32 for the permission to publish this work. In addition, the authors would also like to acknowledge input and constructive review by all who reviewed our work across the organisation.

REFERENCES

Australian Securities Exchange (ASX), 2013a. Chapter 4, ASX Listing Rules: Periodic Disclosure, 13 p. Available from: <https://www.asx.com.au/documents/rules/Chapter04.pdf>

Australian Securities Exchange (ASX), 2013b. Guidance Note 31, ASX Listing Rules: Reporting on Mining Activities, 18 p. Available from: <https://www.asx.com.au/documents/rules/gn31_reporting_on_mining_activities.pdf>

Australian Securities Exchange (ASX), 2016. Chapter 19, ASX Listing Rules: Interpretation and definitions, 36 p. Available from: <https://www.asx.com.au/documents/rules/Chapter19.pdf>

Australian Securities Exchange (ASX), 2019a. Chapter 3, ASX Listing Rules: Continuous Disclosure, 17 p. Available from: <https://www.asx.com.au/documents/rules/Chapter03.pdf>

Australian Securities Exchange (ASX), 2019b. Chapter 5, ASX Listing Rules: Additional reporting on mining and oil and gas production and exploration activities, 21 p. Available from: <https://www.asx.com.au/documents/rules/Chapter05.pdf>

Committee of Sponsoring Organizations of the Treadway Commission (COSO), 2017. *Enterprise Risk Management, Integrating with strategy and performance*. Available from: <https://www.coso.org/SitePages/Enterprise-Risk-Management-Integrating-with-Strategy-and-Performance-2017.aspx>

Deming, W E, 1950. Elementary Principles of the Statistical Control of Quality, Nippon Kagaku Gijutsu Remmei, 103 p.

Institute of Internal Auditors, The (IIA), 2020. The IIA's Three Lines Model. Available from: <https://www.iia.org.au/technical-resources/professionalGuidance/the-iia%27s-three-lines-model>

International Organisation for Standardisation (ISO), 2018. ISO 31000:2018: Risk Management – Guidelines.

JORC, 2012. Australasian Code for Reporting of Exploration Results, Mineral Resources and Ore Reserves (The JORC Code) [online]. Available from: <http://www.jorc.org> (The Joint Ore Reserves Committee of The Australasian Institute of Mining and Metallurgy, Australian Institute of Geoscientists and Minerals Council of Australia).

South32, 2022. Annual Report 2022, South32 Limited. Available from: <https://www.south32.net/docs/default-source/annual-general-meetings/2022/annual-report-2022.pdf?sfvrsn=8b529d95_1>

Telem, D, 2017. The three lines of defense – Making the transition to a mature risk management tool, KPMG LLP. Available from: https://assets.kpmg.com/content/dam/kpmg/ca/pdf/2017/01/three-lines-of-defense-kpmg.pdf

Best practice in Multiple Indicator Kriging (MIK) – importance of post-processing and comparison with Localised Uniform Conditioning (LUC)

G Zhang[1] and I Glacken[2]

1. Consultant Geologist, Snowden Optiro, Perth WA 6000.
   Email: gregory.zhang@snowdenoptiro.com
2. Executive Consultant, Snowden Optiro, Perth WA 6000.
   Email: ian.glacken@snowdenoptiro.com

ABSTRACT

Multiple indicator kriging (MIK) has been used in the minerals industry for some decades. As one of the non-linear estimation methodologies, MIK has advantages related to resolving multiple or mixed populations, high variability data and strongly skewed distributions. Applying the MIK methodology requires a lot more effort when compared to other non-linear methodologies, for example localised uniform conditioning. This is due not only to the variogram modelling for multiple indicators, but the validation is also very important because the panel model is fundamental for follow-up work, including the generation of recoverable resources at the local or SMU scale. Apart from this, only a few reported case studies of MIK have documented post-processing of the estimated modelled distribution, representing the probability distribution at point scale at un-sampled locations. The MIK post processing should include the change of support from the point scale to the panel scale, and then the extraction of quantile values from the panel conditional cumulative distribution function (CCDF) for localisation of grades into the SMUs within each panel, based on a ranking estimate marking the SMUs from high to low. In this paper, we present a case study on a gold deposit with the MIK method with full-post processing implemented for recoverable resources. The MIK point estimate was carried out with an indirect lognormal change of support from point to panel scale, and localisation was done with a custom script. The localised uniform conditioning (LUC) method was also applied for comparison purposes. The comparison shows that MIK has advantages compared to LUC when comes to capturing the high-grades, especially when mixed or varying anisotropy is present within the mineralisation.

INTRODUCTION

Nonlinear methodologies are often considered during the early stage of mineral projects when data is relatively sparse and some notion of mineable resources is required for a given future grade control configuration. The Multiple Indicator Kriging (MIK) approach is advantageous in dealing with mixed data populations, high variability and highly skewed data. Furthermore, it is one of a number of techniques available when resolution at SMU scales is required – for example for mine planning – given that the available data is too sparse, and a direct SMU scale estimate with such data would result in an oversmoothed and locally-inaccurate outcome. MIK is one of a small number of so-called recoverable resource techniques – that is, predicting the available material at the time of mining based upon early-stage or sparse data.

The need for recoverable resource estimates raises the importance of post-processing or *localisation* of the MIK estimates (LMIK) to the future mining scale; however, only a few case studies in the published literature discuss such post-processing (Zhang, 1998; Hardtke, Allen and Douglas, 2011). Thus there is value in sharing our thoughts and experiences on the topic.

This paper starts from a summary of the theory and workflow behind LMIK, follows with a comparison with localised uniform conditioning (LUC), a common alternative nonlinear estimation approach, via a case study, and finally leads to a discussion of the relative strengths and weakness of these two methods in the context of reasonably depicting the high-grade tail of a skewed mineralised distribution.

CASE STUDY – GEOLOGICAL BACKGROUND

The deposit covered in this comparative case study is an unmined post-orogenic Archaean gold deposit. Regionally, there are three north-trending litho-structural domains separated by a series of north-north-east trending strike faults.

At the deposit scale, the host rock is a turbiditic sedimentary sequence, with the mineralisation concentrated along the contacts of the litho-structural domains. The bedding strike direction in the deposit area is 025°, with the strata steeply dipping to the west.

There are three main phases of alteration, and mineralisation has been interpreted as being introduced in the last two:

1. Chlorite-albite alteration and magnetite-hematite-chlorite veins.
2. Quartz-tourmaline-pyrite alteration and veins.
3. Hematite-calcite-pyrite alteration and veins.

Gold mineralisation intensifies from the early high temperature chlorite-albite alteration through to the latter two medium-low temperature phases, where iron facilitated the precipitation of gold. Gold occurs as native gold or as fine inclusions within the base metal sulfides, or in the gangue that consists of quartz, albite, carbonate, muscovite, pyrite and tourmaline.

There are two types of mineralisation:

1. Disseminated gold-arsenopyrite and gold bearing quartz veins.
2. Gold-quartz veins with rare polymetallic sulfides.

The integrated and mixed mineralisation types and directions result in high variance and mixed populations for the gold data. MIK is regarded as a candidate for estimation for this reason.

The drill spacing in the majority of the deposit is 50 m × 50 m (locally 25 m × 25 m), and becomes wider at the deposit's margins. 60 per cent of the drilling is diamond holes and 40 per cent is reverse circulation (RC). The assumed selective mining unit (SMU) size is 5 m × 5 m × 5 m. A direct linear estimate from the drill hole data into SMU size block model would generate an oversmoothed and inaccurate result, and this leads to the consideration of localised MIK to assist in preliminary mine planning.

SUMMARY OF THE LMIK AND LUC WORKFLOWS

In this section, we summarise all the estimates and workflows involved in this study, and the logic behind some parameter selections.

This study workflow includes the following stages:

- MIK point estimate at a larger block size than the SMU.
- Change of support from an MIK point estimate into parent block scale estimate.
- MIK localisation into SMUs.
- Panel Ordinary Kriging estimate with high-yield constraint for the high-grade samples (>4 g/t).
- LUC via in-house software from the panel OK model.

LMIK DETAILS

Workflow

MIK provides a point estimate of the conditional cumulative distribution function (CCDF) at unsampled locations via the estimation of probabilities at various thresholds. 10–15 thresholds are usually selected based on the sample cumulative distribution for each mineralised domain or ore zone. For each threshold, the samples above the threshold are coded as '1' and samples below as '0' in this case. Often the inverse coding is applied, resulting in complementary probabilities. Essentially the samples are treated as binary indicators, variogram models are built leading to an estimate at a relatively large block (or panel) scale. The estimate result at each location is a sequence of probabilities between '0' and '1', forming a CCDF. The total of the probabilities at various thresholds will equal to 1. There are invariably order relations issues, whereby the successive probabilities are not monotonically increasing, and three solutions (as described below) can be used to address these issues.

The MIK post-processing steps involve:

1. Order relations corrections.

2. Change of support of the MIK estimate from a point scale CCDF to panel scale CCDF.

3. Retrieve the grade values from the corrected panel CCDF for various quantiles, corresponding to the number of SMUs.

4. Generate a 'ranking' estimate at the SMU scale using ordinary kriging with the untransformed data, the median indicator variogram model and the same search parameters as used in MIK point estimate.

5. Assign the quantiles from the panel CCDF to the SMUs in each panel according to the ranking estimate.

The disseminated nature of mineralisation within the orebody, combined with the vein type mineralisation, results in multiple mineralisation directions at different grade cut-offs and high variance in the sample data, making direct estimation using ordinary kriging or similar problematic. Moreover, the lack of sufficient drilling density would lead to an oversmoothed estimate if direct estimation were carried out at the SMU size with the sparse data, leading to an incorrect model for subsequent open pit optimisation. The application of MIK with post-processing localisation addresses the above issues.

In the case study the panel estimate had a size of 25 m × 25 m × 10 m and the SMU size adopted for post-processing was 5 m × 5 m × 5 m.

The broad mineralisation shape was interpreted initially using a nominal cut-off of 0.15 g/t, together with the presence of favourable lithology, high intensity alteration, sulfides and fractures. Fourteen indicator thresholds and associated variogram models were generated for the gold data. The indicator variogram modelling started with the median (50 per cent) and progressed towards both ends (27.9 per cent and 99.9 per cent). Considerable effort was expended to maintain smooth changes between neighbouring threshold variogram models (Figure 1).

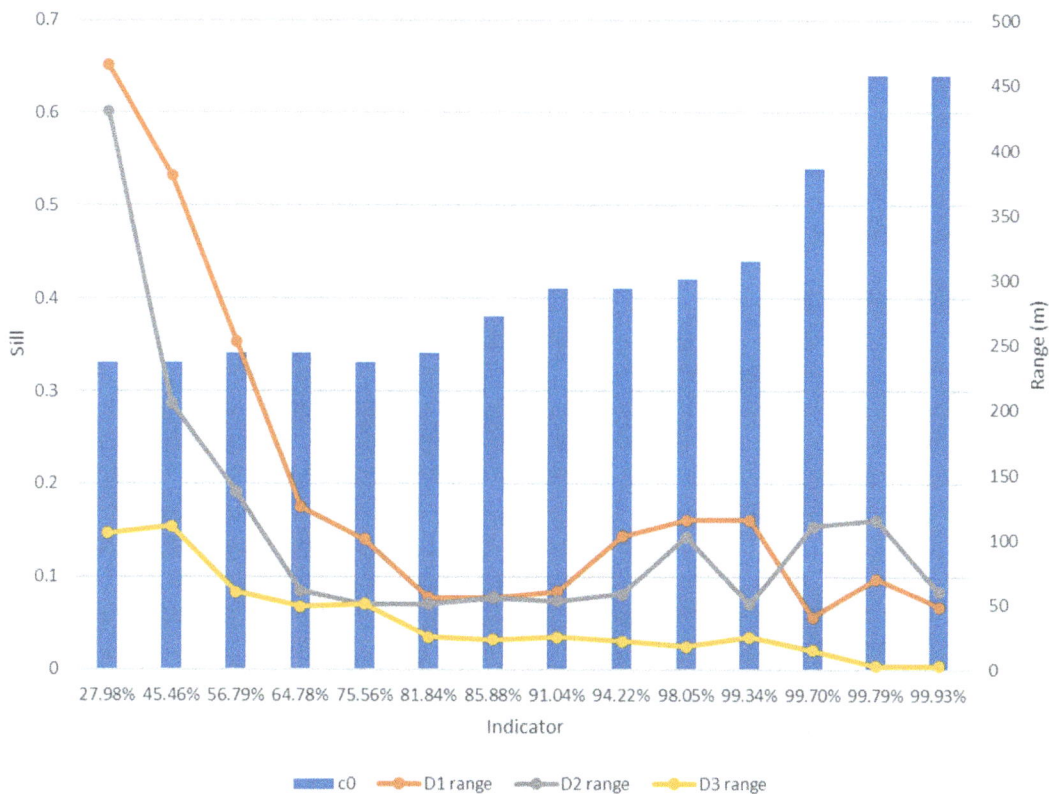

FIG 1 – Variation of nugget, major (D1), semi-major (D2) and minor (D3) directions' ranges across the 14 indicators.

The MIK point distribution estimate comprises a series of probability estimates for 14 thresholds. The average or E-type grades can then calculated by weighting the domain-wide grades of all the intervals by the local probabilities. The outputs of the MIK point estimate are the CCDF at the parent block centroids and the corresponding E-type averages.

Order relations

Order relations violations occur where either the probabilities are not ascending over the increasing thresholds or the CCDF rises above 1; this is because the thresholds are estimated independently. The order relations resolution approach is illustrated in Figure 2. The black solid line represents the uncorrected parent block CDF, where the cumulative probability at a grade of 2 is actually less than the probability at 0.5. This can be addressed by: (1) decreasing the cumulative probability of 0.5 equal to 2 (green dash line); (2) increasing cumulative probability of 2 to 0.5 (blue dash line); or (3) smoothing the cumulative probabilities by averaging the overall minimum and maximum thresholds (black dash line). The last option was selected in this case.

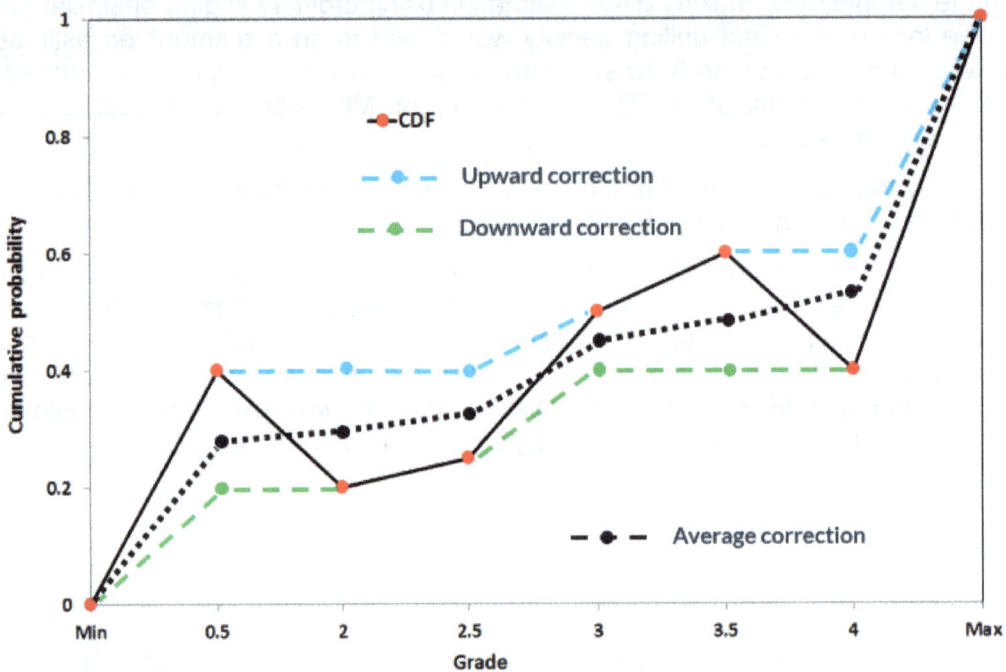

FIG 2 – Schematic illustration of the order relations correction technique.

Change of support

Since the initial MIK distribution as initially estimated is always strictly at point support, a change of support correction needs to be applied to each CCDF to reflect the distribution at the panel scale. A number of options are available for this change of support; given that, in the case study (and indeed in most cases) the CCDFs are positively skewed, an indirect lognormal correction is an appropriate change of support algorithm. This can be carried out using (for instance) the open source GSLIB software routine postik.exe (Daniels and Deutsch, 2014), which is a two-stage process; the first corrects the point CCDF to the panel CCDF using variance reduction based upon the relative point and panel block sizes, and the second process forces the panel support average to match the point CCDF average. The indirect lognormal correction is also available in some generalised mining software or can be implemented using Excel.

An example of the CCDF comparison between parent block and point scales is presented in Figure 3, where the parent block (panel) CDF is 'squeezed' towards to the middle when compared to the point CDF.

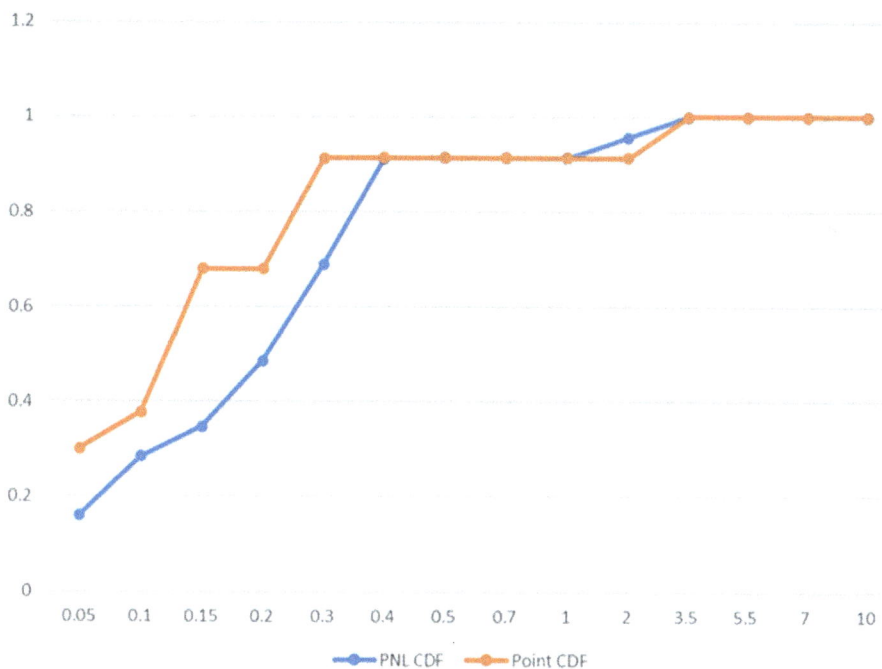

FIG 3 – Comparison of point CCDF (orange) and panel CCDF (blue) for a block (threshold for X axis and probability for Y axis).

It is important to note that if the E-type or average of the CCDF was all that was required, then the change of support may not be necessary; however, since the entire panel support CCDF is required for the localisation process, the change of support is required, and to not implement it would lead to errors in the localised SMU distribution.

Localisation – LMIK

The final stage in the workflow is the localisation of grades from the panel to the SMU scale; this is required to provide mine planners with an SMU-level model for realistic what-if scheduling. The localisation algorithm here is different from LUC, which is discussed below. In essence, however, the localisation for MIK applied in this case starts with the discretisation of the CCDF of each parent block into the number of SMUs.

The processing behind this step includes two parts. The first part is to extrapolate the lower and upper tails of the parent block CCDF, that is above and below the lowest and highest indicators; in this case a power model was selected for the upper and lower tails; the formula is:

$$P = kX^n \text{ when } 0<X<cog_1 \text{ or } cog_{14}<X \tag{1}$$

Where P is the conditional cumulative probability at X, X is the threshold grade, n and k are the coefficients to calculate.

For interpolation between the other thresholds, a linear model was selected, with the formula:

$$P|(cog_m<X<cog_{m+1}) = aX+b \text{ when } |(cog_m<X<cog_{m+1}), 1 <= m <= 13 \tag{2}$$

Where P and X are defined as above. Since quantile P is a known parameter, X for the various quantiles can be calculated using:

$$X=(P/k)^{(1/n)} \text{ when } 0<X<cog_1 \text{ or } cog_{14}<X \tag{3}$$

$$\text{Or } X=(P-b)/a \text{ when } cog_m<X<cog_{m+1} \tag{4}$$

Other models can be used to model the tails of the CCDF; for example, a hyperbolic function can be used for the upper tail, and the logic is the same, albeit with slightly different formulae. In this case study a hyperbolic function has been applied to the upper tail.

After we have the grade X corresponding to every quantile (where the number of quantiles corresponds to the number of SMUs required for the localisation), the ranking estimate is interrogated to determine which grade is to be assigned to each SMU. This part is similar to the LUC approach. In the case study 50 SMUs were generated per panel, and thus the panel CCDF was discretised into 50 quantiles, each corresponding to two-percentile increments. These retrieved quantiles were allocated to the SMU blocks according to the grade trends (per panel) captured in the ranking estimate, in the manner of LUC. Validation was conducted during the whole procedure to ensure that the intermediate results made sense geologically and minimised any edge effects, which are usually caused due to the extended search strategy and lack of enough nearby samples to generate meaningful SMU values.

LUC – LOCALISED UNIFORM CONDITIONING

As perhaps the most commonly used non-linear geostatistical estimation methodology, localised uniform conditioning (LUC) was also conducted for comparison purposes. The LUC workflow here is summarised in its simplest form; for further details on the uniform conditioning stages and underlying theory, refer to Neufeld and Deutsch (2005). For further details of the localisation (LUC) stages, refer to Abzalov (2006).

The workflow of LUC includes the following steps:

- Estimate into the panel with the linear estimation methodology first; for example, by Ordinary Kriging to ensure the minimum conditional bias.

- Test that various supports maintain the multi-Gaussian relationship (theoretically).

- Create a Discrete Gaussian Model (DGM) at point support, panel support and SMU support.

- Transform the cut-offs into Gaussian space via anamorphosis and derive the metal and proportion above each cut-off.

- Translate the corresponding metal values and cut-offs from Gaussian space to the original space.

- Generate the ranking estimate and localise to SMUs on a metal basis, with the grade being back-calculated.

The panel scale estimate was generated using Ordinary Kriging in this case, with a distance constraint for the high-grade samples (>4 g/t). High variability data will cause high conditional bias in the parent block estimate due to the nature of Ordinary Kriging, which requires the input data to have low skewness and moderate variance/coefficients of variation. Either a top-cut or a high yield constraint needs to be applied to the outliers. The UC process is based on a transformation of the samples to a Gaussian distribution. This is different to MIK in that MIK is non-parametric, no distribution is assumed and complex distributions are effectively discretised via the imposition of indicator thresholds.

The process of mapping the distribution to a Gaussian distribution is Discrete Gaussian Modelling through the use of Hermite polynomials. This is not always feasible as some highly – skewed, high variance data is mathematically not suited to a Gaussian transformation (Taleb, 2020). In this case, top-cutting is unavoidable in forcing the variance to meet the normal-score transformation criteria. Another fundamental difference between the LUC method and the LMIK approach described above is that the former relies on a single continuity model per domain, and any changing anisotropy within different grade ranges (as seen in the case study) cannot be captured.

COMPARISON OF LMIK AND LUC

The tonnage-grade curve comparison between LMIK and LUC at the SMU scale is presented in Figure 4, which shows tonnage (as a proportion of the tonnage at zero cut-off) on the X axis, the average grade above cut-off on the Y axis, and cut-offs along the curves. Some observations are as follows:

- The relative tonnage and grade are quite close when the cut-off grade is no larger than 0.3 g/t, which is the declustered composite average. The difference becomes much larger and more obvious at cut-off grades above this.

- The LUC parent block estimate average grade (0.257 g/t) is 7 per cent lower than the MIK parent block estimate (0.274 g/t), so is the metal. This is undoubtedly a function of the different treatment of outlier grades by both ordinary kriging (for LUC) and indicator kriging (for LMIK) The difference gets larger as the cut-off grade increases.

- The comparison shows that the SMUs from the LMIK approach are better able to represent the composite statistics from the perspective of average, skewness and maximum (Figure 5).

- The Gaussian distribution limits the application of the method; when a top-cut or high yield constraint needs to be applied to the composite data to minimise the conditional bias and meet the Gaussian distribution criteria, the global metal or global average will be decreased. MIK does not have this limitation to the same extent, and often (as in this case) the global statistics are better reproduced.

- LUC performs adequately when only considering the global metal, not necessarily the average grade of the composites or the SMUs. The LUC approach is much less effort to achieve when compared to LMIK, although the localisation part of the workflow is conceptually and practically simpler in the LMIK case, being based upon grade quantiles rather than back calculated metal quantiles (as in LUC).

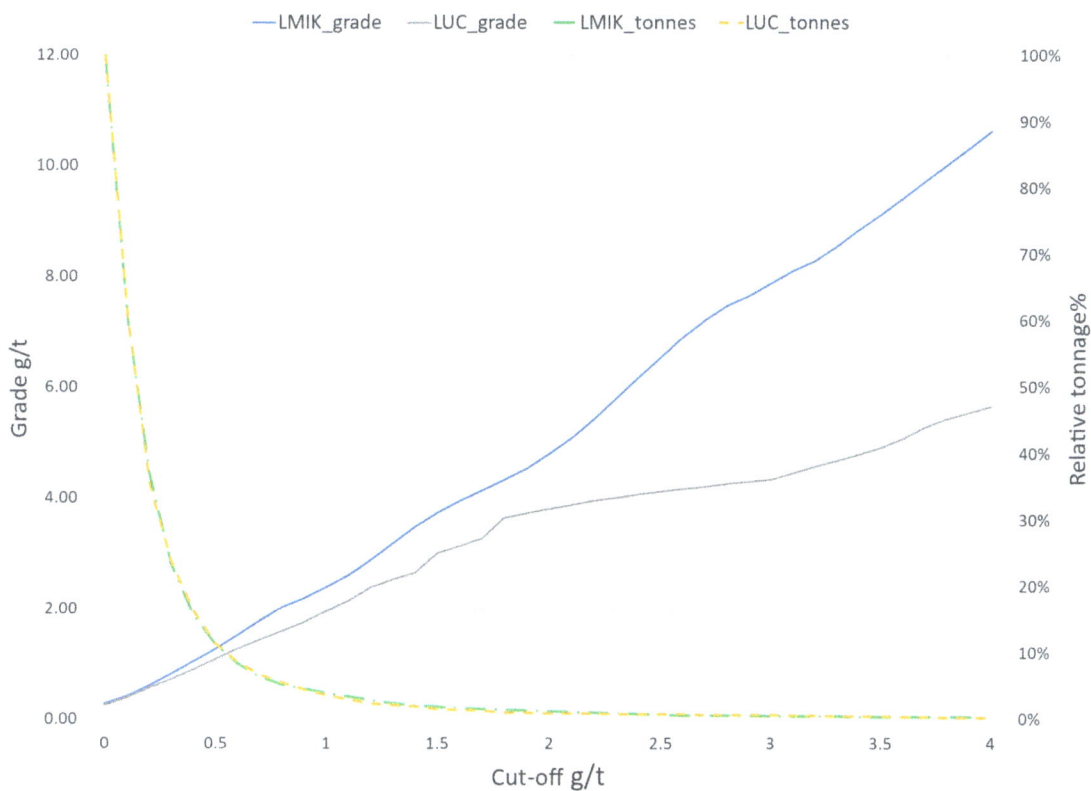

FIG 4 – Comparative tonnage grade curves for LMIK (blue) and LUC (orange) SMU distributions.

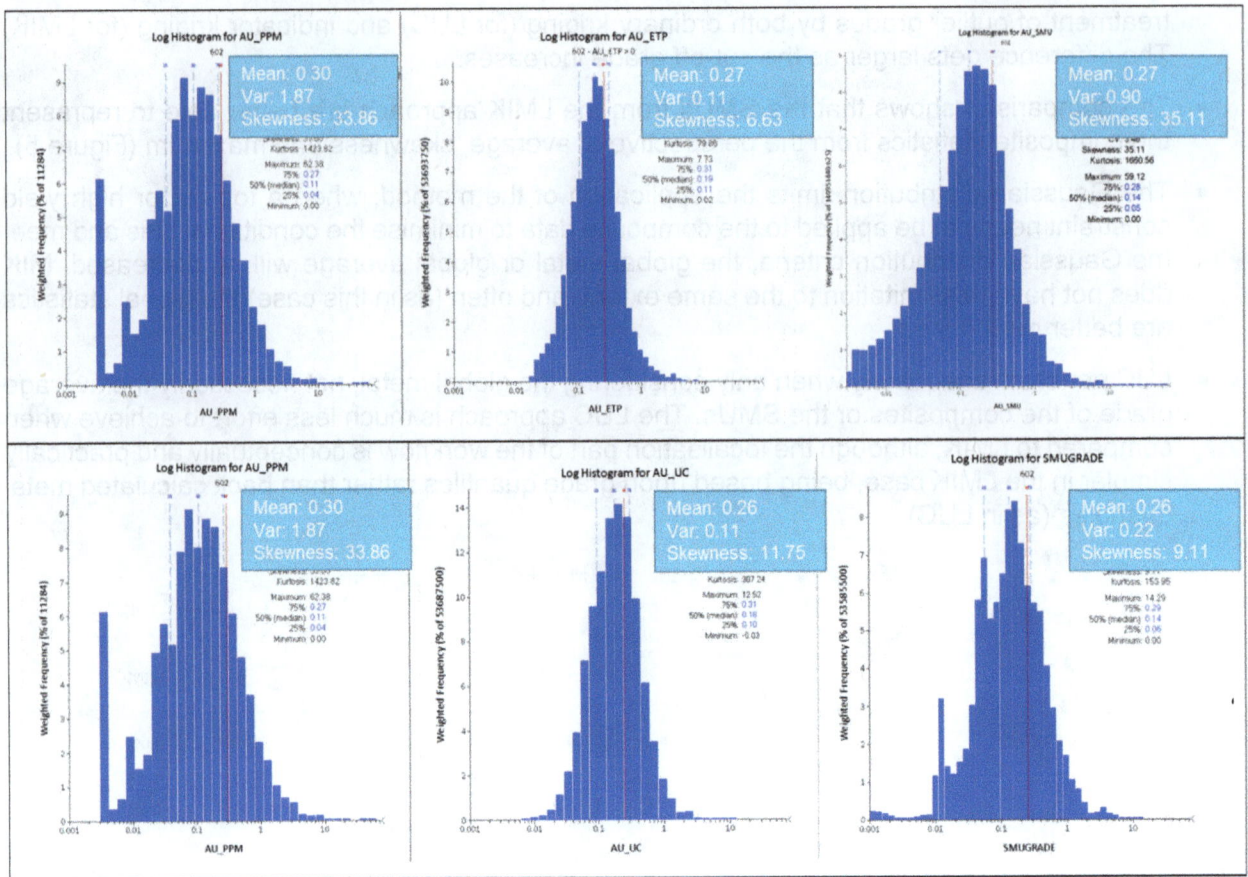

FIG 5 – LMIK (top) and LUC (bottom) statistics comparison for composites (left), panel blocks (middle) and SMUs (right).

CONCLUSIONS

Several conclusions can be drawn from this comparative case study:

- The LMIK SMUs are better able to reproduce the statistics of the composite data, including average, skewness and maximum. The LUC SMUs have a much lower maximum grade than the LMIK approach, indicating that LMIK can better 'capture' the high-grade portion at the SMU level of support.

- High CV data has to be remediated before estimation at the panel scale OK estimate for the application of UC. A high yield spatial constraint has been applied in this case instead of a more conventional top-cut. The metal and average grade therefore decrease by 7 per cent when compared to the MIK parent block. LMIK does not require the top-cut or high yield constraint; rather, the treatment of high-grades is regulated via the modelling of the upper tail of the CCDF, ie that portion of the distribution above the uppermost indicator threshold.

- LMIK is better than LUC when dealing with the high relative variability data and variable mineralisation directions, because LUC makes certain assumptions regarding input data and requires the multi-Gaussian relations between the transformed data (sample, panel and SMU support), which cannot always be met and which can never be verified. The high CV feature of the highly skewed data in this case, leads to more potential for high-grade smearing, which is challenging for LUC.

- At the early stage of the project, the robustness of the change of support is determined by the variogram models. Whether it is affine, lognormal, or indirect lognormal change of support, it depends on the kriged relationship, namely variance at point (sill) is the sum of variance at parent block scale (block variance) plus the variance at SMU scale (gamma bar, which is one of the input parameters for change of support). This can be validated when the project comes to production stage when short range data is available. This issue applies to both LMIK and

LUC; however, the continuity models are more detailed for LMIK as they are derived at a large number of cut-offs, and thus capture any change in anisotropy over different grade ranges.

- Care should be taken when looking at the 'high-grades' in the SMUs of localised MIK or LUC model. As mentioned before, these high-grades are guaranteed to be locally inaccurate in both cases. The SMU grade distribution in each parent block can be inferred via the ranking estimate, and this is only for the convenience of open pit optimisation/planning. The model does not eliminate the need for detailed grade control drilling, and the overall non-linear estimation methodology generated models still need to be 'calibrated' by the close space drillings.

- For a complex, highly variable grade distribution with anisotropy which varies at different grade ranges and which requires localisation to the SMU scale for scheduling, the LMIK approach is arguably superior, since it can handle the complex distribution and the extreme grades better. The localisation algorithm itself, as described above, is simpler than the LUC method; however, the inference of multiple continuity models (14 in this case) is much more demanding than the simple UC approach, which relies on a single variogram model. When the MIK/LMIK post-processing issues, such as order relations corrections and the change of support from point to panel are additionally considered, many practitioners may favour the simpler, but not always appropriate, LUC approach.

- LMIK is an estimation approach, like Ordinary Kriging or Inverse Distance, but it is a tool with more complexity which allows the generation of recoverable resources through localisation to SMU support. During the LMIK workflow there are a number of parameters or opportunities for the practitioner to control the risk and make the estimate as optimistic/conservative as possible (for example the upper tail modelling). In other words, the generation of high-quality models, as ever, requires considerable input and understanding. LMIK is a versatile tool and can be used when it is appropriate for the data complexity and variability.

- For the upper tail modelling, sensitivity testing is recommended using different extrapolation models. Another key decision for LMIK is the selection of the treatment of grades above the top bin value during panel estimate (median versus. mean), which was conducted in this case study and does not have a material effect on the outcome.

- The deposit in this case study is not producing. Therefore, verifying the comparative estimates against actual mining data would be impossible for now. However, workers often use conditional simulation to simulate a 'truth' and then repeat the LMIK and LUC comparative estimates based the dummy holes 'drilled' in the 'truth' model. This can be used as a further test of whether LMIK or LUC estimate result is closer to the 'truth'.

- There are criticisms regarding the inaccuracy during a global change of support (Rossi and Parker, 1993), and general indicator kriging (Emery and Ortiz, 2004), which points to the importance of understanding the limitations of these non-linear estimation methods. Caution should be taken and the estimate results would better be validated with the grade control of production data if possible. In each case the practitioner is advised to consider the benefits of the LMIK approach, in terms of handling rotating anisotropy and extreme skewness, against the criticisms of the methodology.

ACKNOWLEDGEMENTS

The authors are extremely grateful for Paul Blackney, Kahan Cervoj and Oscar Rondon for their helpful technical discussions and advice during the estimation stage. Thanks are also due to Oscar Rondon and Thi Nguyen for their time and patience during the early stages of presentation.

REFERENCES

Abzalov, M, 2006. Localised Uniform Conditioning (LUC); a new approach for direct modelling of small blocks, *Math Geol*, vol 38(4).

Daniels, E and Deutsch, C, 2014. Flexible Uncertainty Representation and localization programs, *Centre for Computational Geostatistics Annual Report*, vol 16, 2014.

Emery, X and Ortiz, J, 2004. Shortcomings of multiple indicator kriging for assessing local distributions, *Transactions of the Institutions of Mining and Metallurgy, Mining Technology, Applied Earth Science (B)*, 113:B249–B259. 10.1179/174327504X27242.

Hardtke, W, Allen, L and Douglas, I, 2011. Localised indicator kriging, in *Proceedings of the 35th Symposium on the Application of Computers and Operations Research in the Mineral Industry (APCOM)*.

Neufeld, C and Deutsch, C V, 2005. Calculating recoverable reserves with Uniform Conditioning, Centre for Computational Geostatistics paper 2005–303, University of Alberta, Canada.

Rossi, M E and Parker, H M, 1993. Estimating Recoverable Reserves, Is It Hopeless?, *In Geostatistics for the Next Century*, (ed: R Dimitrakopoulos), pp 259–276 (Kluwer Academic Publishers: Boston).

Taleb, N, 2020. Statistical consequences of fat tails: Real world preasymptotics, epistemology, and applications, arXiv preprint arXiv:2001.10488.

Zhang, S, 1998. Multimetal recoverable reserve estimation and its impact on the Cove ultimate pit design, *Mining Engineering*, 50(7):73–79 (Littleton: Colorado).

Beyond zero cut-off – validation
at economic cut-offs

Optimising underground resources – a case study

P Boamah[1], G Squissato Barboza[2] and D K Mukhopadhyay[3]

1. MAusIMM(CP), Principal Resource Geologist, Evolution Mining, Sydney NSW 2000.
   Email: paul.boamah@evolutionmining.com
2. MAusIMM(CP), Principal Mining Engineer, Enable Advisory Pty Ltd, Brisbane Qld 4000.
   Email: guilherme.barboza@enableadvisory.com
3. MAusIMM(CP), Senior Manager Resource and Reserve Governance, South32 Limited, Perth
   WA 6000. Email: dk.mukhopadhyay@south32.net

ABSTRACT

Mineral Resource estimates are fundamental to every mining operation or project. It is a JORC Code requirement that the reported Mineral Resource satisfies the criteria of 'reasonable prospects for eventual economic extraction' (RPEEE). While near-surface mineralisation is reported above an economic cut-off within an optimised pit shell, the acceptable practice for deep-seated mineralisation is to report the Mineral Resource estimate above a Reduced Level (RL) using an economic cut-off.

This paper discusses how the RPEEE requirements have been met for South32's Cannington Ag-Pb-Zn underground mine Mineral Resource using the Datamine Mineable Shape Optimiser (MSO) tool. Applying Datamine's Mineable Shape Optimiser (MSO) ensures mining and economic parameters are considered, allowing RPEEE to be assessed and isolated blocks that do not meet the Minimum Mining Volume (MMV) to be excluded.

The result shows a repeatable and less subjective process. The MSO removes isolated blocks and potential errors associated with underground surveys between stopes.

INTRODUCTION

A 'Mineral Resource' is a concentration or occurrence of solid material of economic interest in or on the Earth's crust in such form, grade (or quality), and quantity that there are reasonable prospects for eventual economic extraction. The location, quantity, grade (or quality), continuity and other geological characteristics of a Mineral Resource are known, estimated or interpreted from specific geological evidence and knowledge, including sampling (JORC Code, 2012). All reported Mineral Resources must satisfy the requirement that there is a reasonable prospect for eventual economic extraction (ie more likely than not), regardless of the resource classification (JORC Code, 2012).

Cannington Ag-Pb-Zn underground mine is a longhole open stoping mine and has been operating since 1997. Current mining is mainly from secondary and, in some cases, tertiary stopes. Cannington, being a polymetallic deposit, also uses a Net Smelter Return (NSR) as the grade descriptor. The NSR calculation considers metal prices, metallurgical recoveries, product-associated costs and payable proportions for each product based on the relevant grades for each particular block. Unlike an open pit Mineral Resource estimate (near surface), which is primarily defined by a pit shell using modifying factors (Modifying Factors are considerations used to convert Mineral Resources to Ore Reserves (JORC Code, 2012), many underground operations (deep-seated) define their Mineral Resource estimate as mineralisation above a Reduced Level (RL) and an economic cut-off considering the mining method.

Underground mining is inherently more complex than an open pit. While assessing the resource RPEEE does not replicate the same practices that an Ore Reserve estimation would apply, the best practice is to use stope optimisation software as part of the process (Glacken, 2019). Applying Datamine's Mineable Shape Optimiser (MSO) ensures mining and economic parameters are considered, allowing RPEEE to be assessed and isolated blocks that do not meet the Minimum Mining Volume (MMV) to be excluded.

GEOLOGY OF CANNINGTON

Cannington is a Broken Hill Type Ag-Pb-Zn deposit located within the Proterozoic Mount Isa Inlier, a diversely mineralised, deformed and metamorphosed terrain in north-west Queensland (Figure 1). The inlier is subdivided into three main blocks by major north-striking fault zones, namely the

Western Succession (or Western Fold Belt), the central Kalkadoon-Leichardt Province and the Eastern Succession (or Eastern Fold Belt). The Eastern Succession comprises an Archean-Proterozoic basement of metamorphic rocks variably overlain by three volcano-sedimentary cover sequences, which range in age from 1850 to 1670 Ma (Walters and Bailey, 1998).

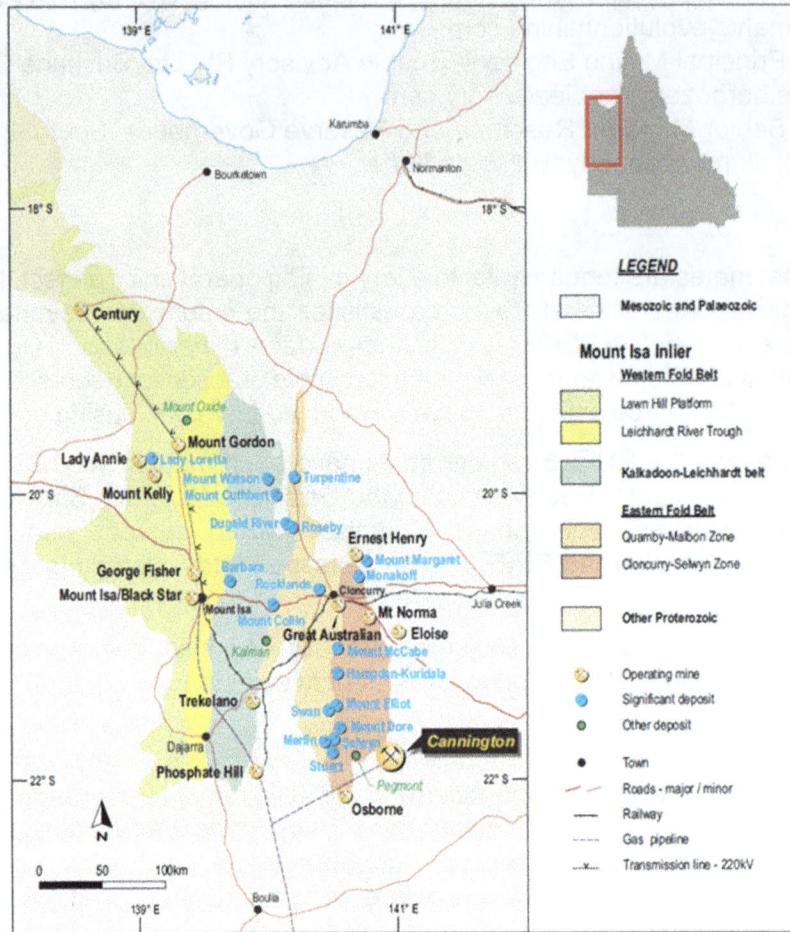

FIG 1 – Simplified geological map of the Mount Isa Inlier after Walters and Bailey (1998).

Cannington mineralisation is hosted within a 70 m to 90 m thick sequence of interbedded garnetiferous psammite and basic volcanic rocks that enclose a fine to medium-grained central amphibolite body up to 100 m thick (Walters and Bailey, 1998). The psammite is intercalated with muscovite-sillimanite schist and enclosed by migmatitic quartzo-feldspathic gneiss. Pegmatite horizons that are semi-conformable within the deposit sequence are thought to be predominantly the result of partial melts of clastic rocks within this package (Walters and Bailey, 1998).

The sulfide mineralogy of the Ag-Pb-Zn mineralisation is dominated by coarse grained galena (PbS), sphalerite (ZnS) and friebergite ($Cu_6(Ag, Fe)_6Sb_4S_{13}$). Silver is predominantly contained in friebergite but also occurs in galena and other silver mineral species. Sphalerite is typically coarse grained. Other sulfide minerals include pyrite, pyrrhotite, chalcopyrite and arsenopyrite (Walters and Bailey, 1998).

Mineralisation is stratiform along the limbs of a tight, northerly striking isoclinal recumbent fold that plunges ~25° to the south. Fold limbs dip between 40° and 70° to the east (Walters *et al*, 2002).

Two major north-west trending faults offset deposit sequence. The Trepell Fault divides the 'Northern Zone' from the 'Southern Zone' and the Hamilton Fault offsets mineralisation in the south of the 'Southern Zone' (Figure 2) (Bailey, 1998).

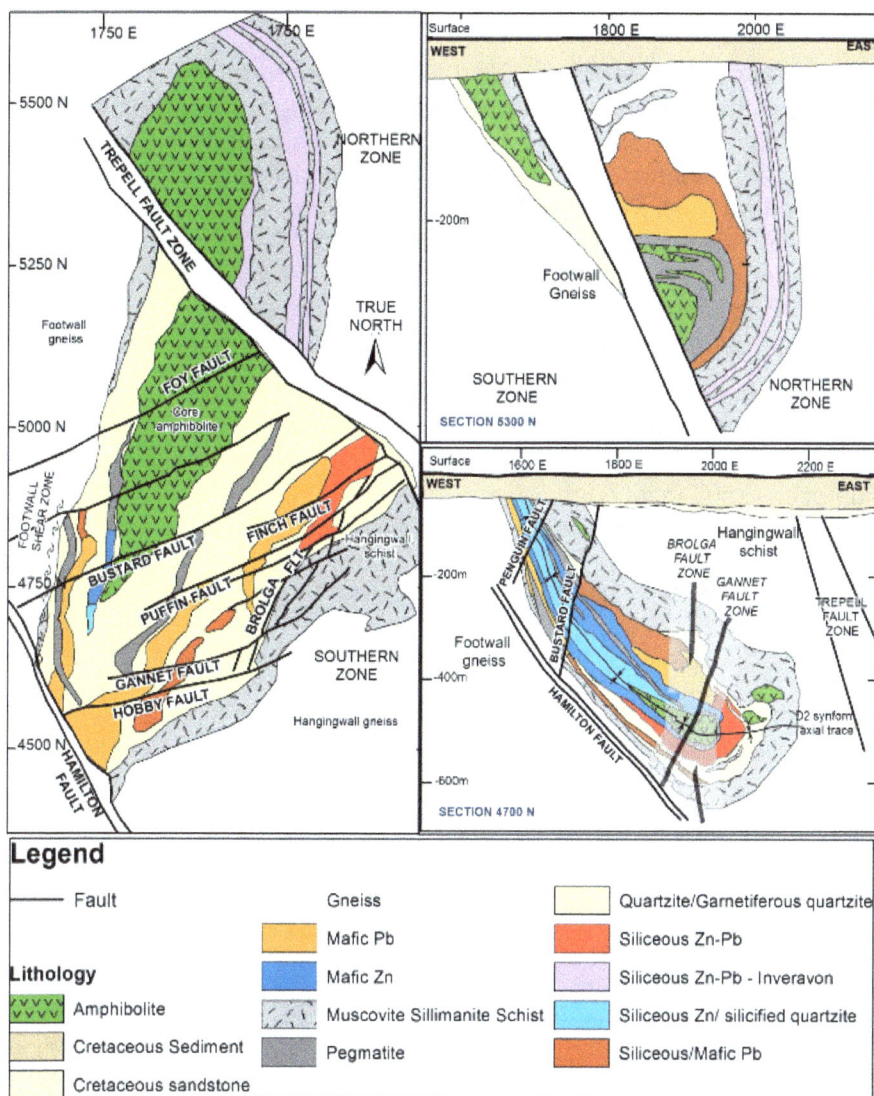

FIG 2 – Cannington geology interpretation and cross-sectional view after Bailey (1998).

ESTIMATION AND MINING

Grades are estimated using ordinary kriging (OK) into block cell dimensions of 5 mE × 5 mN × 5 mRL. These dimensions have been selected to optimise the modelling of grade variability within an average drill hole spacing of 12.5 mN × 15 mRL.

The orebody is mined by conventional longhole open stoping (LHOS) with a backfill. The Selective Mining Unit (SMU) is defined by a stope cross-sectional area of 10 m along strike, 5 m width, and a sublevel height (20 m). Stopes sizes range from a minimum of 2 kt to 200 kt with an average of 34 kt. The stope size is variable as a result of variable orientation due to folding, thickness, grades, and rock mass rating.

Cannington Ag-Pb-Zn underground mine has been in operation since 1997. Current mining is from mainly secondary and, in some cases tertiary stopes. Historical mining has left remnant material that is above the economic cut-off but smaller than the current SMU. Mineral Resource reporting in such an environment is challenging as a resource must satisfy the 'reasonable prospects for eventual economic extraction' (RPEEE) requirement, regardless of the specific estimation approach used or the classification method applied.

Table 1, section 3 of The JORC Code (2012) outlines criteria to be considered for RPEEE. This work proposes the use of an MSO approach to address the resource selection criteria above the economic cut-off.

MSO – A REPORTING TOOL

Mineable Stope Optimiser (MSO) is widely used for generating optimised shapes for underground stopes. MSO automates the design of stope shapes for a range of scenarios and provides optimal shapes designed to maximise an orebody's value by applying constraints detailing mining methods and design parameters (Castanho, 2020). Alford Mining Systems (AMS) first developed the Stope Shape Optimiser algorithm and commercialised the software in 2011. MSO software is currently distributed through major mining software packages such as Datamine, Deswik and Vulcan. MSO has been a game-changer and impacted the underground mine planning process by providing opportunities for improving project value, enabling profitability with marginal projects and identifying future mining areas to target exploration and development (Castanho, 2020).

Although mine planners widely use MSO as the basis of potential stope shapes, MSO can also help report the potential underground Mineral Resource by adjusting the input parameters (including grade cut-off) to reflect criteria that address RPEEE requirements. These input parameters include both mining assumptions (sub level spacing, mining widths, wall angles, expected overbreak, pillars and other exclusion or stand-off requirements) and mining cut-off (grade or NSR).

MINERAL RESOURCE REPORTING – CANNINGTON CASE STUDY

Historical practices

Historically, Cannington's Mineral Resource has been reported using an optimised pit shell with modifying factors to delineate the Mineral Resource for near-surface mineralisation. In addition, the resource model below the optimised pit shell and above the economic cut-off supporting underground mining has been reported to define the underground Mineral Resource. Figure 3 is a pictorial view of the previously adopted reporting process.

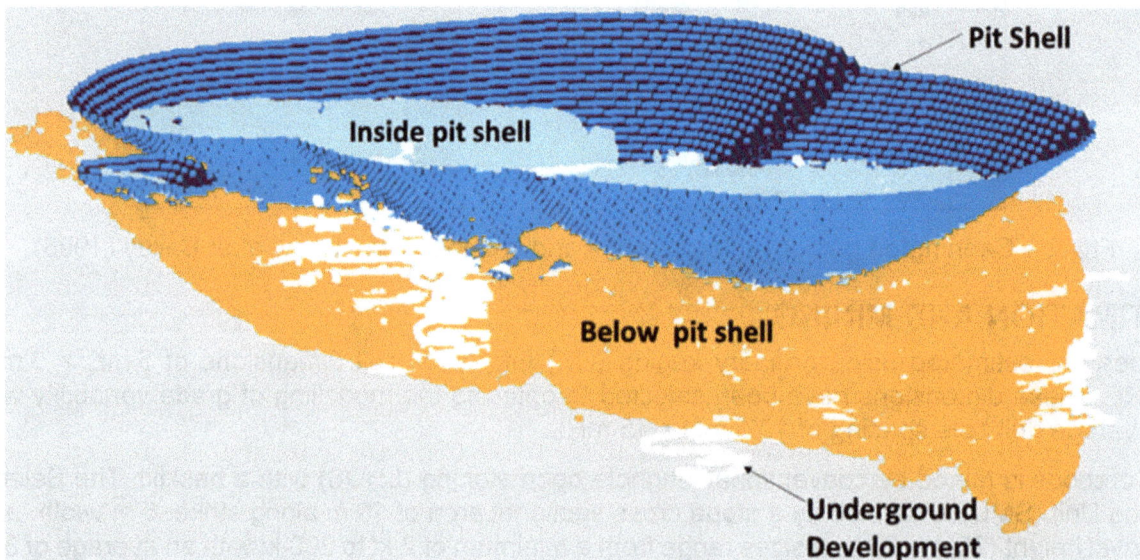

FIG 3 – Historic style of reporting Mineral Resources.

Current practices

Near-surface mineralisation continues to be reported in the same manner above an optimised pit shell. However, MSO is used in the process of defining the mineralisation that supports Cannington's underground operation and is done by optimising material above an economic cut-off supporting underground mining and below the open pit shell. However, the underground stope design which selects material within the open pit shell, is reported as part of the underground Mineral Resource but excluded from the open pit Mineral Resource (Figure 4).

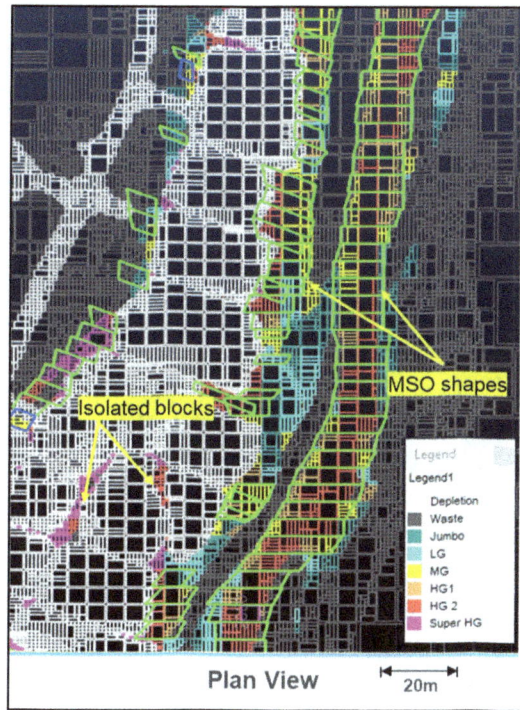

FIG 4 – Current style of reporting Mineral Resources.

MSO process and assumptions

The MSO runs are set-up based on the LHOS mining method with the following assumptions applied based on local experience:

- Grade Descriptor (NSR) based on South32 provided-price protocols for the payable metals; Ag, Pb and Zn.

- Resource cut-off grade.

- Sublevel Spacing based on mined levels and planned development.

- Hanging wall and Footwall angles from 50° to vertical.

- Minimum stoping widths based on Cannington's experience.

- Variable overbreak by mining zone (0.5 m to 1.5 m).

- Stand-off from major faults.

- Limited overlap with mined stopes.

- No designed stope overlaps are allowed ensuring material is not double counted (controlled by order of MSO runs and block model depletion).

The MSO runs are set-up in two steps; Run1 uses the average stope dimension of Cannington with all the assumptions listed above. A second MSO run (Run2) is then performed considering decreased minimum stope dimensions. The second run excludes blocks from Run1 to identify and include smaller stopes. Figure 5 shows the outlines of both Run1 and Run2.

FIG 5 – Run1 and Run2 shapes.

RESULT

South32 adopted the MSO process in reporting the FY2020 Mineral Resource, resulting in a relative reduction in the Mineral Resource, as shown in Figures 6 and 7. The South32 2019 annual report had Cannington's total Mineral Resource at 85 Mt with a Reserve life of 12 years, and the 2020 annual report had Cannington's total Mineral Resource at 79 Mt with a Reserve life of 11 years (mining for FY2019 was 2.7 Mt) as shown in Figure 7. Thus, while the FY2020 Mineral Resource has slightly less tonnes than the historical approach, the application of MSO provides a more robust resource that satisfies RPEEE requirements and less subjective as isolated blocks smaller than the minimum mining volume are excluded, although above the desired cut-off. Figure 6 highlights an example where mineralisation previously reported as resource was excluded from the Mineral Resource as the highlighted areas do not meet the minimum mining volume criteria.

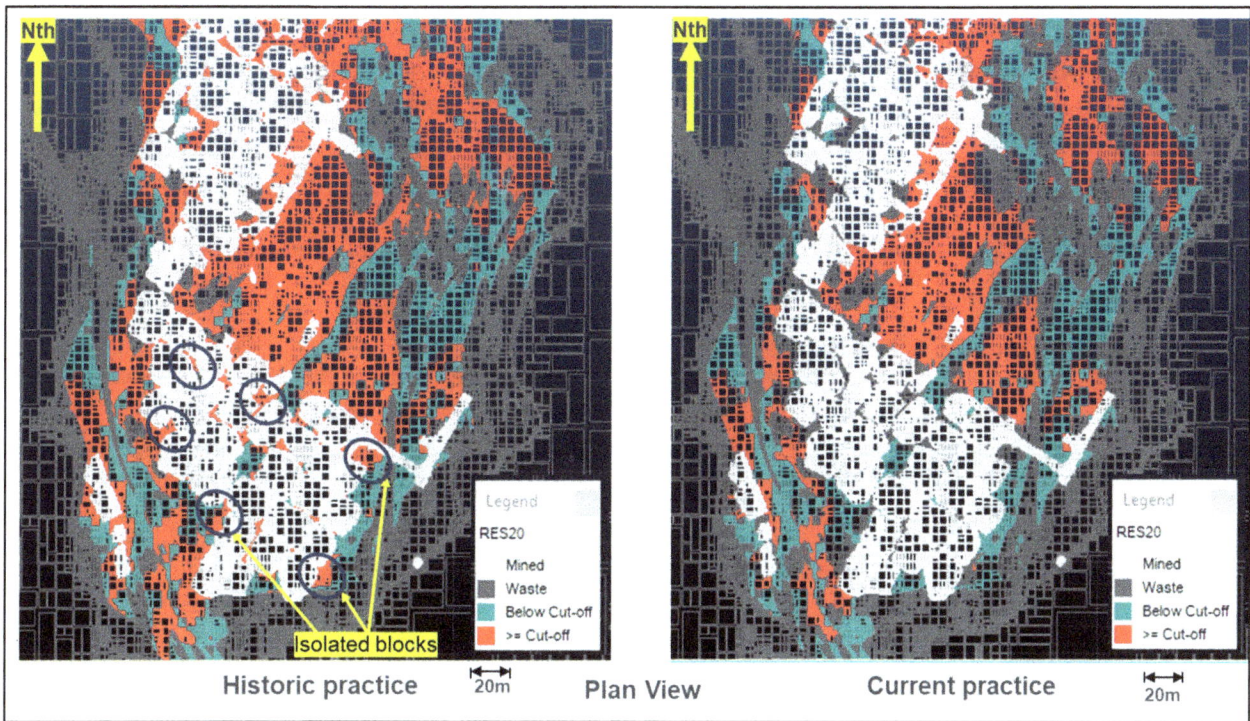

FIG 6 – Comparison of historic and current style of reporting.

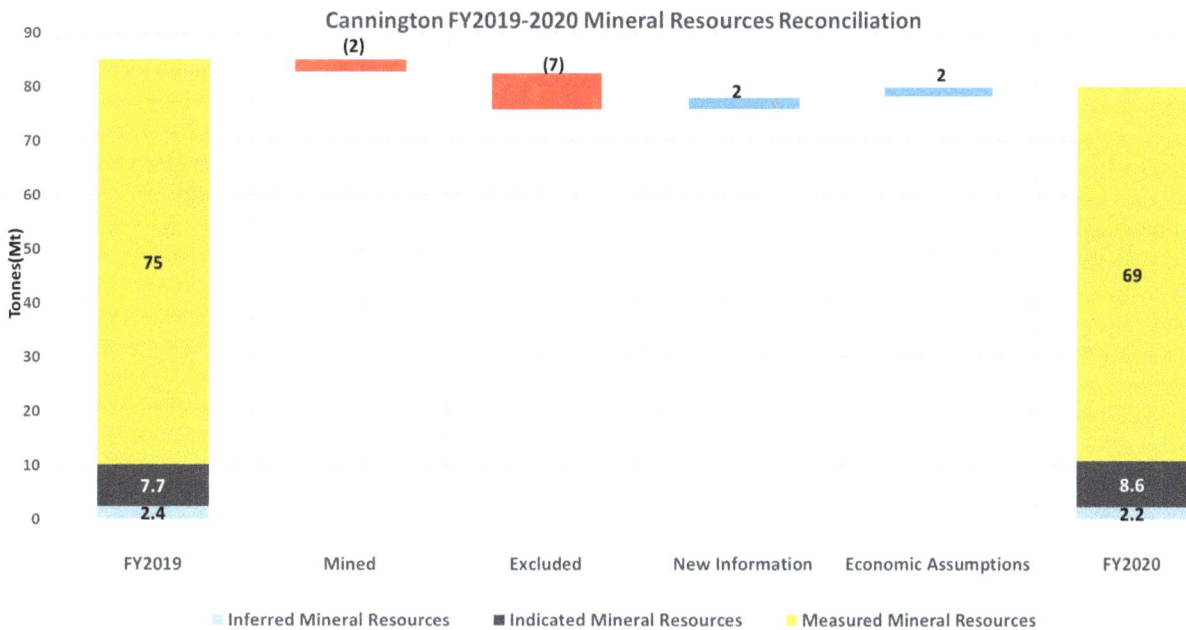

FIG 7 – Cannington's FY2020 Mineral Resource waterfall chart.

CONCLUSION

The MSO tool can be used to refine reporting of an underground Mineral Resource that satisfies the JORC Code (2012) RPEEE requirements. Just as a 'pit shell' is used to delineate Mineral Resource for near surface mineralisation, MSO can be used to define a Mineral Resource supporting underground mining.

Application of the MSO tool ensures:

- RPEEE is satisfied as appropriate assumptions on mining methods, parameters, constraints and cost are applied uniformly.

- A repeatable and less subjective process. It removes isolated blocks and potential errors associated with underground surveys between stopes.

ACKNOWLEDGENTS

The authors would like to thank South32 for granting permission to publish this paper. In addition, the authors acknowledge input and constructive review from David Hope, Manager Tenure, Mining Rights and Resource Geology.

REFERENCE

Bailey, A, 1998. Cannington silver-lead-zinc deposit, in *Geology of Australian and Papua New Guinean Mineral Deposits* (eds: D A Berkman and D H Mackenzie), pp 783–792 (The Australasian Institute of Mining and Metallurgy: Melbourne).

Castanho, N, 2020. Mineable Stope Optimiser (MSO): A Mine Planning Tool More Important Than Ever [online], SRK Consulting. Available from: <https://www.srk.com/en/publications/mineable-stope-optimiser> [Accessed: April 2020].

Glacken, I M, 2019. The highly vexed issue of reasonable prospects for eventual economic extraction (RPEEE) narrowing the range of practice, in *Proceedings Mining Geology 2019*, pp 26–35 (The Australasian Institute of Mining and Metallurgy: Melbourne).

JORC, 2012. Australasian Code for Reporting of Exploration Results, Mineral Resources and Ore Reserves (The JORC Code) [online]. Available from: <http://www.jorc.org> (The Joint Ore Reserves Committee of The Australasian Institute of Mining and Metallurgy, Australian Institute of Geoscientists and Minerals Council of Australia).

Walters, S G and Bailey, A, 1998. Geology and mineralisation of the Cannington Ag–Pb–Zn deposit: an example of Broken Hill type mineralisation in the Eastern Succession, Mount Isa Inlier, Australia, *Economic Geology*, 93:1307–1329.

Walters, S, Skrzeczynski, R, Whiting, T, Bunting, F and Arnold, G, 2002. Discovery and geology of the Cannington Ag-Pb-Zn deposit, Mount Isa Eastern Succession, Australia: development and application of an exploration model for Broken Hill-type deposits, *Integrated Methods for Discovery*, special publication 9, Society of Economic Geologists: Littleton Global Exploration in the 21st Century, pp 95–118.

Case studies

Testing and quantifying geological uncertainty (or a gram of drill data is worth a kilogram of geological interpretation)

C M D Barton[1] and M P Murphy[2]

1. Senior Resource Geologist, IGO, Perth WA 6151. Email: cathy.barton@igo.com.au
2. Manager – Geological Services, IGO, Perth WA 6151. Email: mark.murphy@igo.com.au

ABSTRACT

When developing a new underground mine, the geological interpretation that guides the mineral resource is crucial to the preparation of a reliable ore reserve. In preparing a robust geological model to underpin a resource estimate, practitioners must weigh the comparative risk of accepting a relatively wide-spaced drill hole pattern, versus a more close-spaced pattern that ensures a high local geological confidence for mine planning but may have an exponentially increased time and monetary cost. This paper details the authors' investigations into the geological uncertainty of IGO's Odysseus Deposit, which is approximately 645 km north-east of Perth, Western Australia. At Odysseus, the komatiitic nickel sulfide mineralisation is irregularly cross-cut by nickel-barren pegmatites that have intruded and offset its thick ultramafic orebodies. These pegmatites generate significant uncertainty in the precision of Odysseus' resource volume and tonnage estimates, as well as the subsequent quantification of pegmatite mining dilution in its ore reserve.

To quantify the geological uncertainty in the pegmatite interpretations, IGO's mineral resource practitioners prepared several geological risk studies over the Odysseus North lode, which is the lode most affected by the pegmatites. These studies included categorical conditional simulation of the pegmatites, preparing implicit geological models with alternative algorithms, trialling a machine learning application to prepare a maximum likelihood geological model, and finally assessing the early results of the pre-production close-spaced drilling to these different models. To assess the precision of the different single-outcome models, the authors' compared the estimated pegmatite content of each model to the range of possibilities modelled by categorical conditional simulation within the planned stopes at Odysseus North. An estimation confidence score for each modelling method was then prepared for each alternative single-outcome modelling method. From these studies IGO concluded that a close-spaced drill pattern is essential at Odysseus, supporting the view of the Competent Person in 2017 who proposed a 25 m drill spacing in areas affected by pegmatite which had been partially classified as Inferred.

INTRODUCTION

The Odysseus Deposit (Odysseus) is within a camp of nickel sulfide deposits at IGO's Cosmos Project (Cosmos), which is about 50 km north of the Western Australian regional town of Leinster. Xstrata Plc (Xstrata), a prior owner of Cosmos, discovered Odysseus at approximately 1.0 km below topography using diamond core drilling from surface. Odysseus comprises a northern and a southern domain, with both domains accessed via the Ilias Decline, which is the current entry to the underground reserves. The decline portal is located at latitude 27°36"00' S and longitude 120°34"28' E. At the time of writing, Odysseus is undergoing pre-production development, which includes fitting out a 1.0 km-deep shaft and associated materials handling systems. During this period, the decline is facilitating close-spaced infill resource definition drilling as well as initial mine development activities.

A significant challenge in Mineral Resource estimation (MRE) modelling for the northern Odysseus domain (ODN) has been the three-dimensional (3D) spatial interpretation of the nickel-barren pegmatites that have intruded and displaced the komatiite orebody. The interpretation of the pegmatite geometries is critical to estimating the MRE tonnage accurately, as well as for the dilution effect assumptions applied in the Ore Reserve estimate (ORE). Some of the pegmatites have been found to be sites of significant water ingress into the underground workings, adding further importance to understanding their location for mine planning purposes.

At the time of writing, the prevailing publicly reported MRE for Odysseus was prepared by prior owner Western Areas Limited (WSA) in 2017 and was an update of pre-existing Xstrata models. IGO acquired WSA in June 2022 (IGO, 2022). Drill data used by Xstrata and WSA was surface derived and was at a nominal spacing of 40 m. An independent consultant prepared the 2017 geological interpretation using Leapfrog Geo (Leapfrog) software's standard implicit modelling algorithms. The consultant chose the vein tool to create the 3D interpretation of the pegmatites, referencing previous pegmatite interpretations done by Xstrata geologists during the process. The pegmatite vein model was then used to locally overprint the grade estimation block model. A close spaced drill program of 25 m squares in areas with pegmatite was recommended by the Competent Person in 2017.

After Odysseus was acquired by IGO, several studies were completed to quantify and mitigate the geological uncertainty associated with the pegmatites.

GEOLOGY

The information about Cosmos dates to the discovery of the original near-surface high-grade Cosmos deposit by Jubilee Mines Limited in 1997 (Craven et al, 2000). The following sections of this paper provide a summary of the main regional and local geological features of Cosmos and Odysseus.

Regional

The Cosmos camp of nickel sulfide deposits is located within or adjacent to a local sequence of approximately 2.7 billion-year-old (now metamorphosed) komatiitic lavas that are part of the Agnew-Wiluna Greenstone Belt (AWGB) in the Kalgoorlie Terrane of the Eastern Yilgarn Craton (de Joux et al, 2014). Komatiites are ultramafic rocks that have crystallised from high-temperature, low-viscosity lavas that primarily erupted in restricted linear belts, which are now interpreted to be the signature expressions of early crustal boundaries (Mole et al, 2014). These rocks are rare in post-Archean geology, and their scarcity in the younger rock record is believed to be due to the presence of a much higher internal heat flux in the Archean-age Earth. Today, these ancient margins are the sites of major nickel deposit camps throughout WA, including Cosmos. This geological context has been well documented by past mining, exploration, and academic geologists who have worked in the region over the last 25 years.

Relative to Cosmos, the AWGB extends approximately 115 km north-north-west to the town of Wiluna and approximately 150 km south-south-east to near the town of Leonora. However, the simplified regional geology image depicted in Figure 1, which shows the locations of the major nickel sulfide deposits around Cosmos, only captures the northern half of the AWGB. As annotated in Figure 1, the AWGB hosts multiple world-class high-grade underground nickel deposits in the Cosmos region, such as near Leinster where the Perseverance deposit is reported to have had a pre-mining resource of approximately 50 Mt grading 2.2 per cent Ni (approximately 1.1 Mt nickel in situ). The AWGB also hosts multiple large-tonnage, low-grade nickel deposits that are usually mined using open pit methods, such as Mt Keith, approximately 40 km north of Cosmos, which had a pre-mining resource in the order of 500 Mt grading 0.55 per cent Ni (Duuring et al, 2012).

The Ida Fault demarks the western boundary of the AWGB, and the belt is truncated in the east by the Keith-Kilkenny Lineament. The northern and southern margins are ambiguous, with the northern margin hidden by the Proterozoic-age Earaheedy Group of rocks near Wiluna. Based on geochemical characteristics of the komatiite units, the AWGB extends south-eastwardly to near the town of Leonora. The entire belt has undergone a complex polyphase deformational history, with metamorphism ranging from low temperature pressure prehnite-pumpellyite facies in some rocks near Wiluna, increasing in pressure and temperature to greenschist to lower amphibolite grade in rocks near Agnew, then increasing again to the higher temperatures and pressures of middle amphibolite grade in the rocks around Leinster (de Joux et al, 2014).

The geology of the AWGB is often disrupted by major wrench faults that are traceable over tens of kilometres, and the local geology is often characterised by a steeply dipping stratigraphy, and rocks that can display structural features from up to ten regional deformation events.

Cosmos

On a local scale, Cosmos' geology comprises a metamorphosed sequence of ultramafic, intermediate, and felsic volcanic rocks that contain multiple komatiite-hosted (or associated) nickel sulfide deposits. The mineralised ultramafics have thicknesses of up to 500 m in the Cosmos mine camp, where they dip vertically and face east. However, the komatiites thin towards Lake Miranda, which is just south of Cosmos, and dip more shallowly to the east. The footwall volcanic succession to Cosmos' mineralised and now metamorphosed komatiite lavas is an intercalated sequence of fragmental and coherent extrusive lithologies, ranging from metamorphosed basaltic andesites through to rhyolites, and additionally, younger cross-cutting felsic intrusions and pegmatites. Age dating studies indicate that this footwall volcanic succession has ranged in age from 2.736 to 2.724 billion years (Ga) (de Joux *et al*, 2013).

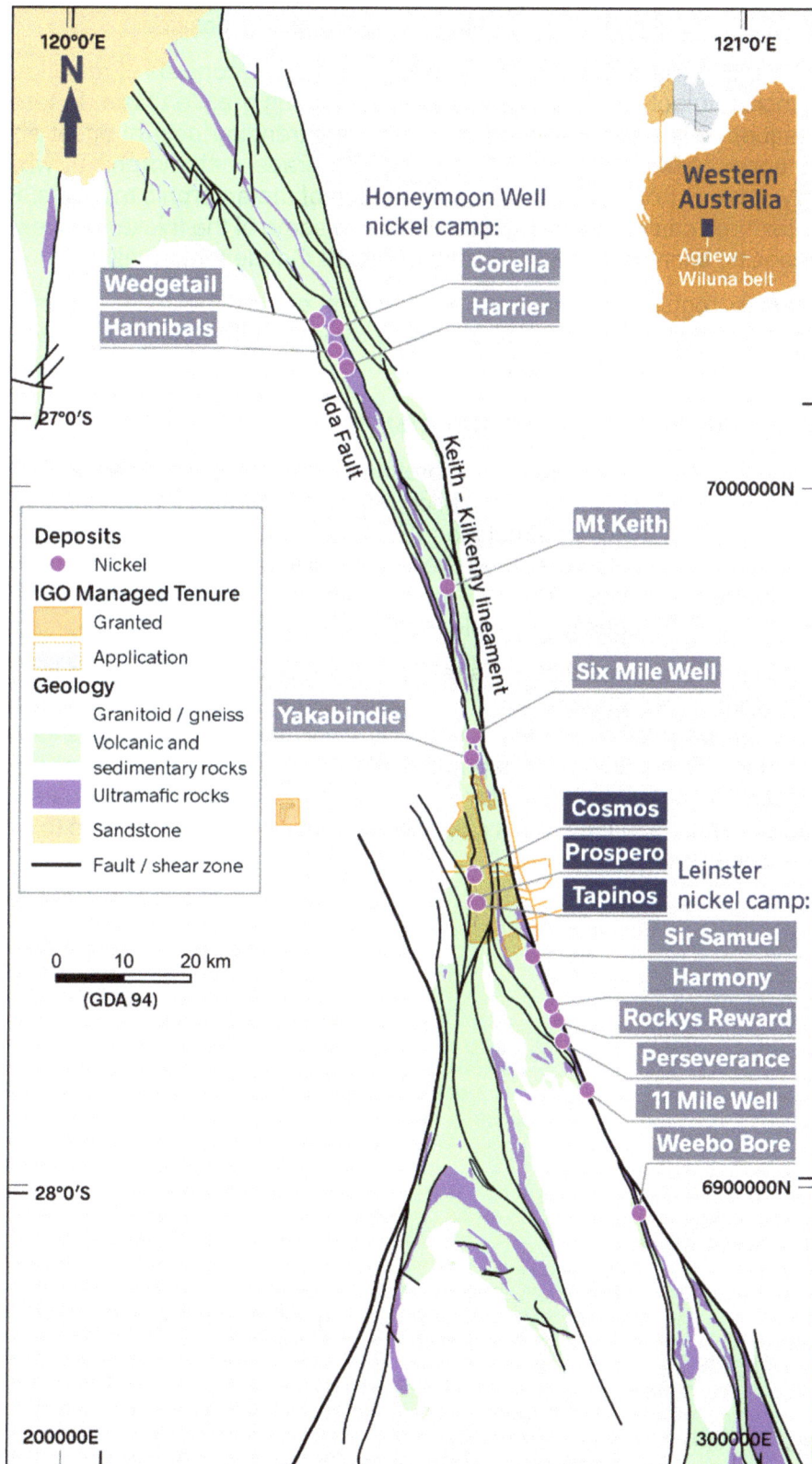

FIG 1 – Northern AWGB simplified geology and nickel deposits.

Cosmos' ultramafic sequence that overlies the footwall metavolcanics has two units identified in mapping and drill hole logging. The first is a basal ultramafic package that is separated from the upper ultramafic by a 2.685 Ga old, discontinuous felsic unit that occurs ~30 to 60 m above the basal contact. This felsic unit is interpreted to signal a hiatus in komatiite lava eruption. Two younger felsic intrusions have also been age dated to be 2.653 Ga and 2.670 Ga, respectively. The younger of these intrusions cross-cuts the upper contact of the upper ultramafic package. Combined with dating from the footwall volcanics, the lower ultramafic package is interpreted to have been deposited inflows from 2.724 to 2.685 Ga, while the upper ultramafic package was deposited from 2.685 to 2.670 Ga. As such, the first phase of komatiite deposition lasted ~50 million years (Ma), followed by a short pause, then a second phase of ~30 Ma of eruption and deposition.

The stratigraphic hanging wall to Cosmos' mineralised komatiites consists of reworked volcaniclastic metasediments, including polymictic conglomerates containing granite clasts. In terms of structure, Cosmos' mine sequence is often disrupted by north-west-trending dextral offset shears. All rocks have undergone upper greenschist to lower amphibolite grade metamorphism, which has usually destroyed primary igneous textures through the formation of metamorphic minerals. However, some areas of primary textures can still be recognised locally in some of the thicker and less serpentinised parts of some deposits, such as in the core zone of the Mt Goode metadunite.

The depth of surface regolith ranges from 40 to 80 m across the local region and comprises transported cover and saprolite clays. The carapace over the ultramafic rocks frequently presents as a siliceous saprock over cavernous clays.

IDENTIFYING GEOLOGICAL UNCERTAINTY

To assess the geological risk involved in interpreting pegmatite geometries at ODN, IGO's MRE practitioners prepared a geostatistical simulation study for the ODN zone using a categorical adaptation of the sequential indicator simulation (SIS) method (Deutsch, 2006). For this study, drill hole data that bounded a 3D volume around the ODN mineralisation were integer coded as '1' for pegmatite and '2' for non-pegmatite. The SIS method was then used to simulate these data over a 1 m × 1 m × 1 m grid, with 50 equally probable realisations of the pegmatite geometry prepared.

Analysis of the SIS results showed that at Xstrata's drill spacing of ~40 m, there is significant scope for materially different interpretations of the spatial distribution of the pegmatites (Figure 2). Additionally, while the pegmatite indicator continuity analyses (variography) indicated there is a nominal 100 m range 3D continuity of pegmatite before samples at this spacing are no longer correlated, if selecting a 50 per cent indicator semivariance as a metric that equates to moderate to strong correlation of nearby samples, pairs can only be separated by 15 to 20 m to achieve this high connectivity confidence level.

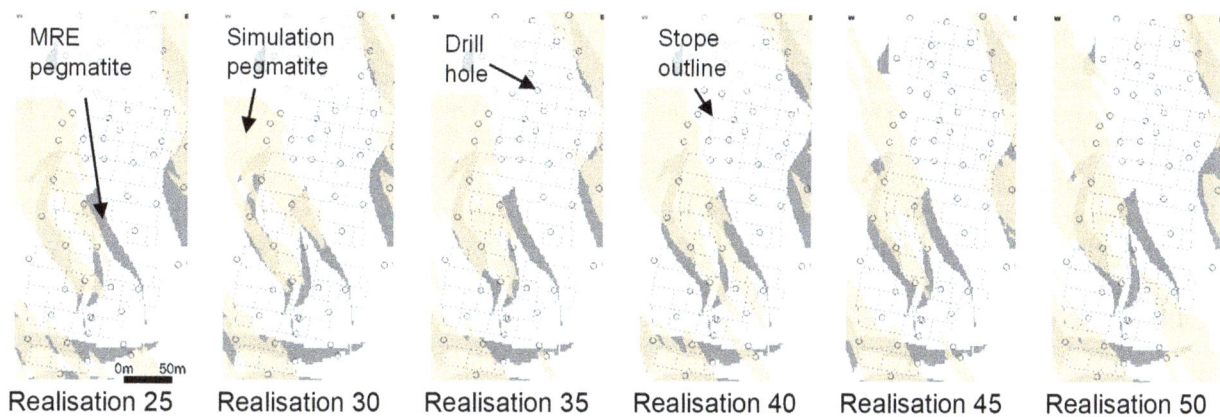

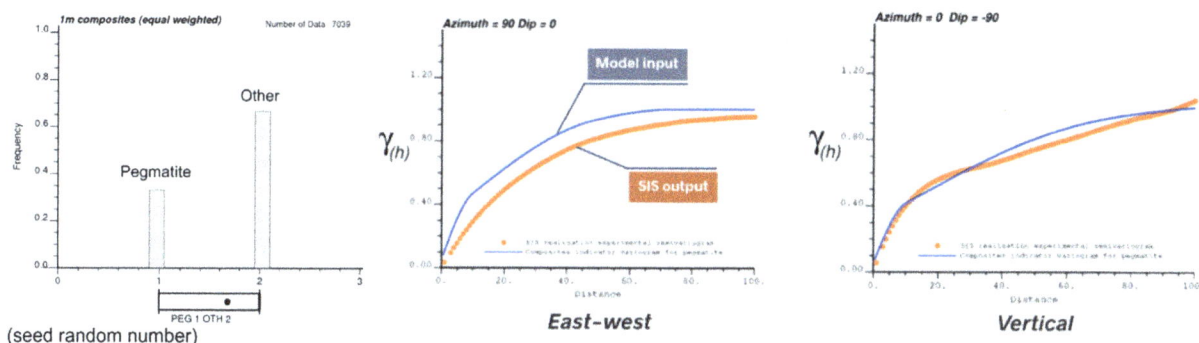

FIG 2 – Six example SIS realisations of ODN pegmatite at 9400.5 m elevation ±2 m.

Based on these findings, the recommendations from the pegmatite SIS study were that closer-spaced drilling would be necessary before mining, and that a grade control 'drill out' should aim for an orebody drill hole pierce point spacing of 20 m or less. In some areas, holes should be drilled even closer to confirm the short-range character of the pegmatites that disrupt the economic mineralisation. From these recommendations, a drill program was planned to test the short-range continuity of pegmatites and mineralisation at ODN, along with the assessment of the modelling methods to be applied in the final pre-mining MRE as discussed in the following sections of this paper.

TESTING UNCERTAINTY WITH ALTERNATIVE MODELS

Table 1 of the current JORC Code (2012) includes a section prompting Competent Persons to discuss the uncertainty in their MRE models. However, MRE practitioners rarely provide objective quantitative assessments of the uncertainty in their estimates and typically only report a subjective opinion on the confidence of their geological interpretations (McManus *et al*, 2022). There are several reasons why quantifying uncertainty is difficult, including the fact that it may not be economically feasible to have multiple geologists model the same deposit to produce alternative geological interpretations or a lack of in-house skills or time to prepare simulated geology models. Additionally, for ORE work, mining engineers prefer to have only one model for mine planning purposes for the life of the mine, and stochastic models can then be used to assess the uncertainty in the mine plan.

As an alternative to SIS, IGO decided to test three alternative 'single outcome' geological modelling methods for the pegmatites at ODN using commercially available software systems to provide at least three alternatives to assess uncertainty. These methods were completed in 2022 and are discussed below. The three models were then compared to the bandwidth of uncertainty derived from the categorical SIS results discussed above to test which method might be more reliable for future modelling of the pegmatites for mine planning purposes.

Implicit vein and intrusion modelling

In revising ODN's pegmatites models, IGO's resource modeller used the same software, method, and data that WSA adopted for its MRE in 2017. IGO prepared the new 3D implicit models from first principles. IGO's new vein interpretation consisted of 41 individual veins, each requiring checking and editing of six Leapfrog software control elements every time a vein was created or updated. These elements included the vein hanging wall and footwall surfaces, segments, pinch-out locations (if required), midpoints (controlling the reference surface), and a boundary (if in use). When modifying a vein system, the interaction between veins and vein priorities, also required checking during the revision process. IGO's modeller found that the vein tool algorithm was effective for the modelling of narrow, laterally continuous pegmatite bodies. However, the resulting interpretation was subjective, particularly when multiple cross-cutting veins of the same lithology occurred, and manual intervention was required during the interpretation process to remove local unrealistic software artefacts. Specifically, the trends of anastomosing pegmatite veins were challenging to predict where they intersected. Where the drill spacing was closer in the ODN data set, the WSA and IGO pegmatite models were generally similar, but where the drill spacing was wider, the models differed significantly. Notwithstanding, the complexity of this process, IGO's modeller found that during sectional interpretation of the pegmatites using the vein tool, a potential relationship between the pegmatites and water flow sites was identified. This useful result demonstrates the value of practitioners being aware of the multiple utility of geological models and being alert to ancillary emergent relationships in the modelling process.

ODN's pegmatites are also amenable to geological modelling using Leapfrog's intrusion implicit algorithm, which is principally designed for the modelling of the more rounded geometries of large bulbous intrusives. However, although the method is deemed less suitable for modelling curviplanar and anastomosing veins, most of ODN's pegmatites are thick enough, and drilled out at a close enough spacing, for the intrusion tool to produce a geologically realistic alternative model. A significant advantage of the intrusion tool over the vein algorithm is that there is no need to allocate each pegmatite interval to a specific vein name, as a simple grouping of pegmatite lithology codes is sufficient to define that all veins are part of a single intrusive. Moreover, the intrusion volume is automatically revised when the drill hole database is updated, unlike the vein tool, which requires manual manipulation of the vein selections when new data arrives.

Artificial Intelligence (AI) modelling

The third alternative model that IGO assessed for ODN's pegmatites was prepared using Maptek's Domain MCF (DMCF) application of machine learning artificial intelligence (AI) software. IGO has tested DMCF previously on its Nova-Bollinger Deposit and the method delivered good results on that deposit's many and varied mineralisation geometries (Pym *et al*, 2022). DMCF requires only two items as inputs. These are firstly a block model definition covering the modelling space and sub-blocking parameters which, if needed, can be prepared by the DMCF interface, and secondly, a comma-separated-values (CSV) text file that contains lithology domains and sample data. IGO used Vulcan software's Geology Core tools to prepare the comma-delimited text file and block model definition file, but other commercially available software, such as Excel, could equally have been used for the task. DMCF prepared a model for 13 lithological domains, including pegmatite, in less than 36 minutes, and its output additionally included grade trend information for eight elements that were input with the geology codes to better inform the AI algorithm.

Model comparison

The plan views through ODN in Figure 3 illustrate the results of the different pegmatite models prepared in this study. Figure 3a–3c are example SIS realisations, and Figure 3d–3f are respectively the implicit vein and intrusion models, and then the AI model. Note that many of the planned mining envelopes in the lower left and centre left of these images are at the main concerns for pegmatite uncertainty in the current MRE model.

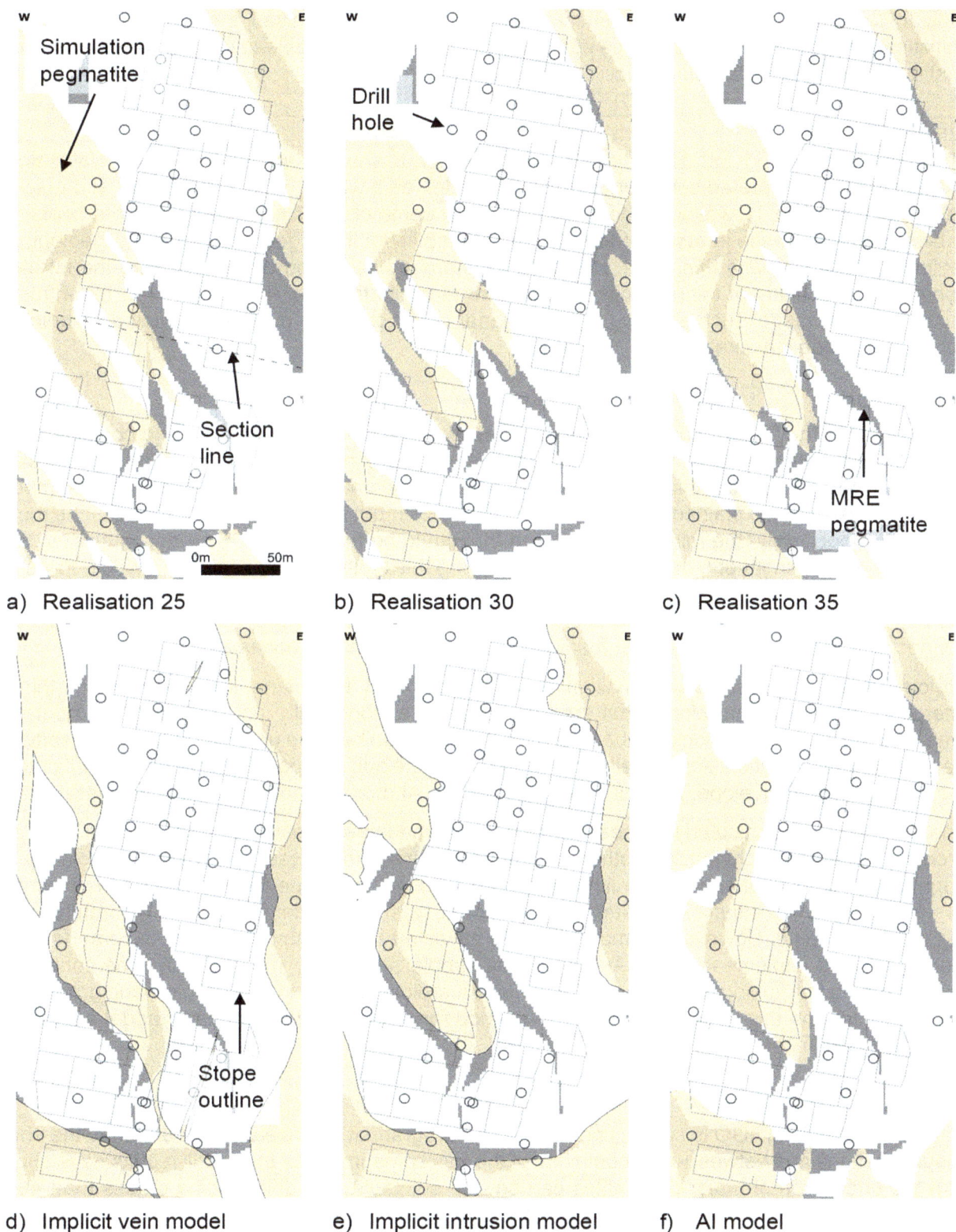

a) Realisation 25　　b) Realisation 30　　c) Realisation 35

d) Implicit vein model　　e) Implicit intrusion model　　f) AI model

FIG 3 – Model compared at the same local elevation through Odysseus North.

Vein model

While IGO's modeller found that the implicit vein modelling revealed relationships between the pegmatites and the water flow sites, IGO concluded that the method was time-consuming and may not always be practical to prepare as an alternative model. While the vein modelling tool is beneficial when modelling thinner units, its interpretation can be subjective and biased towards overly enhancing lateral continuity. Specifically, the vein tool is designed to model laterally continuous strata, like coal seams. There is a risk that its results can encourage modellers to enforce lateral geological continuity and connectivity where none exists, particularly when the drill spacing is too wide to know the short-range continuity. Therefore, IGO emphasised the importance of using statistical assessments of short-range continuity and alternative models to challenge the interpreted continuity while using the vein modelling algorithm.

Intrusion model

The implicit modelling intrusion algorithm applied for this study was found to be valuable in challenging the interpreted continuity of the vein model, as it was simple to implement and quick to set-up and maintain. However, it should be noted that some of the thinner pegmatite intercepts had to be excluded from the intrusion model due to the inability of this algorithm to produce geologically realistic results for these thin veins. However, to improve interpreted continuity using the intrusion approach, particularly of thinner units or where larger units thin out, modifications to the intrusion model could be made, such as using a trend (planar or structural) or manually adding data points to control the shape if desired.

DMCF model

IGO found that the DMCF model of the pegmatites was geologically reasonable and consistent with the data, and overall, the AI modelled the pegmatites more realistically than Leapfrog's intrusion model algorithm. While the thinnest pegmatite veins in the data did not appear at all in the AI model, their absences were explained by IGO's choice of the minimum sub-block size specified in the model definition. Specifically, the minimum block size that could be made in the AI model was set to be a cube with a side length of 5 m, and pegmatites having thicknesses of less than 5 m would therefore not be expected to be modelled due to the relatively coarse block resolution. However, given that the drill spacing at Odysseus, is on average much wider than 5 m, the modelling of volumes and connections of thin veins at sub 5 m resolution is a highly uncertain endeavour.

Although the AI model block definition could be modified to ensure the modelling of thinner intercepts, the risk is that blocks that are too small can exponentially increase the AI computation time, and the numbers of blocks in the output models can also become unwieldy for end-use software transfer and visualisation. Additionally, in this test case, the sub 5 m pegmatite veins are well outside the mineralisation envelope, so their omission from the AI model is not a material consideration.

The AI model was particularly successful in modelling the large, anastomosing pegmatite units and predicting orientation changes. Another advantage of the AI model was that it modelled not only the pegmatites but also all other major lithological units, as well as grade trends. At the time of writing this paper, there are insufficient water data points to use the AI model for predicting potential water ingress to the mine. However, as more data are collected in the future, this may become feasible. Setting up and implementing the AI model was simple and quick, with the only time-consuming process being fine-tuning of the input CSV file, and it rapidly generated a lithological and grade trend model. IGO found this to be a very efficient method for producing an alternative model and testing geological assumptions.

DRILLING RESULTS

At the time of writing, IGO is conducting an infill drill program at Odysseus to better define the short-range locations and between-hole continuity of the barren pegmatites. Initial drill results have shown that the orientation of major pegmatites can be unpredictable at a drill spacing of less than 40 m. The Odysseus MRE is based on diamond holes drilled from the surface that were primarily designed to intersect the mineralisation envelopes at high angles to best gauge the ultramafic-hosted mineralisation's true thickness. However, this strategy often results in shallow-angle pegmatite intersections, and there is uncertainty in the interpretation of their orientations and thicknesses. While each modelling technique discussed above has its benefits, none can predict the presence or absence of pegmatites in all areas of ODN without close-spaced drilling. For instance, Figure 4a is a north-facing cross-section of the MRE model, illustrating the pegmatite drill hole intercepts and resulting MRE modelling of mineralisation and pegmatite in the MRE block model. In contrast, Figure 4b is the same cross-section with results from infill drilling from underground platforms, revealing a significant change in understanding due to the close-spaced information. Figure 4c–4f depict the same section again, respectively overlaying the implicit vein model, the implicit intrusion model, the AI model and an example SIS realisation.

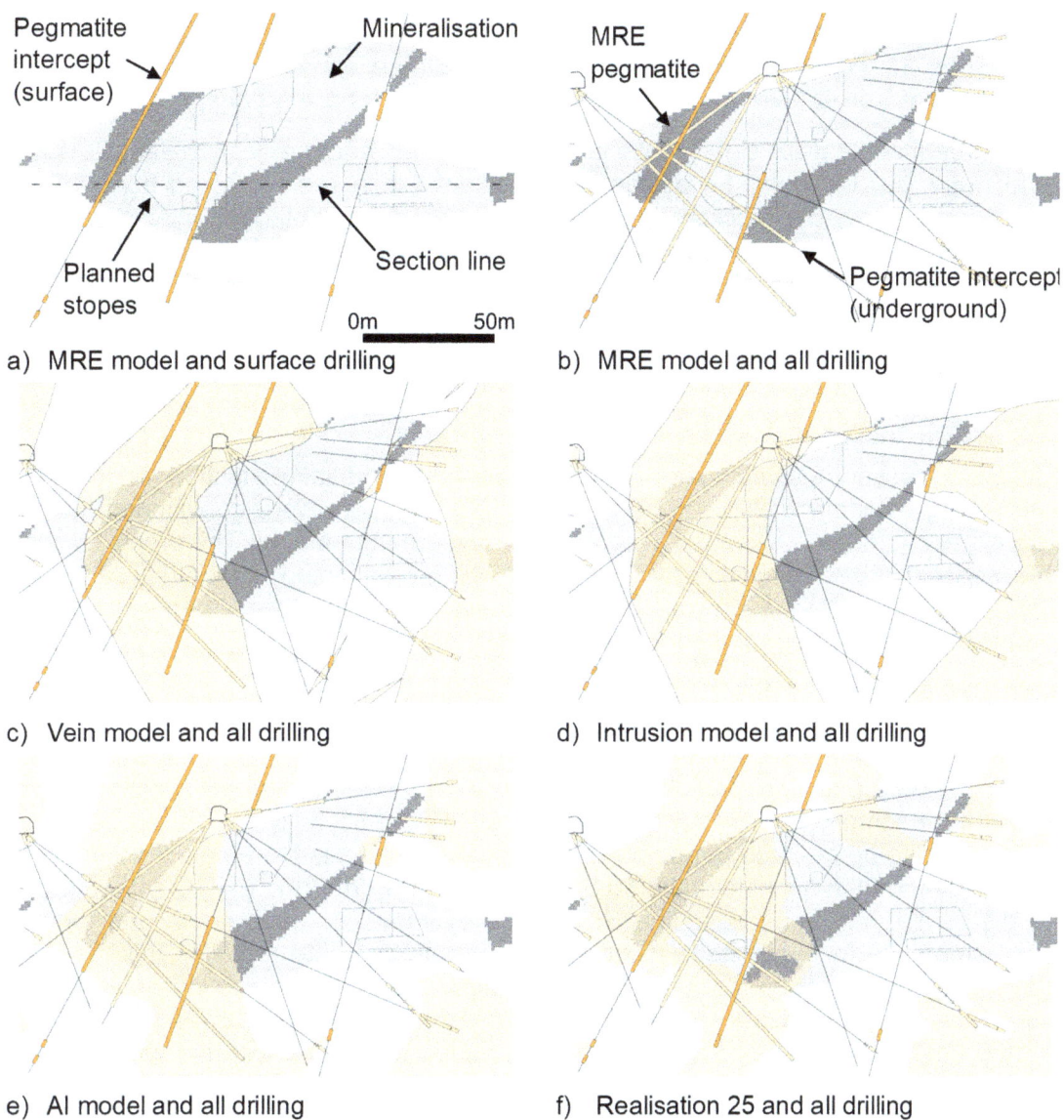

FIG 4 – Comparative cross-sections for Odysseus North drilling and models, superimposed on the 2017 planned stopes.

QUANTIFYING UNCERTAINTY

IGO's goal in developing alternative pegmatite models for ODN is to determine the most suitable approach for the planned revision of the MRE and the subsequent update of the ORE. To assess the confidence in the chosen pegmatite interpretation, a semi-qualitative criterion based on the agreement between alternative interpretations and a quantitative distance-to-data metric will be used. These criteria will be combined into a geological confidence score ranging from zero to one for each MRE block. This will enable end-users of the model to obtain an average confidence score for specific volumes of interest, such as complete mining stopes. The result will provide a pseudo probability metric indicating the likelihood of achieving the planned stoping target based on its pegmatite content.

Geological confidence scoring

IGO has developed a system for assigning geological confidence scores to the MRE blocks of Odysseus in future, based on the results of three alternative pegmatite models. The planned process is as follows:

- If all three models (vein, intrusive and AI) indicate the presence of pegmatite at a block location, and the nearest drill hole is within 10 m, the block will be assigned a high-confidence score. The actual value of the score (0.8, 0.9 or 1.0) will be determined by practitioners who will also consider local risk factors such as proximity to mineralisation boundaries and data quality.

- For blocks where all three models indicate the presence of pegmatite, but the nearest drill hole is between 10 and 20 m away, a medium-confidence score ranging from 0.5 to 0.7 (in increments of 0.1) will be assigned, again with flexibility for discretionary choice.

- If the three models differ significantly in their predictions and the nearest sample is more than 20 m away, the block will be assigned a low-confidence score ranging from 0.0 to 0.4 (in increments of 0.1). For blocks with low-confidence scores in planned mining areas, IGO will recommend infill drilling at an appropriate spacing and time before mining commences in those areas.

This suggested method is yet to fully assessed and will be cross-checked with the simulation results as discussed below.

So, what's the best model?

As previously discussed, IGO simulated 50 equally probable spatial realisation models of the ODN pegmatites using categorical SIS. These categorical SIS models were conditioned to the Xstrata drilling data that informed Odysseus's prevailing MRE pegmatite model. The precision of the MRE pegmatite model and the three alternative pegmatite models discussed in this study can be assessed using the results of ODN's SIS work. Through these precision assessments, the best 'single outcome' modelling approach for planned MRE revisions can be identified.

To assess the precision of the different models considered in this study, IGO performed an intersection analysis of ODN's 2017 planned ORE stopes with the SIS model and derived several pegmatite estimation precision related metrics for each stope in the ORE. These metrics included the expected value, or 'e-type' percentage volume of pegmatite per stope, which represents the average percent pegmatite across the 50 SIS realisations. Additionally, the 90 per cent confidence interval of pegmatite percentage per stope was calculated, which is the range of percent pegmatite between the 5th and 95th quantile of the 50 SIS realisations. Furthermore, IGO computed the mean absolute difference (MAD) between the SIS e-type estimate for each stope and the estimated pegmatite percentage for each stope using the four single outcome models. This resulted in four MAD metrics, one for each model under consideration. The authors caution here that while the supposition that the SIS e-type provides the best average or expected result, this is clearly a modelling assumption. However, this assumption is supported by the fact that the SIS approach is the only modelling method that accounts for an objectively derived pegmatite continuity model (variogram) resultant from the data, which is not the case for the four single outcome test models.

Out of the total 198 stopes planned in 2017 into ODN, 50 are predicted to contain some pegmatite within their volumes in at least one of the 50 SIS realisations. For these 50 stopes the average MAD for each test model was calculated across all stopes to provide an overall measure of precision for each model type. As a second measure of precision, IGO tallied the instances where a model's estimated mean percent pegmatite for a stope fell within the 90 per cent confidence interval (CI) of the SIS results for that stope. This metric aims to evaluate the usefulness of each model estimate, as the estimates are only meaningful if they predict a percentage value that has a 90 per cent chance of occurrence based on the likelihood supported by the SIS results.

Figure 5 contains four panels of box-and-whisker plots with the following features:

- Each box-and-whisker graphic represents the distributions of pegmatite percentage derived from the SIS modelling and intersection of the ORE stopes with the SIS model.

- The box-and-whisker plots are sorted from left to right in order of increasing e-type mean percentage pegmatite to allow easy assessment of trends in relation to percentage fill.

- The box-and-whisker plots denote the minimum, maximum, 5th and 95th quantiles (p05 and p95, respectively), median quantile (p50) and e-type mean of each stope's pegmatite distribution – refer to the figure legend for details of these graphic elements.

- Each panel includes an additional graphic element, which is the stope mean for each of the respective test models – refer to the direct labels in each panel to identify the model plotted.

- Each panel shows the average MAD for all stopes from the e-type mean for all stopes and the tally of stopes falling inside the SIS 90 per cent CI, as discussed above.

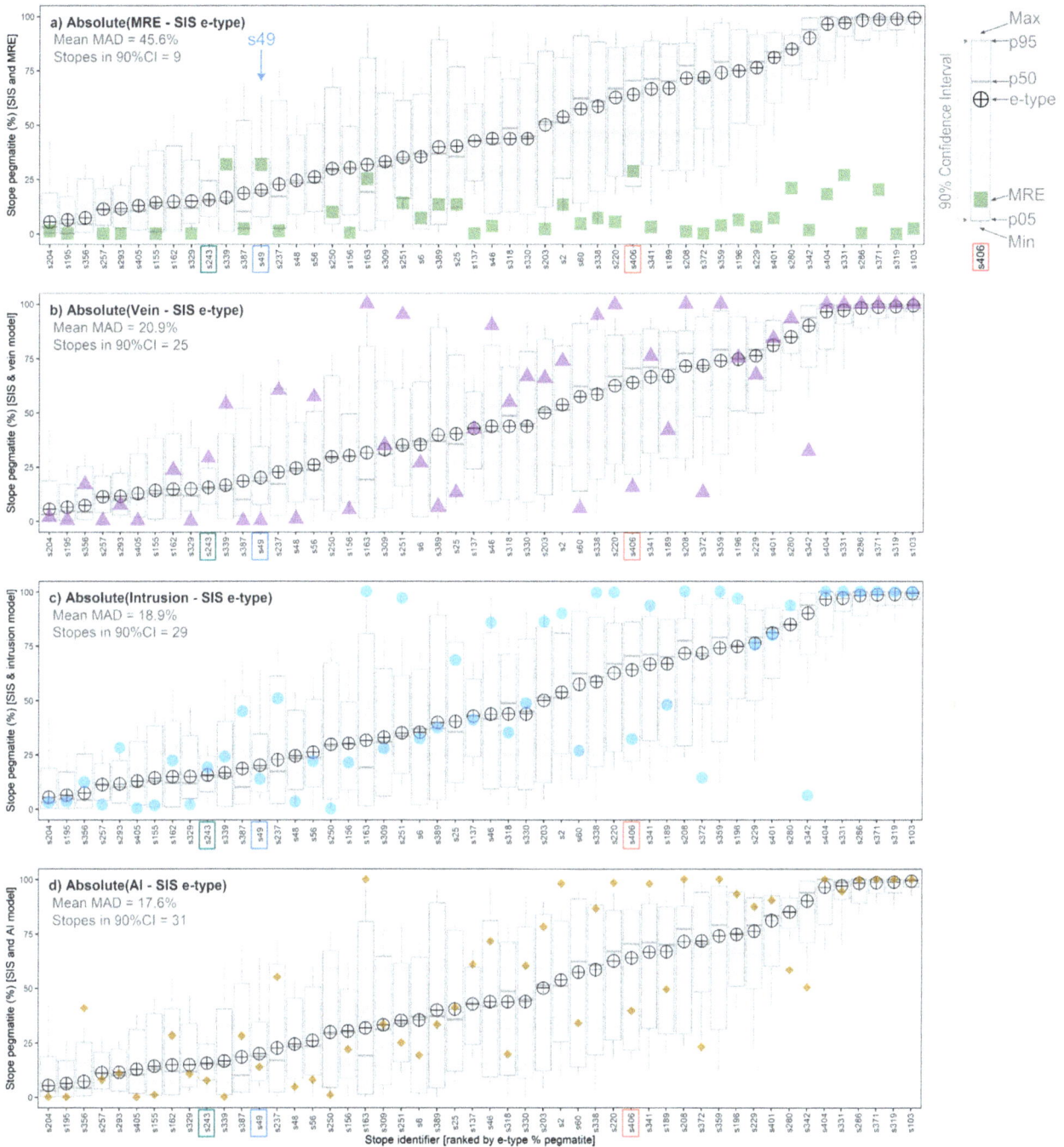

FIG 5 – SIS stope distributions of percent pegmatite compared to MRE, vein, intrusion and AI estimates.

To illustrate what is depicted in each panel of Figure 5, consider the stope identified as 's49' in the study which is in the southern part of ODN and intersects the 9400 m level. This stope is highlighted on the horizontal axis and used as a legend in Figure 5, using the MRE as the legend example. Table 1, combined with Figure 5's legend, provides detailed information on the value associated with each graphic element.

TABLE 1

Stope s49 analysis using stope percent pegmatite results.

Stope ID	Min	90% confidence interval			e-type	MRE	MAD	In 90%CI
		p05	p50	p95				
s49	4.3	7.7	18.7	34.6	20.1	31.7	11.6	Yes

As demonstrated in Table 1 and Figure 5, stope s49 is one of the stopes in which the 2017 MRE model's estimate of percent pegmatite, 31.7 per cent, falls within the 90 per cent CI predicted by the SIS results. The SIS results range from 7.7 per cent at the p05 to 34.6 per cent at the p95, as shown in the figure legend. Although the MRE model's estimate of percentage pegmatite in stope s49 has a MAD of 11.6 per cent from the SIS e-type mean of 20.1 per cent, the fact that it falls within the 90 per cent CI of the SIS results suggests that the MRE result is reasonable. However, the reasonableness of the bandwidth of the 90 per cent CI is another matter for consideration, and further comments on what a 90 per cent CI bandwidth should be for a stope are provided below.

With the two overall summary precision metrics explained above in detail, Table 2 is a summary listing of the overall MAD and the number of stopes falling within the SIS 90 per cent CIs for each test model used, as plotted in the panels of Figure 5.

TABLE 2

Ranking of selected inputs and output of each model produced for Odysseus North.

Model	Mean MAD% pegmatite from SIS e-type	Stope means in 90% confidence
2017 MRE	45.6%	9
2022 Vein implicit	20.9%	25
2022 Intrusion implicit	18.9%	29
2022 Artificial intelligence	17.6%	31

Several observations and interpretations can be drawn from Figure 5 and Table 2 as follows.

The box-and-whisker plots in Figure 5 reveal a correlation between the 90 per cent CI bandwidth and the upper and lower tails of the percentage of pegmatite in the stopes. For example, the stopes with e-type means up to around 20 per cent pegmatite percentage have a narrower 90 per cent CI bandwidth than stopes with e-type means ranging from 25 per cent to 75 per cent. Above 75 per cent, the bandwidth decreases markedly, with the five stopes with e-type means near 100 per cent indicating near certainty that these stopes are all pegmatite. The important conclusion here is the confidence in the volume of pegmatite in stopes having a near all-mineralisation or near all-pegmatite content is very high, and as such, not much infill drilling is needed to resolve uncertainty. However, the many stopes with near zero pegmatite content as indicated by the prevailing MRE model (Figure 5a) deserves the most attention for follow-up drilling and analysis in the ORE schedule, given the SIS results are indicating many of the 2017 planned stopes are likely to have significant pegmatite content, in some cases with high certainty the content is near 100 per cent, such as stope s286 and s319 in Figure 5a.

The 2017 model has the highest stope average MAD from the SIS e-type for each stope. The authors attribute this to an optimistic application of the implicit vein modelling method in places. The veins were modelled under the assumption that they are relatively thin and/or subvertical, which is not an unreasonable interpretation as geological context choices are up to the practitioner. However, given the issues with the data spacing and drill hole intersection angles, as depicted in the example section of Figure 4a, this highest MAD for the MRE demonstrates the risks of committing to an optimistic model for a single outcome. The authors also suggest that this modelling approach explains why only nine stopes in the MRE have estimates of pegmatite inside the bandwidth of the 90 per cent confidence intervals of the 50 stopes deemed likely to contain some internal pegmatite.

The new IGO models are a significant improvement on the 2017 model with much lower MADs and many more of the stopes having the estimated value within the 90 per cent CI of the SIS results for those stopes. The intrusion implicit model and the AI model are very close in terms of the two precision metrics, with the AI model marginally superior. Given that both these most precise models require minimal practitioner intervention, it is reasonable to conclude that minimising the human intervention in the model seems to be more reliable, and using modern algorithms to address the data rather than preconceived geological interpretation is a better approach.

In Figure 5, besides the low and high percent pegmatite tails, the uncertainty bandwidth in most planned stopes is wide, with some stopes such as s163 ranging from possibly containing nearly zero pegmatite to almost 100 per cent pegmatite. On the other hand, stope s243, which is positioned tenth from the left side of Figure 5, has a much narrower bandwidth of uncertainty. Additionally for this lower uncertainty stope, both the intrusion and AI models fall within the 90 per cent CI, suggesting either method being reliable for modelling this stope, and by inference for situations where the uncertainty bandwidth is minimised by having closer spaced information.

What's an acceptable uncertainty?

While stope s49 was discussed as an example of lower uncertainty bandwidth, its e-type pegmatite estimate is 20.1 per cent, with a 90 per cent CI ranging from 7.7 per cent to 34.6 per cent. This equates to a 90 per cent CI e-type relative uncertainty bandwidth of approximately ±33 per cent. Although this is narrower than the uncertainty bandwidth of other stopes in the middle sections of Figure 5, a ±33 per cent uncertainty may be too wide for mine planning purposes. The question arises then as to what level of uncertainty may be considered reasonable?

For several years, experts in the field have proposed that mines should aim for a relative uncertainty bandwidth for their primary product(s) of no more than ±15 per cent on a quarterly production basis. When this level of uncertainty is achieved, the mineralisation being mined can be considered Measured Mineral Resources/Proved Ore Reserves. Similarly, an annual production ±15 per cent level of uncertainty is commensurate with mineralisation classified as Indicated Mineral Resources/ Probable Ore Reserves (Murphy *et al*, 2005). The concept behind this proposal is that as larger 'samples' are taken from the orebody, the uncertainty decreases, with the mining of the entire orebody having no uncertainty but mining a single stope or model block having much higher uncertainty. In practice, applying this concept to quarterly and annual production periods involves having a simulation study to model uncertainty for each production unit.

A worked example helps to illustrate the concept. Consider a hypothetical mine where the production rate requires the mining of four stopes per quarter. SIS modelling has determined that all stopes in the mine have, on average, a relative 90 per cent confidence interval (R90 per cent CI) ±49.4 per cent for their e-type estimate of contained barren pegmatite. Assuming that the distribution of the uncertainty metric in each stope is normal (bell-shaped), the uncertainty metric's standard deviation (SD) can be calculated by dividing the ±R90 per cent CI value by 1.645. This is because ±1.645 SD about the mean of the standard normal distribution represents the 90 per cent CI. Therefore, the uncertainty R90 per cent CI value of ±49.4 per cent equates to an SD of ±30 per cent. Note that the assumption that the uncertainty metrics from SIS results are normally distributed is usually reasonable as long as 50 or more realisations have been prepared.

Continuing with the hypothetical example, if four stopes are mined per quarter, the uncertainty reduces by a factor of the square root of the number of mining units, according to the standard error of the mean (Wikipedia, n.d.). Therefore, the quarterly relative 90 per cent confidence interval (QtrR90 per cent CI) would be as follows:

$$QtrR90\%CI = \frac{Stope\ SD \times 1.645}{\sqrt{number\ of\ stopes\ mined\ per\ quarter}} = \frac{30\% \times 1.645}{\sqrt{4}} = \frac{49.35}{2} = \pm 25\%\ (aprox)$$

This result indicates that stopes having an ±30 per cent SD uncertainty (or ±49.4 90 per cent CI) will result in a quarterly uncertainty that is too high for assigning Measured Mineral Resources/Proved

Ore Reserves and meeting any production forecasts that call for better precision than ±25 per cent. Reducing the uncertainty would require more close-spaced drilling.

Continuing again with the hypothetical example, the annual production requires mining 12 stopes, and this information can be used to estimate the annual relative 90 per cent confidence interval (AnR90 per cent CI) uncertainty as follows:

$$AnR90\%CI = \frac{Stope\ SD \times 1.645}{\sqrt{number\ of\ stopes\ mined\ per\ year}} = \frac{30\% \times 1.645}{\sqrt{12}} = \frac{49.35}{3.46} = \pm14\%\ (approx)$$

Therefore, the annual uncertainty level is acceptable for Indicated Mineral Resources and Probable Ore Reserves. However, there are several technical details in implementing this approach in practice because stope tonnages and the 90 per cent CI uncertainty distributions of individual stopes usually vary. This is part of the ongoing investigation at Cosmos at the time of writing.

In summary, the SIS-based approach involves modelling the uncertainty of the variable of interest, and then aggregating the results into mining units on a quarterly or annual basis to confirm (or not) that the uncertainty measure for the mining units is commensurate with the resource/reserve classification applied and industry/market expectations for the deposit under consideration.

CONCLUSIONS

In this study, various methods were used to test the uncertainty of pegmatite geology at ODN, which helped challenge assumptions made by geologists throughout the project. The results from the categorical SIS showed that the prevailing MRE had room for multiple alternative interpretations, which was confirmed by the materially different interpretations produced using an implicit vein modelling tool, an implicit intrusion modelling tool, and artificial intelligence machine learning. Implicit and AI modelling software has made it easier to produce alternative models quickly, and the alternative models produced in this study clarified where geological uncertainty exists spatially. At ODN, the use of modelling methods which require limited intervention are effective at modelling the pegmatites.

Communicating geological uncertainty remains a challenge for mineral resource estimation practitioners, especially where upfront drilling costs need to be justified to assess a deposit's complexities. IGO geologists plan to trial the use of a geological confidence category in the next Odysseus MRE. This category will be built into the model by assigning a confidence score to each block after comparing the three alternative geological models at each block, along with the distance to the nearest sample. The SIS and alternative models will also be used to quantify geological uncertainty in the planned mining envelopes by evaluating the range in percentage pegmatite volumes produced by each model in individual stopes.

The key to reducing risk around geological uncertainty is to have sufficient close spaced drill testing in place well in advance of mine planning to minimise delays and mistakes. One of the main learnings from this study is that although the costs of a close-spaced drilling program may be difficult to justify in the initial stages of a project's development, the costs of not having sufficient data to manage geological uncertainty can be significantly greater. Specifically, not mitigating important geological risks prior to mining can confound mine planning and production targets, result in wasted development into areas that are found to be unviable, lead to the installation of infrastructure or capital development in sub-optimal locations, or worse, result in the mining and processing of waste under the false belief that it was ore, or unknowingly sterilising economic mineralisation.

Finally, based on the results of this study, Dr Harry Parker's famous adage that 'an ounce of geology is worth a pound of geostatistics' could be paired with a new aphorism that 'a gram of drill data is worth a kilogram of geological interpretation'.

ACKNOWLEDGEMENTS

The authors would like to sincerely thank:

- IGO for providing permission to publicly release the technical information included in this paper and providing the time and tools to complete the task.

- Shane Taylor and Chris Johnson from IGO for image drafting of introductory images.

- Jacqueline Murphy for her technical edits and often tedious and iterative drafting of plans and sections of the SIS and model results.

- The belated Dr Harry Parker for his mentorship, insight, inspiration, wry humour, and ability to always distil the crux to the obvious reality of the situation.

REFERENCES

Craven, B, Rovira, T, Grammer, T and Styles, M, 2000. The role of geophysics in the discovery and delineation of the Cosmos nickel sulphide deposit, Leinster area, Western Australia, *Exploration Geophysics*, 31(2):201–209. Available from: <https://doi.org/10.1071/EG00201>

de Joux, A, Thordarson, T, Denny, M, Hinton, R W and de Joux, A J, 2013. U-Pb dating constraints on the felsic and intermediate volcanic sequence of the nickel-sulphide bearing Cosmos succession, Agnew-Wiluna greenstone belt, Yilgarn Craton, Western Australia, *Precambrian Research*, 236:85–105. Available from: <https://doi.org/10.1016/j.precamres.2013.06.008>

de Joux, A, Thordarson, T, Fitton, J G and Hastie, A R, 2014. The cosmos greenstone succession, Agnew-Wiluna Greenstone Belt, Yilgarn Craton, Western Australia: Geochemistry of an enriched Neoarchaean volcanic arc succession, *Lithos*, 205:148–167. Available from: <https://doi.org/10.1016/j.lithos.2014.06.013>

Deutsch, C V, 2006. A sequential indicator simulation program for categorical variables with point and block data: BlockSIS, *Computers and Geosciences*, 32(10):1669–1681. Available from: <https://doi.org/10.1016/j.cageo.2006.03.005>

Duuring, P, Bleeker, W, Beresford, S W, Fiorentini, M L and Rosengren, N M, 2012. Structural evolution of the Agnew-Wiluna greenstone belt, Eastern Yilgarn Craton and implications for komatiite-hosted Ni sulfide exploration, *Australian Journal of Earth Sciences*, 59(5):765–791. Available from: <https://doi.org/10.1080/08120099.2012.693540>

IGO, 2022. IGO ASX Release – 20 June 2022 – Completion of Western Areas Scheme of Arrangement, in IGO ASX release, pp 1–1. Available from: <https://www.igo.com.au/site/PDF/3a2d943e-8bbc-4118-a868-6226c14c7044/CompletionofWSATransaction>

JORC, 2012. Australasian Code for Reporting of Exploration Results, Mineral Resources and Ore Reserves (The JORC Code) [online]. Available from: <http://www.jorc.org> (The Joint Ore Reserves Committee of The Australasian Institute of Mining and Metallurgy, Australian Institute of Geoscientists and Minerals Council of Australia).

McManus, S, Rahman, A, Coombes, J and Horta, A, 2022. Measuring spatial domain models' uncertainty for mining industries, in *Proceedings of the 12th International Mining Geology Conference*, pp 114–120 (The Australasian Institute of Mining and Metallurgy: Melbourne).

Mole, D R, Fiorentini, M L, Thebaud, N, Cassidy, K F, Campbell McCuaig, T, Kirkland, C L, Romano, S S, Doublier, M P, Belousova, E A, Barnes, S J and Miller, J, 2014. Archean komatiite volcanism controlled by the evolution of early continents, in *Proceedings of the National Academy of Sciences of the United States of America*, 111(28):10083–10088. Available from: <https://doi.org/10.1073/pnas.1400273111>

Murphy, M P, Parker, H, Ross, A and Audet, M, 2005. Ore-Thickness and Nickel Grade Resource Confidence at the Koniambo Nickel Laterite (A Conditional Simulation Voyage of Discovery), in *Geostatistics Banff 2004, Quantitative Geology and Geostatistics* (eds: O Leuangthong and C V Deutsch), 14:469–478 (Springer). Available from: <https://link.springer.com/chapter/10.1007/978-1-4020-3610-1_47#citeas>

Pym, F A, Crook, K E, Hetherington, P M and Murphy, M P, 2022. Machine learning in resource geology – why data quality is critical, in *Proceedings of the 12th International Mining Geology Conference,* pp 149–170 (The Australasian Institute of Mining and Metallurgy: Melbourne).

Wikipedia, n.d. Standard Error [online], *Wikipedia,* Retrieved 12 April 2023, Available from: <https://en.wikipedia.org/wiki/Standard_error>

An evolution of drill hole spacing studies at Newmont Corporation

A Jewbali[1] and L Allen[2]

1. FAusIMM, Group Executive, Resource Modeling, Newmont Mining Corporation, Denver CO 80237, USA. Email: arja.jewbali@newmont.com
2. Senior Director, Value Assurance, Newmont Mining Corporation, Denver CO 80237, USA. Email: lawrence.allen@newmont.com

ABSTRACT

At Newmont Corporation drill hole spacing studies are done to support the business in understanding the cost of collecting additional information (with a specific focus on drill hole sampling density) versus the value of additional drilling in terms of resource risk reduction. For the last 20 years drill hole spacing studies have been executed in an effort to answer the following question: *What spacing should the deposit be drilled at to achieve a certain level of resource confidence so that when mining occurs the resource model delivers as predicted ie what is the optimal sampling density?* The optimal sampling density supports a resource model that provides a sound foundation for mine planning purposes and resource/reserve reporting. At Newmont Corporation drill hole spacing studies are now a requisite for understanding drill data density needs for new deposits and near mine expansions. This paper details the evolution of drill hole spacing studies at Newmont Corporation in terms of methodology and usage.

INTRODUCTION

Public reporting codes for mineral reserves and resources are nonspecific concerning confidence thresholds for determination of Measured, Indicated and Inferred categories. Many mining companies, including Newmont Corporation, have adopted the practice of defining Indicated resources as having a confidence range on tonnage, grade and metal of ±15 per cent at 90 per cent confidence on an annual production basis (Deutsch and Begg, 2001). The Measured category is often defined as ±15 per cent on a quarterly basis. Drill hole spacing studies are used to determine the spacing requirements necessary to achieve these confidence levels for Measured and Indicated classification. The Inferred category typically relies on geological understanding and benchmarks from similar deposit types. The drill hole spacing for Inferred material is typically too wide for calculation of a robust variogram model that is necessary for the drill hole spacing study and thus the approaches discussed in this paper are mainly used to define Indicated and Measured spacings.

At Newmont Corporation the two main methods used over the last 20 years for executing drill hole spacing studies are either estimation variance calculations or the conditional simulation approach. The uses and limitations of both methods have been widely documented by many including Davis (1992); Verly, Postolski and Parker (2014); Dimitrakopoulos, Godoy and Chou (2009). This paper discusses the evolving methodology within Newmont Corporation for drill hole spacing studies along with comparisons between methodologies and conclusions.

EVOLUTION OF METHODOLOGY

Prior to 2004, Newmont Corporation used the estimation variance method or benchmarks against similar deposits to determine appropriate drill hole spacings for classification. Newmont Corporation's classification methodology in use at the time did not directly tie drill hole spacing with classification into Measured and Indicated categories. Historically, the classification methodology included various techniques that made use of distance from drill holes and kriging variance.

While these drill hole spacing and classification methods worked well in the majority of deposits that Newmont Corporation mined at the time (mostly open pits with large estimation domains and low cut-off grades) and with the further evolution of the simulation methodology, it was recognised that additional benefit could be gained from the use of conditional simulations. These benefits included the ability to examine areas of greater local variability, assessment of geological versus grade uncertainty, and assessment of varying cut-off grade strategies.

Beginning in 2004, Newmont Corporation began exploring conditional simulation methodologies for drill hole spacing studies. Initially only grade was simulated but as categorical simulation methods matured those were incorporated in the workflows. In the following sections several applications of the simulation methodology are discussed.

SINGLE SIMULATION FOR GENERATION OF MULTIPLE DRILLING PATTERNS

An early application of the simulation methodology was on a disseminated gold/copper deposit that was mined by open pit. Given the computationally intensive nature of multiple conditional grade simulations in 2004, Newmont Corporation experimented with a methodology for drill hole spacing analysis that used one, well validated, grade simulation from which multiple simulated drill hole patterns could be extracted. Kriged models were then generated from these multiple patterns.

The process for this method is described in a Newmont Corporation memorandum Newmont Corporation (2004) and shown in Figure 1:

- Develop and validate a single conditional grade simulation of the main intrusive in the deposit. While other domains contain economic mineralisation, the main intrusive has the highest variability and is of the greatest economic importance. Also, only gold values were simulated as they are more variable than copper and would likely require a closer drill spacing. The simulation was completed on a regular 6.25 m grid.

- Extraction of simulated drill holes. The 6.25 m node spacing allowed for 16 different drill hole patterns for each of the drill hole spacings to be used in the analysis (50, 75, 100, 125, 150 m).

- Estimation of multiple models. For each drill hole pattern at each spacing, a model was estimated using the parameters developed for the resource model. This resulted in 80 different models.

- Run each model through the annual mine plans and plot variability for each drill spacing by year. Use these plots to check for compliance with ±15 per cent at 90 per cent confidence requirement.

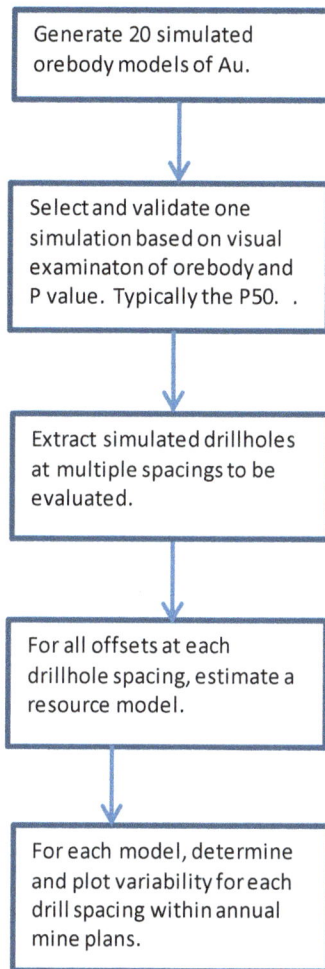

```
┌─────────────────────────┐
│ Generate 20 simulated   │
│ orebody models of Au.   │
└─────────────────────────┘
            │
            ▼
┌─────────────────────────┐
│ Select and validate one │
│ simulation based on     │
│ visual examinaton of    │
│ orebody and P value.    │
│ Typically the P50.   .  │
└─────────────────────────┘
            │
            ▼
┌─────────────────────────┐
│ Extract simulated       │
│ drillholes at multiple  │
│ spacings to be          │
│ evaluated.              │
└─────────────────────────┘
            │
            ▼
┌─────────────────────────┐
│ For all offsets at each │
│ drillhole spacing,      │
│ estimate a resource     │
│ model.                  │
└─────────────────────────┘
            │
            ▼
┌─────────────────────────┐
│ For each model,         │
│ determine and plot      │
│ variability for each    │
│ drill spacing within    │
│ annual mine plans.      │
└─────────────────────────┘
```

FIG 1 – Single simulation method workflow.

An example of the presentation of results from this analysis for one year are shown in Figure 2. This shows plots for the various drill spacings compared to the expected value based on the simulation for multiple models. The ±15 per cent boundaries are shown. Examination of the results for the nine years of mine life resulted in a drill hole spacing of 100 m being selected. At the time of this study, this methodology facilitated efficient computer execution as estimation of multiple models was much faster than performing multiple simulations.

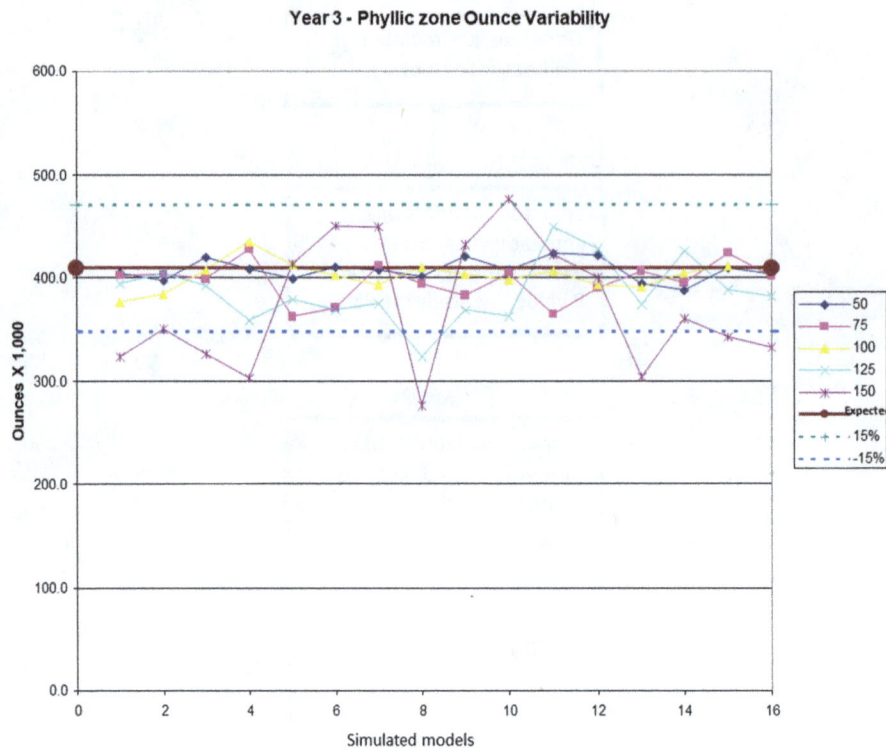

Year 3 - Phyllic zone Ounce Variability

FIG 2 – Disseminated gold/copper project – annual variability for year 3 of mine life (on the horizontal axis are the simulated models with each line representing a different drill hole spacing).

EXTENDING THE 'SINGLE SIMULATION FOR GENERATION OF MULTIPLE DRILLING PATTERNS' APPROACH TO AN UNDERGROUND SCENARIO

After the previous successful implementation of the simulation methodology Newmont Corporation (2011) applied the workflow discussed above to determine the drill hole spacing necessary for Indicated classification for a proposed underground project. This volcanic deposit is composed of multiple breccia types associated with different episodes of gold mineralisation. For this underground case, multiple sets of simulated drill hole fans were extracted from a single grade simulation of the deposit as shown in Figure 3. As in the open pit analysis discussed above, multiple estimations were made and run through the annual mine plan.

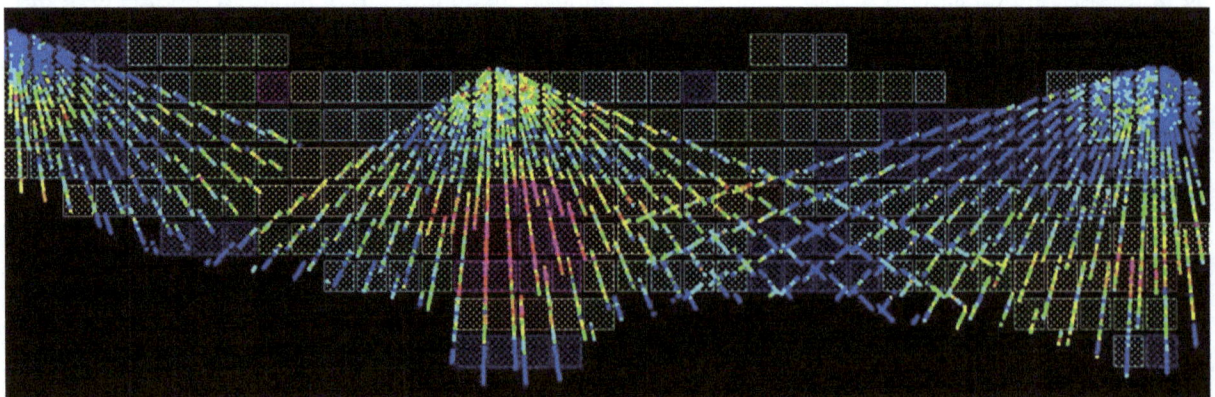

FIG 3 – Simulated drill hole fans.

Using this approach, a drill hole spacing of 8 m was determined as necessary for Indicated classification. This spacing appeared to be unreasonable based upon the nature of the deposit mineralisation and known geological controls. Careful examination of the deposit by a cross functional team of subject matter experts (Geology, Geostatistics and Mine Engineering) concluded that a 20 m spacing would be more appropriate for classification based on observed geological and grade continuity.

A later drill hole spacing study using multiple simulations by Newmont Corporation (2013) produced a similar result to that of Newmont Corporation (2011). Upon further investigation and analysis it was concluded that these two studies did not account for location uncertainty and the flexibility to adjust the future underground design as a function of grade variability. Every simulation positions the underground stopes in a different location as a function of grade variability. This is typically not an issue in the open pit scenario as the exact location of the ore within a large open pit is not as important as all material, ore and waste, will be mined. The exception to this being the pit bottom, where location is important and thus is often drilled to a tighter spacing. This was a key learning for Newmont Corporation in the application of drill hole spacing studies for deposits that will be mined with underground methods. Any approach that holds the designed stopes static will not allow for a reliable definition of uncertainty(usually the designed stopes are derived from an estimated resource model). A second more obvious insight was that drill hole spacing study workflow needs to be in-step with the mining methodology as described below.

To account for this location uncertainty, an analysis was conducted by Jewbali, Wang and Johnson (2015). As part of this analysis, an uncertainty profile was constructed using stope optimisation tools to generate mine designs and production profiles for individual simulations of the deposit. This workflow is shown in Figure 4. At the time of this study, stope optimisation was a computer intensive process, therefore only 50 of the 100 simulations were selected for uncertainty characterisation.

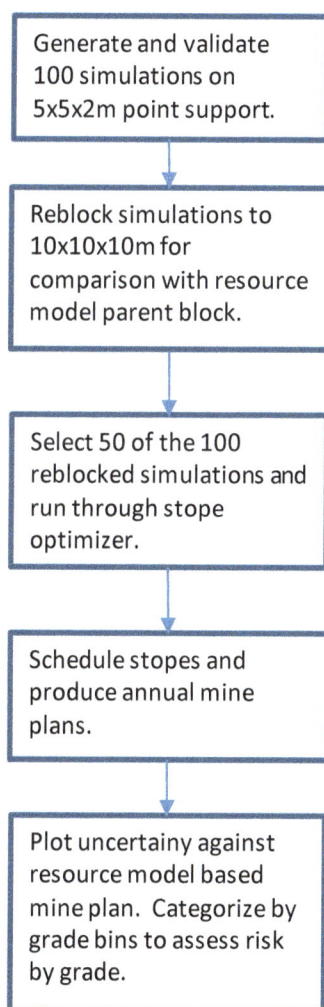

Generate and validate 100 simulations on 5x5x2m point support.

↓

Reblock simulations to 10x10x10m for comparison with resource model parent block.

↓

Select 50 of the 100 reblocked simulations and run through stope optimizer.

↓

Schedule stopes and produce annual mine plans.

↓

Plot uncertainty against resource model based mine plan. Categorize by grade bins to assess risk by grade.

FIG 4 – Multiple simulation method with stope optimiser workflow.

Conclusions from this study suggested the selected drill spacing of 20 m was adequate for estimation of tonnes, grade and ounces at the economic (ore/waste) cut-off. However, comparisons between simulations and the resource model indicated that the high-grade portion of the deposit was likely overestimated and that high-grade stopes were at risk. To improve project economics, mining engineers wanted to mine higher grade stopes in the earlier years and were in effect implementing a 'high-grade cut-off strategy'. The drill hole spacing study did not take this into account and did not make recommendations of what drill hole spacing was required for material above a higher cut-off. The link between mining methodology and drill hole spacing studies was another key insight for Newmont Corporation. To be effective and set the right expectations for model performance drill hole spacing studies need to be in-step with how the deposit will be mined.

MULTIPLE SIMULATIONS FOR GENERATION OF MULTIPLE DRILLING PATTERNS – A SIMULATION FOLLOWED BY ESTIMATION METHODOLOGY

With improvements in computing speed and more efficient simulation algorithms incorporating multiple simulations in the drill hole spacing workflow became possible. This also removed the requirement to select one single well validated simulation, which had always been difficult to justify given the variability encountered between different simulated models that reproduce the same statistical properties. Newmont Corporation (2012) completed a drill hole spacing study using multiple simulations for an orogenic type deposit mined using open pit methods.

Fifty simulated orebody models for Gold were generated and validated using a 2 × 2 × 2 m node spacing. These simulations were then sampled at seven different spacings: 10 × 10, 20 × 20, 30 × 30, 40 × 40, 50 × 50, 60 × 60 and 70 × 70 m, extracting drill hole data sets. These artificial drill holes were extracted at a 50 degree dip towards the south-west to match existing drilling. This resulted in 350 different data sets.

Using each of these data sets an estimated model for gold was created resulting in 350 different models (7 drilllhole spacings times 50 simulated models). Estimation for these models was done on a 10 × 10 × 6 m block size basis. Estimation parameters and methods were adjusted for the different drill hole spacings in an attempt to maintain consistent change of support. This was important because estimated resource models (using linear methods) for open pit mines at Newmont Corporation were built to reproduce the required SMU (selective mining unit) grade distribution and any estimated model that is smoother will display differences due to estimation parameter choice. Simulated models were then re-blocked to the estimated models block size.

The difference between the estimated models and the re-blocked simulations (considered to be the ground truth) was tabulated and quantified using relative precision (Equation 1):

$$Relative\ Precision = \frac{|Estimated\ model - Reblocked\ simulation|}{Reblocked\ simulation} \tag{1}$$

The Relative Precision was tabulated and plotted for annual and quarterly production periods to determine drill hole spacing requirements for Indicated and Measured material according to the Newmont Corporation requirements. The workflow for this methodology is shown in Figure 5 and example plots at one drilling spacing for tonnes, grade and ounces are shown in Figure 6. The figure shows that at the 50 × 50 m drill hole spacing the Indicated compliance criteria is met (ie within ±15 per cent at 90 per cent confidence requirement for annual production volumes) except for years 1, 7 and 8 which mine smaller volumes of material. The key insight from this approach was that if this methodology (simulation followed by estimation) was to be used; how the estimated models were built was important else the drill hole spacing would artificially be driven by the differences between simulated and estimated models. An acknowledgement of this issue led to the methodology discussed subsequently termed the 'simulation followed by re-simulation' approach.

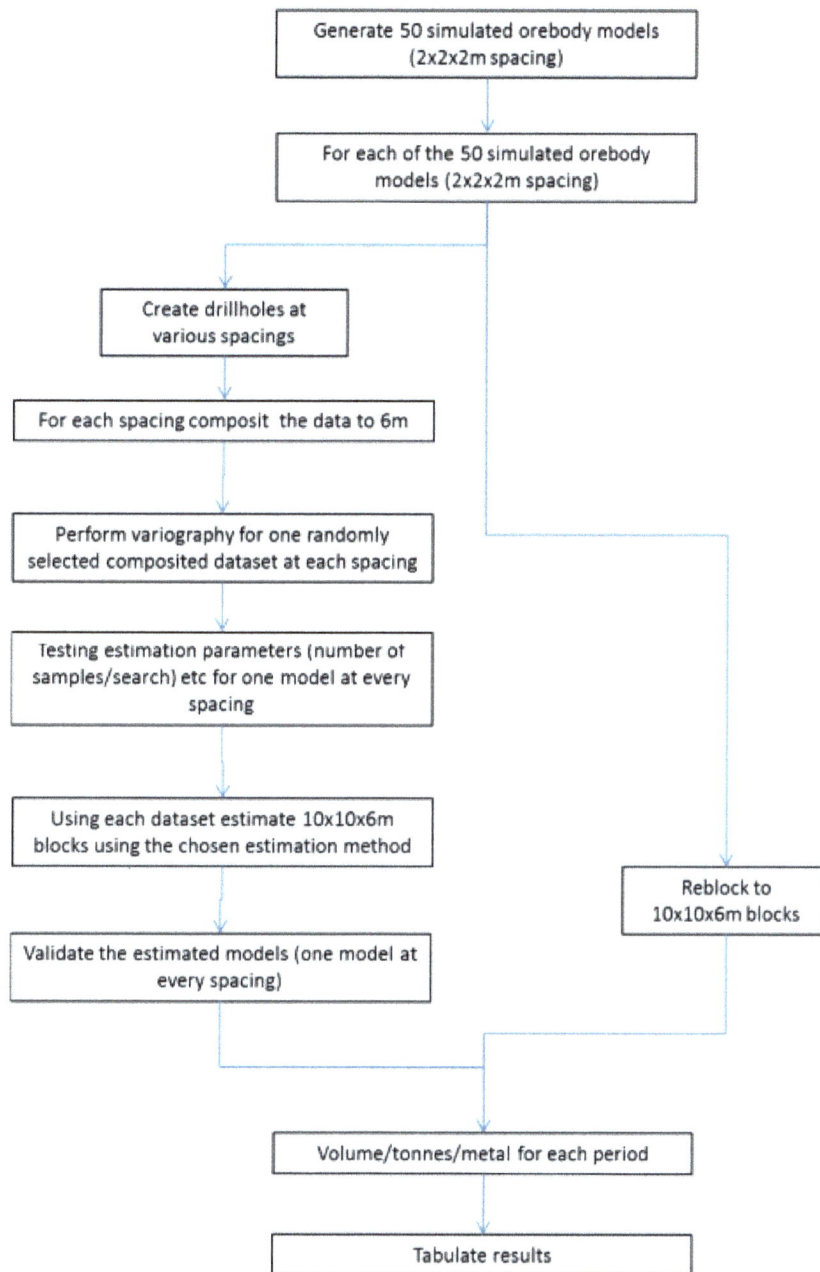

FIG 5 – Multiple simulation workflow.

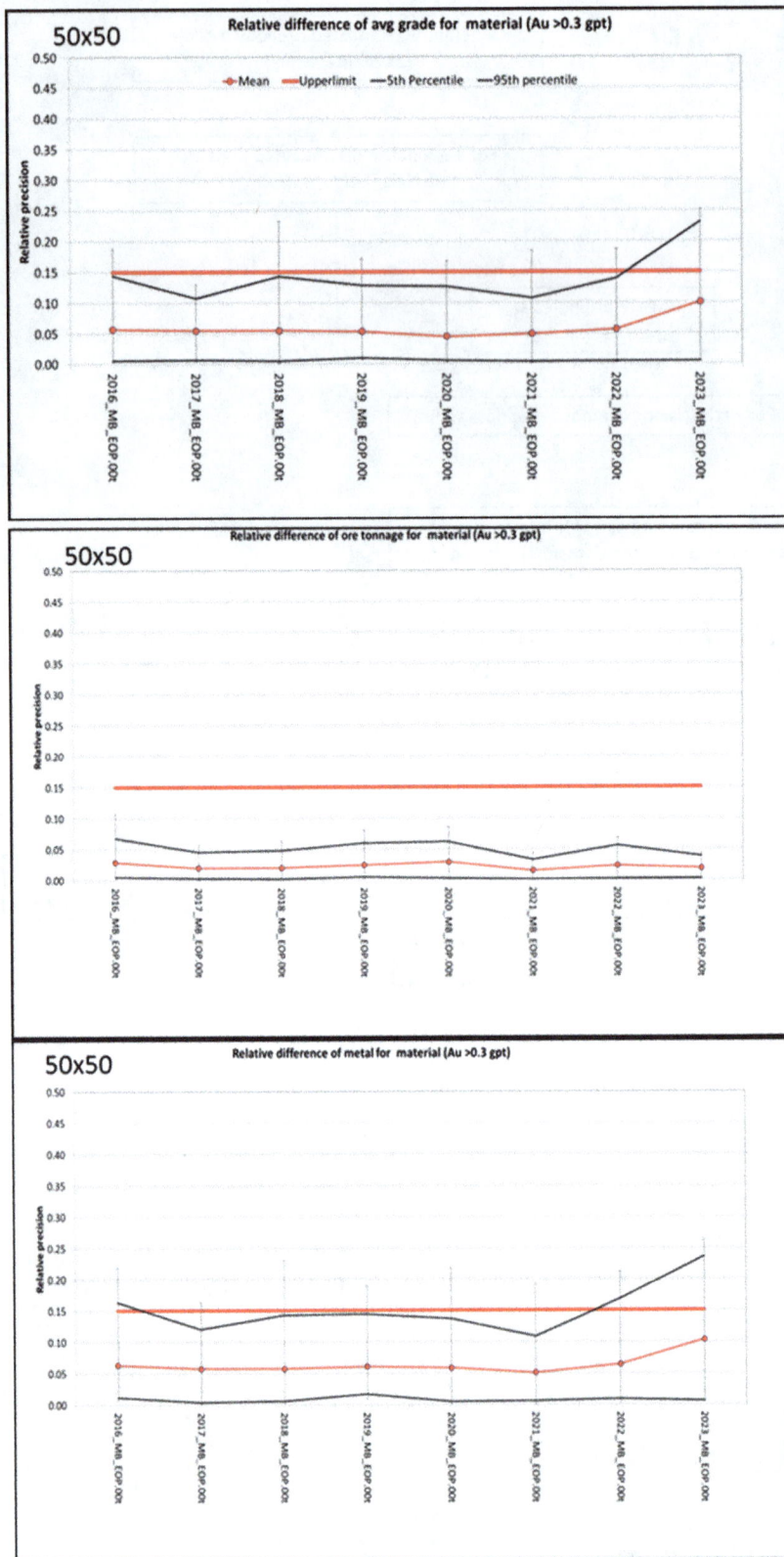

FIG 6 – Relative precision for Au grade, Ore tonnes and Au metal at 50 m × 50 m spacing for annual mine plan (horizontal axis are annual periods).

MULTIPLE SIMULATIONS FOR GENERATION OF MULTIPLE DRILLING PATTERNS – A SIMULATION FOLLOWED BY RE-SIMULATION METHODOLOGY WITH INCLUSION OF GEOLOGICAL UNCERTAINTY

By 2020, drill hole spacing workflows in Newmont Corporation evolved to include simulation for geological controls followed by simulation of the grade variables within the simulated geological controls. An example for a narrow vein deposit is discussed by Silva and Deutsch (2020) who developed the workflow used subsequently by Newmont Corporation on other deposits. Since the vein thickness had a significant influence on the contained metal it was decided that simulation of vein volumes was critical for this drill hole spacing study.

The spacing study included a total of 32 veins which were then consolidated into four groups based on orientation and panes of continuity. The vein simulation workflow entails simulation of the vein top or bottom followed by vein thickness. The subsequent grade simulations for each vein simulation follows a typical continuous variable workflow as detailed in Figure 7. For this analysis the ore tonnage was calculated within panels matching quarterly and yearly production (tonnes above cut-off for each production period). Panels containing less than 80 per cent of the target tonnage (as defined by the Mining Engineers) were discarded for the analysis. The re-simulation production plans at each drill hole spacing were then compared to those from the reference simulation to determine the drill hole spacing that would be necessary to meet Newmont Corporation classification criteria.

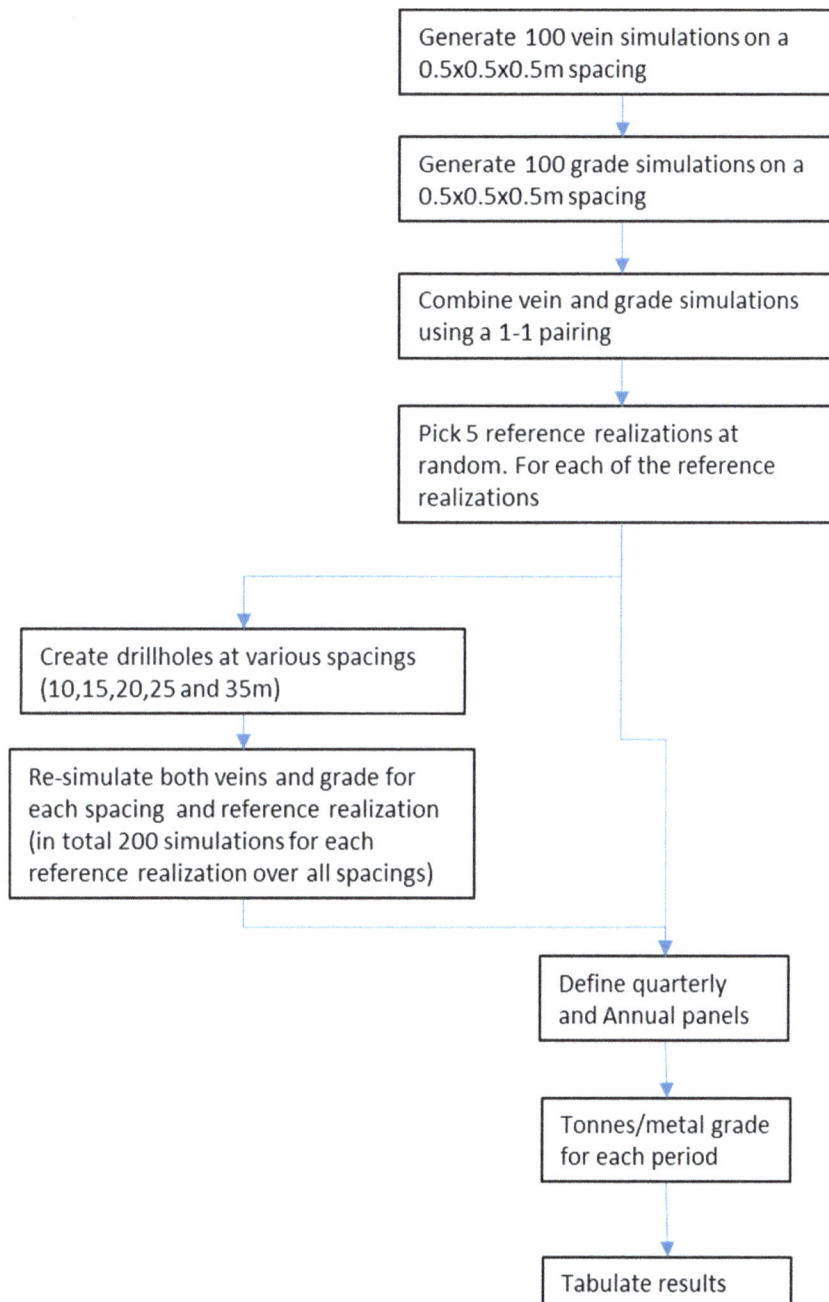

FIG 7 – Simulation followed by re-simulation workflow for a narrow vein deposit.

Figure 8 shows the Per cent Relative Precision for vein group 1 for annual production volumes. It shows that in general, at 25 m the Indicated compliance criteria is met (ie within ±15 per cent at 90 per cent confidence requirement for annual production volumes). This result matched an earlier, preliminary analysis, using the estimation variance method. While both methods produced the same result, the simulation based method provided more benefit including uncertainty analysis over the mine life and identification of areas with higher variability. One of the key insights for Newmont Corporation from executing this drill hole spacing study was that this approach is not trivial to implement; there are numerous validation steps required and re-simulating geological and grade variables at wider spacings is challenging.

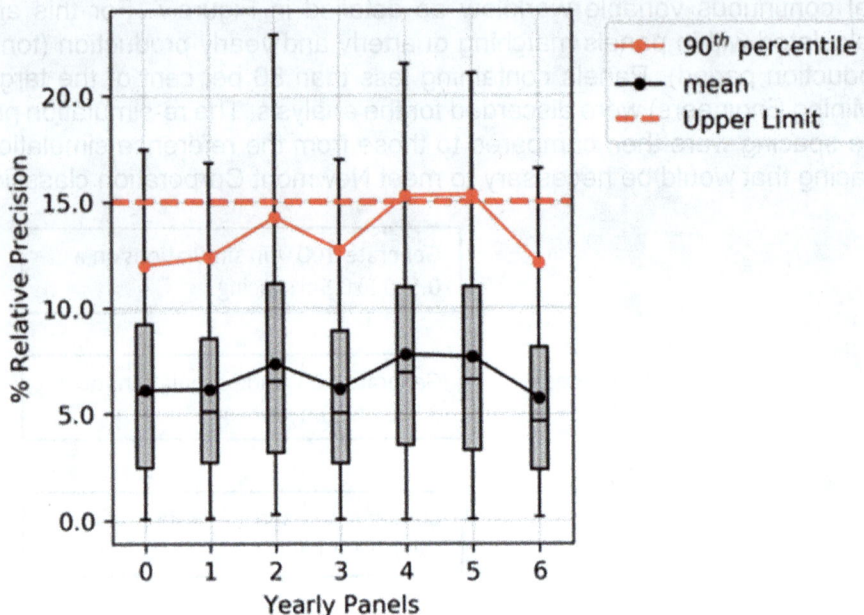

FIG 8 – Relative yearly ore tonnage precision for 25 m spacing for Group 1.

Notice that in this case the definition of ore is applied over reference and re-simulated realisations (Equation 2). There are other approaches as displayed in Figure 9 which shows a generalised workflow for combining geological and grade simulations in the simulation followed by re-simulation workflow. Three primary options for defining ore can be considered:

1. Best estimate (E-type model).

2. Grade control estimates.

3. Re-simulated realisations.

$$\%Relative\ Precision = \frac{|Re-simulated\ model - Reference\ simulation|}{Reference\ simulation} \times 100\% \qquad (2)$$

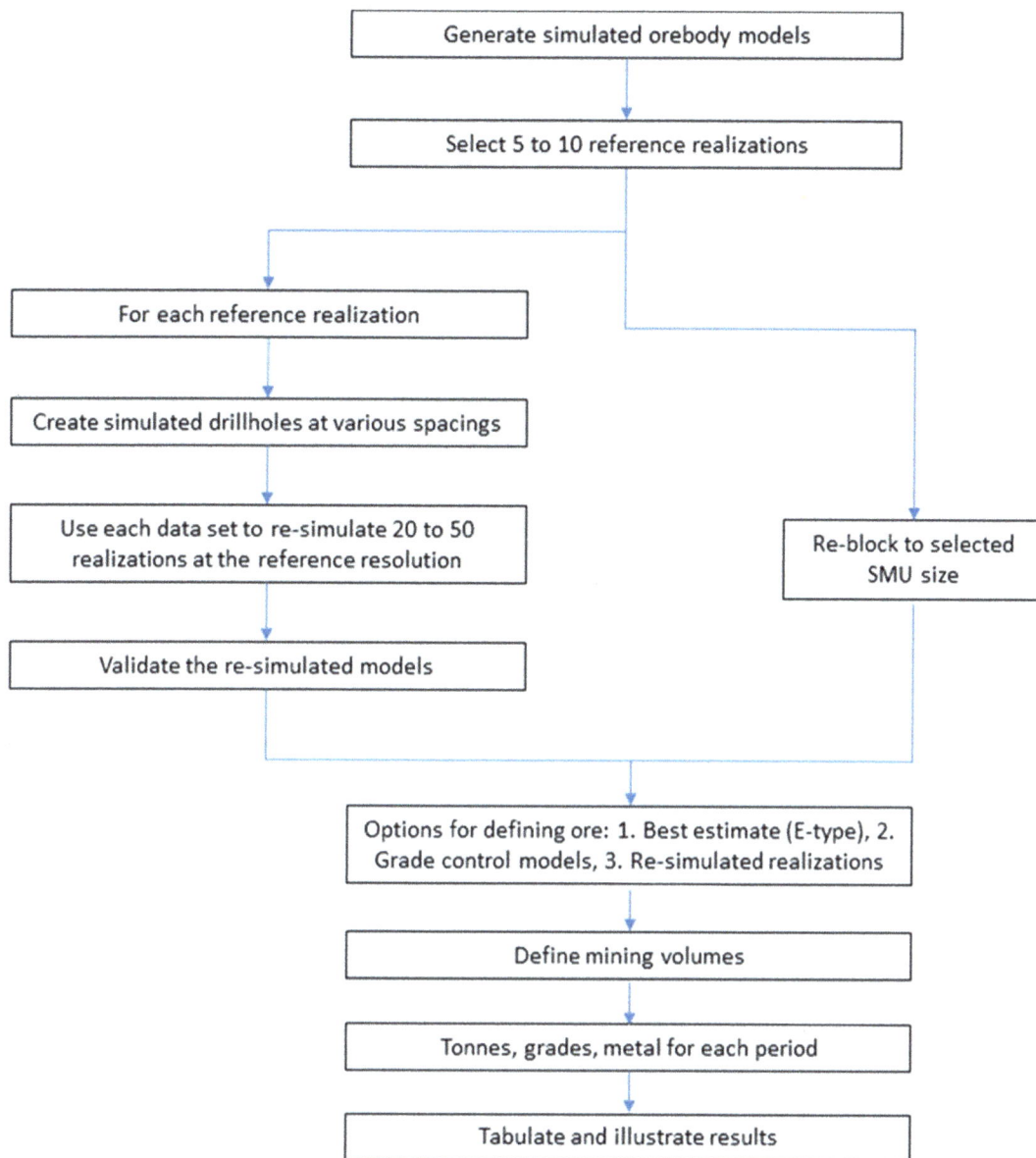

FIG 9 – Generalised Conditional re-simulation-based drill hole spacing study workflow.

Option 1 could introduce some degree of smoothing as a result of averaging the re-simulated realisations. Option 2 allows accounting for the information effect (more drilling available at the time of mining). It involves re-drilling each re-simulated realisation at a measured spacing and obtaining an estimate (eg Ordinary Kriging) used as a grade control model; a single kriging plan can be used for all the drill hole spacings. Option 3 may result in overly optimistic results since it implies no errors in the ore/waste definition (no misclassification since ore/waste is defined on the full re-simulation).

CONCLUSIONS AND CONSIDERATIONS

This paper presented Newmont Corporation's journey with drill hole spacing studies over an approximate 20 year period and the snapshot of various methods during this evolution. The single simulation, multiple estimation method, was discarded in favour of multiple simulations with multiple estimations to eliminate the possible risk in selection of a single simulation representative of the deposit. Over time, the simulation plus estimation and the simulation plus re-simulation methods were used on various deposits, along with the comparison to the simple estimation variance calculations method. Also, Newmont Corporation has trialled various categorical simulation methods for characterisation of geological uncertainty as well as grade uncertainty for drill hole spacing studies as discussed by Jewbali *et al* (2014). The vein simulation method described above is one of the most recent efforts in that area. This experience gathered over using these methods for Newmont

Corporation's operations and projects have resulted in recommendations as to when the use of each method is considered appropriate. These are shown in Table 1.

TABLE 1

Selection of a drill hole spacing methodology-Newmont Corporation guidance.

Estimation variance based method	Simulation based method	
	Conditional simulation plus re-simulation	Conditional simulation plus estimation
Consider using when: • There is insufficient data to support a simulation based study. • It is early in the project (Stage 0 or earlier). This method does not take cut-off grades into account. For OP, ore is defined at generally low cut-offs. For UG, the whole volume is mined (eg the whole vein).	Consider using when: • There is enough data to support a simulation based study. • The study can be completed within a reasonable time. This is a primary method for both OP and UG. This method is especially important to use for studies at multivariate deposits where multivariate relationships between variables need to be maintained in simulated models.	Consider using for OP when: • Estimation allows for reproduction of the proper change of support. • There is enough data to support a simulation based study. If estimates are too smooth, the simulation plus re-simulation method must be used. If multiple variables are present, the assumption is that the reproduction of multivariate relationships between them is not necessary for correct ore/waste definition. Consider using for UG when: • The size of simulated models is too large for a re-simulation study to be completed within a reasonable time. • It can be demonstrated that the estimated models validate well in terms of global bias, swath plots and change of support. • It can be proven that optimised stopes are realistic and reasonably reflect the way a particular deposit is mined.

In addition during the last 20 years, Newmont Corporation has evolved the resource classification processes to ensure that the selected drilllhole spacings for Indicated and Measured material derived from the drill hole spacing studies are translated into classification. Towards this end, Newmont Corporation has standardised on a three-hole rule geometric process that ties the drill hole spacing directly to classification.

Drill hole spacing studies, when completed, result in an appropriate spacing at time of the study. Spacing is dependent of the inherent variability and continuity of the mineralisation, the cut-off grade, and the production rate. Each of these may change over time and require a re-evaluation of the spacing necessary for classification:

- **Variability and continuity of mineralisation.** Drill hole spacing for a project or operation is completed within a specific mine design. As the mine design expands because of economics or additional drilling at the margins of the deposit, these extension areas may have different mineralisation characteristics and require a different spacing.

- **Cut-off grade.** Cut-off grades may change over time and require re-analysis. Or a different cut-off grade strategy may be developed in the production schedule than what was used in the original drill hole spacing study. A strategy that incorporates cut-off grades higher than the economic cut-off in particular time periods may require a closer drill spacing to ensure production expectations are achieved. This would not affect classification, as that is typically done using the economic cut-off grade.

- **Production rate.** The production rate may change over time. Or, if mining multiple deposits, the proportion of contribution from individual deposits may change as the mine plan evolves. Drill hole spacing may need to be re-evaluated for each deposit to ensure production expectations.

REFERENCES

Davis, B, 1992. *Confidence Interval Estimation for Minable Reserves*, SME Preprint 92–39, 7 p.

Deutsch, C V and Begg, S H, 2001. How Many Realizations Do We Need?, CCG Paper 2001–112, Centre for Computational Geostatistics, University of Alberta, Edmonton, Canada.

Dimitrakopoulos, R, Godoy, M and Chou, C, 2009. Resource/Reserve Classification with Integrated Geometric and Local Grade Variability Measures, in *Proceedings Orebody Modelling and Strategic Mine Planning: Old and New Dimensions in a Changing World Conference*, pp 207–214 (The Australasian Institute of Mining and Metallurgy: Melbourne).

Jewbali, A, Perry, B, Allen, L and Inglis, R, 2014. Applicability of Categorical Simulation Methods for Assessment of Mine Plan Risk, in *Proceedings Orebody Modelling and Strategic Mine Planning: Integrated Mineral Investment and Supply Chain Optimization*, pp 85–97 (The Australasian Institute of Mining and Metallurgy: Melbourne).

Jewbali, A, Wang, H and Johnson, C, 2015. Resource model uncertainty analysis and stope optimizer (MSO) evaluation for an underground metal mine project, in *Application of Computers and Operations Research in the Mineral Industry (APCOM)*, pp 625–640 (Society for Mining, Metallurgy and Exploration (SME): Fairbanks).

Newmont Corporation, 2004. *Drill hole Spacing Study*. Unpublished internal company document.

Newmont Corporation, 2011. *Drill hole Spacing Study*. Unpublished internal company document.

Newmont Corporation, 2012. *Drill hole Spacing Study*. Unpublished internal company document.

Newmont Corporation, 2013. *Drill hole Spacing Study*. Unpublished internal company document.

Silva, H and Deutsch, C V, 2020. Drill hole Spacing Study. Unpublished consulting report.

Verly, G, Postolski, T and Parker, H, 2014. Assessing Uncertainty with Drill Hole Spacing Studies – Applications to Mineral Resources, in *Proceedings Orebody Modelling and Strategic Mine Planning: Integrated Mineral Investment and Supply Chain Optimization*, pp 109–118 (The Australasian Institute of Mining and Metallurgy: Melbourne).

Maximising the value of a drilling program – case study in a challenging environment

A A Latscha[1] and D O'Connor[2]

1. MAusIMM, Principal Geoscience, Rio Tinto, Perth WA 6000.
   Email: anne-audrey.latscha@riotinto.com
2. Specialist Geologist, Rio Tinto, Perth WA 6000. Email: darragh.oconnor@riotinto.com

ABSTRACT

Increasing orebody complexity, restrictions in ground access, and longer lead times for disturbance approvals, have generated the need for the Resource Development Team within Rio Tinto Iron Ore (RTIO), to explore optimised drill design techniques that deliver resource conversion value while maximising program outcomes. In response to this challenge, the Team initiated a case study to produce an AI assisted/designed drill program using a non-machine learning AI.

The case study involved the assessment and implementation of an AI generated drill program at one of RTIO's Pilbara iron ore deposits, located in the Hamersley Province of Western Australia. The Mineral Resource is currently classified as 100 per cent Inferred. Access constraints inherent to topography, and the presence of exclusions zones have prevented adequate drilling coverage, resulting in lower resource confidence and a poor definition of the orebody extents and geometry. The drill campaign, planned in Q1 2023, needs to consider access constraints, complex topology and the limitations related to additional ground clearance, while addressing the requirement of improving orebody knowledge.

The drill plan had to address two objectives. The first and main aim of drilling was to upgrade the Resource Classification to allow for possible conversion to Ore Reserve. The input for this objective was a volumetric representation of the areas currently not meeting pre-defined quantitative Resource Classification upgrade criteria. The second objective was to better define the orebody extents and geometry. For the second objective, the input to the drilling optimiser was a volumetric representation (blocks) of the areas of high uncertainty in the mineralisation/waste boundary, which were defined using an indicator conditional simulation technique.

The drilling optimiser output was a range of compliant drill plans. A set of KPIs based on economic and orebody knowledge criteria were used to assess the plans and to select the most optimal drill design to be drilled in 2023. The geology and management teams were able to make better informed, objective decisions using the trade-off between cost and orebody knowledge gained. As part of the assessment, the AI generated plans were compared to our current planning practises. The AI assisted plans, when gauged against the project KPIs, showed clear and measurable improvements in both meeting project objectives and in managing operational risk.

INTRODUCTION

Traditionally, drilling designs are defined using a regular grid pattern approach, with spacing dependent on required orebody knowledge based on studies requirements. In this specific case study, the authors have been working on a deposit (called the Deposit) where mining is scheduled to start within the next two years, and 100 per cent of the Mineral Resource is still classified as Inferred.

Historically, drilling at the Deposit has not been based on the traditional grid pattern approach, due to access constraints and steep topography. A drilling campaign is planned in 2023 to increase the confidence in the orebody knowledge and to support the definition of Ore Reserves. Moving away from the grid pattern and using reduced angled holes (RAD drilling), with varying numbers of holes, drilling locations, depth, angle, means that an infinity of alternative drilling plans is possible. Once a drilling plan is designed by the geologist, it is difficult to assess the performance of this plan until the deposit is mined, through to reconciliation, and it is then too late to alter the plan. The drilling plan produced by the geologist will be drilled without considering the range of other options which are likely to include better performing plans.

For this case study, the DRX™ AI based drilling optimiser (the optimiser), was used to maximise the value of the program. DRX™ is a trademark of Objectivity.ca. Based on the given inputs, constraints and objectives, the optimiser produces ranges of drilling plans which can be assessed holistically, allowing the geologist to make informed decisions on resource conversion, drilling volumes, programme duration and budget. Trade-offs between a range of plans can also be clearly communicated to management to help align drilling execution with corporate strategy.

The first part of this paper describes the project's context: the deposit geology, constraints as well as drilling objectives. The second part outlines the drilling optimiser workflow. The third part presents the methods used to create a volumetric representation of the drilling targets. And finally, the performance indicators utilised to compare the drilling plans between each other and against the baseline (manual) plan will be described, as well as the process to select the final plan which will be executed in 2023.

Geology setting

The Deposit is considered a typical, high phosphorus, supergene iron deposit hosted in the Lower Proterozoic Brockman Iron Formation within the Hamersley Basin of the Pilbara region, Western Australia. The Hamersley Group Iron Formations are comprised of low energy chemical precipitates of alternating iron oxide and silicate rich bands, ranging from micro to macro scale, interbedded with cherts, shales, dolomites and volcaniclastics. Deep supergene enrichment of precursor banded iron formations (BIF) occurred during the Mesozoic to Tertiary, with this style of mineralisation accounting for most economic Iron Ore deposits within the Pilbara region. The Hamersley Group has been intruded by a regional set of dolerite dyke swarms, with dominant WNW-NNW trends that cross-cut the Deposit at the local scale stopping out mineralisation.

Stratigraphy within the Deposit area dips to the south between 20 and 35 degrees. A variably mineralised hydrated cap blankets the Deposit and is typically 10 to 15 m thick. Supergene iron enrichment occurs to depths of greater than 100 m in some areas. Locally, ore/waste boundaries exhibit a complex geometry; in some instances, mineralisation preferentially follows more iron-rich, BIF dominant units resulting in fingering patterns adding to the overall complexity. A typical long section of the Deposit is shown in Figure 1, and a typical cross-section of the Deposit is presented in Figure 2.

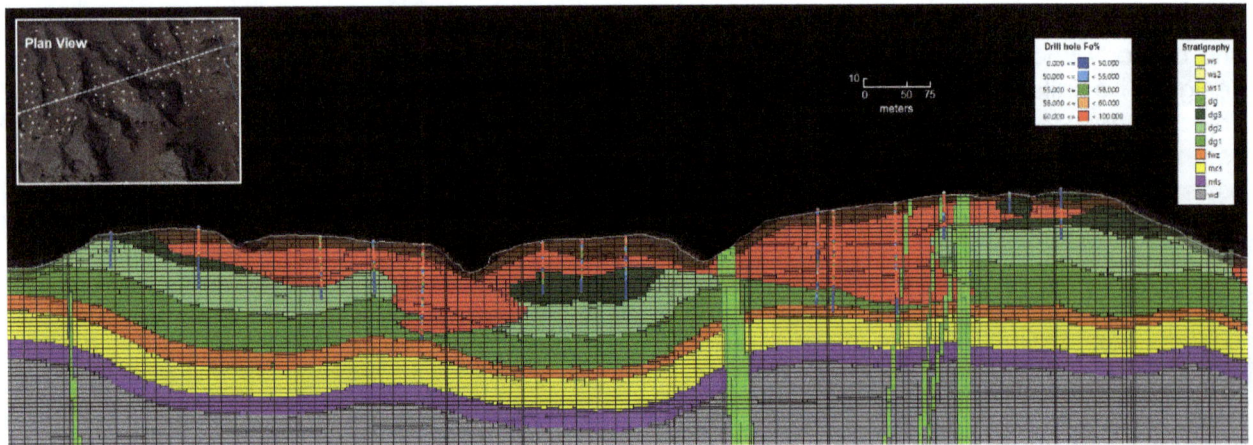

FIG 1 – ESE-WNW typical long section; mineralisation using a 50 per cent Fe cut-off is shown with the red blocks. The hydrated surface blanketing the fresh mineralisation is in brown.

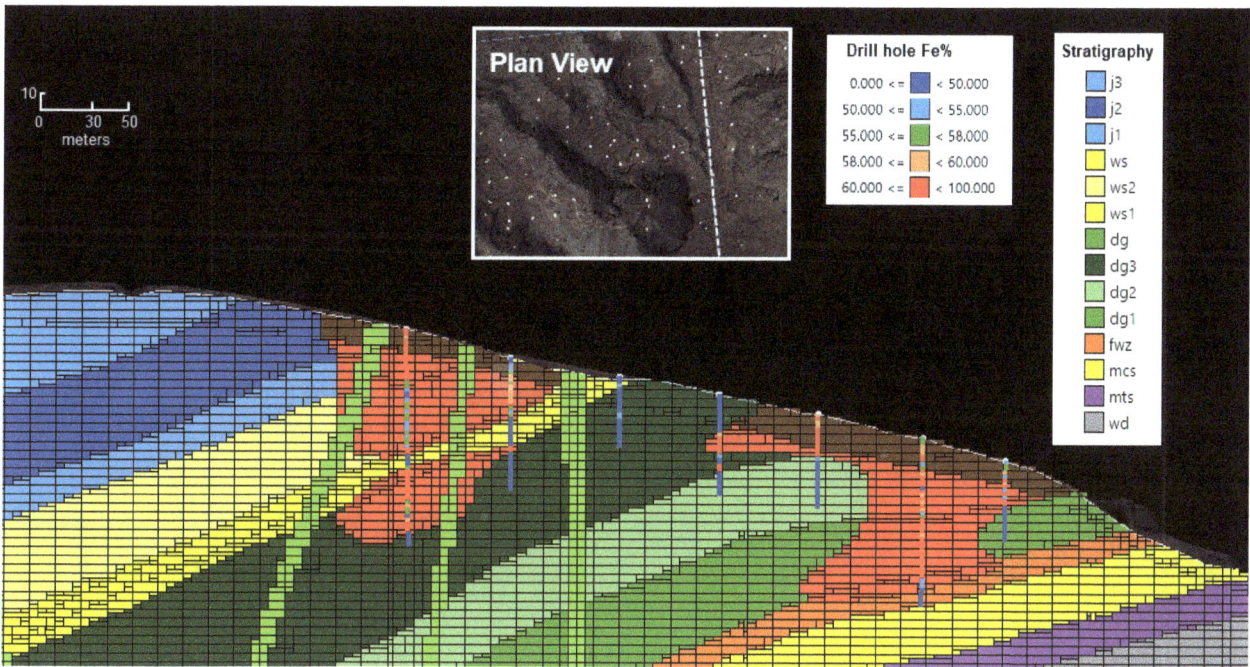

FIG 2 – NE-SW cross-section; main crusher feed sources are the Joffre Member (in blues) and the Dales Gorge Member (in greens).

Current work

The Deposit area is covered by 1:5000 scale mapping to assist with geological interpretation and modelling. Drilling has been completed over numerous campaigns from the late 1970s to present. Drilling for grade, metallurgical, density and structural purposes has been undertaken to determine the mineralogical and material handling properties of the Deposit and to provide geotechnical parameters for pit design. Drilling metrics for the Deposit, as of January 2023, are detailed in Table 1.

TABLE 1

Drilling complete within the Deposit by drill rig type.

Drill type	No. holes	Total metres
Diamond	31	2151.30
Percussion	5	648.50
Reverse Circulation	156	10 256.00
Total	**192**	**13 055.80**

Progressive iterations of geological modelling and resource estimation have been undertaken with the latest modelling and estimation completed in 2021. The Deposit currently sits within the Inferred category and requires additional drilling to increase resource classification and allow conversion of resource to reserve prior to commencement of mining. Drilling is scheduled to commence in Q1 2023, and will utilise Reduced Angle Drilling (RAD) to target previously inaccessible areas of the deposit restricted by topographical and environmental constraints. RAD drilling utilises a modified Reverse Circulation (RC) drill rig, allowing low angle drilling to be completed with a range of inclinations from -35 degrees to 90 degrees, thus allowing greater optionality when generating drill designs.

Constraints, assumptions, and objectives

Significant constraints exist within the Deposit area; the main constraint hampering drilling activities can be attributed to topography. Steep terrain has hampered drill rig access; the Deposit is incised by numerous steep sided gullies which pose significant restrictions to drill hole locations, leaving large sections of the Deposit poorly drilled and tested. Furthermore, the Deposit area is covered by numerous biological exclusions which require further study prior to the granting of mining approvals.

This, coupled with the complex nature of ore/waste boundaries has resulted in the deposit being classified in the Inferred category.

Given the above constraints, it was decided to trial a drill hole optimisation solution to improve resource confidence (at a minimum converting inferred resource to indicated to then allow conversion to reserve) and to better define the geometry of the deposit to allow informed pit optimisation and design, while honouring the abovementioned constraints.

DRILLING OPTIMISATION PROCESS

Drilling optimisation workflow

The purpose of the drilling optimiser is to maximise the value of every metre drilled while addressing the orebody knowledge requirements. The drilling optimisation process workflow followed in this case study is shown in Figure 3.

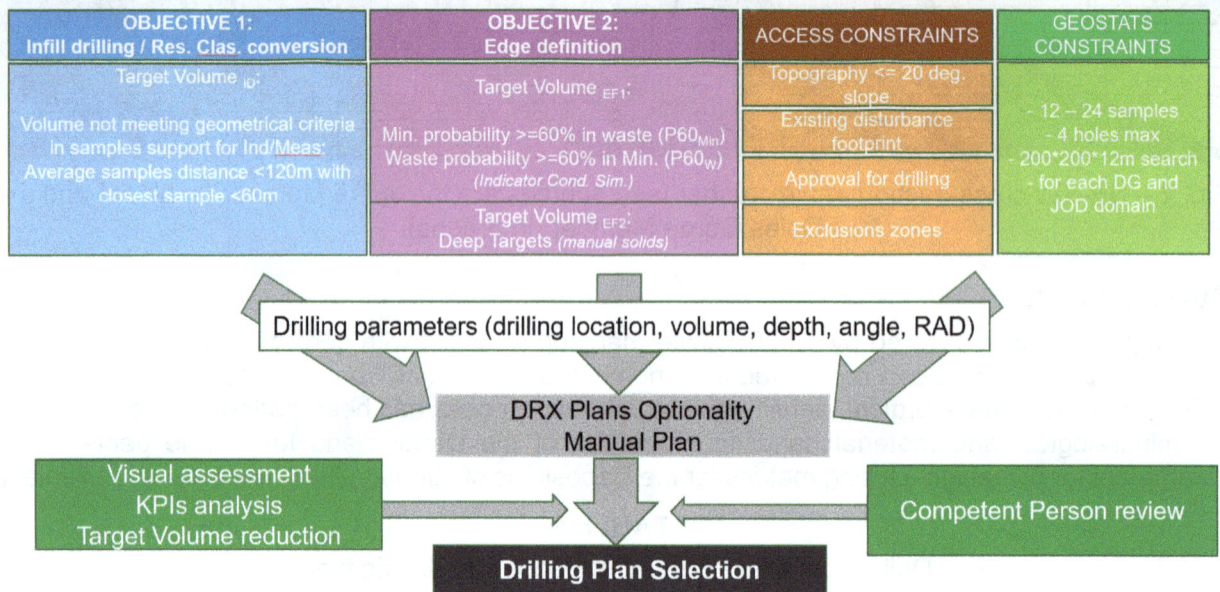

FIG 3 – Drilling optimisation process workflow.

For this case study, the drill plan had two major objectives. The first objective (objective 1) was to increase the confidence in the definition of the Mineral Resource, and this was expressed in upgrading the classification from Inferred to Indicated or Measured. Objective 1 was focused on infill drilling within the existing mineralisation envelope. The second objective (objective 2) was to better define the orebody geometry and to close off the mineralisation envelope which is still open at depth and laterally. Objective 2 was important given the location of the initial planned pits. The optimisation tool requires physical inputs corresponding to volumetric representations of the targeted areas. For each objective, a solid (volume of blocks or enclosed triangulation) created around the targeted areas was provided to the optimiser.

Flexible variables utilised to run the optimisation included drilling parameters such as number of drill holes, collar location, drilling depth, and angle. Based on the set of inputs, constraint, and objectives, the optimiser produced a range of compliant drill plans. The drill plans were assessed and cross-compared using a set of project specific performance indicators. The optimiser plans were also compared to a manual plan produced by an experienced geologist (the baseline plan).

The need to infill a volume, while also testing the extents of this volume, produce two competing criteria for the optimisation process. The former is a volumetric or density of drilling problem, while the boundary problem is a geometric constraint. The optimiser used in the study can perform multi-objective optimisation to address these competing requirements.

Objective 1 – Infill drilling for resource classification conversion

Definition of Target Volume ID

Assumptions and inputs

As explained above, a target volume (called Target Volume $_{ID}$) was required to address objective 1. This volume was linked to objective criteria used for resource classification conversion. The target volume represented blocks that did not meet the resource classification criteria for Indicated/ Measured resource classification based on existing sample support.

The definition of objective criteria for resource classification was based on the analysis of existing data from an operating mine site (the mine) abutting the Deposit. It was assumed the conditioning data at the Deposit and at the mine show a similar behaviour, and that no difference in distribution or spatial variability was observed during the analysis of the combined data set for the production of the latest resource model. This operating site was covered by a drill spacing of 60 × 60 m and the Mineral Resource classification was either Inferred, Indicated or Measured.

The first criterion used for the analysis was the distance between a block, and the closest sample used to estimate that block, and was called DCS (distance to closest sample). The second criterion was the average distance between all the samples used to estimate that block and was called ADAS (average distance between all samples). This analysis of the data at the operating mine allowed the generation of thresholds which were then applied to the Deposit.

The analysis of the mine data set revealed that 94 per cent of the blocks classified as Measured were estimated at the first estimation pass, using a search ellipsoid of 200 × 200 × 12 m, with a number of samples between 12 and 24 and using four holes minimum. The analysis also showed that 99 per cent of those Measured blocks have a DCS of less than 63 m and an ADAS of less than 125 m. Tables 2 and 3 give the details of the analysis.

TABLE 2

Percentage of mineralised blocks estimated per estimation pass at the mine.

Resource classification	Estimation pass	Percentage estimated blocks
Measured	1	94
	2	6
Indicated	1	85
	2	15
Inferred	1	62
	2	38

TABLE 3

Maximum (99th percentile) distance to closest sample (DCS) and average distance between samples (ADAS) per resource classification category for the blocks estimated at the first estimation pass at the mine.

Resource Classification	DCS (m)	ADAS (m)
Measured	63	125
Indicated	69	134
Inferred	107	154

Based on this analysis, thresholds were set to 60 m for DCS and 120 m for ADAS and were applied to the Deposit. Mineralised blocks which had a DCS greater than 60 m, or a DCS less than 60 m but an ADAS greater than 120 m, within a search ellipse of 200 × 200 × 12 m, were considered not to qualify as either Indicated or Measured. This rule was extended to Indicated as well and no separate threshold was defined between Indicated and Measured, using a conservative approach.

The blocks not meeting the above-mentioned criteria were extracted and combined to create a solid which was used as the target volume for objective 1, and called Target Volume $_{ID}$. A summary of the neighbourhood criteria utilised to define Target Volume $_{ID}$ is shown in Table 4.

TABLE 4

Minimum, resource classification, conversion geometrical criteria based on samples' neighbourhood restrictions. All blocks not meeting those criteria were combined and formed the Target Volume ID solid, which was used as an input to the optimiser to address objective 1.

Data quantity			Data configuration		
Min. # of samples	Max. # of samples	Min. # of holes	Search ellipsoid	DCS	ADAS
12	24	4	200 × 200 × 12 m	<60 m	<120 m

Limitation

It needs to be acknowledged that the above-mentioned objective criteria are not solely attributable to enabling the upgrade of Mineral Resource classification from Inferred to Indicated or Measured, as there are other criteria that need to be considered by the Competent Person, who prepares and signs off the public reporting of Mineral Resources in accordance with the JORC Code (2012). This means that the classification of a block meeting those neighbourhood criteria will not necessarily be converted, however, those criteria have been set as a minimum requirement.

Visual representation of Target Volume $_{ID}$

The target volume defined and utilised as input for objective 1 (Target Volume $_{ID}$) is shown in Figure 4.

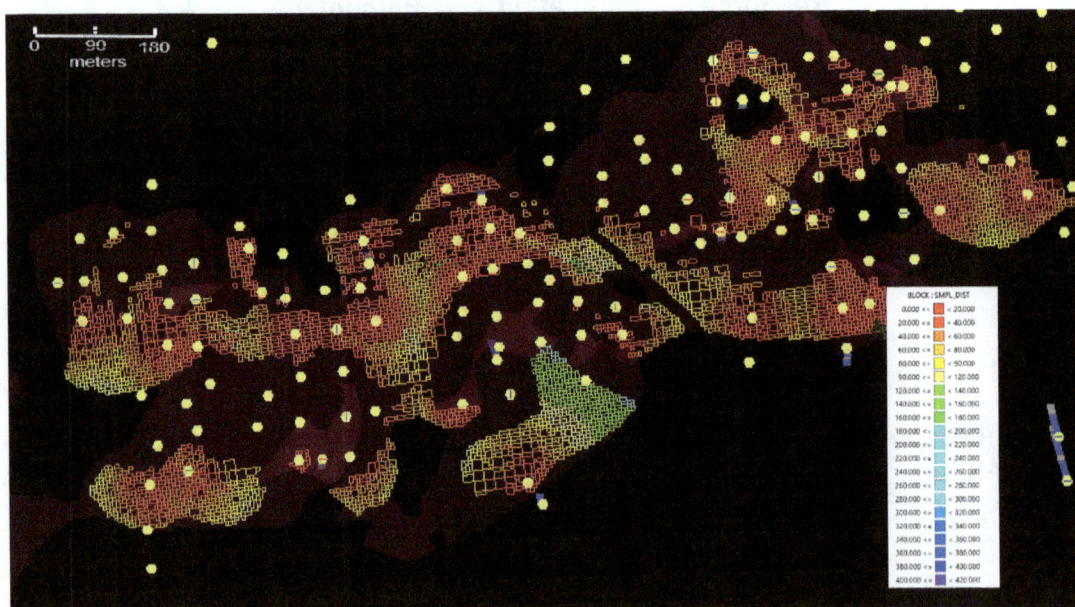

FIG 4 – Target Volume $_{ID}$ (blocks with DCS >=60 m or (DCS <60 m and ADAS >=120 m)).

Objective 2 – Orebody edge definition

Similarly to objective 1, a target volume was required to address objective 2. Given the sparse and irregular drilling distribution, the orebody has not been closed off laterally and/or at depth. Objective 2 targeted areas with a low level of confidence within the current interpreted mineralisation boundary, due to either poor drilling coverage or to unsupported interpretation.

Two target volumes were created: Target Volume $_{EF1}$ and Target volume $_{EF2}$.

Target Volume $_{EF1}$ was a volume representation of the areas of high uncertainty in the current mineralisation geometry and was created using an indicator conditional simulation technique. Target Volume $_{EF2}$ captured blocks of mineralisation, which had been interpreted at depth with no or very limited sample support, and could also correspond to an artefact of the mineralisation modelling technique utilised. Target Volume $_{EF2}$ was created manually.

Definition of Target Volume $_{EF1}$

Target Volume $_{EF1}$ is represented by a solid encompassing the blocks which have a high uncertainty of being misclassified regarding their mineralisation category. The Target Volume was created using an indicator conditional simulation technique, and reflects the uncertainty present in the current mineralisation/waste boundary. It includes blocks currently interpreted as waste, but which have a probability greater or equal to 60 per cent of being mineralised, as well as blocks currently interpreted as mineralised, which have a probability greater or equal to 60 per cent of being waste based on geostatistical analysis. The simulation was performed using Isatis.Neo™ v.2022, which is a Trademark of Geovariances SAS. The software follows the methodology described in Desassis *et al* (2015).

Pluri-Gaussian Simulation workflow

The simulation was performed using the Pluri-Gaussian Simulation technique, which is a stochastic technique used for spatial modelling of categorical variables, represented by the truncation of one or more Gaussian random fields. This simulation technique was selected as it honours the relationship in contacts between mineralisation categories which is an important consideration for banded iron formation hosted deposits.

Based on the spatial relationships and contacts between mineralisation categories, a number of Gaussian random fields and a truncation rule have been defined. The thresholds of the truncation rule vary vertically and laterally according to the proportions of each facies.

Then the spatial correlation structure of the Gaussian random fields is modelled. A set of Gaussian values corresponding to the categories is first generated at the data locations by Gibbs sampling. The remainder of the simulation was completed in Isatis.Neo™ by using a turning bands process with conditioning to simulate the Gaussian values, which are then converted back to categories, using the truncation rule, for the final simulation output (Maleki and Emery, 2014).

Application to the case study – Categorical variables

A volume of interest (VOI) was defined around the relevant area, in order to reduce the computation time and was limited by the topography, the lateral extents of the deposit and the 500 m elevation level at the footwall of the deposit.

The input variable was a categorical variable corresponding to the mineralisation with four categories. Each sample's value is assigned with one of the four categories: mineralisation, waste, detritals and hydration. The indicators are created for each sample by assigning a 0 or 1 value for each of the four categories.

Proportions

The proportion of each category was modelled for each block using an estimation algorithm. The proportions were variable and the estimation of the proportions used a method based on a maximum likelihood. In Isatis.Neo™, the estimation of the proportions is done using the SPDE (Stochastic Partial Differential Equation) method, which allows handling local anisotropies. An example of the proportions for each of the four categories on cross-section is shown in Figure 5. The proportions are used to work out thresholds on Gaussian values for conversion to categories. The thresholds for the conversion will be different on different blocks. The thresholds are calculated from the standard Gaussian distribution.

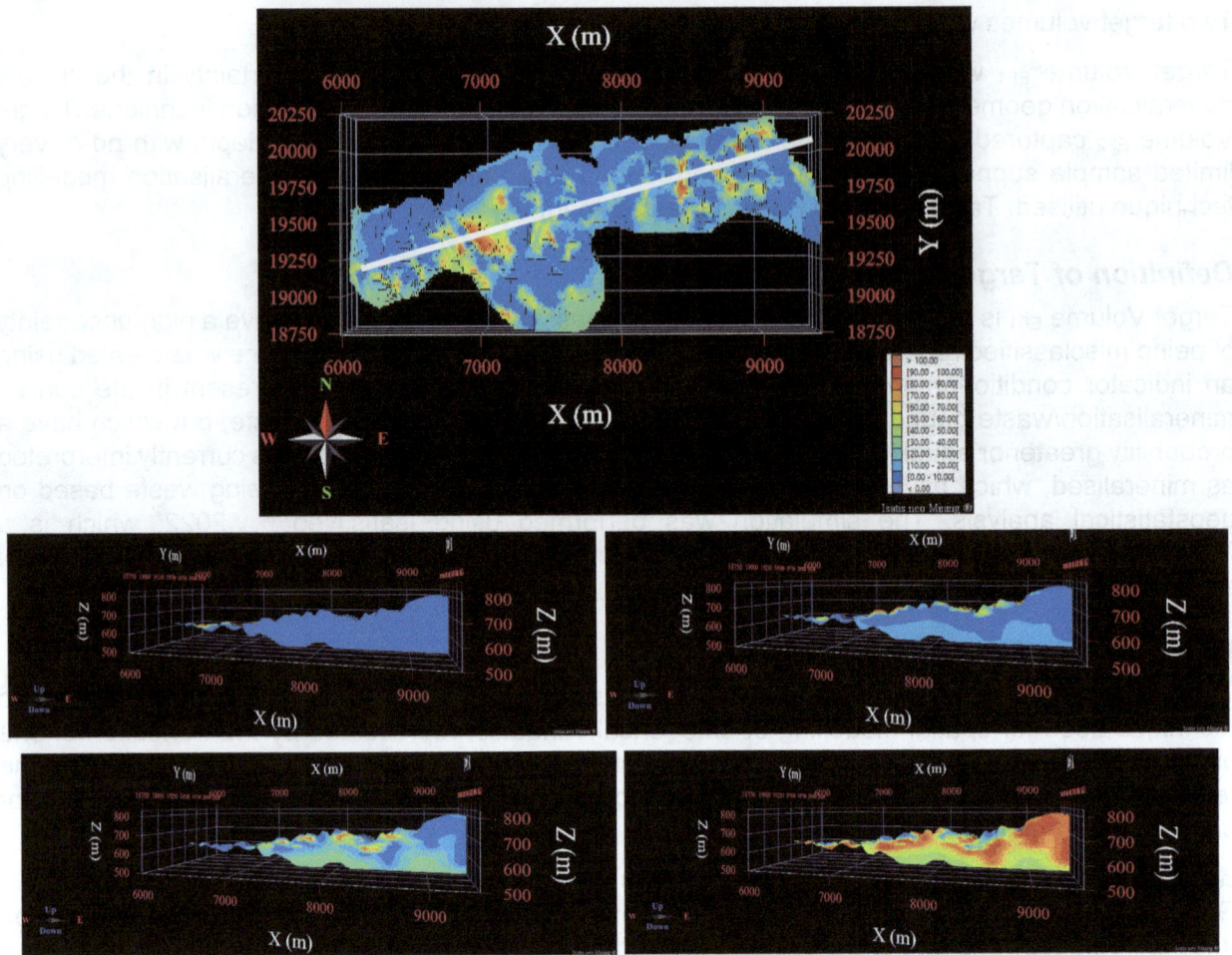

FIG 5 – Proportions (top: plan view and cross-section location; from top left to bottom right: detritals, hydration, mineralisation, waste proportions).

Truncation rule

The lithotype (truncation) rule reproduces the possible contacts between the different mineralisation domains. In the case of the Deposit, the possible transitions are as follow: the mineralisation domain underlies hydration, detritals can overly either hydration or waste, and waste can contact any of the other domains. Following the application of those rules, the truncation rule suggested by the possible transitions is shown in Figure 6.

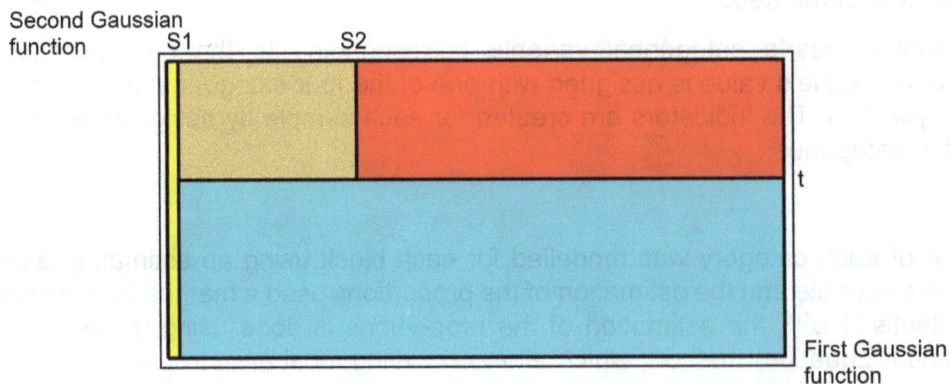

FIG 6 – Truncation rule; yellow: detrital; brown: hydration; red: mineralisation; blue: waste. S1 and S2 are the thresholds for the first Gaussian function and t for the second Gaussian unction.

The first Gaussian variable controls the contact between the detritals and hydration or waste, and the second one controls the contact between the waste and the hydration or mineralisation.

The thresholds of the truncation rule vary vertically and laterally according to the local proportions of each category. As we know the proportions at each location and using the lithotype rule, the Gaussian values can be converted to categories.

Simulation results

The 100 realisations produced 100 possible interpretations of the mineralisation. The results of the first and tenth simulations are shown in Figure 7 as examples.

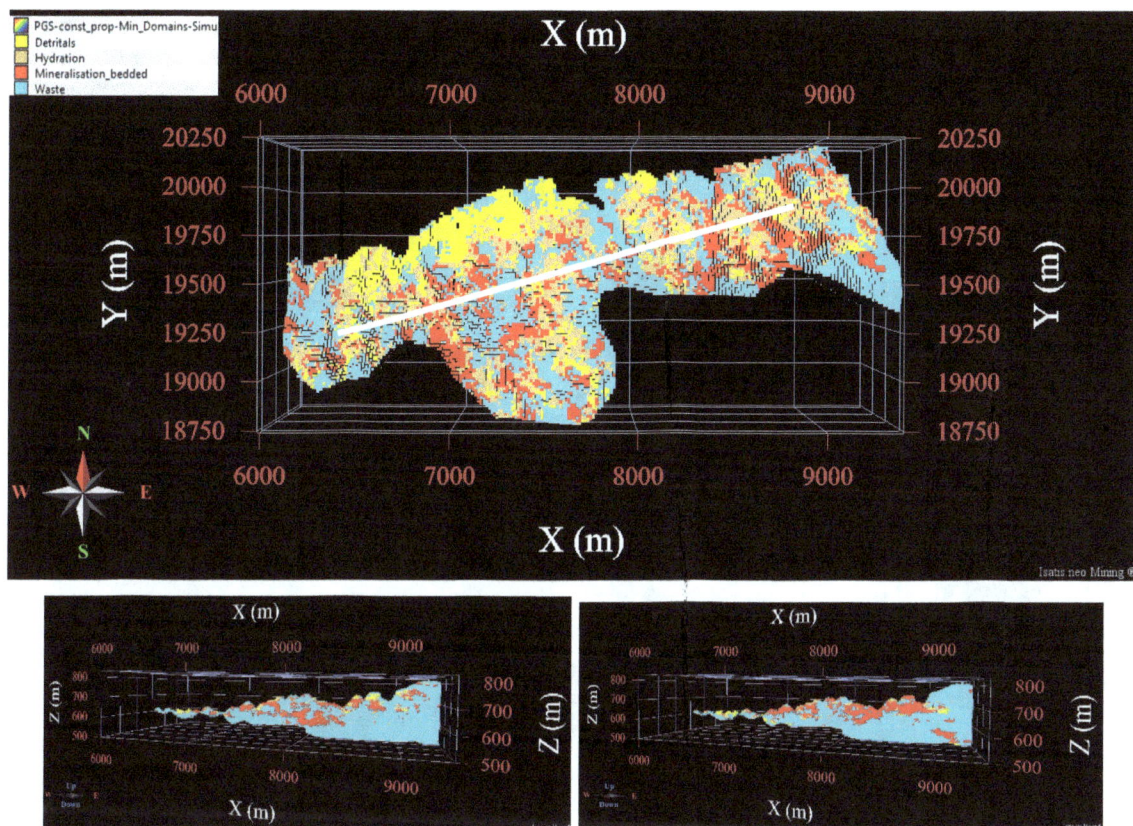

FIG 7 – Pluri-Gaussian simulations results (top: plan view and cross-section location; left: first simulation; right: tenth simulation).

To validate the simulation results, the variograms of the indicators of the simulated categories were compared with the variograms of the indicators of the input data, and the comparison was deemed acceptable.

Post-processing and use input to the drilling optimiser

The sum of the indicators for each category were calculated for each block and for the 100 realisations. The resultant number divided by the number of simulations (100) gave the probability of occurrence of each category for that block. Each block's category is defined by the category having the highest percentage from the 100 realisations.

The model was exported from Isatis.Neo^{TM} as a csv format file to be used as an input to the drilling optimiser. Two numeric (integer) variables 'mineralisation' and 'waste' were created and captured the likelihood of those two categories based on the outputs from the simulation.

A cut-off of 60 per cent was selected to provide a slight preference in selecting blocks that were currently miscategorised (eg waste blocks within a mineralised volume and vice versa). This defines blocks which had a high uncertainty on their current interpreted mineralisation boundary. They are shown in Figure 8 against the current interpretation of the mineralisation. Those blocks show some correlation to the solid Target Volumes $_{EF1}$ which have been used for the drilling optimiser as an input target to address the edge definition uncertainty.

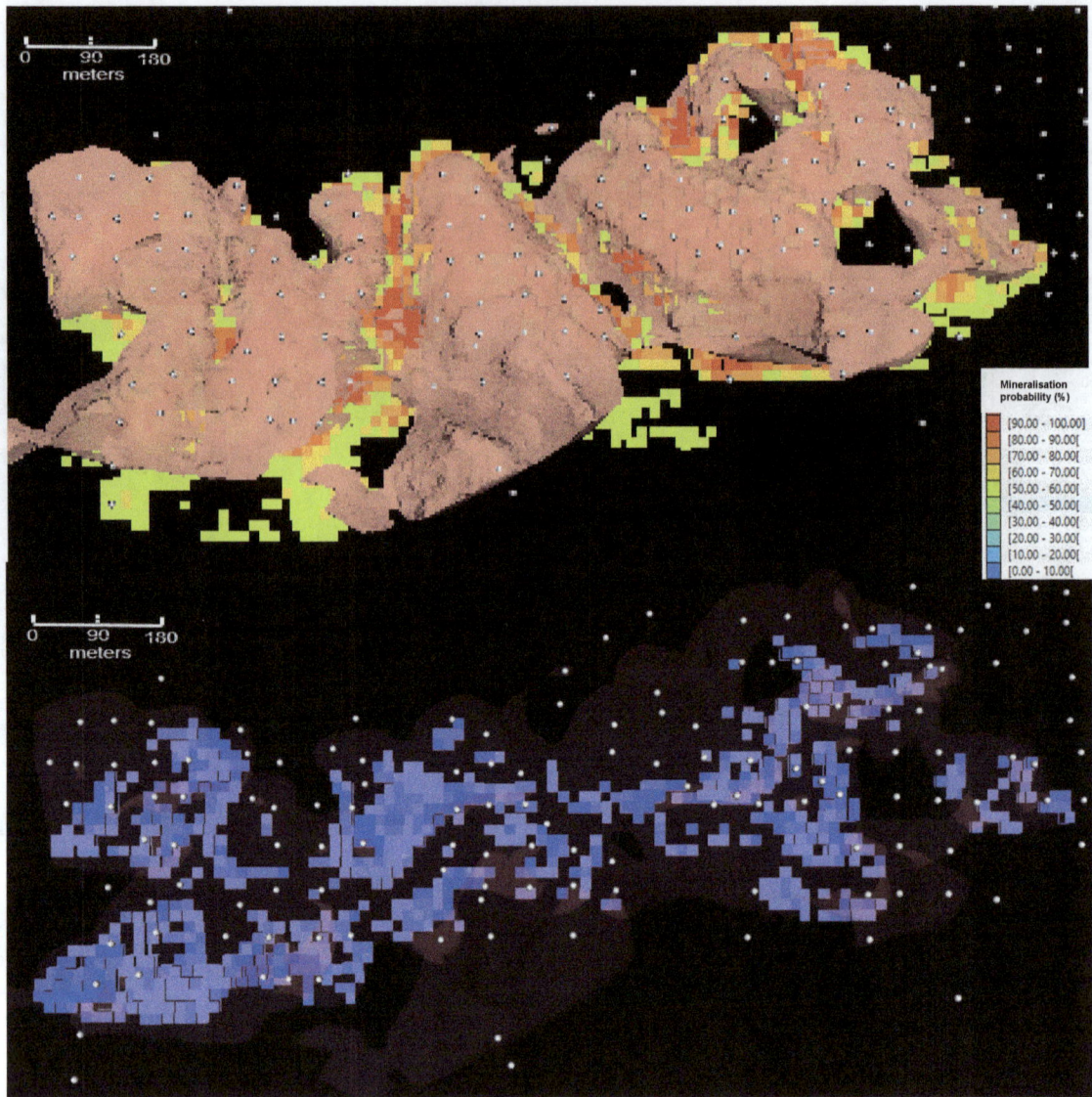

FIG 8 – Blocks with a high uncertainty on mineralisation boundary, with a probability greater or equal than 60 per cent to be misclassified. Volumes of blocks used as inputs into the drilling optimiser (top: Target Volume EF1 P_{60Min} showing potentially misclassified waste blocks; bottom: Target Volume EF1 P_{60W} showing potentially misclassified mineralised blocks; pink solid: current mineralisation interpretation).

Definition of Target Volume $_{EF2}$

Target Volume $_{EF2}$ captured mineralisation which has been interpreted at depth with no or limited sample support. This unsupported mineralisation interpretation may be attributed to an artefact in the geological modelling process, inherent from the interpolation parameters of the implicit modelling engine utilised to model the orebody geometry. An example of such a non-supported 'finger' of interpreted mineralisation at depth is shown in Figure 9. Drilling is then required to test this deep interpreted mineralisation.

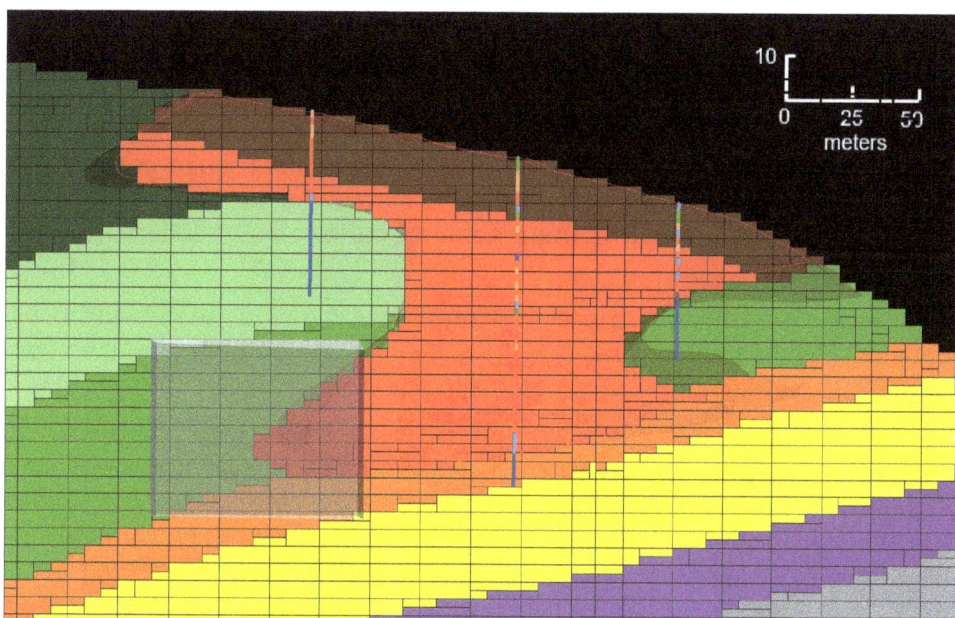

FIG 9 – Example of section view showing Target Volume $_{EF2}$ (grey solids), manually created around deep fingers of mineralisation (red solid/blocks) with no or poor samples support.

These deep fingers, which are not supported by drill hole data, were not necessarily identified with the conditional simulation, as the latter uses drill hole data independently from the interpreted mineralisation solid. Therefore, manual shapes were added to the targeting model. These shapes, designated as Target Volume $_{EF2}$, are represented in Figure 10. The optimiser was then provided with the solid volumes as a second set of targets for objective 2. Targeting the above manually generated solids allows testing the presence of mineralisation and refining the orebody geometry at depth.

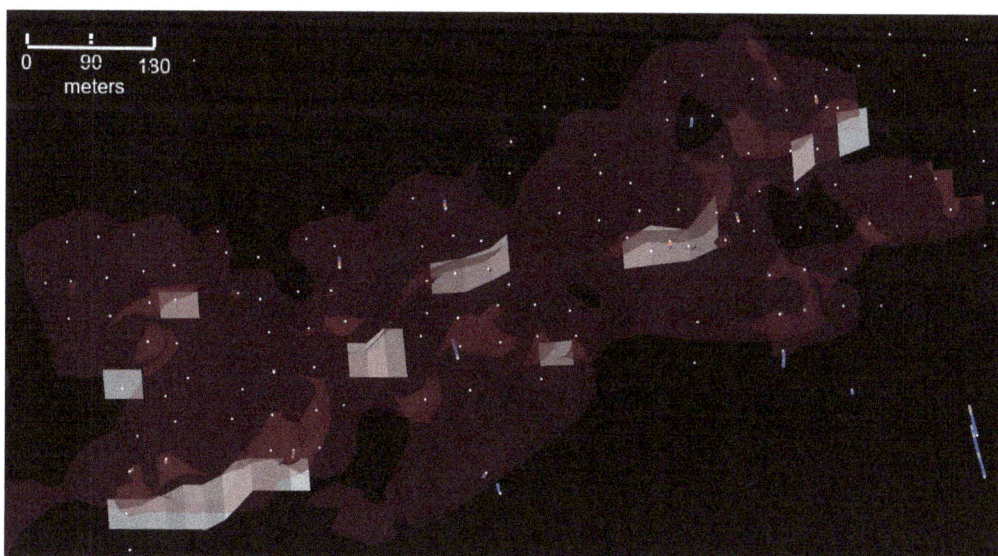

FIG 10 – Solids capturing blocks with unsupported deep 'fingers' of mineralisation (grey solid: Target Volume EF2), red solid: current interpreted mineralisation).

DRILLING PLANS OPTIONALITIES AND ASSESSMENT

Process iteration

The drilling optimiser was run using the set of inputs, constraints and objectives. The first iteration used a specific set of constraints which were subsequently flexed and adjusted with each run based on output review and discussions.

Further iterations consisted in having 75-degree-angled holes and seeing how this affected the optimisation. Then the angle was increased down to 35 degrees. Those holes would have to be drilled by a specific rig. Other iterations looked at opening the access not only to the same pads than the baseline plan but to any ground where the topography was favourable (<20 degrees) and outside any exclusion zones. However, the execution of such plans would require considerable earthwork leading to an increased cost and a lengthier field program. As part of this iterative process, it appeared that the main value driver was the amount of earthworks, resulting from the creation of new pads.

Another iteration added a restriction to preferentially target direct crusher feed from higher value geological domains. This helped constrain the volume given the amount of drilling being budgeted for the program.

The iterative process allowed the geology team to understand the drilling program's value drivers. One of these significant value drivers was the cost of earthworks to develop new drill pads. Given the fact that opening the ground required an increase in the earthworks, resulting in a lengthier and more expensive execution program, in addition to the limitation in clearing more ground, it was decided to restrict the collaring to the existing ground disturbance footprint – the major constraint for the optimisation.

With limited surface disturbance settled, the major flex became drilling location and angle. The use of RAD drilling provided reach underground but also added new variables for cost and time constraints. Ultimately, the program was again constrained to decrease RAD drilling given the limited availability of a RAD rig.

The KPIs were used to assess the plans and to compare them against each other. The outcome was to find the right trade-off between OBK requirements and costs/execution of the drilling program. The iterative process not only defined an initial set of KPIs but then also refined those KPIs as an understanding of the value drivers and the real world constraints were simulated within the optimiser.

Drilling plans assessment

The assessment of the plans followed a three step process: visual assessment, objective assessment using key performance indicators (KPIs), and analysis of the investment curves. The involvement of the Competent Person was a key component of the assessment process.

Visual assessment

A visual assessment of the produced plans against the geological model was performed. This step involved the verification of the validity of every single hole. The holes criteria (location, angle, length) were criticised and adjusted to produce a new iteration. Following this first step, the optimiser was re-run taking into consideration those changes. This iterative process was used to tune both the team's expectations and the algorithm to produce plans that were executable. An example of visual assessment is shown in Figures 11 and 12. Visual assessment was also important in helping the team gain confidence in the output being produced by the optimiser.

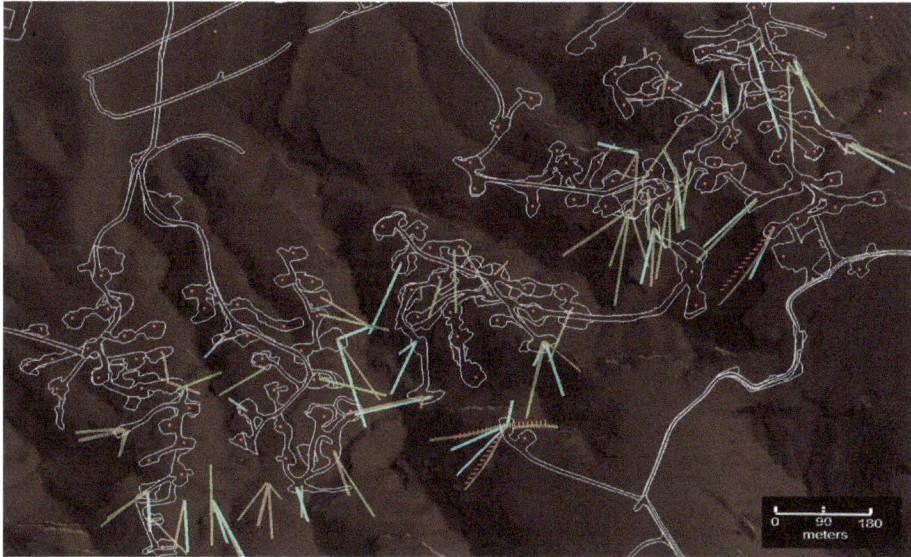

FIG 11 – Visual assessment of a range of drilling plans – plan view (red dots: existing drill holes, green lines: manual (baseline) plan, red, blue and brown lines: DRX™ produced plans).

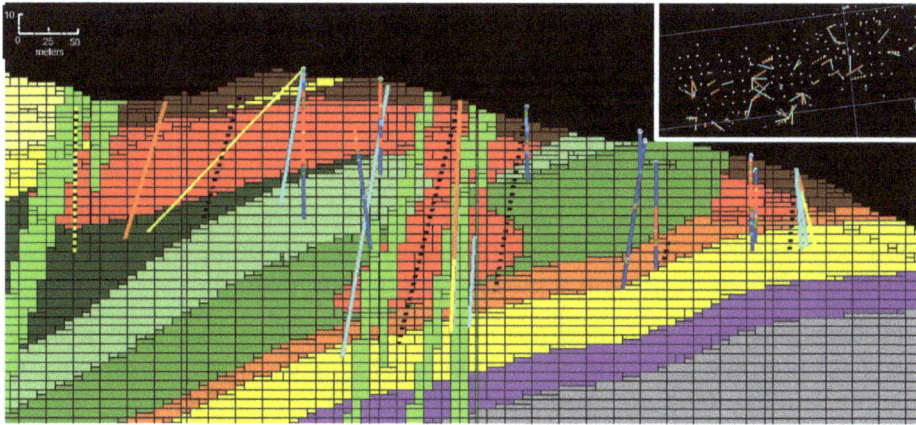

FIG 12 – Visual assessment of a range of drilling plans against geological model and existing drill holes, loaded by iron grades.

Following the visual assessment, a more objective assessment of the plans was done using investments curves and key performance indicators (KPIs).

Objective assessment using Investment Curves

To objectively compare a wide range of plans, using different assumptions, each of the plans was plotted on a DRX™ investment curve. The investment curve plots cost against KPIs. The simplest KPI is the expected volume being converted. However, in more complex cases, the curve allows a range of drilling plans to be compared. The geology and management teams can make better informed, objective decisions using trade-off between cost and orebody knowledge gained. The X axis is a measure of the amount of drilling (expressed in this case by the cost of the campaign including earthworks and drilling) and the Y axis is a measure of value generated (expressed by volume of mineralisation resource classification conversion, based on the neighbourhood criteria). The investment curve which was produced for objective 1 is shown in Figure 13. The brown curve corresponds to all possible drilling plans produced by the optimiser using the given set of inputs and constraints. Some specific plans have been extracted from the curve and represented by the small brown triangles (minRad6H 39, minRad 6H 45 and minRad6H 68). Two other plans expressed by the orange triangles (minRad 34 and minRad 43) were produced using a slightly different set of inputs or constraints. The pictures also show the manual (baseline) plan produced by the geologist (Protrack v2+6) for comparison.

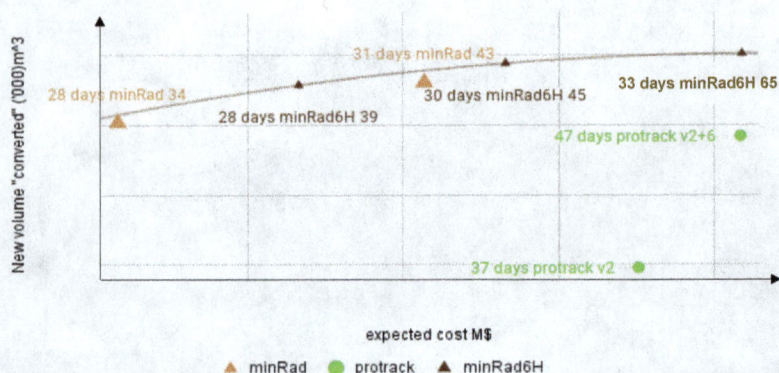

FIG 13 – Investment curve showing drilling plans produced by the optimiser. The X axis represents the costs of the drilling campaign, and the Y axis represents the volume of mineralisation conversion based on neighbourhood criteria. The green dots represent the manual plans.

This curve shows the baseline plan would produce a similar resource conversion volume as the optimiser's minRad 34 plan but for an additional cost, representing a roughly 29 per cent decrease of the total program budget by using the optimiser plan. The optimiser plan could potentially be drilled quicker saving time and bringing the resource to reserve status quicker.

For a similar cost, the minRad6H 65 plan would produce a higher resource conversion volume than the baseline plan. The baseline plan is less optimal than any of the plans produced by the optimiser.

Objective assessment using KPIs

The KPIs used to assess the plans were based on the cost of the program (drilling + earthworks), the orebody knowledge (OBK) gain, the drilling efficiency (expressed as the ratio between OBK gain and cost), the field program duration and a measure of ground disturbance.

The assessment was aiming at selecting the most optimal drilling plan with a trade-off between minimising cost, field program duration and ground disturbance while maximising OBK gain and efficiency. A summary of the KPIs and their assumptions is shown in Table 5.

TABLE 5

Description of the KPIs used to assess the drilling plans optionality.

KPI	Measure	Assumptions
Cost (AU$)	Minimising cost of earthworks + drilling. Maximising budget to be freed up to use and add value elsewhere.	RAD pad preparation is five times more expensive than standard pad preparation. Using existing pad has no cost.
OBK gain	Maximising resource classification conversion volume (objective 1). Minimising P_{60Min} and P_{60W} volumes (objective 2). Maximising drilling metres in deep targets (objective 2).	Geometrical samples support criteria as a metric for resource classification conversion from Inferred to Indicated/Measured (objective 1).
Efficiency	Maximising Mineralisation volume converted per $ spent ($m^3/\$$).	Metric valid only for objective 1.
Field Program duration	Minimising duration for drilling campaign. Maximising time to be freed up to use and add value elsewhere.	RAD hole drilling rate is 0.67 standard hole drilling rate (m/day). RAD pad preparation is five times longer than standard pad.
Ground disturbance	Minimising number of new pads/tracks. Minimising RAD drilling volume	Pads/tracks within existing ground disturbance footprint. RAD pad surface is 2.25 bigger than standard pad surface.

The KPIs were produced by the optimiser for each of the plans assessed, including the manual plan, and are shown in Table 6. Those plans were addressing objective 1 only at this stage. Comparing the baseline plan with minRad43, for an identical drilling volume, the baseline plan would generate less OBK conversion, would be more costly, involve a longer drilling campaign and include more RAD drilling, meaning also more ground disturbance.

TABLE 6

KPIs values for a selection of plans assessed for objective 1, including the baseline plan. The relative values are calculated against the baseline plan.

	Drilling		Cost	OBK		Efficiency
Plan	Drilling volume (m)	% RAD	Relative cost (AUD M$)	Resource volume conversion (%)	Relative volume conversion ($106*M^3$)	Relative efficiency (M^3/$)
Baseline plan	4394	86	-	86.1	-	-
minRad6H-39	4112	55	-20%	87.9	+15%	+44%
minRad34	3616	62	-29%	86.5	+4%	+45%
minRad43	4432	55	-15%	88.0	+16%	+35%

	Duration	Ground disturbance	
Plan	Drilling days	New RAD pads	New standard pads
Baseline plan	47	20	4
minRad6H-39	28	14	3
minRad34	28	13	2
minRad43	31	15	3

Based on this analysis, the minRad6H-39 plan was selected for further refinement to address objective 2. Relatively to the other plans, the cost of this plan was intermediate, but it provides a high resource conversion rate (87.9 per cent) at a lower percentage of RAD drilling (55 per cent) and a lower program duration (28 days), with a limited need for the creation of new pads for the RAD Rig (15).

Following an iterative process, the criteria for objective 2 were added as a second step in the inputs to the optimiser. The resulting plans were designed to address both objectives 1 and 2. The KPIs for the final plan selected are shown in Table 7. This plan will allow a conversion of an additional 15 per cent of the resource volume in comparison to current state (if no additional drilling). It will target 88 per cent of the P_{60Min} blocks, 69 per cent of the P_{60W} blocks and will also include 675 m of drilling within the manually defined, deep target volumes.

TABLE 7

KPIs related to the final selected plan (CondSimP1+9/10) addressing objectives 1 and 2. The relative cost is calculated against the starting point.

Plan	Drilling		Cost	OBK (Objective 1)	Efficiency (Objective 1)
	Drilling volume (m)	% RAD	Relative Cost (AUD M\$)	Resource volume conversion (%)	Absolute efficiency (M³/\$)
Starting point (Current – if no additional drilling)	-	-	-	74.0	-
CondSimP1+9/10	4443	57	+0.7%	89.1	1.8

Plan	Edge definition (Objective 2)			Duration	Ground disturbance	
	% P$_{60Min}$ blocks impacted by drilling	%P$_{60W}$ blocks impacted by drilling	Drilling volume in Deep Targets (m)	Drilling days	New RAD pads	New standard pads
CondSimP1+9/10	88	69	675	32	19	5

Engagement with the Competent Person

An important component of the project has been the continuous engagement of the Competent Person, to assist with the assessment of the plans regarding the parameters for resource classification, other than sample support and neighbourhood criteria. Reviewing the plans, the KPIs and the trade-offs is an important means of ensuring that there will be few surprises once the program moves to execution. Competent Person engagement also helps identify areas of concern that can be added either as targeting criteria or additional constraints.

CONCLUSION

The use of the drilling optimiser allowed a range of drilling optionality and an informed decision to be made based on objective criteria, while addressing the two priority objectives – infill drilling for resource classification upgrade (objective 1) and edge definition drilling to close off the mineralisation (objective 2 – based on the given constraints and inputs.

For each objective, the optimiser required the input of quantitative volumetric and geometric criteria. These were based on estimation, support sample, neighbourhood criteria to address the first objective, and on a mineralisation/waste boundary uncertainty definition, using indicator conditional simulation technique to quantify the second objective.

Using an iterative approach, the value drivers for this specific deposit were identified. It appeared the location of the collars and the angle of the drilling were key criteria in adding value to the drilling. Regarding the location of the collars, a constraint was set to restrict the collaring locations to the existing ground disturbance footprint.

The inclusion of low angled drill holes in the plan increased the value of the drilling considerably; however, it requires the use of a specific reduced angled drilling (RAD) rig, implying additional earthworks (to increase the size of the pads), as well as a longer and a more costly field program.

Once the value drivers were understood, KPIs were defined to assess the drilling plans produced by the optimiser, as well as the manual (baseline) plan produced by the geologist. Based on the KPIs, the baseline plan was performing less efficiently that the plans produced by the optimiser. When directly compared to the current practice baseline plan, the optimiser solution, meeting all the identified constraints decreased costs by a third.

In addition to this cost decrease, the optimiser added OBK by optimising the geometric placement of holes to the target volumes defined, using conditional simulation and manually defined geological targets.

The assessment of the plans, and the selection of the final plan which will be used for the execution of the program, were achieved using a trade-off between addressing the orebody knowledge objectives and the constraints of the drilling program (costs, duration, ground disturbance, earthworks), and based on the inputs (access and statistical constraints as well as geological context). The selection of the final plan was also performed alongside the engagement of the Competent Person.

The use of an Artificial Intelligence based algorithm helped the geology and management team to develop and select a drill program that, given the site's complexity, maximised the expected orebody knowledge and resource conversion while also decreasing time and costs. The assessment demonstrated how emerging tools can be integrated into our current processes quickly and effectively. AI may not be a threat, but rather a means to better understand each project's unique value drivers, and to create programs to specifically maximise the final outcomes from late-stage drilling programs.

ACKNOWLEDGEMENTS

This paper described work undertaken by the authors while employed by Rio Tinto Iron Ore and is presented with the Company's permission.

The authors wished to express their appreciation to the Rio Tinto Members who reviewed the paper: J Dvorak, D Sackers, L Olssen, J O'Connell, A A Jambaa and acknowledge D Vreugdenburg and M Nelson for their technical input and support. Special acknowledgement is made to J Templeton-Knight for the support in initiating the project and the engagement with Objectivity.ca, as well as N Richards and P Savory for their ongoing support leading to the completion of the project as well as their review of the paper. The project has highly benefited from the experience, ideas and advise from K Cole Rae, J Heim, R Gold and J Pocoe whose technical inputs have been extremely valuable. The authors would like to thank as well D Barry and M Woligroski from Geovariances for their technical support. The ongoing and thorough engagement of Andrew Dasys from Objectivity.ca has been a significant contribution in the successful outcome of the project.

REFERENCES

Desassis, N, Renard, D, Beucher, H, Petiteau, S and Freulon, X, 2015. A pairwise likelihood approach for the empirical estimation of the underlying variograms in the plurigaussian models. Available from: <https://arxiv.org/abs/1510.02668>

JORC, 2012. Australasian Code for Reporting of Exploration Results, Mineral Resources and Ore Reserves (The JORC Code) [online]. Available from: <http://www.jorc.org> (The Joint Ore Reserves Committee of The Australasian Institute of Mining and Metallurgy, Australian Institute of Geoscientists and Minerals Council of Australia).

Maleki, M and Emery X, 2014. Joint simulation of Grade and Rock type in stratabound Copper Deposit, *International Association for Mathematical Geosciences*, pp 471–495.

Schrödinger's kittens – lifting the lid on resource drill hole data after mining

J Moore[1], M Grant[2], D Corley[3] and W Randa[4]

1. Group Manager Resources, Oceanagold, South Brisbane Qld 4101.
   Email: jonathan.moore@oceanagold.com
2. Senior Resource Geologist, Oceanagold, Waihi 3610, NZ.
   Email: matthew.grant@oceanagold.com
3. Group Geologist, Oceanagold, South Brisbane Qld 4101. Email: doug.corley@oceanagold.com
4. Principal Geologist, Oceanagold, South Brisbane Qld 4101.
   Email: wesly.randa@oceanagold.com

ABSTRACT

Resource estimates are the corner stone of technical and investment decision-making. Prior to mining, resource estimation uncertainty has the greatest potential to lead to poor investment decisions, despite a significant component of resource estimation uncertainty being unknowable at this critical stage in the mining cycle. This presents a conundrum to the resource geologist in terms of risk evaluation and resource classification.

Ahead of mining, there is normally considerable focus on drill hole spacing analysis to determine the 'optimal' drill hole spacing, taking grade and geological continuity into account; cost-benefit analyses balance improved resource definition against cost; the law of diminishing returns and the exponential cost increases acting together as spacing is reduced. Typically, a lower limit to the drill hole spacing is identified. After completing the resource drilling to an agreed spacing, resource estimation is then undertaken. At this point it is common to attempt to bracket the resource estimation uncertainty. In many cases conditional simulation is used, based on a modelled, but largely assumed, variogram model. It is also assumed that the histogram of the available drill hole sample data is representative of the in-ground mineralisation.

The purpose of this paper is not to diminish the importance of drill hole spacing and simulation studies, but rather to illuminate an important aspect of estimation uncertainty that, although previously recognised, is typically overlooked – the question of how representative the available data set is in characterising the true (but unknown) distribution of mineralisation. To do this, we lift the lid on one of OceanaGold's former operations, the mined-out Globe Progress Mine, by resurrecting a high quality, close-spaced, reverse circulation (RC) grade control data set. The Globe Progress Mine is within the West Coast Region of New Zealand, it was closed and transitioned to rehabilitation in 2016 and is now known as the Reefton Restoration Project.

The exhaustive Globe Progress grade control data was used to repeatedly 'redrill' the deposit by extracting 35 m × 35 m spaced subsets from the original 5 m × 5 m spaced grade control data. Utilising closely spaced grade control data removed the need for assumptions regarding short range continuity, which are necessary with forward-looking analyses that are based upon broader spaced resource drilling. The extraction process used a nearest neighbourhood algorithm, repeatedly moving the origin in 5 mE, or 5 mN increments. Individual resource estimates were then completed for each of the extracted drill hole data sets (49 in total) as well as a grade control estimate based upon the exhaustive data set. Whilst the data unpinning each of the 49 estimates changed, the geological assumptions, variography, and modelling parameters remained constant. This approach was taken to isolate the impact of changing the input data. The 49 estimates were then compared against each other and the grade control estimate.

The mean of all 49 sensitivity estimates was close to that of the grade control estimate in terms of contained gold, tonnes and grade. Whilst it is acknowledged that the grade control estimate itself is subject to some degree of estimation uncertainty, the close match between the average of the 49 estimates and the grade control estimate suggests that the resource estimation methodology is reasonable and appropriate. Whilst this comparison is important, the focus of this study is on the component of estimation uncertainty related to the underlying data, and this is reflected in the spread across the estimates. The spread across the 49 global estimates (highest to lowest) for this particular case study was found to be significant (approximately 20 per cent in grade and metal) and that is attributable solely to the underlying data. This exercise quantifies a component of the estimation uncertainty that is inherent to all drill hole data and is distinct from the uncertainties associated with modelling methodology choices, sample and subsample quality and drill hole spacing-related interpolation uncertainty. Importantly, this uncertainty is unknowable prior to mining; we can only directly compare the resource drill hole data against grade control data after mining has taken place.

For the case study, approximately 65 per cent of the estimates fell within 5 per cent of the grade control estimate for contained-gold, suggesting that for many projects the histogram of drill hole data is unlikely to differ noticeably from that of the in-ground resource. However, about 15 per cent of the estimates in this study differed by more than 7.5 per cent, suggesting that a not-insignificant proportion of estimates will be materially compromised. Whether or not the available drill hole data is representative comes down to the 'luck of the draw' and cannot be known at the time of resource estimation.

Given the challenge of attempting to evaluate forward-looking estimation uncertainty, a component of which that can only be quantified retrospectively, what should we do as resource geologists? What are the implications for risk evaluation, resource classification and reconciliation? Furthermore, without rigorous post-mining data checks, resource geologists may conflate suboptimal modelling with the shortcomings of the underlying data. An example of this problem is discussed in the following section.

INTRODUCTION

This paper seeks to raise awareness of a component of resource estimation uncertainty inherent to drill hole data underpinning all estimates. The uncertainty being evaluated in this paper is not associated with sampling, or subsampling quality or locational uncertainty of the drill holes, nor is it directly related to drill hole spacing. Rather the uncertainty being evaluated is that drill hole data available for estimation might not be globally representative of the resource being estimated, even at drill spacings commonly used to classify Indicated Resources. By its very nature, this uncertainty can only be quantified once the resource has been mined and presents a conundrum to the resource geologist who is attempting to evaluate forward-looking estimation uncertainty, prior to mining, not retrospective uncertainty.

A case study is presented to demonstrate how significant this uncertainty can be and to remind resource geologists of the limitations of forward-looking data-based risk analysis. Data from OceanaGold's former Globe Progress Mine within the West Coast Region of New Zealand was chosen because of the size and quality of the reverse circulation (RC) grade control sample data set.

But before we look at the case study, it is worth sharing an experience that led to the writing of this paper.

In 2014, mining commenced at OceanaGold's Coronation open pit at its Macraes operation in the South Island of New Zealand. Coronation was a low angle, mineralised shear zone that pinched and swelled. The broad geology was not complex, but the erratic gold distribution, particularly at the economic cut-off (0.5 g/t Au), presented considerable challenges.

By late 2016, grade control had defined approximately 30 per cent (33 koz) more gold than estimated by the resource model, despite using resource estimation approaches well established at Macraes. The default assumption was that the poor reconciliation reflected poor modelling (either geological interpretation or estimation). Nonetheless, a comprehensive reconciliation review was undertaken, including data comparisons and reviews of the interpretation and estimation parameters. Nearest neighbour blasthole sample grade versus twin RC sample – grade quantile-quantile (QQ) comparisons revealed no evidence of sample bias. However, when the resource drill hole locations were superimposed over the post-mining contained-gold contours (from grade control model), the cause became clear; the resource drilling had fortuitously missed many of the high-grade shoots (see Figure 1).

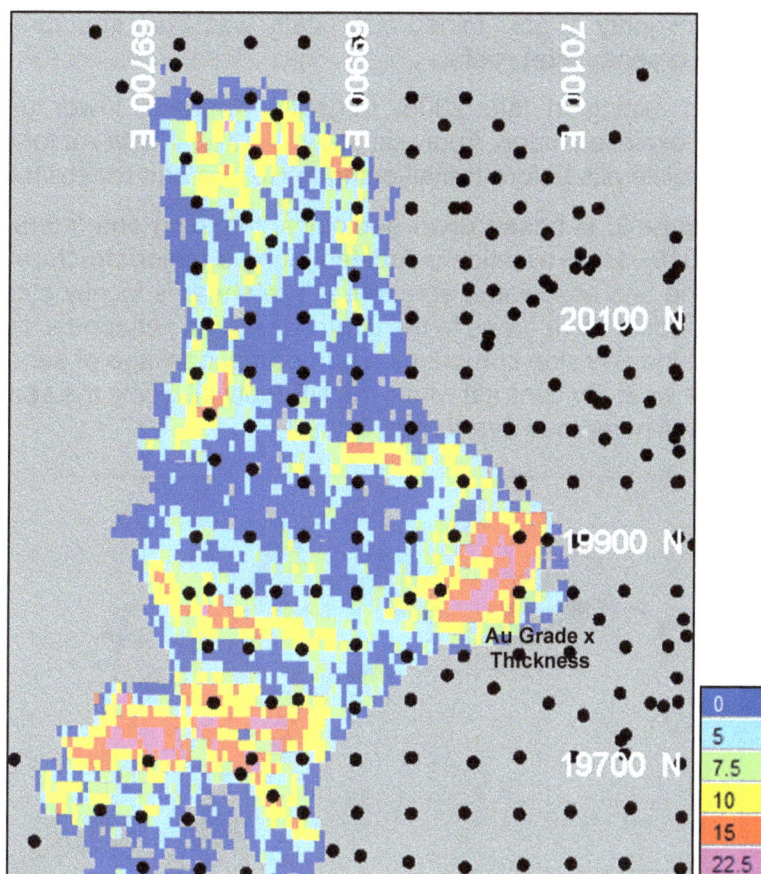

FIG 1 – Coronation Deposit, drill hole locations versus contained-gold contours from grade control.

The story however, does not end here. Before the underlying causes of poor model to mine reconciliation were understood, significant changes had already been made to the resource estimation domaining wireframes. This over-reaction, based on a conflation of causes, resulted in overly optimistic grade estimates for the subsequent Coronation cut-back.

Whilst the drill hole spacing at Coronation was broader than was typical at Macraes, the experience raised concerns as to whether this uncertainty was more widespread in the industry than realised, even for deposits drilled with tighter patterns. A few years later, when time permitted, we decided to commence some case studies.

LEVERAGING OFF PRODUCTION DATA

Each year around the world, vast swaths of grade control data are accumulated by mining companies. After being integrated into ore definition and reconciliation processes, these troves of closely spaced data are often retired to company databases, becoming sunken treasures.

This paper uses grade control data resurrected from OceanaGold's former Globe Progress Mine, now known as the Reefton Restoration Project, after transitioning to rehabilitation in 2016. The Globe Progress data set was chosen because of the high quality and closely spaced RC grade control sample data, as well as the interesting challenges posed by the geological complexity, and the locational uncertainty of historical mining depletion.

GLOBE PROGRESS DEPOSIT

The Globe Progress deposit is situated approximately 7 km south-east of the township of Reefton, within the West Coast Region of New Zealand. Located within the Reefton Goldfield, the deposit is hosted within Silurian-Devonian, lower greenschist facies-metamorphosed sandstones and argillites (Cooper, 1974; Adams, Harper and Laird, 1975).

Gold bearing quartz lodes were discovered on Globe Hill around 1876 and by 1920 an estimated 1.1 Mt at 12 g/t Au for a total of 0.41 Moz gold (gold recovery ~80 per cent) had been mined from 11 underground levels. The majority of gold mined historically was from structurally controlled, high-grade quartz (± stibnite, ± arsenopyrite) reefs.

OceanaGold commenced mining in late 2006, targeting remnant, lower grade mineralisation enveloping the historically exploited lodes. Approximately 11 Mt @ 1.8 g/t Au for 646 koz of gold was mined from Globe and satellite pits before transitioning to closure and rehabilitation in 2016.

The schematic map in Figure 2, is based upon geological mapping and interpretation by Andrew Allibone (Allibone *et al*, 2018). Mineralisation is located along a major structural break, the arcuate Globe-Progress shear zone and associated structures. The shear is locally cataclastic, is concave upwards, dips on average about 45 degrees to the south-west, and juxtaposes two distinct domains. The hanging wall block on the SW side comprises a homoclinal package of sandstone and argillites dipping ~40 degrees to the west. By contrast within the NE footwall block, the sedimentary sequence is tightly folded and includes numerous overturned limbs.

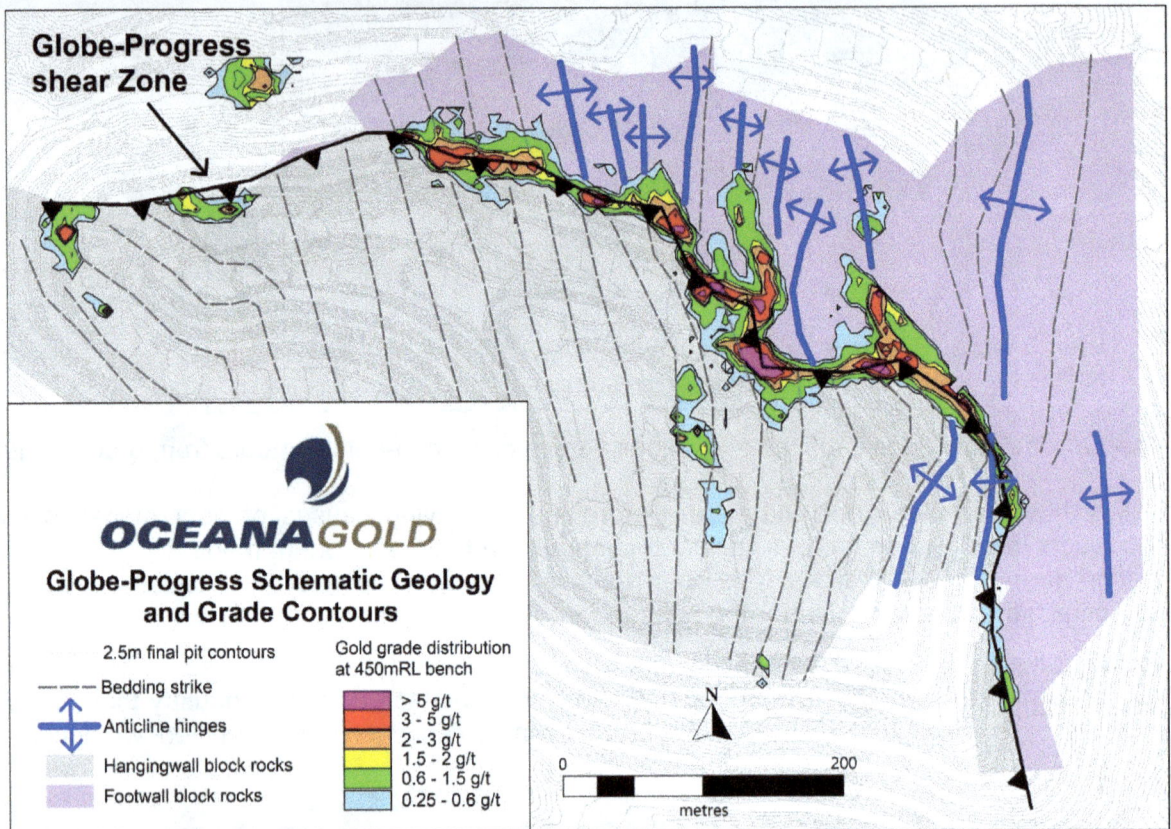

FIG 2 – Plan slice at 435.25 mRL of Grade Contours, and Geology. Simplified from Allibone *et al* (2018).

Grade control data

During open pit mining at Globe Progress, 7.5 m benches were mined on three 2.5 m flitches. Grade control used approximately 9 m long, RC holes, inclined at 60 degrees, on an approximate 5 m × 5 m pattern. Grade control coverage extended beyond the mineralised envelope except locally where mineralisation trended into pit walls/floor. Samples were collected at 1.5 m downhole intervals, and each interval was subsampled via a rotary splitter (and when not available, a three-tier riffle splitter). A nearest neighbour comparison (5 m E × 5 mN × 1 mRL search) between 2847 RC and diamond twin samples, revealed no significant bias. Sample recovery was often lost in historical stopes, and in these cases, no sample was collected.

There were 363 292 samples available for the study, providing a large reference data set of closely spaced (5 m × 5 m) data.

35 m × 35 m drill hole data extraction

A nearest neighbour algorithm was used to extract unique 35 m × 35 m × 1.5 m subsets from the reference 5 m × 5 m × 1.5 m spaced grade control sample data. The resulting 35 m × 35 m × 1.5 m extracted sets were, like the original RC grade control drill holes, inclined at 60 degrees. Extracting to a 35 m × 35 m grid meant that 49 unique subsets could be extracted (ie seven 5 m spacings in the east direction and seven 5 m spacings in the north direction =>7 × 7 = 49). For each successive subset extraction, the origin of the grid was shifted in 5 m east or 5 m north increments (moving point-of-origin) until all 49 possible permutations were extracted.

A 35 m × 35 m spacing was chosen for this study for practical reasons. Drill spacings much greater than 35 m × 35 m were likely to preclude sensible estimates, given the complexity of the mineralisation and locational uncertainty of historical mining depletion. Using spacings significantly less than 35 m × 35 m would have resulted in a smaller number of moving point drill set permutations and would have lessened the statistical rigour of the analysis. For example, a 25 m × 25 m drill hole spacing would have resulted in only 25 estimates (cf 49 at 35 m × 35 m), given the 5 m × 5 m grade control spacing.

Each of the 49 unique, equiprobable, 35 m × 35 m drill hole sets were then used as input data for a separate sensitivity estimate. Note that combining the 49 unique drill hole data subsets reproduces the original exhaustive grade control data sets. This suggests that the average of the 49 resource estimates should match the grade control estimate – discussed in later sections.

Resource estimation

Resource estimation for the Globe Progress deposit faces a number of challenges. The deposit exhibits complex structural controls and gold distribution of mineralisation. In many cases mineralisation locally transgresses lithological boundaries. Many of boundaries remain ambiguous in terms of soft versus hard estimation decisions. Insufficient historical records remain to volumetrically deplete historically mined mineralisation. For this reason, it has been difficult to distinguish between drill hole intercepts that represent *in situ* mineralisation versus remnant pillars. Conversely, it is not possible to accurately determine the 3D extent of mined out voids when intersected by drill holes.

Despite the challenges above, it has been possible to produce resource estimates that reconcile well against grade control data, both spatially and quantitatively. Given the time required to construct 50 independent estimates (ie 49 sensitivity estimates and one grade control estimate), a streamlined methodology was adopted:

- Large panel (20 mE × 20 mN × 2.5 mRL) recoverable estimates were made using multiple indicator kriging in GS3 proprietary software. Fourteen indicator thresholds were defined.

- A composite length of 1.5 m was used. Of the 355 298 composite grades, the highest eight values were 569, 357, 301, 130, 110, 110, 102 and 96.9 g/t Au. A top cap of 100 g/t Au was applied.

- Three broad estimation domains were created. Figure 3 shows the two mineralised estimation domains surrounding grade contours of the mineralisation (East–west Trend Mineralisation – Black Dashes, North-west Trend Mineralisation – White Dashes.

- The domains were estimated as soft boundaries but with domain-specific search parameters, indicator variograms and indicator thresholds/means.

- Reasonable indicator variograms were modelled for the resource estimates from the 35 m × 35 m drill hole data. This confirmed grade continuity at 35 m spacing, an important assumption.

- Only input data and indicator thresholds/means were changed for each successive estimate. All other parameters remained constant.

- The estimated blocks were classified as 1, 2 and 3, corresponding approximately to Measured, Indicated and Inferred confidence categories, but these categories are not externally reportable due to absence of Reasonable Prospects for Eventual Economic Extraction (RPEEE).

- The 49 estimates were reported within the mined pit volume at a 0.5 g/t Au cut-off grade.

- The grade control estimates were based upon ordinary kriged estimates into 5 mE × 5 mN × 2.5 mRL blocks. Ore dig line polygons were manually designed at a 0.5 g/t Au cut-off grade. A 60 cm rind of edge dilution and an additional 2 per cent of ore loss and dilution was applied to approximate extraction effects.

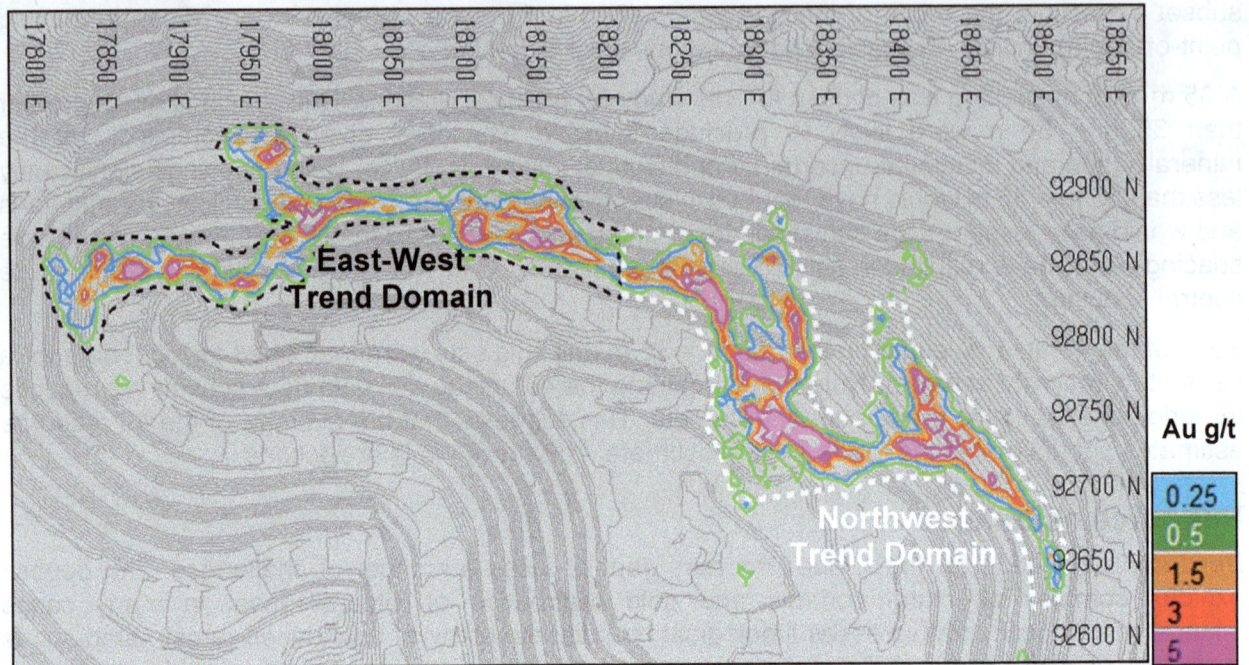

FIG 3 – Plan slice at 435.25 mRL of Grade Contours, Estimation Domains and Mined Pit.

RESULTS

The average of the 49 estimates based on 35 m × 35 m spacing was 10.7 Mt @ 1.63 g/t and 561 koz gold for combined classes 1, 2 and 3 compared to the grade control estimate of 10.4 Mt @ 1.69 g/t and 564 koz gold (NB: there were no reportable resources remain at Globe Progress due to mine closure and restoration). Tonnes, grade and contained-gold are within +3 per cent, -4 per cent and 0 per cent respectively of the grade control estimate. The close comparison suggests that the resource estimation process is reasonable.

Figure 4 compares each resource estimate against the grade control estimate using ratios of estimate/grade control. A ratio >1.0 indicates over-estimation by the resource estimate. Whilst this comparison against the grade control estimate is an important metric, the focus of this study is on the component of estimation uncertainty related to the underlying data, and this is reflected in the spread across the estimates. The scatterplot of ratios shows an approximate spread of 20 per cent (±10 per cent) for contained-gold and 15 per cent (+5 per cent to -10 per cent) for grade. The differences are driven by the input data because only the input data (with associated indicator thresholds/means) vary between estimates. This presents a dilemma. If the available input data is not representative of the resource being estimate (even for reasonable drill hole spacings), prior to mining how can we be sure that our estimate is reasonable? And how can we evaluate the range of possible estimates? The latter requires access to a sample grade histogram that is representative of the resource being estimated.

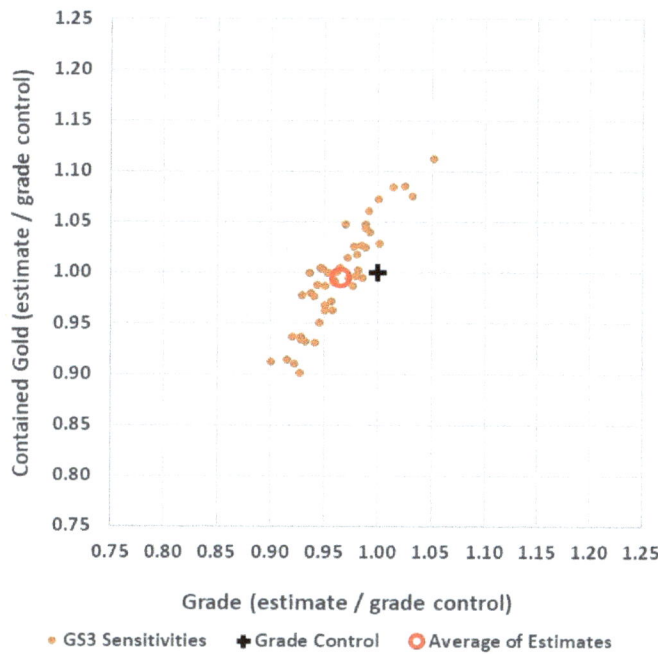

FIG 4 – Scatterplot of comparisons between estimates and grade control for grade and contained-gold.

Figure 5 includes two images. The left-hand image shows two of 49 possible drilling grid origins, points A and B on the entire set of 49 origins (black crosses). Point A and point B origins are only 10 mN apart. The two alternative 35 m × 35 m drill hole patterns propagated from point A and point B are superimposed on the right-hand image as black squares and white circles respectively. These are superimposed over a high-resolution gram-metre contour map for the Globe Progress orebody based upon actual RC grade control sample data (NB: 'gram-metre' equals grade (g/t) multiplied by true length (m)). The view is looking at -60 degrees to the NE, which is the downhole direction. Each pixel (spaced 5 m × 5 m) summarises a string of grade control drill holes, stitched end on end, at -60 degrees through the entire thickness of the orebody. The drill hole gram-metres were accumulated over the full length of the string (thickness of the deposit).

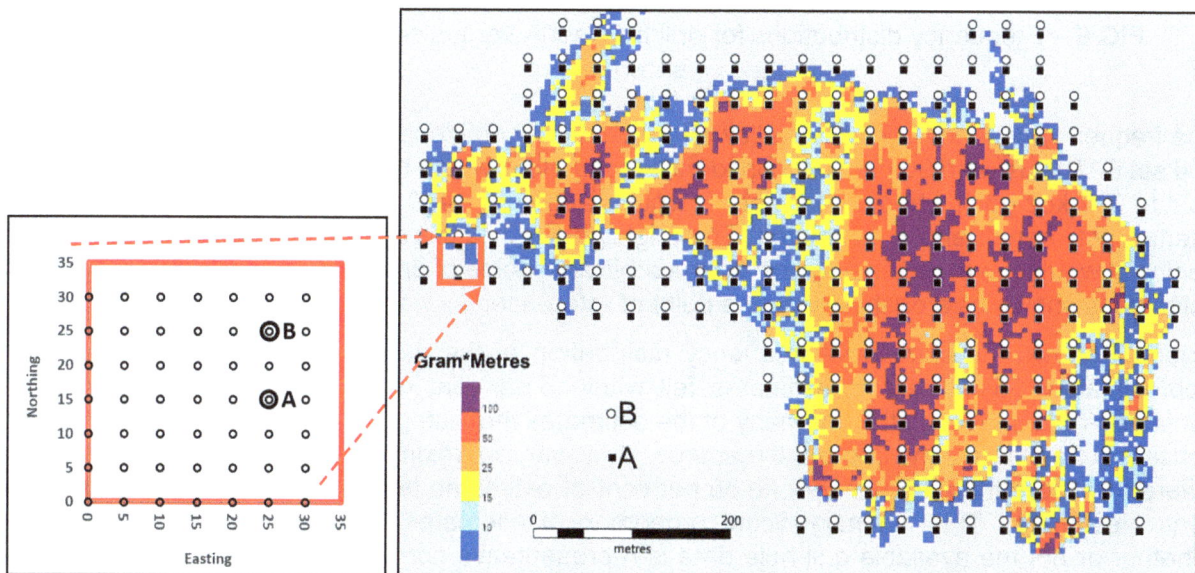

FIG 5 – The right-hand image shows two of 49 possible drilling grids, sets A and B, over RC Grade Control Gram-metres. The grid of possible origin points in the left-hand image shows the relative location each set.

The drill hole patterns both appear to provide similar coverage of the mineralisation; both patterns share 35 m × 35 m spacing and are offset by only 10 m. However, the mean grades of the two sets are markedly different. Above a 0.4 g/t Au cut-off, Set A has 2093 samples @ 2.249 g/t Au versus set B with 1978 samples @ 2.178 g/t Au. Set A has a higher mean grade and a larger number of samples above 0.4 g/t Au. Resource estimates based upon each set, yield significantly different outcomes.

The estimate based on the drilling propagated off point A is 11.0 Mt @ 1.78 g/t for 627 koz gold

The estimate based on the drilling propagated off point B is 10.1 Mt @ 1.57 g/t for 509 koz gold.

This large difference in the estimates is entirely related to the data available for resource estimation, which reflects the arbitrary choice of the drilling grid origin.

Not only are the estimates materially different, but the expected ranges of potential estimates differ, depending on whether set A or B is used. One hundred realisations were completed for each of set A and set B using Turning Bands Conditional Simulation. These realisations were then used to produce frequency distributions for the contained gold estimates (see Figure 6).

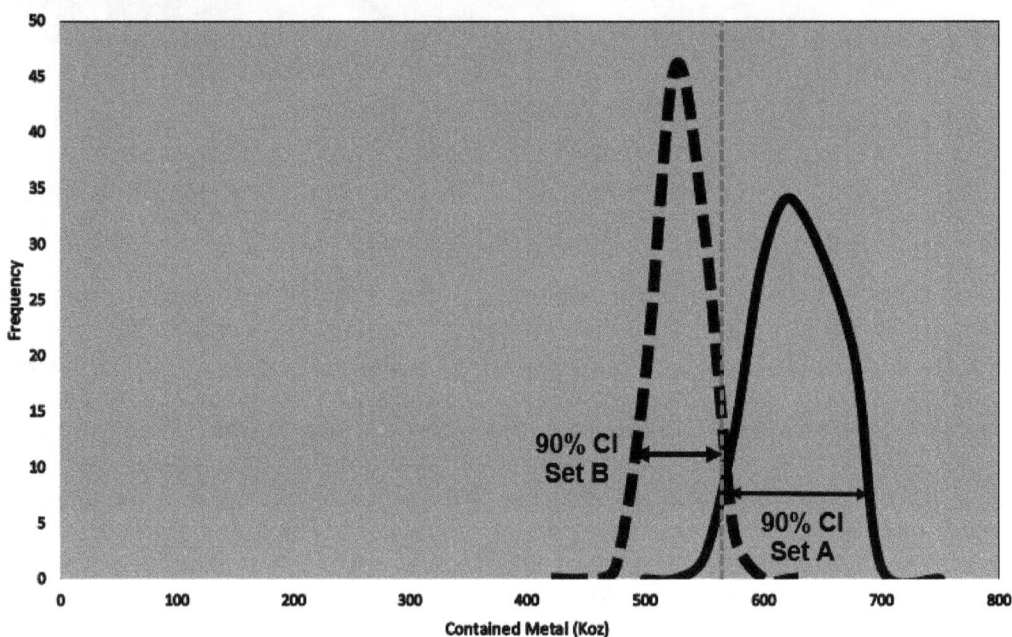

FIG 6 – Frequency distributions for drill hole set A versus set B. Grade control estimate shown in red.

The frequency distributions above, allowed 90 per cent confidence intervals to be defined for set A and set B. That is, with 90 per cent confidence, estimate A would be expected to range from 580 koz gold to 674 koz gold, whereas estimate B would range from 479 koz gold to 539 koz gold, with no overlap between A and B estimates. This highlights very different perceptions that the resource geologist would have, depending only on the arbitrary choice of drilling grid origin. The grade control estimate is shown as a dashed line as a point of reference.

Figure 7 shows the cumulative frequency distribution of the 49 estimation errors for this study. Approximately 65 per cent of estimates fell within 5 per cent of the grade control estimate for contained-gold, reflecting that for many of the estimates the histogram of drill hole data did not differ noticeably from that of the in-ground resource. However, approximately 15 per cent of the estimates differed by more than 7.5 per cent (ie 85 per cent of estimates fell within 7.5 per cent of the grade control estimate). This not-insignificant proportion of estimates were significantly compromised. Whether or not the available drill hole data is representative comes down to the 'luck of the draw' and cannot be known at the time of estimation.

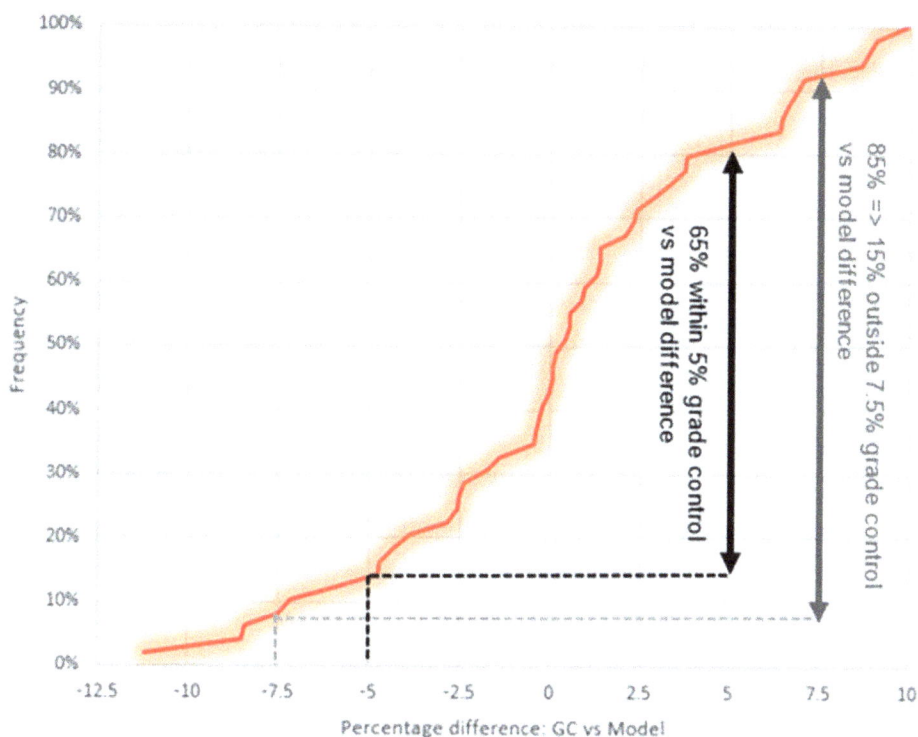

FIG 7 – Cumulative probability plot of estimate/grade control gold ratios.

If further case studies generalise these findings, the implication would be that many projects will be developed and mined without noticeable impact. Some proportion of them however, despite best efforts, may not fare so well.

CONCLUSIONS

The objectives of this paper are for OceanaGold to share its learnings from this case study and to raise awareness of the inherent uncertainty related to representativity of the available drill data. With estimates completed prior to mining, there is no means of determining whether the available data are representative of the resource being estimated or whether the data are an outlier set. It is however crucial to be aware that your data set for estimation has the potential to be an outlier. As mining progresses, it is important to compare the declustered block means of resource against grade control sample data. This step is essential in understanding whether the drivers of reconciliation disparities are data-related or modelling-related and to avoid conflating the two. Future work will include scenarios that quantify the extent to which reduced drill hole spacing mitigates uncertainty, although there are already studies into the relationship between drill hole spacing and estimation uncertainty (for example, Collier *et al*, 2022).

In the view of the authors, the approach undertaken for this study was a rigorous one. The RC data used provides an exhaustive and high-quality data set. Being closely spaced, no short-range continuity assumptions were required (*cf* forward-looking simulation studies based upon broader spaced resource drill hole data). The reconciliations suggest that resource estimation methodology was appropriate, and that by changing only the input data (and associated indicator thresholds/means), the spread across the 49 estimates can only be attributed to the input data. This paper however, includes only one case study.

Additional examples are required. OceanaGold hope to complete further case studies using operational data from other mineralisation styles and including other drill hole spacings. Additional case studies might provide useful insights and semi-quantitative rules-of-thumb for forward indicators of uncertainty. For example, deposit type, spatial partitioning of gold (the extent to which gold content is disproportionately distributed at resource drilling and grade control scales), structural controls, mineralogy, coefficient of variation, number of samples etc. Other companies with operating mines are encouraged to contribute.

Ultimately, as we gain a broader and deeper understanding of this uncertainty this may become a consideration during resource classification. But for now, a footnote might suffice, acknowledging that there is some risk that the resource estimate may be based on an outlier drill set.

Many companies have databases archiving large quantities of accumulated production data. These invaluable troves have the potential to provide answers to problems that can't be provided with broader spaced resource drill hole data.

ACKNOWLEDGEMENTS

The authors would like to thank Mike Stewart for his thoughts and comments. Thank you to Sean Doyle for assisting with the root cause analysis at Coronation in 2016. Thanks to Andrew Allibone for supplying the geological map for Globe Progress.

Many of the approaches underpinning this paper were either directly or indirectly gleaned from Neil Schofield over the years. Thanks also to Neil for feedback on the draft paper.

Finally, the authors would like to thank the mining industry for under-drilling our deposits. Without them, there would be less uncertainty.

REFERENCES

Adams, C J D, Harper, C T and Laird, M G, 1975. K-Ar ages of low-grade metasediments of the Greenland and Waiuta Groups in Westland and Buller, New Zealand, *New Zealand Journal of Geology and Geophysics*, 18:39–48.

Allibone, A, Blakemore, H, Gane, J, Moore, J, MacKenzie, D and Craw, D, 2018. Contrasting Structural Styles of Orogenic Gold Deposits, Reefton Goldfield, New Zealand, *Society of Economic Geologists, Inc. Economic Geology*, 113(7):1479–1497.

Collier, P, Rees, I, Pocoe, J and Espinoza, D, 2022. Mineral project risk and orebody knowledge – quantifying the value of drilling using decoupled net present value analysis, *International Mining Geology Conference 2022* (The Australasian Institute of Mining and Metallurgy: Melbourne).

Cooper, R A, 1974. Age of the Greenland and Waiuta Groups, South Island, New Zealand, *New Zealand Journal of Geology and Geophysics*, 17:955–962.

Classification

Assembling the geological complexity in mineral resource classification

B C Afonseca[1]

1. MAusIMM, Resource Geologist, Glencore, Brisbane Qld 4000.
   Email: brunocafonseca@yahoo.com.br

ABSTRACT

Mining companies must clearly specify the criteria used for mineral resource classification in their public reports. The classification aims to categorise the resources based on the level of uncertainty associated with modelling and estimates. However, reporting codes lack standardised rules for categorising mineral resources, making this classification challenging. Decisions should be based on the geological, technical and operational factors unique to each deposit. This lack of objective guidance leads to subjectivity in uncertainty categorisation, especially when evaluating the risk in geological modelling. Greater geological complexity increases the risk of misinterpretation. To address this issue, this study proposes modifications of conventional geometrical, geostatistical and mathematical techniques to enable the objective assessment of uncertainty in geological models. The anisotropic continuity approach is based on functions such as variograms to quantitatively analyse local continuities. Calculating the indicator variogram in different geological domains can determine a continuity's magnitude. Complexity is inversely proportional to the anisotropic range. The second technique involves morphological operations on image scanning. Multiple erosion iterations are performed on a grid until the entire domain is eroded. This process provides contrasting results for simple and complex images and assesses local uncertainties by counting the number of iterations required to erode a grid node. The third approach is based on the signed distance function, which is widely used to assess geological model uncertainty. Multiple realisations of the distance function are simulated to produce equiprobable boundaries for geological contacts. The dispersion of realisations measures the risk of misinterpretation arising from geometrical complexity. The angular gradient approach maps sudden anisotropy variations by converting anisotropy fields into angular gradients. The purpose of transforming angular measures into numerical intervals is to create a spatial variable that accounts for local variations in anisotropy. Highly deformed regions generate greater variability in gradient estimates. This study applies these techniques to both a hypothetical case and a real structurally complex deposit. The results show that resource classification will benefit from these techniques, which can be easily implemented and automated in most geostatistical packages, supporting risk assessment in mineral resource classification.

INTRODUCTION

In resource evaluation, predictions of material tonnes and grade inevitably contain errors. The differences between predicted estimates and the actual recovered material arise from errors that accumulate during different stages in the mining chain. Harmful errors mostly occur during three stages of the mining process: (i) sample collection, (ii) geological interpretation and (iii) grade estimation.

Sources of uncertainty

The first source of uncertainty is that related to data gathered during sample collection. This is the most critical element in the entire mining chain and may affect decisions based on assay values obtained from sampled material. Therefore, logging inconsistencies, poor topographic surveys, biased chemical analysis and inappropriate sampling protocols will lead to misinterpretation and estimation flaws.

The second source of uncertainty relates to the decision of stationarity and geological interpretation. Stationary domains, in which observations share similar geological and statistical properties, represent a physical constraint for resource estimation. Hartman (1992) defines these domains as consolidating all available geological knowledge. There are various ways to produce geological shapes, from traditional sectional wireframing to implicit algorithms based on radial basis function interpolation (Cowan *et al*, 2003). Irrespective of the chosen technique, the model will contain intrinsic errors; that is, there will inevitably be some uncertainty in the interpretation, especially for unsampled locations or complex geological shapes.

The third source of uncertainty in resource predictions relates to the grade estimation process. This is possibly the most well-documented topic in the entire resource evaluation framework. Novel methodologies and technologies for improving grade estimation models have long been published by both the scientific community and mining companies in academic papers and technical reports.

Classification metrics

International organisations such as the Australasian Institute of Mining and Metallurgy (JORC, 2012) and the Canadian Institute of Mining, Metallurgy and Petroleum (CIM) have established strict guidelines for guaranteeing technical competency and transparency in mineral resource reporting. An essential criterion for technical reports is to communicate all risks and how they may affect the accuracy of and confidence in disclosed numbers. Thus, the uncertainties related to each evaluation step must be comprehensively outlined to justify the risk categorisation. However, reporting codes do not provide objective criteria for the classification of resources. Therefore, rather than being prescriptive, risk categorisation relies on the conditions specific to each mineralisation system and varies from one deposit to another.

The lack of clear generic rules for risk classification makes the competent person (CP) crucial in resource reporting. The CP must have demonstrated relevant experience in the mineralisation style being evaluated and is responsible for the declared decisions and results. CPs of companies have applied a variety of classification approaches over the years. Many papers on this topic have been published, presenting geometrical, statistical and mathematical solutions to assess grade uncertainty (Arik, 1999; Souza, 2007; Afonseca and Miguel-Silva, 2022). A traditional means of classifying mineral resources is to measure the quality of kriging estimates by assessing parameters such as the slope of regression, kriging efficiency or kriging variance (Matheron, 1963). A more sophisticated approach to assessing uncertainty is to create a set of equiprobable realisations. Simulation techniques are used to generate local distribution functions in which dispersion is a direct measure of uncertainty. Some companies use the 90/15 rule to bring a degree of objectivity to the classification criteria. In this approach, based on a 90 per cent confidence level, materials for which the error is below 15 per cent for a production scale equivalent to three months are classified as 'measured', while regions with an error of less than 15 per cent at an annual productional scale are classified as 'indicated' (Parker and Dohm, 2014). The error is calculated as follows:

$$\varepsilon = \frac{\left[\frac{m - p_5}{m}\right] + \left[\frac{p_{95} - m}{m}\right]}{2}$$

where p_5 and p_{95} are the 5th and 95th quantiles, respectively, and m is the mean of the distribution drawn from the realisations. In this formulation, the error (ε) is dependent on the mean, while the 5th and 95th quantiles are not. Therefore, in extreme scenarios (eg strongly skewed distributions) the simulated mean may lie outside of the confidence interval, resulting in inconsistent error values. One limitation of this simulation-based methods is that it accounts exclusively for uncertainty in the grade distribution (ie it fails to provide an objective measure of geometrical uncertainty). In this paper, geometrical uncertainty refers to how accurately a three-dimensional (3D) representation of a geological domain aligns with its actual *in situ* volume and shape.

Confidence in geological interpretation

In different geological systems, the mineralisation style may have strong structural controls, among other factors. Depending on the existence, number and severity of deformational events, the mineralised orebody can present a complex morphology, making modelling challenging. In such cases, the primary source of uncertainty may be the ore shape and its contacts rather than grade distribution.

Nevertheless, uncertainty in geological interpretation remains an overlooked and subjective aspect of resource classification. Generally, mining companies have extensive guidelines dedicated to monitoring the performance of estimates and data quality. However, these guidelines are not equally applied to confidence in geological interpretation. Therefore, this study presents quantitative metrics to assess the risks embedded in 3D representations of complex geological shapes. The following section reviews and suggests adaptations to geometrical, geostatistical and mathematical techniques to achieve objective metrics for risk assessment. Finally, the proposed workflows are demonstrated in a real structurally complex deposit.

RISK METRICS FOR GEOLOGICAL COMPLEXITY

Mineral deposits are created by a combination of physical, chemical and geological phenomena. The more complex these natural events, the more challenging it is to build an accurate representation of the mineralisation geometry. For example, Cenozoic bauxite and alluvial deposits are essentially placed under tabular horizons related to their weathering profiles. In such cases, the risk involved in representing ore continuity is related to understanding whether the ore contact is sharp or transitional. In contrast, modelling mineral deposits that have been structurally affected by multiple deformational events may be highly confusing and laborious, even for experienced modellers. Srivastava (2005) argues that failing to appropriately address model uncertainty may result in a naive quantification of uncertainty in grade estimates. Therefore, morphology complexity is a risk factor that should be considered when classifying mineral resources.

This section presents several objective metrics to assess uncertainties arising from the geometrical complexity of mineral domains. To this end, a hypothetical two-dimensional (2D) scenario is created to demonstrate how the approaches perform in different structural domains and anisotropy trends.

Anisotropic continuity

Geostatistical methods use functions such as variograms to quantify the continuity of geological properties. For example, erratic or nuggety ores, such as those found in precious metal deposits, are typically more discontinuous compared with well-behaved stratiform coal ores. Thus, the experimental variogram calculated for each deposit will provide information on their spatial variability differences. The same concept may be used for geometrical continuity. In this sense, geological complexity is expected to be negatively related to the experimental variogram range.

As an alternative to traditional deterministic modelling, indicator-assisted domaining can be used to represent mineralisation characterised by different spatial trends (Abzalov and Humphreys, 2002). In this approach, an indicator probability map is estimated, and the practitioner may decide whether the location is inside or outside a given domain based on defined pseudo-probability thresholds. Similarly, local geometrical complexity can be expressed by the continuity of indicator variograms calculated for each subdomain. Given the structural domain K, the location u is coded to be inside or outside K according to the indicator formalism:

$$I_K(\mathbf{u}) = \begin{cases} 1, \text{if u} \in K \\ 0, \text{otherwise} \end{cases}$$

The indicator variogram is used to determine the spatial continuity of I related to domain K:

$$\gamma_k(\boldsymbol{h}) = \frac{1}{2}Var\{I_k(\boldsymbol{u}) - I_k(\boldsymbol{u} + \boldsymbol{h})\}$$

To illustrate the procedure, a hypothetical 2D example was created to emulate strain partitioning. Figure 1 shows a complex structural condition in which deformation evolves from a low – to a high-strain stage (where the axial surface becomes parallel to the shear plane). These types of structurally controlled domains often require the construction of anisotropy fields. This step is necessary because mineralisation continuity varies locally depending on the structural context. Thus, variograms are calculated for each subdomain and its respective local orientation angles (see Figure 1b). Applications of locally varying anisotropy in the presence of deformed ores have been well documented by Boisvert, Manchuk and Deutsch (2009), Martin *et al* (2019) and Caixeta (2020).

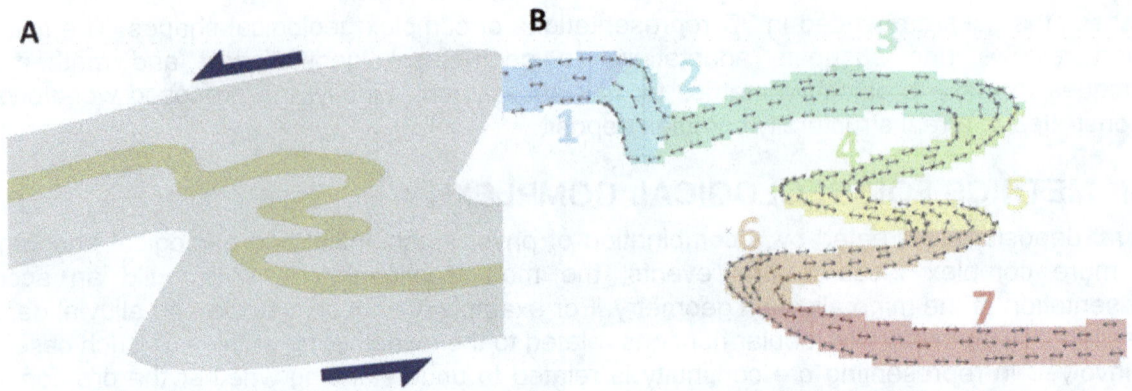

FIG 1 – Hypothetical shear zone with (a) an ore vein (yellow) at different deformation degrees and (b) respective anisotropy field. The colours represent the structural subdomains 1 to 7.

The continuity of each subdomain measured by the variograms (see Figure 2) can be used as an objective criterion to rank the risk of misinterpretation because it provides direct information on geological complexity. As observed, the range of anisotropy varies, with the locations having the tightest folds being those with shorter anisotropy ranges. In contrast, undeformed flat regions such as those at the long limbs (Domains 1, 3 and 7 in Figure 1) have increased continuity values. Variations in anisotropy magnitudes from one domain to another can be expressed as continuity ratios. For classification purposes, the risks arising from geological complexity can be matched to local ratios ranging from 0 to 1, with 1 being the highest level of geometrical confidence. When applying the ratios approach, it is important to be aware that these measures are relative (ie the approach disregards absolute anisotropy ranges).

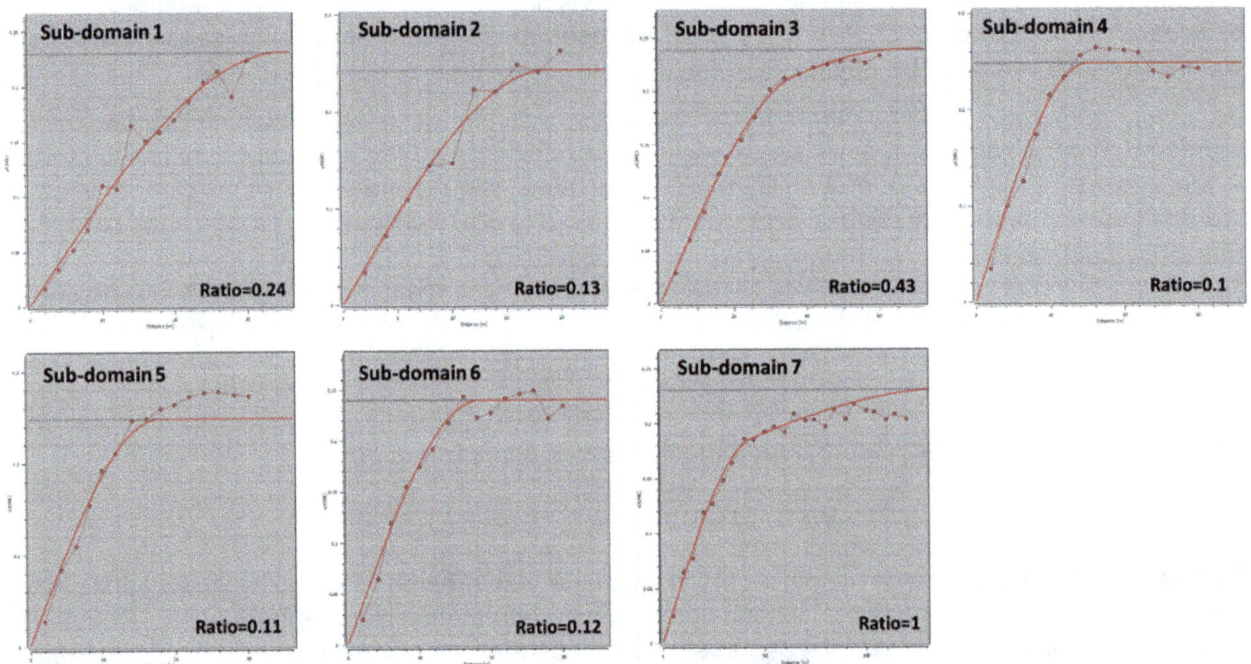

FIG 2 – Local variograms for each structural domain.

In the presented example, continuities are defined for seven structural domains. However, the analysis may be as detailed as necessary. For example, to produce a dense grid of local anisotropies, limb subdivisions, hinges and small-scale variations in continuity may be considered. However, one drawback of this workflow is the requirement for local anisotropy fields. Building a reliable vector field that appropriately represents geological conditions may be time consuming and requires a modeller with previous experience in geological control of mineralisation. This may not be the case in the early exploration stage of mineral deposits. An oversimplified interpretation of local continuities will result in understated geometrical risks.

Morphological operations

Morphological operations (MOs) are a set of techniques used for image processing. Initially developed by Georges Matheron and Jean Serra (Serra, 1982), MOs involve the application of a structuring element to scan a binary input image, creating an output based on the correspondence between the pixels in the input image and those in the structuring element (see Figure 3).

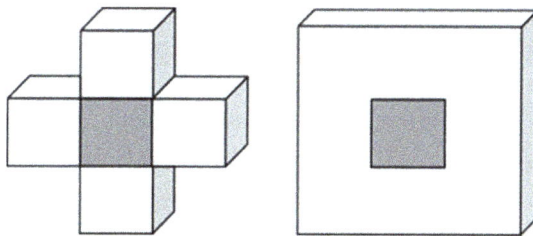

FIG 3 – Examples of isotropic structuring elements. The morphological operation considers the centre pixel (grey) as the origin during image scanning.

Traditional applications of MOs include morphological dilation and erosion. In the dilation process, an image is scanned, and voids beneath the structuring element are filled if their origin matches an equally filled cell in the input image. In contrast, morphological erosion removes filled cells from the input image if the origin of the structuring element matches a void cell.

In the mining sector, MOs have been employed to restore missed continuity in geological models. Abzalov and Humphreys (2002) and Afonseca and Costa (2021) used the dilation algorithm to fill gaps in indicator probability models. In contrast, geological risk can be assessed by repeatedly applying the erosion transformation to an input grid with increasing dimensions of the structuring element (see Figure 4). The iteration stops when the entire grid is eroded. Structuring elements commonly assume simple geometrical shapes such as crosses or squares; thus, MOs will produce contrasting results when applied to simple or complex images.

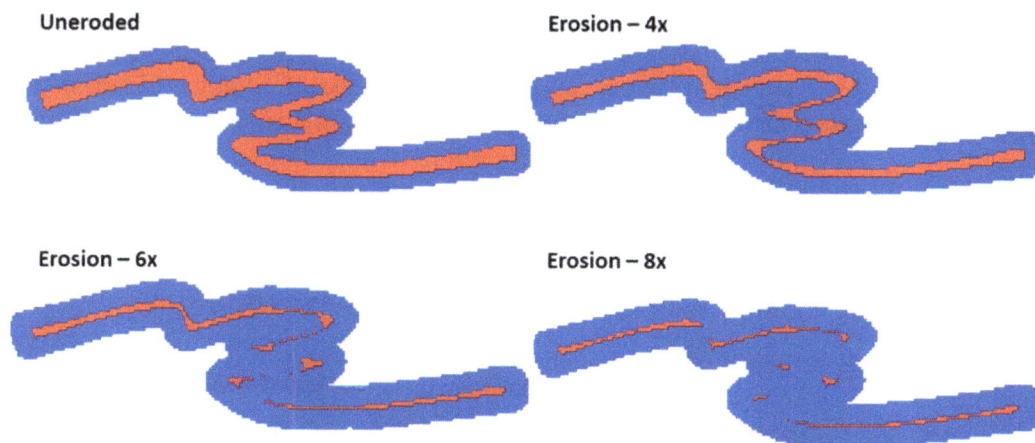

FIG 4 – Results of the erosion transformation as the dimensions of the structuring element increase.

After running morphological erosion, each cell of the output image will contain binary information about whether the grain was transformed into a pore. After multiple MO scenarios, it is possible to identify local frequency functions that account for the probability of a node being eroded. The number of erosion scenarios must be sufficient to produce meaningful local histograms. For example, a large structuring element may result in complete image erosion in a single pass.

The evaluation of results can be done by creating probability maps. Maps such as that shown in Figure 5 can be easily produced from the multiple erosion scenarios. This is achieved by visiting each cell and determining the local erosion frequencies. The local probability of being eroded is calculated by counting the number of eroded scenarios and dividing by the total number of iterations. Analysing MO maps is comparable to analysing those produced by probabilistic geostatistical techniques such as multiple indicator kriging or simulation algorithms. However, unlike multiple indicator kriging, MO probability maps provide an objective measure of geometrical complexity rather than a distribution of an interpolated grade variable. This information may be useful for models with sharp anisotropy variations, such as those found in highly deformed orebodies. The probability map (see Figure 5) generated in this study confirmed the expectation of geological complexity: the most continuous and/or undeformed regions were the last that were eroded. Using the MO approach, these areas were assigned a lower geometrical risk score (blue and green areas in Figure 5). In contrast, the edges and tightly folded regions of the model are generally the most complex parts (yellow and red areas) to be represented in geological designs. An MO probability map could capture this risk.

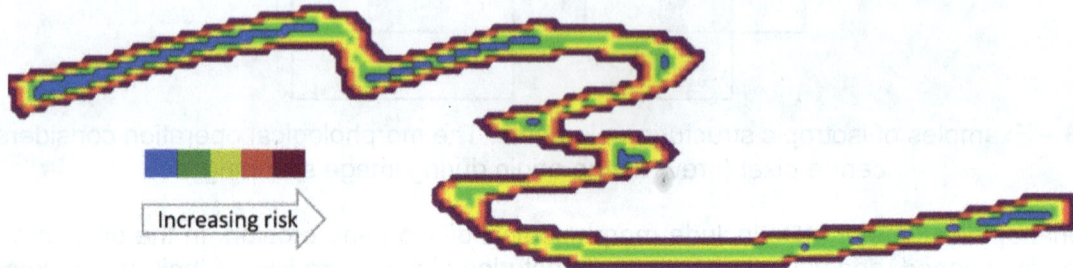

Increasing risk

FIG 5 – Geometrical complexity map designed by the morphological operations probability approach.

When applying MOs, one critical issue is the anisotropy of the structuring element used in the transformations. Particular attention should be paid to rotation of the element and its dimension in each direction. Applying simplified isotropic geometries may produce artefacts and worthless erosion scenarios. Keeping the element features consistent with the overall anisotropy (ie continuity ratios and rotation angles) is recommended.

The signed distance function

The complexity of geological shapes can also be quantified by simulating the position of the boundary between two contacting domains. This approach requires calculating the distance function, defined as the Euclidean distance, between a point and the nearest point located at the domain boundary. The interpolation of the signed distance function is the mathematical basis of some implicit modelling techniques. Moreover, many authors have adapted this method to account for uncertainty in geological contacts (Cáceres *et al*, 2011; Wilde and Deutsch, 2011; Silva and Deutsch, 2012; Amarante, Rolo and Costa, 2019).

The first step in the calculation of the distance function is to code all the nodes as either outside or inside the domain according to the following:

$$i(\mathbf{u}) = \begin{cases} 1, \text{if } u \in K \\ 0, \text{otherwise} \end{cases}$$

where u is a given point in the space, and K is the domain of interest. Considering one location vector from u_a to u_k the Euclidean distance between these two locations is calculated and positively signed if the point is outside the domain; otherwise, the distance function is negatively signed (see Figure 6):

$$i(\mathbf{u}) = \begin{cases} + (u_a - u_k) \text{ if } i(\mathbf{u}) = 0 \\ - (u_a - u_k) \text{ if } i(\mathbf{u}) = 1 \end{cases}$$

where u_k is the closest boundary point from the location u_a. The function module decreases as a location approaches the domain boundary and increases as a location moves away from the boundary. A distance function of 0 represents a location exactly at the interpreted boundary. The presence of anisotropy must be considered when calculating distance. This can be addressed by applying local rotation angles and anisotropy ratios to the distance vector. Wilde and Deutsch (2011) warn that the signed distance function is designed for binary systems (ie representing locations inside or outside the domain of interest). Thus, it does not permit the modelling of many interlaced domains, although multiple domains could be hierarchically taken in pairs.

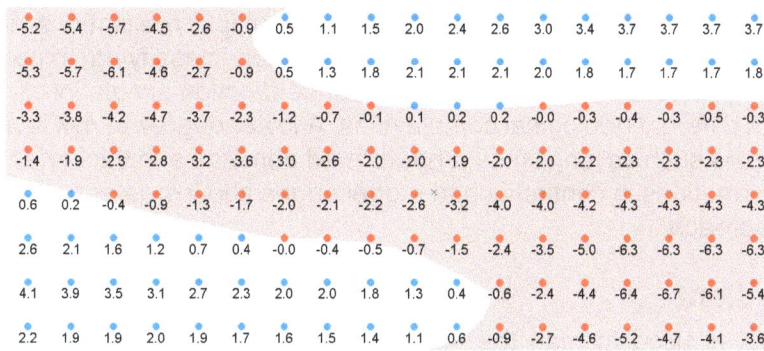

FIG 6 – Calculated distance function (m). Positive distance functions (blue dots) are located outside the domain of interest, while negative values (red dots) are inside the domain of interest (pink shadow area). Distance functions with a value of 0 represent the boundary location.

Once the distance functions of the entire grid have been calculated, the model can be simulated to create multiple scenarios of the boundary location. The sequential Gaussian simulation workflow produced 100 realisations of the distance function (see Figure 7). Therefore, unique equiprobable shapes can be inferred from each sequential Gaussian simulation realisation. In this situation, a small positive nugget effect and Gaussian variograms are recommended to maintain high short-range continuity and mathematical stability (Wilde and Deutsch, 2011). Unlike geological variables, the distance function is not a direct measure of natural phenomena. Thus, the modelling of its spatial continuity may be slightly rambling and less intuitive. Hence, the consistency of the simulated boundaries must be checked with respect to the geological expectations, data observations and expert judgements.

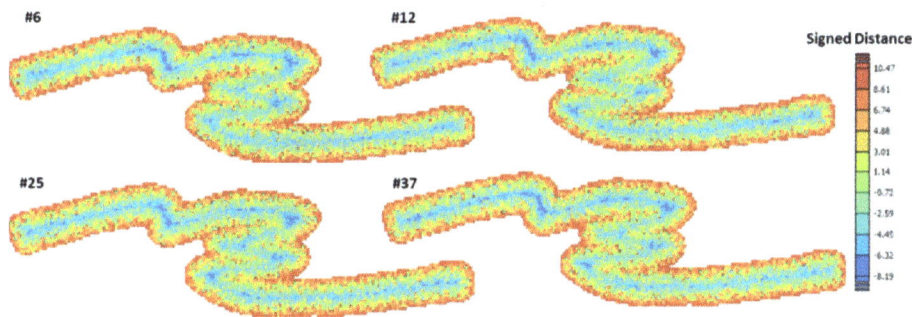

FIG 7 – Four sequential Gaussian simulation realisations of the distance function.

From multiple realisations, distribution functions can be drawn at each grid node. The local distance function histograms provide information on the node's probability of being inside, outside or at the domain's boundary. For instance, if a location shows a high frequency of positive simulated values, it is unlikely that it belongs to the domain of interest. Consequently, local uncertainty regarding model geometry is assessed by examining the dispersion of histograms. The greater the variance, the more considerable the uncertainty about whether the point is inside or outside the domain (see Figure 8).

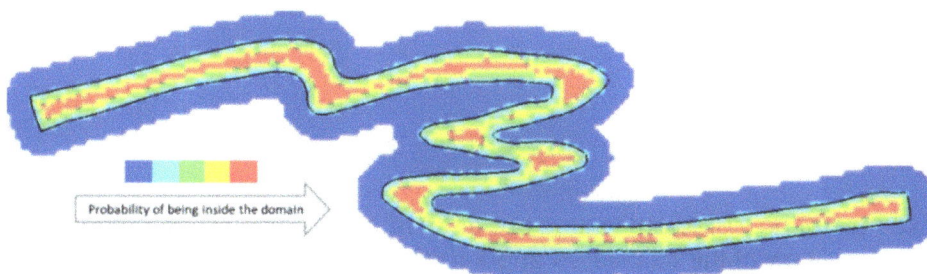

FIG 8 – Probability map of being inside the domain.

The outcome of the distance function simulation is comparable to that of conventional grade simulations used in uncertainty quantification; therefore, risk maps can be produced for the purpose of visualisation. Moreover, statistical summaries of the multiple realisations enable the objective analysis of many locations to manage uncertainties when classifying mineral resources. This hypothetical example provides evidence that the distance function simulation approach is useful for identifying areas with different geological complexities. A lower degree of risk is assigned to the most continuous subdomains such as long limbs, undeformed domains and thickened hinges. In contrast, regions with abrupt variations in continuity and model edges show greater geometrical risk using the distance function approach.

The angular gradient

The angular gradient is used to map the intensity of anisotropy variations in a given geological property. Generally, strongly deformed domains or mineral systems formed under complex geological conditions may have significant short-scale variations in continuity. In contrast, continuous, undeformed or gently folded domains will better preserve global anisotropy. The method was designed to regionalise these variations to support risk assessment.

The initial step involves identifying the anisotropy axis that suffers from the greatest rotation compared with the original undeformed state. Depending on the deformation mechanism, this can be especially challenging. The inference of 3D rotation is possible by considering a full 3D rotation matrix such as the unfolding functions (Deutsch, 2005) provided in mining packages. However, using a unique axis, the one with the most outstanding contribution to the anisotropy changes will improve the ease and speed of the method without reducing its effectiveness.

With the direction defined, the local anisotropy vectors must be converted into angular gradients. The purpose of transforming angular measures into numerical intervals is to create a spatial-related variable that accounts for local variations in anisotropy. Figure 9a presents the criteria for converting azimuth readings into angular gradients. Depending on its azimuth, a gradient value ranging from 0 to 1 will be assigned to each vector. Zero values are given to the regions understood as 'undeformed' (ie those for which the anisotropy is equivalent to the initial state). In this example, ground zero would be in the east–west direction. As the anisotropy trend becomes locally rotated compared with the east–west direction, the gradient increases to 1, which represents the most discordant anisotropic trend. Figure 9b presents a map of the calculated gradient ranges.

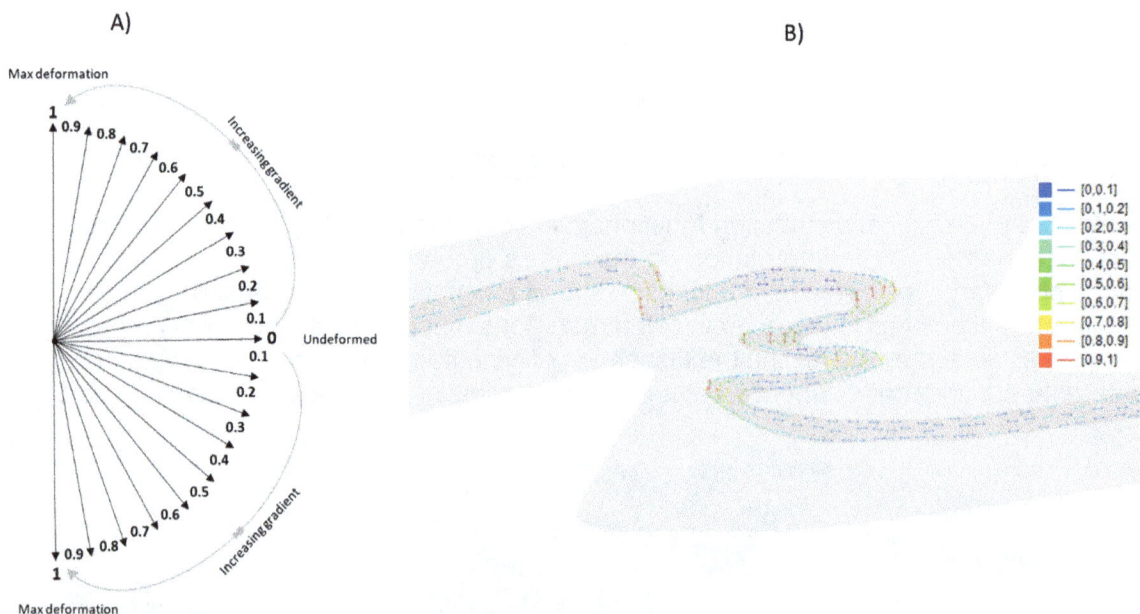

FIG 9 – (a) Example of how gradient values are defined based on local anisotropy trends; (b) plan view of the calculated angular gradient at each location.

From this conversion, we can see that the gradient intervals show a spatial pattern and can be geostatistically treated to produce numerical models that account for geological complexity. For this purpose, this work applies indicator kriging estimates (Journel, 1982, 1988). The 10 gradient intervals are converted into binary variables following the indicator formalism and then kriged with the variogram model of the median gradient. A complete distribution function is drawn from the indicator estimates in each grid node. The dispersion of the local histograms is a proxy of geological complexity (ie areas with high variance values represent an increased probability of variation in anisotropy trends). The results may be visually or numerically analysed. Dispersion measurements and maps like that presented in Figure 10 can be used for risk analysis to assist resource categorisation.

FIG 10 – Risk map produced by applying the angular gradient approach.

Other probabilistic approaches such as simulation-based methods may also be employed as an alternative to indicator kriging. However, one concern is the transformations applied to the gradient distribution. For example, during normal score transformation, despiking techniques are used as a means of reordering duplicate values. This may result in an artificially increased gradient variability arising from despiking in regions in which low variability is expected.

APPLICATIONS

This section presents the four approaches for modelling geometrical risk applied to a real mineral deposit. For the sake of confidentially, information about location, geological settings and grades has been omitted from this paper. However, the disclosure of some structural features is necessary to provide context. In this case, the mineralisation in the deposit is structurally controlled by folding in the host rock. The deformation mechanism creates a strain partitioning in which moderate deformation domains may laterally transit to highly folded ones. The primary grade trend follows the folding axis, while the secondary continuity direction is the strike, orthogonal to the global axial surface. The strike direction better highlights the strain partitioning; therefore, it is used to generate all cross-sections presented hereafter.

Historically, the magnitude of changes in geological interpretation is related to local structural complexity. Reconciliation works show that the most deformed areas suffer from the greatest variations in tonnes and metals with the addition of new drilling. The methods are aimed to highlight those areas with a greater chance of misinterpretation.

The simplest way to perceive local variations in continuity is by calculating the directional indicator variograms for each structural sector. Figure 11a shows a cross-section of five subdomains defined according to the local dominant anisotropy. The magnitude of continuity along the strike is informed by the variogram range in each sector (Figure 11b). The greater the range, the lower the probability of being in a discontinuous or complex folded region. As expected, the less-deformed portion, termed Domain 1, showed greater continuity, while Domains 2–5 produced short-range variograms.

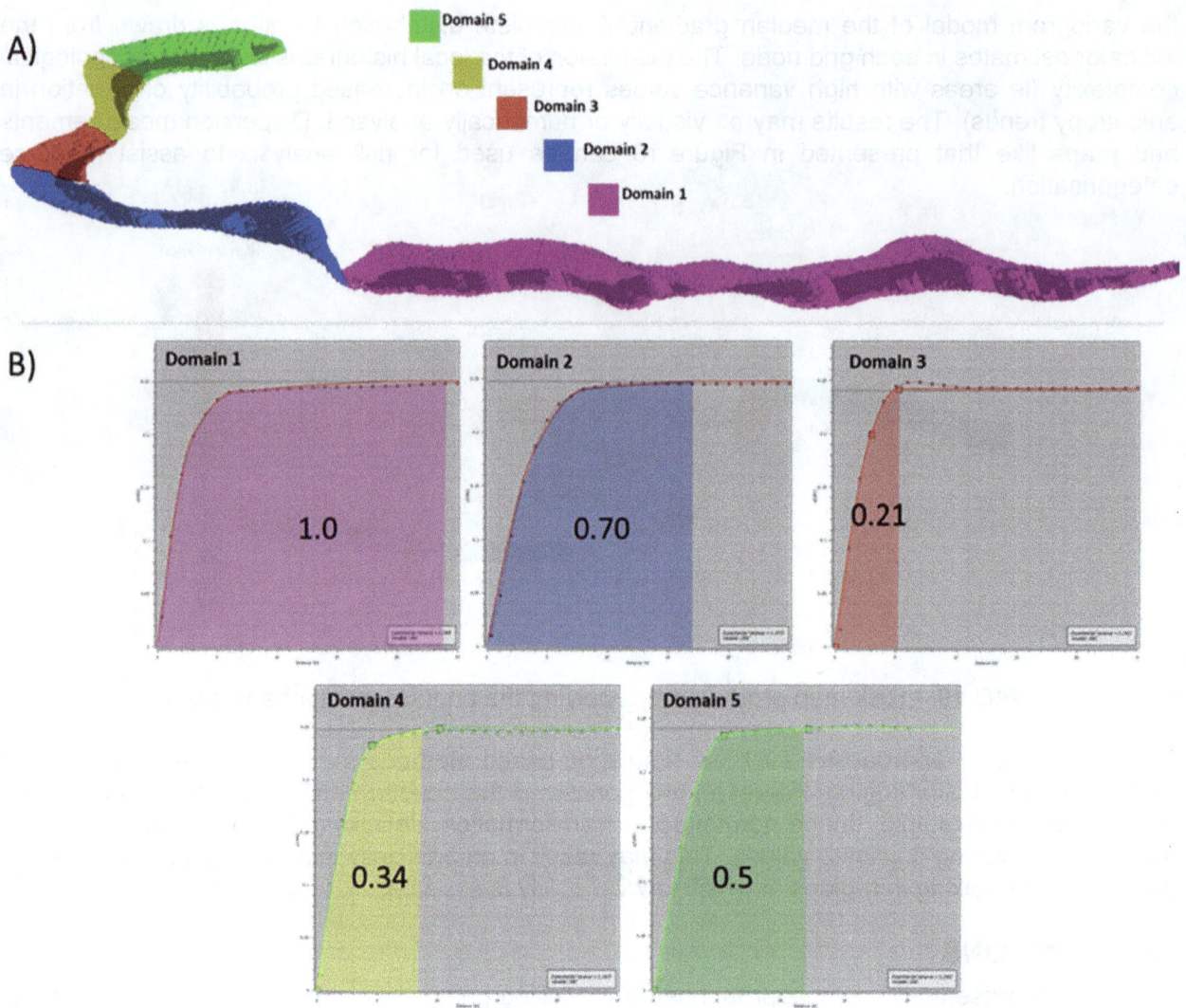

FIG 11 – (a) Anisotropy domains defined at different structural contexts; (b) the respective continuity functions and indexes.

MOs were shown to be another helpful alternative in quantifying modelling risk in complex shapes. By iterating the erosion algorithm with increasing cell dimensions, the highly deformed portion was the first to be depleted from the model (see Figure 12a). Consequently, a risk map (see Figure 12b) can be produced. The method locally accounts for uncertainty in geological interpretations by assigning higher risk values to the model's contacts and most structurally complex regions. This is an exciting outcome for modelling risks that lie mainly in defining the contacts and representing complex geometries in strongly deformed areas.

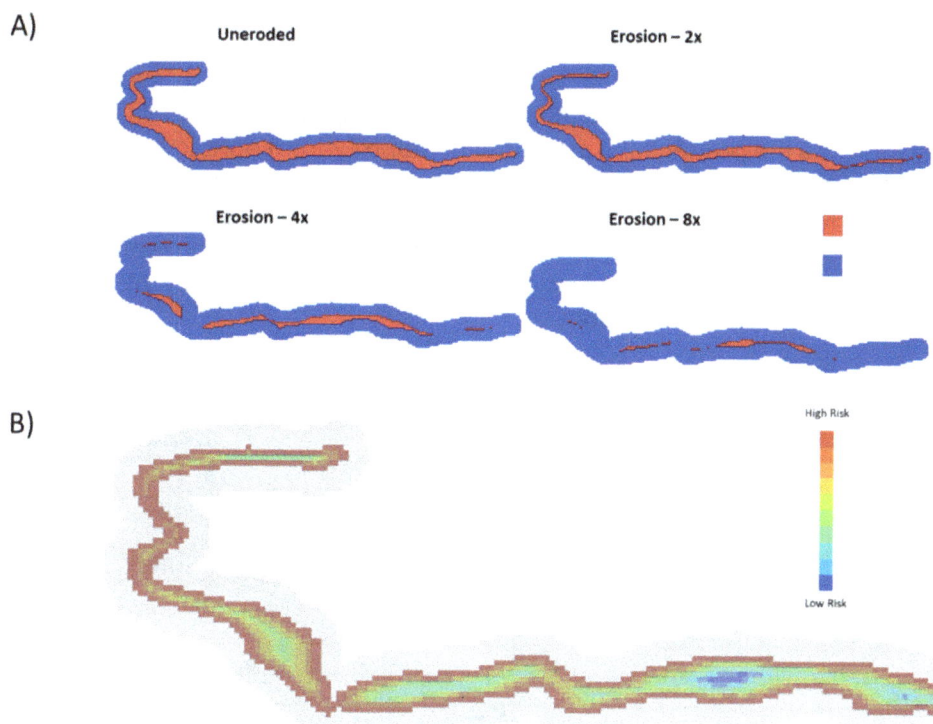

FIG 12 – (a) Mineralisation depletion from increased erosion; (b) risk map produced using the morphological operations approach.

Simulation of the signed distance function (see Figure 13a) produced 100 equiprobable scenarios of the mineralisation contacts. In the presence of hard boundaries, uncertainty in the contact position is measured as the frequency that a given point falls inside or outside the simulated boundary. Figure 13b shows the variability in the boundary location. Using this approach, the areas with increased geometrical uncertainty are the model borders and areas with greater complexity such as the folded and pinched areas. In contrast, the thickened portions and those with smooth variation in the anisotropy directions receive a greater probability of being inside the domain, thus can be assigned as low risk in this context.

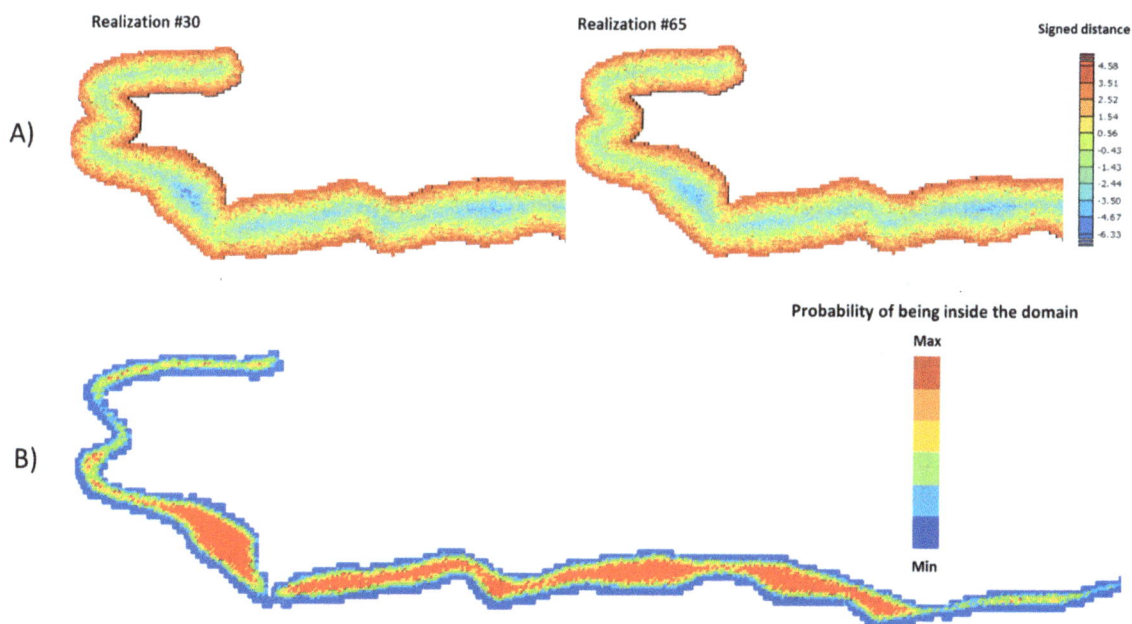

FIG 13 – (a) Two equiprobable realisations of the signed distance function; (b) probability map of being inside the interpreted domain.

Varying anisotropy fields are currently in use for implicit modelling and estimates. However, one non-traditional application for these vectors is to integrate them into the risk assessment framework. In this real-world example, we can see that the angular gradient approach successfully captures the variations in anisotropy and may be used as a proxy of geological complexity. The gradient approach appropriately highlights the gradation in structural complexity from right to left. Figure 14a shows the anisotropy vectors converted into gradient intervals, and Figure 14b shows the final block model with the calculated variance for the local gradient distribution. The results of the angular gradient approach are transferred to each block as local variances to assist in conventional risk assessment. The approach is highly efficient in capturing the variations in anisotropy by assigning higher risk levels to more structurally complex regions.

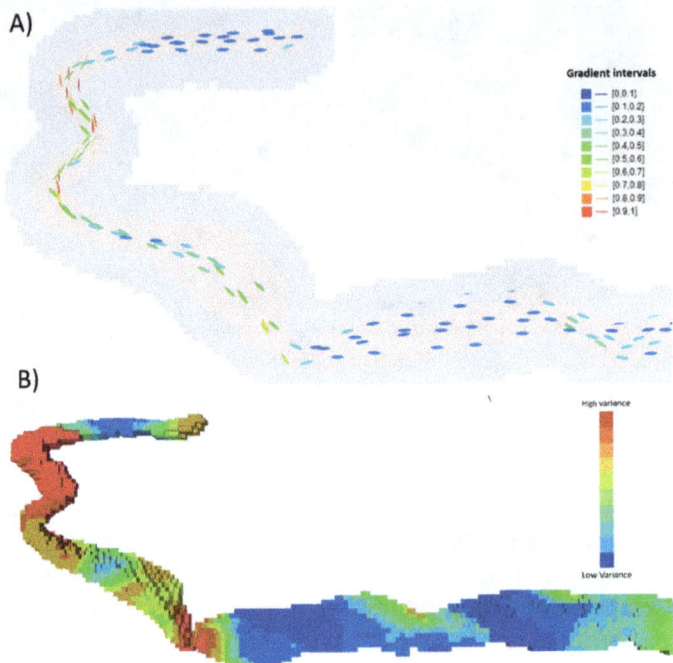

FIG 14 – (a) Structural vector data converted into gradient intervals; (b) block model of geological risk assigned by the angular gradient method.

DISCUSSION

Mineral resource estimate uncertainty has long been a concern for geostatisticians and geologists in mining, leading to improved methodologies. Simulation-based workflows for defining sample spacing are widely used for resource classification; however, the complexity of uncertainty requires a more comprehensive approach. The case study presented in this paper reveals that the main source of discrepancy between long-term models and grade control models is geological misinterpretation, particularly in highly deformed areas. Thus, in this specific case, the focus should be on structural settings rather than on grade variability when classifying resources, which was the inspiration for this study.

All approaches (apart from anisotropic continuity) show probabilistic behaviour at a certain point. For example, the morphological erosion method creates local frequency distributions that ultimately represent the probability of erosion. Similarly, the distance function method creates multi-equiprobable model boundaries. By kriging the multiple gradient indicators, local histograms show the variance in the anisotropy directions. One practical outcome of probabilistic methods is the possibility of obtaining dispersion measures such as variance or standard deviation. This enables the combination of uncertainties arising from independent variables such as simulated grade uncertainty and interpretation uncertainty using the distance function workflow. Then, consider two random variables: x being the grade and y being an indicator of whether the node belong to a given domain. If the variables are independent, the combined variance between them is given by:

$$Var_{(xy)} = [E(x)]^2 \, Var(y) + [E(y)]^2 \, Var(x) + Var(x) \, Var(y)$$

The choice of method and its efficiency will depend on the deposit's physical characteristics. These methods are effective in detecting areas of varying geometrical complexity in folded environments, as demonstrated in both the hypothetical example and a real-world deposit. However, they may not be suitable in cases where fault systems cause lateral or vertical displacement. Moreover, the signed distance function approach may only be appropriate for hard boundaries. Adaptations may be needed for soft boundaries to accurately capture uncertainty in stationarity transitions. When selecting a method or combination of methods, the limitations and conditions for applicability must be considered.

CONCLUSION

Resource classification is a contentious aspect of resource evaluation. Confidence in the categorisation of mineral resources is subjective and based on the judgement of the qualified person rather than on an objective uncertainty assessment. Methods such as probabilistic schemes and geometrical drilling patterns are commonly used to reduce subjectivity because they are easily understood and enable a comparison of results from different deposits. However, these standard approaches often rely on grade continuity and may overlook other geological or structural complexities. Many geologists and resource modellers recognise that one of the most significant risks when reporting ore volumes arises from 3D representations of complex geological shapes. Further, this work reminds practitioners of the relevance of geology fundamentals in evaluating mineral resources. This paper argues that careless geological interpretation may be more detrimental compared with naive geostatistical techniques and proposes objective approaches to quantify geological interpretation risks.

In conclusion, the techniques developed in this work are shown to benefit resource classification through the adoption of two practical examples. All methods can be easily integrated and automated in commonly used geostatistical mining software, enhancing risk assessment in mineral resource evaluation.

REFERENCES

Abzalov, M Z and Humphreys, M, 2002. Resource estimation of structurally complex and discontinuous mineralisation using non-linear geostatistics: case study of a mesothermal gold deposit in northern Canada, *Exp Min Geol J*, 11(1–4):19–29. Available from: <https://doi.org/10.2113/11.1-4.19>

Afonseca, B C and Costa, J F, 2021. Dynamic anisotropy and non-linear geostatistics supporting short term modelling of structurally complex gold mineralization, *REM–Int Eng J*, 74(2):199–217. Available from: <https://doi.org/10.1590/0370-44672020740034>

Afonseca, B C and Miguel-Silva, V, 2022. Defining optimal drill-hole spacing: a novel integrated analysis from exploration to ore control, *J South Afr Inst Min Metall*, 122(6):305–316. Available from: <https://doi.org/10.17159/2411-9717/2024/2022>

Amarante, F A N, Rolo, R M and Costa, J F C L, 2019. Assessing geologic model uncertainty—a case study comparing methods, *REM–Int Eng J*, 72(4):643–653. Available from: <https://doi.org/10.1590/0370-44672019720037>

Arik, A, 1999. An alternative approach to resource classification, in *APCOM'99: Computer Applications in the Mineral Industries: Proceedings* (eds: K Dagdelen, C Dardano, M Francisco and J Proud), pp 45–53 (Colorado School of Mines, Golden).

Boisvert, J B, Manchuk, J G and Deutsch, C V, 2009. Kriging in the presence of locally varying anisotropy using non-Euclidean distances, *Math Geosci*, 41(5):585–601. Available from: <https://doi.org/10.1007/s11004-009-9229-1>

Cáceres, A, Emery, X, Aedo, L and Galvez, O, 2011. Stochastic geological modeling using implicit boundary simulation, in *2nd International Seminar on Geology for the Mining Industry: Geomin 2011* (eds: J Beniscelli, R Kuyvenhoven and K O Hoal), p 21 (Gecamin: Santiago).

Caixeta, R M, 2020. Contribuições para o uso de anisotropias locais na geostatistica [Contributions to the use of local anisotropies in geostatistics], PhD thesis (unpublished), Universidade Federal do Rio Grande do Sul, Porto Alegre.

Cowan, E J, Beatson, R K, Ross, H J, Fright, W R, McLennan, T J, Evans, T R, Carr, J C, Lane, R G, Bright, D V, Gillman, A J, Oshust, P A and Titley, M, 2003. Practical implicit geological modelling, in *Proceedings of the Fifth International Mining Geology Conference*, pp 89–99 (The Australasian Institute of Mining and Metallurgy: Melbourne).

Deutsch, C V, 2005. Practical unfolding for geostatistical modeling of vein type and complex tabular mineral deposits, in *Application of Computers and Operations Research in the Mineral Industry: Proceedings of the 32nd International Symposium on the Application of Computers and Operations Research in the Mineral Industry (APCOM 2005)* (ed: S D Dessueault), pp 197–202 (A A Balkema: Leiden).

Hartman, H L (ed), 1992. *SME Mining Engineering Handbook*, p 333 (Society for Mining, Metallurgy & Exploration, Englewood, CO).

JORC, 2012. Australasian Code for Reporting of Exploration Results, Mineral Resources and Ore Reserves (The JORC Code) [online]. Available from: <http://www.jorc.org> (The Joint Ore Reserves Committee of The Australasian Institute of Mining and Metallurgy, Australian Institute of Geoscientists and Minerals Council of Australia).

Journel, A G, 1982. The indicator approach to estimation of spatial data, in *Proceedings of the 17th International Symposium on the Application of Computers and Operations Research in the Mineral Industry (APCOM 1982)* (eds: T B Johnson and J Randal), pp 793–806 (Port City Press: New York).

Journel, A G, 1988. New distance measures: the route toward truly non-Gaussian geostatistics, *Math Geol*, 20(4):459–475. Available from: <https://doi.org/10.1007/bf00892989>

Martin, R, Machuca-Mory, D, Leuangthong, O and Boisvert, J B, 2019. Non-stationary geostatistical modeling: a case study comparing LVA estimation frameworks, *Nat Resour Res*, 28(2):291–307. Available from: <https://doi.org/10.1007/s11053-018-9384-5>

Matheron, G, 1963. Principles of geostatistics, *Econ Geol*, 58:1246–1266.

Parker, H M and Dohm, C E, 2014. Evolution of mineral resources classification from 1980 to 2014 and current best practice, Finex 2014 Julius Wernher lecture [online]. Available from: <https://www.crirsco.com/docs/H_Parker_Finex.pdf>.

Serra, J, 1982. *Image Analysis and Mathematical Morphology* (Academic Press: London).

Silva, D A and Deutsch, C V, 2012. Modeling multiple rock types with distance functions: methodology and software [online]. *CCG Annual Report 14*, Paper 307. Available from: <https://www.ccgalberta.com/ccgresources/report14/2012–307_multiple_rt_modeling_with_distance_functions.pdf>.

Souza, L E, 2007. Proposição geoestatística para quantificação do erro em estimativas de tonelagens e teores [Geostatistical proposition for quantifying error in tonnage and grade estimates], PhD thesis (unpublished), Universidade Federal do Rio Grande do Sul, Porto Alegre.

Srivastava, R M, 2005. Probabilistic modelling of ore lens geometry: an alternative to deterministic wireframes, *Math Geol*, 37(5):513–544. Available from: <https://doi.org/10.1007/s11004-005-6670-7>

Wilde, B J and Deutsch, C V, 2011. Simulating boundary realizations [online], *CCG Annual Report 13*, Paper 106. Available from: <https://www.ccgalberta.com/ccgresources/report13/2011–106_boundary_simulation.pdf>.

Comparison of two quantitative mineral resource classification methods – a case study from a large copper porphyry-skarn deposit

C Artica[1]

1. MAusIMM, Principal Advisor – Resource Geologist, Rio Tinto, Brisbane Qld 4007. Email: cecilia.artica@riotinto.com

ABSTRACT

Two quantitative methods for Mineral Resource classification have been applied to a copper skarn deposit beneath a large open pit that is mining a world-class porphyry complex. A drill hole spacing study (DHSS) using the single block kriging method (SBK) and a conditional simulation method (CS) were applied to the skarn deposit and the results compared. Inputs to both processes included the proposed mining rate and stope designs to estimate the likelihood of achieving grade and metal production targets.

Mineral Resource classification of the deposit considers the relative precision of tons, grade and metal predictions over quarterly and annual production periods. These relative precision estimates are calculated at 90 per cent confidence interval (CI).

The SBK method represented the production period as a single large block. Kriging estimates were performed in the large block for various synthetic drill hole grids, and the estimation errors derived. The error was assumed to be normally distributed due to the large block size and quantity of samples, allowing CIs to be deduced.

The CS method followed a conventional sequential Gaussian simulation workflow where grades were simulated with a 5 ft × 5 ft × 5 ft grid node spacing inside mineralisation domains. Each realisation was then averaged up to monthly production volumes represented by a combination of designed stopes, and the coefficient of variation (CV) for the averaged grade was calculated. CI on the relative precision of grade predictions were then deduced from the CVs.

The two methods produced similar results; despite being calculated independently. CS results were used to report uncertainty in the proposed life-of-mine plan and highlight areas for additional drilling. SBK results were used as the basis for Mineral Resource classification as this method has been previously applied to the skarn deposit.

INTRODUCTION

International reporting codes, such as JORC and NI 43–101, imply that quantitative methods for Mineral Resource classification should be applied. The JORC Code requires that the Competent Person considers the relative confidence in tonnage/grade estimations in the classification and comments on the relative accuracy and confidence level in the Mineral Resource estimate. The CIM's Estimation of Mineral Resources and Mineral Reserves Best Practice Guidelines, referenced in NI 43–101, states that *'The confidence category selection should consider uncertainty and risk existing in the Mineral Resource estimate'* and provides examples of geostatistical drill hole spacing studies to assist in the process, including Froidevaux (1982), Deutsch, Leuangthong and Ortiz (2007), and Verly, Postolski and Parker (2014).

Although, JORC and NI 43–101 do not prescribe how to apply quantitative resource classification methods; one recognised industry practice is that for a Measured Resource, the drill hole spacing should be sufficient to predict tonnage, grade and metal on quarterly production with ±15 per cent relative precision at the 90 per cent CI. In the case of an Indicated Resource, the ±15 per cent relative precision should be achieved on an annual production volume. For the Inferred Category, data can be inadequate for assessing confidence intervals and knowledge from analogous deposits can be applied (Verly, Postolski and Parker, 2014). For a summary of the historical precedents for quantitative mineral resource classification, the reader is referred to Verly and Parker (2021). Other qualitative considerations may also influence the classification as noted in the discussion section below.

This paper has assessed two quantitative methods in the classification of a copper skarn deposit, which is the focus of a proposed underground stoping area beneath a large open pit. A DHSS, applying the SBK method described by Verly, Postolski and Parker (2014) was compared with a CS method that has been summarised by Verly *et al* (2017) and also largely coincides with the method described by Abzalov and Bower (2009).

The SBK method implemented here considers the production rate, the variogram model and sample CV to assess the variability of grade estimation over various production periods for various drill hole grids. It is referred to as the single block kriging (SBK) method because the production volume is represented as a single large block, which may be equivalent to a month or quarter of production. A kriging estimate is performed in the large block for each drilling grid, and the estimation error stored. The error is assumed to be normally distributed due to the large block size and many samples, which then allows the practitioner to deduce the 90 per cent CI (Verly, Postolski and Parker, 2014).

There are numerous and diverse examples in the literature of CS being used in the quantification of uncertainty in a mineral deposit (eg Ravenscroft, 1994; Schofield, 2001; Murphy *et al*, 2004; Abzalov and Bower, 2009; Dimitrakopoulos, Godoy and Chou, 2009; Silva and Boisvert, 2014; Cortes, 2017). The variables input to the CS, the CS algorithms, post-processing of the outputs, and interpretation of the results are equally diverse, and no single method stands out as being generally preferred. The CS method described by Verly *et al* (2017), however, is attractive in that it is directly comparable with the SBK method. In both methods, the final post-processing step applies the same formula in the derivation of CIs, and therefore, it has been the preferred application for this paper.

The skarn deposit that is the subject of this case study contains various lithologies and alteration types that can be related to the tenor of mineralisation. Unfortunately, at the time the current study was prepared, these lithologies and alteration types had not been adequately represented in a three-dimensional (3D) geological model. Therefore, stationarity was achieved by domaining based on a set of grade shells whose geometry was locally guided by the orientation of the skarn boundaries, and whose anisotropy was guided by the copper variography. These grade shells were modelled at >=2 per cent Cu (high-grade), >=1 per cent <2 per cent Cu (low-grade) and very low-grade (<1 per cent Cu) thresholds.

METHODS APPLIED

Drill hole spacing study

The author applied the current DHSS SBK study by updating the inputs to the previous SBK study of the skarn deposit (internal company report, 2012). The current mine design production rate for the underground stoping area is 2740 short tons (tn)/day, or ~83 000 tn/month. This production rate has been represented in the current SBK study as a production panel with dimensions 200 ft × 75 ft × 50 ft, which at a tonnage factor of 0.108 tn/ft³ gives 81 000 tn. The author considered that this was sufficiently close to the mine design monthly production to serve as the basis for further calculations. The inputs to the current SBK study included the following:

- Mine Design production/day: 2740 tn
- Mine Design production/month: 83 000 tn
- Tonnage factor: 0.108 tn/ft^3
- SBK Monthly production panel dimensions: 200 ft × 75 ft × 50 ft
- SBK Monthly production panel volume: 750 000 ft^3
- SBK Monthly production panel tons: 81 000 tn

For reference, a tonnage factor of 0.108 tn/ft^3 is equivalent to a bulk density of 3.45 g/cm^3.

The DHSS SBK method has been applied as follows:

- The 5 ft composites used in the study were selected inside the high-grade domain and declustered. The high-grade domain coincides with approximately 90 per cent of the currently designed stopes for the deposit.

- The CV was calculated from the declustered and grade-capped composites.

- The data and domain wireframes were rotated so that the tabular deposit was orientated to be approximately horizontal. This facilitated the construction of the production panel and drilling grids.

- A variogram (correlogram) was modelled from the rotated data set.

- The dimensions of a block equivalent to one month of production were defined:

  o Two scenarios were tested in the DHSS comprising three horizontally adjacent stopes and three spatially independent stopes.

- Artificial drilling grids at various spacings were constructed.

- Artificial drill hole grids were generated at 25 ft, 50 ft, 75 ft, 100 ft, 125 ft, 150 ft spacings. Some examples are provided in Figure 1.

- A kriging estimate was run in the large single block for each artificial drilling grid.

- Relative standard error (RSE) at 90 per cent CI was calculated for the various artificial drilling grids.

- Results were scaled up to represent quarterly and yearly volumes assuming independence of the mining areas and production periods.

- Results were plotted and risk-based Mineral Resource category definitions were applied.

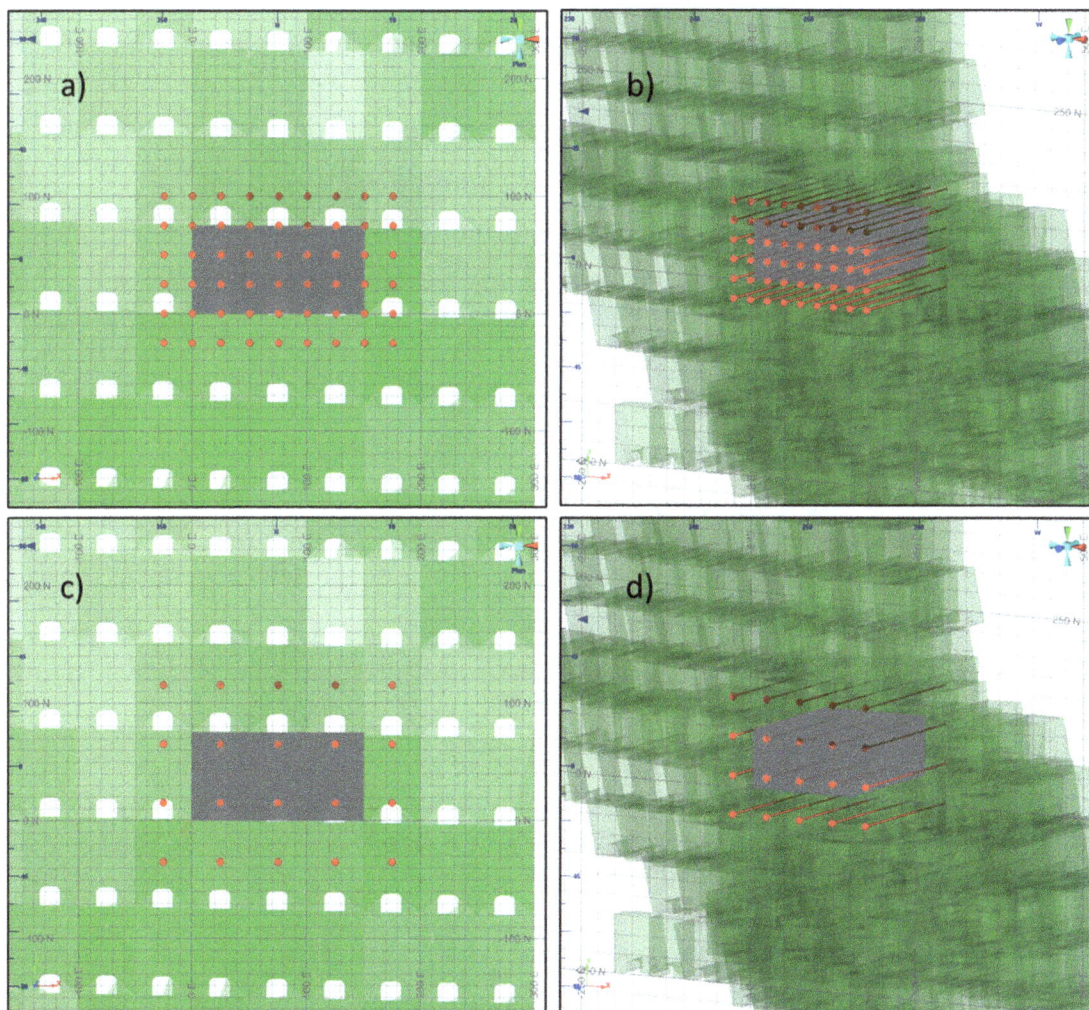

FIG 1 – Views showing stope designs (green), monthly production panel, assuming three horizontally adjacent stopes (grey) and artificial drilling grids (red). (a) 25 ft × 25 ft drilling grid: long section; (b) 25 ft × 25 ft drilling grid: perspective view; (c) 50 ft × 50 ft drilling grid: long section; (d) 50 ft × 50 ft drilling grid: perspective view.

The 90 per cent CI is calculated according to the method described by Verly, Postolski and Parker (2014). When using a unit-sill variogram model (USVM), the estimation variance or kriging variance is denoted by $\sigma^2 Er_{USVM}$. The relative estimation standard deviation, or Relative Standard Error (RSE) can also be expressed:

$$\sigma Er_{USVM\_rel} = sqrt(\sigma^2 Er_{USVM}) \times CV$$

Assuming the error of estimation is normally distributed, which is reasonable for large blocks estimated with many samples, a 90 per cent relative confidence interval on the error for a month of production is given by:

$$90\%CI\_rel = 1.645 \times sqrt(\sigma^2 Er_{USVM}) \times CV$$

Assuming independence of the errors of each month, the estimation variance for N periods of production is the production period estimation variance divided by N:

$$\sigma^2 Er_{USVM} / N$$

Therefore, based on a production panel representing one month of production, the RSE over a quarter or a year at 90 per cent confidence interval is given by:

$$QR90CI = 1.645 \times RSE / sqrt(3)$$

$$YR90CI = 1.645 \times RSE / sqrt(12)$$

Conditional simulation study

The author applied a conventional conditional simulation method that included the following steps:

- A grid with 5 ft × 5 ft × 5 ft node spacing was constructed without rotation and the high-grade, low-grade and very low-grade domains were coded into the grid.

- The 5 ft composites input to the simulation were selected and declustered by domain.

- The declustered composites were normal-score transformed.

- A variogram of the transformed composites was modelled for each domain.

- Sequential Gaussian Simulation (SGS) was run independently for each domain. Search orientation and anisotropy were related to the variogram anisotropy. Fifty realisations were generated.

- The simulated realisations were back-transformed to Cu% grades.

- Simulation results were then validated against input composites and variograms by domain.

- Post-processing included: (a) reblocking the simulated block model to 200 ft × 75 ft × 50 ft; (b) calculating 90 per cent CI; and (c) converting the grid to a 5 ft × 5 ft × 5 ft block model and reporting directly from the simulated block model. The process details were as follows:

  o 200 ft × 75 ft × 50 ft reblocked simulated block model:

    - Each realisation in the 5 ft × 5 ft × 5 ft simulated block model was reblocked to 200 ft × 75 ft × 50 ft.

  o Calculation of the 90 per cent CI:

    - The 90 per cent CI was calculated from the grades of each reblocked realisation.

    - The results were then scaled up to represent quarterly and yearly volumes and allow the calculation of Measured and Indicated resource classifications respectively using the calculations given below.

  o 5 ft × 5 ft × 5 ft simulation block model:

    - Stope designs grouped by year were coded into the 5 ft × 5 ft × 5 ft simulated block model.

    - Tons and average grades from each realisation were then reported by year. Average grades for each realisation and each year were weighted by tons.

- 90 per cent CI was calculated from the grades for each realisation after they had been averaged into the designed stopes grouped by year using the calculations given below.

The formulas applied followed (Verly *et al*, 2017), which are similar to those provided by Abzalov and Bower (2009). First find the standard deviation of the realisations in the 200 ft × 75 ft × 50 ft blocks:

$$SD_{csim} = sqrt((\sum(Real^i - Etype)^2) / N_{(realisations)} - 1)$$

Where Real^i is each realisation number and Etype is the mean of the realisations. Then the 90 per cent CI in the block is given by:

$$BK90CI = (SD_{csim} / Etype) * 1.645$$

As the block dimensions represent a month of production, the result can be rescaled to represent a quarter (QR90CI) or year (YR90CI) of production, assuming the independence of each month:

$$QR90CI = BK90CI / sqrt(3)$$

$$YR90CI = BK90CI / sqrt(12)$$

The validation of the simulation output comprised:

- Statistical comparison of the declustered input composites mean grades against the mean grades of each realisation.

- Q-Q plots comparing the grade distribution of the declustered input composites against each realisation.

- Checking for acceptable reproduction of the input variogram in selected realisations.

- Visual comparison of the declustered input composites with selected realisations in plan and cross-section, such as the examples provided in Figure 2.

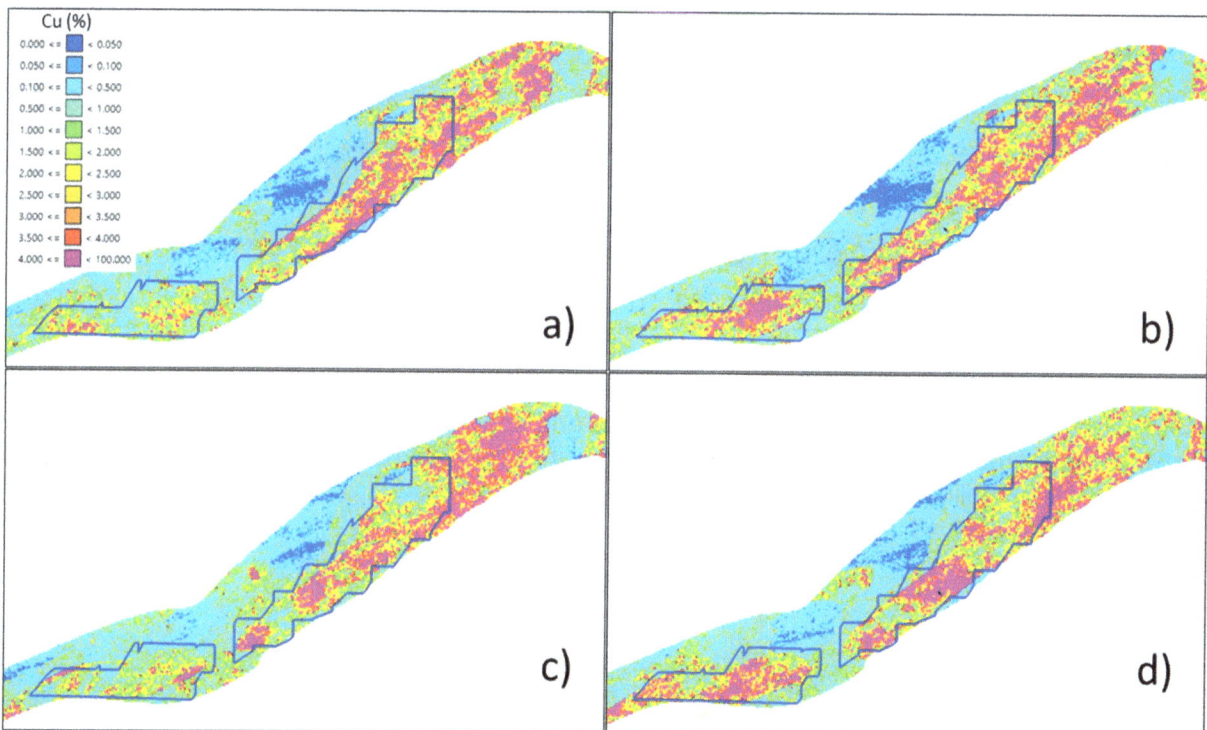

FIG 2 – West–east cross-section at 3350 ftN showing copper grades in four randomly selected realisations through the approximate centre of the skarn deposit: (a) Realisation 1; (b) Realisation 25; (c) Realisation 33; (d) Realisation 50. Copper legend displayed, Production period Year 1 – Year 18 combined stope outline displayed as a blue line. Section is aligned with the dip direction of the skarn.

The validation results were considered acceptable by the technical team and the CS study progressed to the post-processing phase.

RESULTS

Drill hole spacing study

The results of the two SBK scenarios show that as the dimensions of the drill grid increase, the kriging variance, RSE, QR90CI and YR90CI also generally increase (Table 1). Conversely, the slope of regression and the number of samples are observed to decrease. The CV is calculated from the input composites, the block variance from the variogram and block dimensions, which are constant for each drilling grid. The table also shows that in the 'Three horizontally adjoining stopes' scenario, the QR90CI values exceed 15 per cent between the 50 and 75 ft grids. Meanwhile, in the 'Three spatially disconnected stopes' scenario, the QR90CI values do not exceed 15 per cent.

TABLE 1

SBK results.

Scenario	Drill grid (ft)	Kriging variance	CV	RSE	QR90CI	YR90CI	Block variance	Slope of regression
Three horizontally adjoining stopes	25	0.002	0.787	0.031	5.0%	2.5%	0.055	0.998
	50	0.006	0.787	0.061	10.0%	5.0%	0.055	0.985
	75	0.022	0.787	0.118	19.4%	9.7%	0.055	0.933
	100	0.019	0.787	0.109	18.0%	9.0%	0.055	0.895
	125	0.026	0.787	0.126	20.7%	10.4%	0.055	0.871
	150	0.034	0.787	0.145	23.9%	11.9%	0.055	0.824
Three spatially disconnected stopes	25	0.003	0.787	0.043	4.1%	2.1%	0.080	0.999
	50	0.012	0.787	0.087	8.3%	4.1%	0.080	0.994
	75	0.032	0.787	0.141	13.4%	6.7%	0.080	0.939
	100	0.028	0.787	0.132	12.5%	6.3%	0.080	0.908
	125	0.029	0.787	0.134	12.8%	6.4%	0.080	0.899
	150	0.030	0.787	0.137	13.0%	6.5%	0.080	0.893

The 'Three horizontally adjoining stopes' scenario is plotted in Figure 3 where it can be seen that the QR90CI line crosses the 15 per cent threshold at a grid spacing of approximately 60 ft, which could meet the requirements for a measured resource using the '±15 per cent relative precision at the 90 per cent CI' guideline.

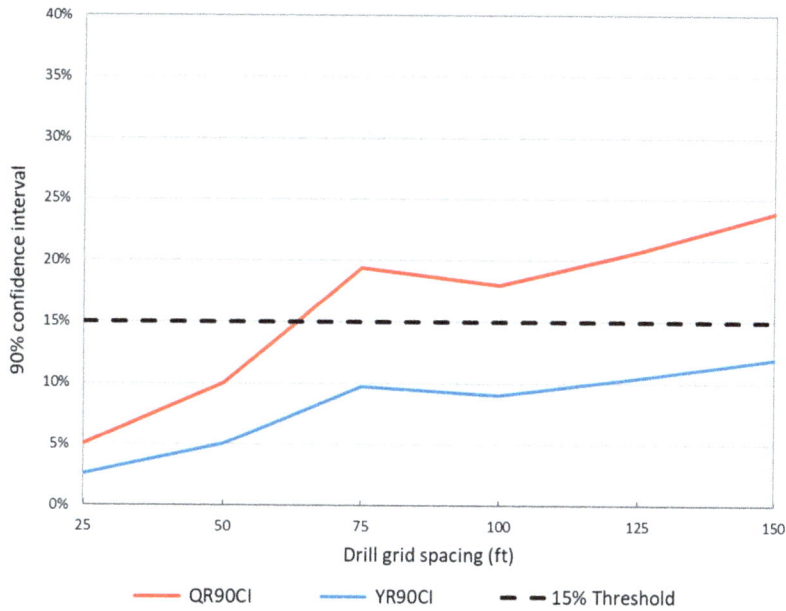

FIG 3 – SBK scenario most aligned with the mine plan.

This scenario was selected by the technical team as being the most representative of the proposed mining plan and therefore, the block model was coded with SBK resource categories based on this scenario. This was achieved by:

- Estimating the average distance to the three nearest drill holes.

- Coding each block as Measured where the average distance was <50 ft, Indicated >=50 ft <140 ft, and Inferred >=140 ft. These values have a 3D cartesian correction applied to the distances interpreted from the SBK results (H Parker, personal communication).

- Applying a smoothing routine in Vulcan to ensure coherent and contiguous resource categories.

Resulting SBK categories are displayed in Figure 4.

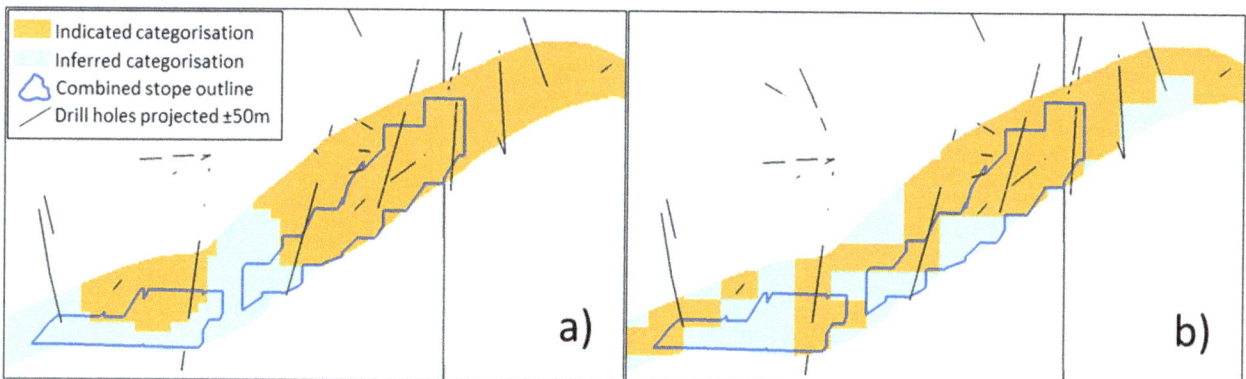

FIG 4 – West–east cross-section at 3350 ftN showing: (a) SBK resource categorisation coded into the block model based on average distance to the three nearest drill holes; (b) CS resource categorisation coded into the block model based on 90 per cent CI.

Conditional simulation study

Reblocking the simulation block model to a 200 ft × 75 ft × 50 ft block dimension, approximately representing a month of mine production, enabled the calculation of the monthly grade and metal 90 per cent CI. This could then be rescaled to represent quarterly and yearly volumes, enabling the resource categorisation of each block (Figure 4).

It is acknowledged that the CS resource categorisation of such large blocks may not be practical for mine planning. Nevertheless, the exercise was completed to provide a comparison with the SBK resource categorisation. A small amount of material meeting the criteria for the Measured category was also categorised by both methods, however, for the purposes of the comparison, the Measured category has been recoded as Indicated. The Measured category will not be discussed further in this paper.

Post processing of the 5 ft × 5 ft × 5 ft simulation block model enabled reporting of the 90 per cent CI by year. This could then be compared with the resource categorisation using the CS and SBK methods. Results of these comparisons are provided in Figure 5 where the 90 per cent CI is displayed for all categories and for the Indicated and Inferred categories separately.

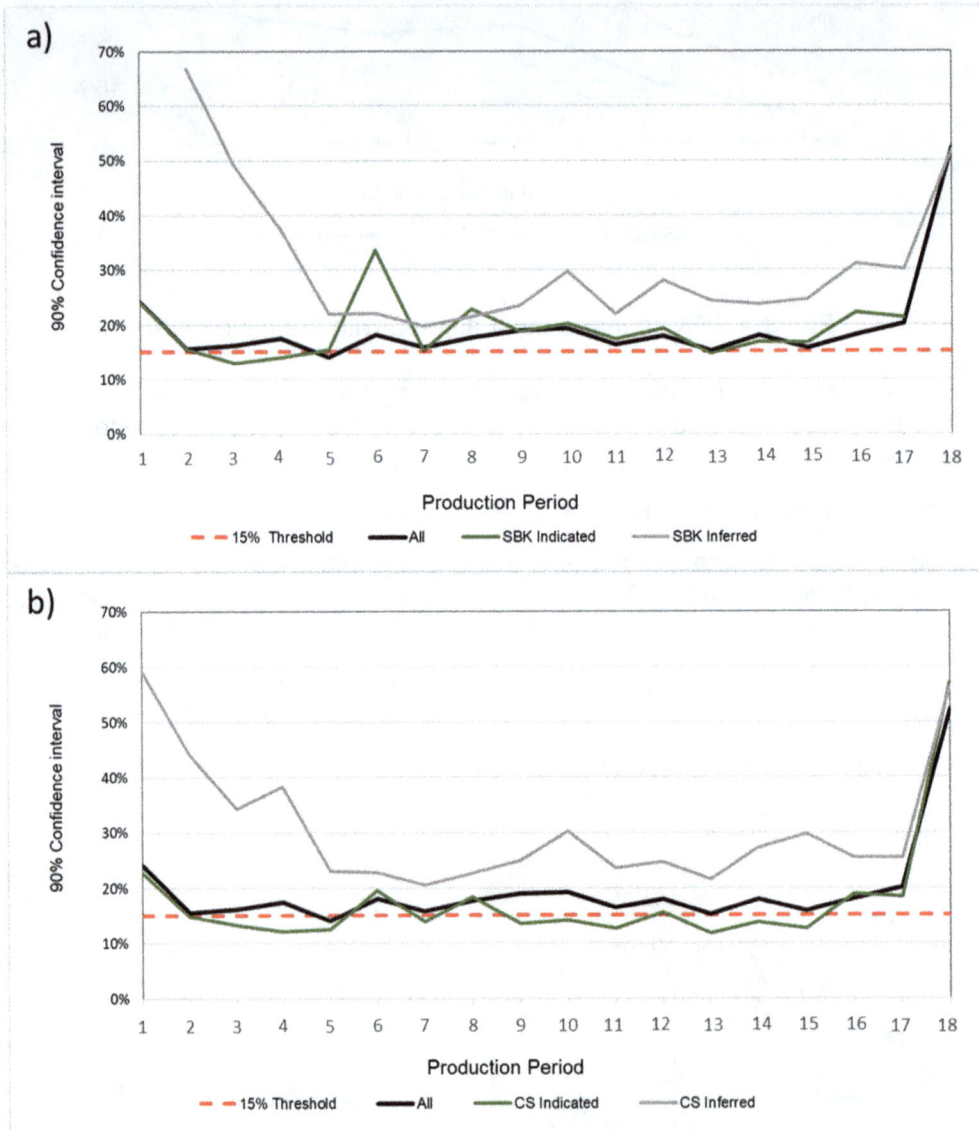

FIG 5 – CS 90 per cent CI reported by: (a) SBK categories and production period; (b) CS categories and production period.

Additionally for comparison of the SBK and CS methods, the breakdown of the tons attributed to each category according to each method is compared with the average drill spacing in each category (Figure 6). This comparison is further developed in Figure 7 where the relative proportion of Indicated tons according to the SBK and CS methods are compared.

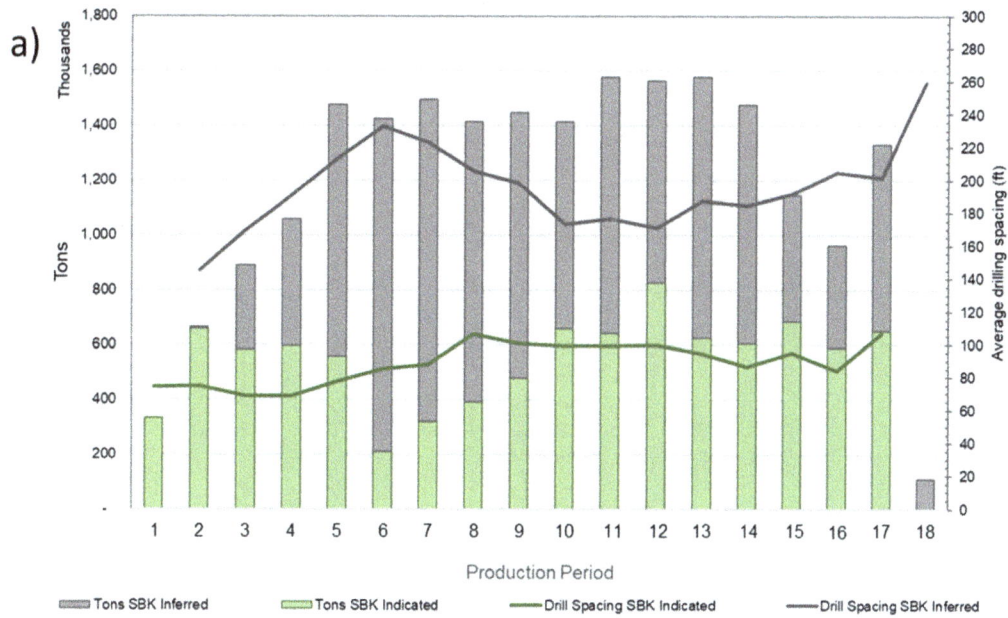

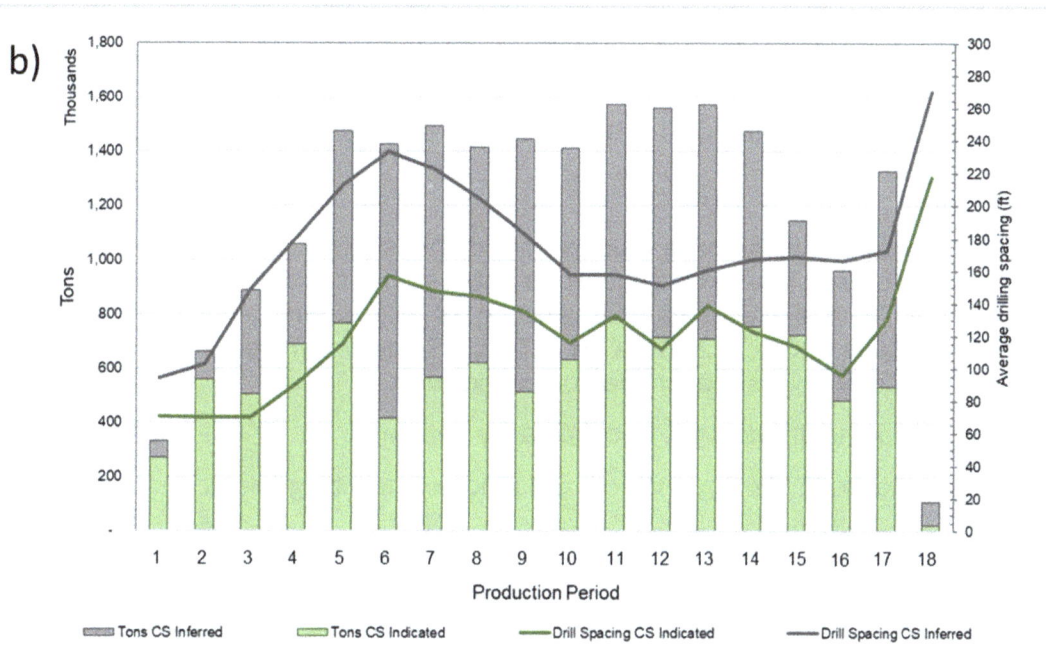

FIG 6 – Tons and drill spacing reported by: (a) SBK categories and production period; (b) CS categories and production period.

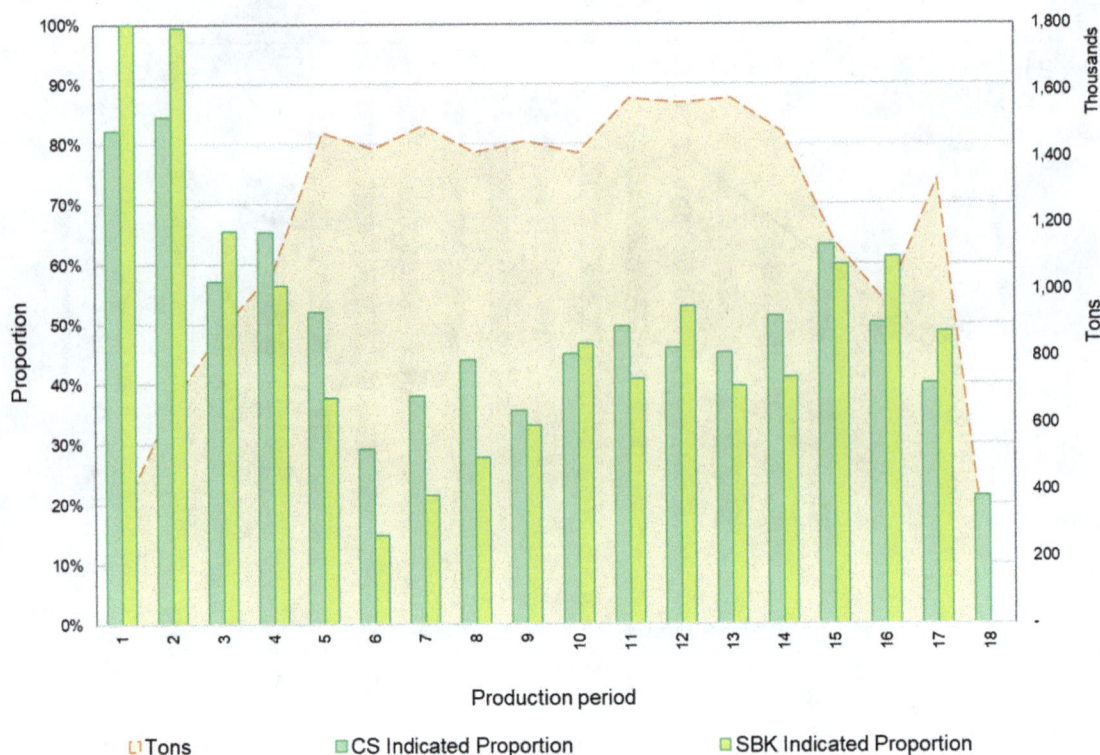

FIG 7 – CS indicated and SBK indicated proportion by production period.

DISCUSSION

The author has presented these case studies as a demonstration of two quantitative methods for resource classification and to allow comparisons between them. Additional qualitative criteria also informed the final Mineral Resource classification of this skarn deposit, in accordance with recommended practice (JORC, 2012; CIM, 2019). Some examples of qualitative criteria to supplement a quantitative classification method can include:

- The reliability of the drilling data, and the assay database.

- Reliability or certainty of the geological and grade continuity, geological model interpretation.

- Structural and geological model interpretation.

- Reliability of inputs to assess reasonable prospects of eventual economic extraction and cut-offs (eg Metallurgical test work, geotechnical data, social license to operate etc).

- Legal and land tenure considerations (CIM, 2019).

Notwithstanding the above, the implementation and comparison of the quantitative methods are discussed below.

Drill hole spacing study

The kriging variance in a SBK study is typically computed on a smaller initial block compared to the intended production block. The outcome is adjusted to accommodate the larger block, assuming the errors are independent (Verly, Postolski and Parker, 2014). In practice, the SBK production volume can represent independent mining volumes in time and/or space. For example, as in the current SBK study, the monthly production volume was represented as three spatially disconnected stopes (statistically independent) and three adjoining stopes (statistically dependent). Therefore, due to the spatial and temporal independence assumption, the calculations for the two scenarios are different such that in the 'three spatially disconnected stopes' scenario, the calculation for the quarterly 90 per cent CI is: QR90CI = 1.645 × RSE/sqrt(9), while in the 'three adjoining stopes' scenario the calculation is: QR90CI = 1.645 × RSE/sqrt(3). Naturally, the kriging variance calculated on the smaller, single stope volume was higher than that of the larger, three stopes volume, however, this was more than offset by the independence assumption (compare the RSE with the QR90CI values between the two scenarios, Table 1).

Mineral Resource Estimation Conference 2023 | Perth, Australia | 24–25 May 2023

There is a limit to how small the independent volumes can be, however, as the risk increases that the normal distribution of errors assumption could be violated (Verly, Postolski and Parker, 2014). Additionally, it becomes increasingly difficult to include sufficient drill holes to provide meaningful results at the larger-spaced drilling grids as the volume decreases. The technical team believed that this theoretical 'small volume limit' was exceeded in the case of the 'three spatially disconnected stopes' scenario, and that the results might not be reliable. Additionally, the theoretical limit may have also been exceeded for the larger drilling grids in the 'three adjoining stopes' scenario as the RSE is not observed to monotonically increase with increased drill spacing as would be expected. Considering this possible limitation of the SBK method, and also considering the relevant qualitative information, the technical team agreed to place the boundary between the Indicated and Inferred categories at 140 ft.

Generally, the smaller production volumes typically encountered in selective underground mining methods relative to bulk mining methods tends to limit the applicability of the SBK method. Conversely, as the production volume increases, it becomes easier to justify the assumptions required in SBK.

Conditional simulation study

The CS workflow applied in the current study was fairly simplistic in that it did not consider uncertainty in the input data or parameters, neither a trend in the copper grades, nor the uncertainty in the domain boundaries. A more complete workflow would have provided a more robust result, however, the processing and professional time required could not be justified for the current study. Nevertheless, validation of the simulation provided sufficient confidence in the results to continue to the next phase of the study.

The goal of the study was to provide ±15 per cent relative precision at the 90 per cent CI on tons, grade and metal. For the current study, the tons were assumed to be fixed as the cut-off grade is implicit in the mine design. Therefore, no cut-off grade was applied when reporting each realisation. As a result, the metal CI was equal to the grade CI and the tons did not need to be considered in this study.

Plotting the yearly 90 per cent CI based on reporting the 5 ft × 5 ft × 5 ft simulation block model showed that the 90 per cent CI of all resource categories generally fluctuates between 15 per cent and 20 per cent (Figure 5). The exceptions to this are in the first and last years of the mine plan when the production volumes are significantly smaller (Figure 6). This is an expected outcome that demonstrates the relationship between volume and variance.

When the 90 per cent CI is split out by resource category, the Inferred category is observed to sit significantly above the 15 per cent threshold, while the Indicated category generally sits slightly below this threshold. This outcome is also in line with expectations as the Indicated category should be predictable within ±15 per cent over a yearly production volume at 90 per cent CI (Verly, Postolski and Parker, 2014). Exceptions are observed in years 6 and 8, which contain a larger proportion of domain boundaries where higher variance can be observed.

Method comparisons

The main finding of this study was that despite being derived from distinct and independent methods; the categories generated by the two quantitative approaches were largely comparable (Figure 7). The categorisation using the SBK method, however, appears to contain more mixing of lower and higher variance material, which is not surprising as the method purposely included a smoothing step in the workflow. The strong correlation between the more robust CS method and the more presumptive SBK method enhanced the reliability of the SBK method as a classification tool for future use by the technical team.

In addition to the CS method's reduced dependence on assumptions, other significant departures of the CS method from the SBK method include:

- The CS inherently considers the local variability.

- The CS incorporates the proportional effect, where higher grades display higher variability.

- The CS allows volume-variance relationships to be investigated for any proposed mine production rate and the issues regarding the small volume limit of selective mining methods are irrelevant.
- The CS method requires considerably greater processing and professional time to produce a meaningful result.
- The CS uses and honours the actual drill hole locations and grades, and does not assume a regular drilling grid.

Considering the above, the CS model will be used by the technical team as the basis for optimising future resource drilling. The team expects that this will lead to closer-spaced drill hole designs in the higher-grade zones, where uncertainty is observed to be higher.

CONCLUSIONS

The practical implementation of two quantitative Mineral Resource classification methods on the same deposit has demonstrated that the outcomes were largely comparable. This provided confidence that the relationship between volume and variance had been adequately modelled in both the SBK and CS methods, with the result of each method providing support for the other.

The SBK approach to drill hole spacing studies is a quick and straightforward method for evaluating uncertainty, making it useful for simple scenarios. Although not as thorough as conditional simulation, it is a viable option for generating reasonably reliable confidence intervals for estimated variables over quarterly to annual time periods, particularly when there are limitations in professional time or funding. Nevertheless, the resulting confidence intervals should be viewed as approximate rather than exact.

In contrast, the CS approach provides a more robust and complete representation of the uncertainty, which enables the investigation of a wide range of potential issues related to variability, of which Mineral Resource classification is but one example. Although requiring considerably more time to prepare, the wide range of uses can make the investment worthwhile.

ACKNOWLEDGEMENT

The author wishes to thank Rio Tinto for permission to publish this paper. Peer reviews by Ana Chiquini, Antonio Cortes, Graham Crook, Andrew Fowler, Cael Gniel, and James Pocoe were much appreciated.

REFERENCES

Abzalov, M Z and Bower, J, 2009. Optimisation of the drill grid at Weipa bauxite deposit using conditional simulation, in *Proceedings Seventh International Mining Geology Conference*, pp 247–251 (The Australasian Institute of Mining and Metallurgy: Melbourne).

CIM, 2019. CIM Estimation of Mineral Resources and Mineral Reserves Best Practice Guidelines, Prepared by the CIM Mineral Resource and Mineral Reserve Committee, Adopted by the CIM Council on November 29, 2019. <https://mrmr.cim.org/media/1129/cim-mrmr-bp-guidelines_2019.pdf>

Cortes, A, 2017. Comparison of two multivariate grade simulation approaches on an iron oxide copper gold deposit, in *Geostatistics Valencia 2016* (eds: J Gómez-Hernández, J Rodrigo-Ilarri, M Rodrigo-Clavero, E Cassiraga, and J A Vargas-Guzmán).

Deutsch, C V, Leuangthong, O and Ortiz, J M, 2007, Case for Geometric Criteria, in *Resources and Reserves Classification: Society for Mining, Metallurgy and Exploration*, 2007 Transactions, vol 322, pp 1–11.

Dimitrakopoulos, R, Godoy, M and Chou, C L, 2009. Resource / reserve classification with integrated geometric and local grade variability measures, in *Proceedings Orebody Modelling and Strategic Mine Planning 2009* (ed: R Dimitrakopoulos), pp 207–214 (The Australasian Institute of Mining and Metallurgy: Melbourne).

Froidevaux, R, 1982. Geostatistics and Ore Reserve Classification, *CIM Bulletin*, 75(843):77–83.

JORC, 2012. Australasian Code for Reporting of Exploration Results, Mineral Resources and Ore Reserves (The JORC Code) [online]. Available from: <http://www.jorc.org> (The Joint Ore Reserves Committee of The Australasian Institute of Mining and Metallurgy, Australian Institute of Geoscientists and Minerals Council of Australia).

Murphy, M, Parker, H, Ross, A and Audet, M A, 2004. Ore-Thickness and Nickel Grade Resource Confidence at the Koniambo Nickel Laterite (A Conditional Simulation Voyage of Discovery), in *Geostatistics Banff 2004*, pp 469–478 (Springer: Netherlands).

Ravenscroft, P J, 1994. Conditional simulation for mining: practical implementation in an industrial environment, in *Geostatistical Simulations: Proceedings of the Geostatistical Simulation Workshop* (eds: M Armstrong and P A Dowd), pp 79–87.

Schofield, N A, 2001. Determining optimal drilling densities for near mine resources, in *Mineral Resource and Ore Reserve Estimation—The AusIMM Guide to Good Practice,* AusIMM Monograph 23, pp 293–298.

Silva, D S F and Boisvert, J B, 2014. Mineral Resource classification: a comparison of new and existing techniques, *Journal of the Southern African Institute of Mining and Metallurgy*, 114(3):265–273.

Verly, G, Parker, H M, Artica, C, Kim, H and Bortoletto Machado, E L, 2017. Classification and Dilution Study by Simulation of a Large Copper Deposit, Peru, in *Proceedings of the 38th APCOM International Symposium* (ed: K Dagdelen), pp K-11–K-18.

Verly, G and Parker, H M, 2021. Conditional simulation for mineral resource classification and mining dilution assessment from the early 1990s to now, *Mathematical Geosciences*, 53(2):279–300.

Verly, G, Postolski, T and Parker, H M, 2014. Assessing Uncertainty with Drill Hole Spacing Studies – Application to Mineral Resources, in *Orebody Modelling and Strategic Mine Planning*, pp 109–118.

Drill hole spacing analysis for classification and cost optimisation – a critical review of techniques

I Glacken[1], O Rondon[2] and J Levett[3]

1. FAusIMM(CP), Executive Consultant, Snowden Optiro, Perth WA 6000.
   Email: ian.glacken@snowdenoptiro.com
2. Principal Geostatistician, Snowden Optiro, Perth WA 6000.
   Email: oscar.rondon@snowdenoptiro.com
3. Principal Consultant, Snowden Optiro, Perth WA 6000. Email: jane.levett@snowdenoptiro.com

ABSTRACT

Reporting Codes are not prescriptive on methodologies to report or classify Mineral Resource Estimation results, but the assessment of risk and uncertainty is required, and is likely to be increasingly important in future Code updates. The 2012 JORC Code indicates that relative confidence in tonnage-grade estimations should be considered for classification, and that there should be a discussion of relative accuracy and confidence levels in the Mineral Resource Estimate. These considerations are also relevant to the drill hole spacing required for various classification categories, along with other geological and data quality criteria.

This paper summarises the theory, principles and current practice of various drill hole spacing analysis techniques for both achieving a desired level of resource categorisation and for optimising grade control or resource drill-out costs. The authors provide practitioners with a guide in their use to assess uncertainty and relative accuracy, as illustrated by a number of case studies. While all drill hole spacing studies rely on the assumption of consistency of the continuity model from current to future drilling, the range of methods available provide much more objective outcomes than previously adopted 'rules-of-thumb'.

INTRODUCTION

The motivation for drill hole spacing analysis (DHSA)

Prior to the capital expenditure associated with building a mine and a processing plant, one of the largest costs associated with the mine value chain relates to drilling. There is an ever-increasing emphasis on the consideration of Mineral Resource classification categories, whether via the JORC Code (Australia) the CIM Guidelines (Canada), the SAMREC Code (South Africa), S-K 1300 (USA), the PERC Code (Europe) or other CRIRSCO-based schemes. As a consequence of this, explorers and miners wishing to achieve production recognise the criticality of attaining an Indicated level of classification, since this is the lowest confidence level which can be converted to reserves and thus contribute to a forecast cash flow via the pre- and feasibility process. While all of the aforementioned Codes emphasise the prerogative of the individual Competent or Qualified Person to allocate the various resource categories (and especially the Indicated Resource category), for various reasons there is an increased search for objectivity in Mineral Resource classification. These reasons may include:

- the need to assess projects competing for scarce capital.

- the desire, despite the individual CP/QP decision, to achieve an objective standard for classification across what may be a significant number of projects in a large mining company.

- in the case of external funding, the need to achieve a certain minimum level of higher-confidence resources and reserves in the early years of production.

As a result, increased emphasis is now placed on drill hole spacing analysis (DHSA) to achieve pre-determined classification levels. However, there is also a strong motivation to optimise the costs of drilling throughout the exploration-development-production cycle.

This paper presents a critical assessment of the relatively wide range of approaches to DHSA, whether to achieve a certain classification level (usually Indicated) or whether to optimise the costs of grade control drilling. Many of the techniques are common across both motivations. A review of the relatively simple and straightforward group of methods based upon estimation variance and its derivatives is presented, along with reminders of the key assumptions inherent in the application of these techniques. The connection between a Kriging variance (KV) or confidence interval (CI) -based approach and the very common '90:15' standard for resource classification (Verly, Postolski and Parker, 2014; Parker and Dohm, 2014; Owusu and Dagdelen, 2019) is emphasised. A case study using these approaches to quantify drill spacing in Mineral Resource development is presented for a nickel-cobalt laterite project.

The increasing use of conditional simulation in DHSA is detailed, both for classification and for general cost optimisation. The benefits of using conditional simulation and the drawbacks (increased workload, quantification of uncertainty) are emphasised. The combination of simulation and cost optimisation for DHSA is presented, with an overview of the benefits of considering the cost versus revenue dimension for simulated drill holes. Finally, a case study of grade control drilling optimisation for a nickel-copper project, using multivariate simulation and value optimisation, is presented.

Overview of techniques and previous work

The description and documentation of DHSA techniques, whether for achieving a desired resource classification level, or for general grid optimisation, has become a relatively popular topic in the past decade. DHSA studies reinforce the themes of compliance with increasingly prescriptive classification Codes, the capture and quantification of uncertainty through simulation, and the use of risk profiles to optimise the risk-reward balance associated with various levels of drilling.

Rivoirard *et al* (2016) proposed a two-step approach for classification. First, a measure of spatial sampling density of the deposit is derived. Second, a resource classification is obtained by thresholding a coefficient of variation derived from the measure of spatial sampling density. This approach has the advantage of comparing objectively different drill hole spacings across the volume of interest to derive a classification category.

Verly, Postolski and Parker (2014) describe a production-linked DHSA approach designed to achieve a certain classification level. The principles underlying this approach are simple:

- define the set of drill hole spacings to evaluate.

- compute the relative error of an attribute within a large production block corresponding to a period of interest using the specified drill hole spacings.

- assume that the relative error has a Gaussian distribution and deduce relative confidence intervals accordingly.

- assign a classification based on whether the confidence interval is within a defined error tolerance.

- select the widest drill hole spacing which achieves less than the desired error.

The relative error is determined as the ratio of the Kriging variance σ_{OK}^2 and the square of the mean value m of the composited drill hole data. Therefore, under the Gaussian assumption, the $\alpha\%$ confidence interval is:

$$\alpha\% \, CI = F^{-1}\left(\frac{1 + \alpha}{2}\right)\frac{\sigma_{OK}}{m}$$

where F^{-1} is the inverse of the standard Gaussian distribution function. This function can easily be evaluated using Excel.

Industry practice (Parker and Dohm, 2014; Verly, Postolski and Parker, 2014) is to consider a 90 per cent confidence interval, a 15 per cent error tolerance and classification as follows:

- Measured resources – material which has a 90 per cent confidence of being within 15 per cent of the true value for parcels reflecting quarterly production.

- Indicated resources – material which has a 90 per cent confidence of being within 15 per cent of the true value for annual production parcels.

The 90 per cent confidence means that $F^{-1}(1 + \alpha/2) = F^{-1}(0.95) = 1.645$. Therefore, it is only required to compute the Kriging variance σ_{OK}^2 for the different drill spacings and check if $1.645 * \sigma_{OK}/m$ is less than the error tolerance of 15 per cent. Hereafter, this approach will be referred as to the 90:15 method.

The authors have recently observed that a wider error margin, such as 30 per cent, is being used to separate Inferred areas from unclassified areas and endorse this refinement. Verly's co-author, the late H M Parker, presents a historical perspective of the development of this technique in the 2014 paper and also in a presentation with C E Dohm (Parker and Dohm, 2014). The 90:15 method was also first addressed by Parker and others in Murphy *et al* (2004) and by Wawruch and Betzhold (2004).

Considerable effort has been devoted to the development of DHSA techniques which align with this standard. Nowak and Leuangthong (2019) describe a number of approaches, based upon KV and a 90 per cent CI but also including simulation and the idea of resampling a reference realisation, a method covered by the authors in a case study below.

Usero, Misk and Saldanha (2019) also detail the use of conditional simulation of grades, which are then resampled and scaled up to the various production units required by the 90:15 method; the optimal drill spacing is a presented as a function of the accuracy, as measured at a 90 per cent CI with a 15 per cent simulation risk (error). In this work the different aspects of grade risk and thickness risk (in a nickel laterite case study) are compared, and it is important to acknowledge that grade uncertainty (as provided by conditional simulation) is often only one aspect of the total risk.

Neves *et al* (2023) also examine the role that simulation plays in optimising a drilling grid, again with a case study on a nickel laterite deposit. In this case, the simulation examines three dimensions of uncertainty; nickel grade, thickness, and ore type (as simulated via a categorical indicator approach). Using the 90:15 approach for quarterly panels, the drilling grid required to achieve a maximum 15 per cent error is defined. This study does consider geological uncertainty (via the categorical ore type simulation) along with thickness and grade, which is not available in most cases due to the relative complexity of the geological controls on mineralisation.

Alfonseca and Silva (2022) present two alternative DHSA workflows, both based upon simulation. One approach involves the sampling of a dense simulation to generate pseudo-drill holes (as in the case study below) followed by estimation and comparison with the original simulation using a number of misclassification measures. The second documented workflow involves re-simulation of pseudo-drill holes from a prior simulation based upon the existing drilling data, followed by calculation of the error of the group of realisations. In each case the chosen drill spacing is a function of the desired level of error.

VARIANCE-BASED METHODS

Overview

Kriging variance (KV) based methods of DHSA are simple and relatively straightforward to undertake. These DHSA approaches allow for the basic assessment of estimation uncertainty, where risk assessment requirements are not complex, and the time available to complete a simulation is lacking.

KV-based methods calculate the Slope of Regression (SR) and Kriging Efficiency (KE) for a range of conceptual drill grid spacings, where it is assumed that the overall grade continuity model remains constant with increased (conceptual) information. These methods use project-specific variograms and estimation block parameters, although it is important to note that the quality of the variogram will impact the result of the study, and that the variogram models may change with increased physical information. The theoretical estimation quality can be compared for each grid, and the results can provide a general indication of risk and uncertainty with each potential grid spacing. The KV based methods can also be used to optimise expenditure, given some simple cost assumptions.

Variance-based methods of DHSA do not consider potential risks, including variations in geology, deleterious elements, or mining considerations. The methods, as described in Verly, Postolski and Parker (2014), link to the 90:15 assumptions as discussed above, and provide a reasonably objective framework for the definition of drilling for Measured or Indicated Resources. The 90:15 approach is best applied to large production situations; there is an attempt to quantify error via a confidence interval, but there is no consideration of grade or geological risk.

Case study – Wilconi nickel laterite

This case study describes the use of a KV-based approach to define a drill spacing intended to give Measured Resources. A-Cap Energy's majority-owned Wilconi Project is situated 700 km from Perth within the Norseman-Wiluna Greenstone Belt. The nickel-cobalt mineralisation has developed through lateritisation of the Perseverance ultramafic sequence, extends for around 20 km along strike and is up to 1500 m wide. Nickel and cobalt mineralisation have been formed by intense weathering and have concentrated in a saprolite clay layer that overlies the ultramafic rock unit. The depth to significant (>0.5 per cent Ni) mineralisation ranges from 2 m to 60 m. The nickel mineralisation zone can be up to 30 m thick, averaging around 4 m in thickness (Figure 1).

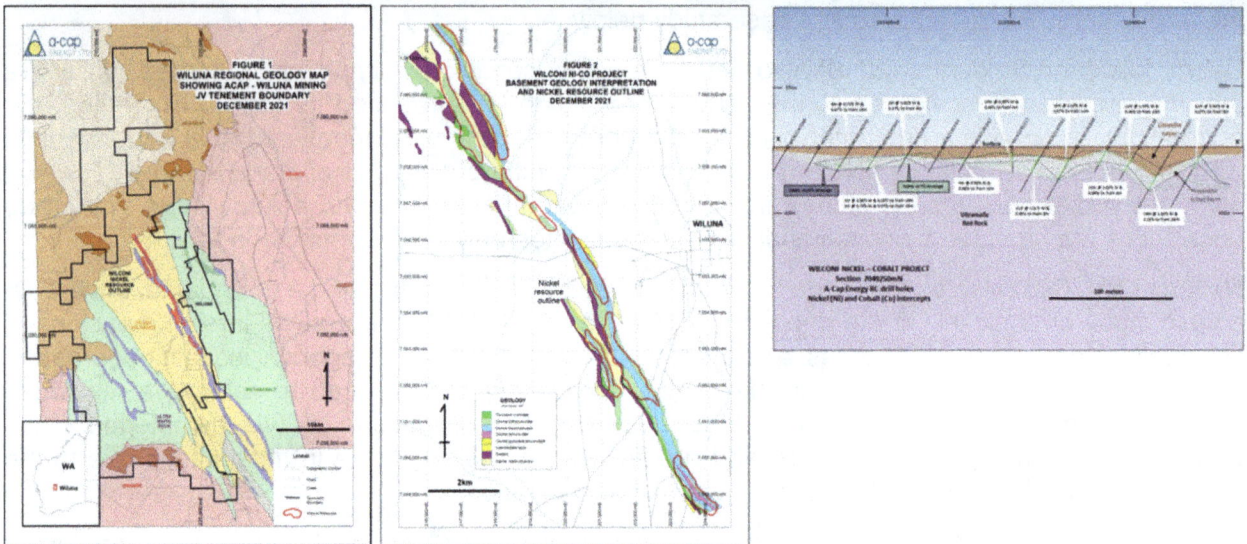

FIG 1 – Location and geology of the Wilconi nickel laterite project.

Two variance-based analyses were undertaken to recommend appropriate drill spacing for Measured Resources. A very basic DHSA, using the project variograms and drilling data and the Mineral Resource Estimate parent cell size, was undertaken. The mineralisation domain selected to undertake the analysis was chosen as it had the 'worst' variography – that is, the highest nugget and shortest range, and it also wholly encompassed another higher-grade mineralisation domain (Figure 2, where Y is north and X is east). The logic is that any drill spacing recommended for this 'worst' domain would be more than adequate for zones of better continuity. A range of conceptual drill grids with vertical holes was tested down to a 10 m × 10 m spacing, and KE versus drill costs were graphed. The optimal drill grid spacing was selected where the improvement in Kriging quality (KE and SR) plateaus, and the drill costs start to increase significantly (Figure 3).

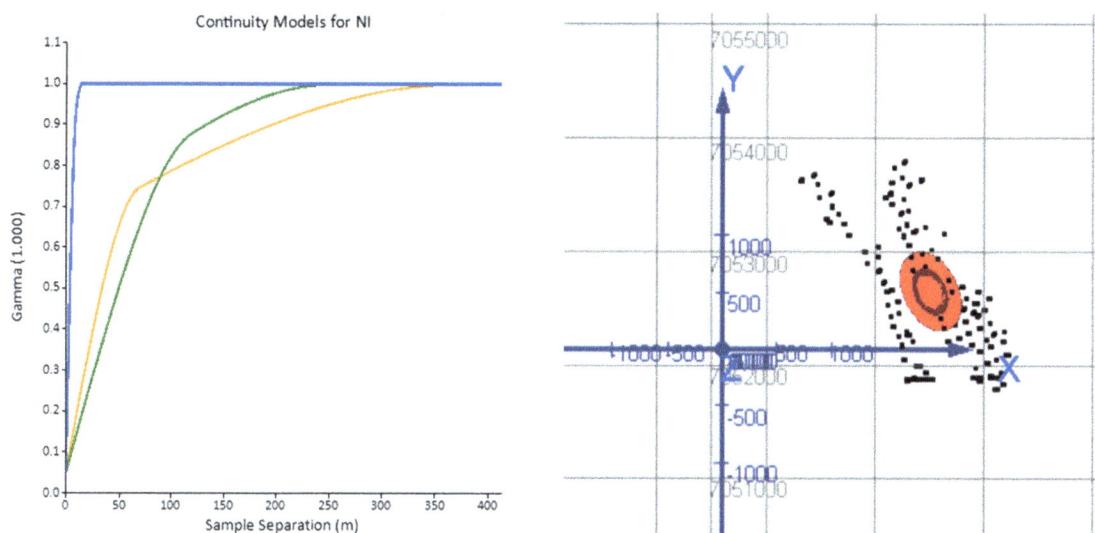

FIG 2 – Variogram model (left) and search ellipse with existing drilling data (right) for mineralisation domain.

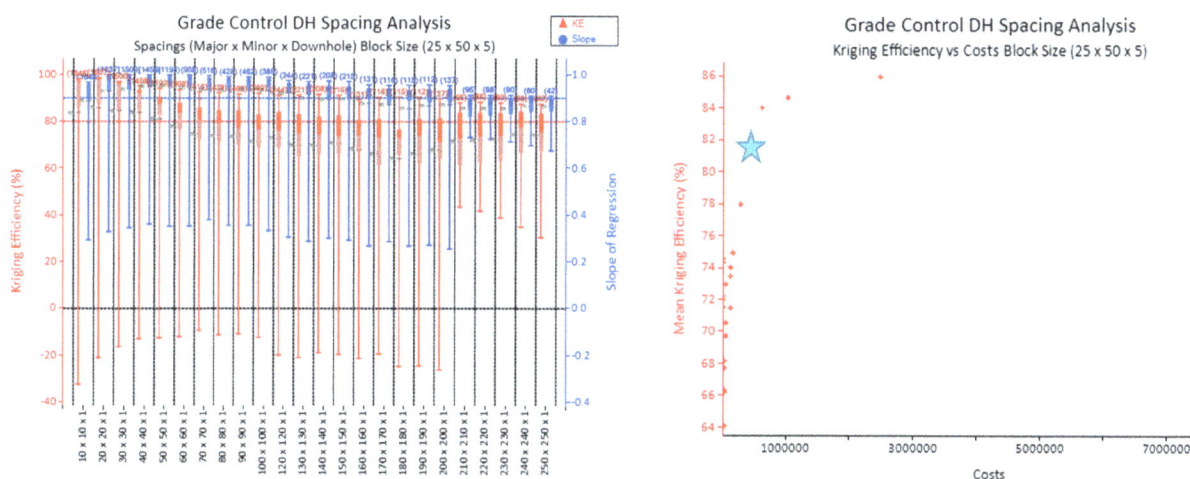

FIG 3 – Outputs of the DHSA at the Mineral Resource block size.

The same process was undertaken to encapsulate the 90:15 approach; however, in this case, a quarterly production volume was applied to the block size (250 m × 250 m × 5 m) rather than the parent cell volume (Figure 4). The same mineralisation domain variography and conceptual drill grids were used, although a large number of samples were used to offset the potential impact positively-skewed populations may have on confidence interval calculations. Post processing for the 90:15 requirements, using the method of Verly, Postolski and Parker (2014) suggests that a drill grid of 50 m × 50 m is appropriate (Figure 5). This shows the 15 per cent error (orange line) and the scaled 90 per cent CI based upon the Kriging variance as the blue line. As a comparison, a nearby advanced nickel laterite project defines Measured Resources based upon a 50 m × 50 m drilling grid.

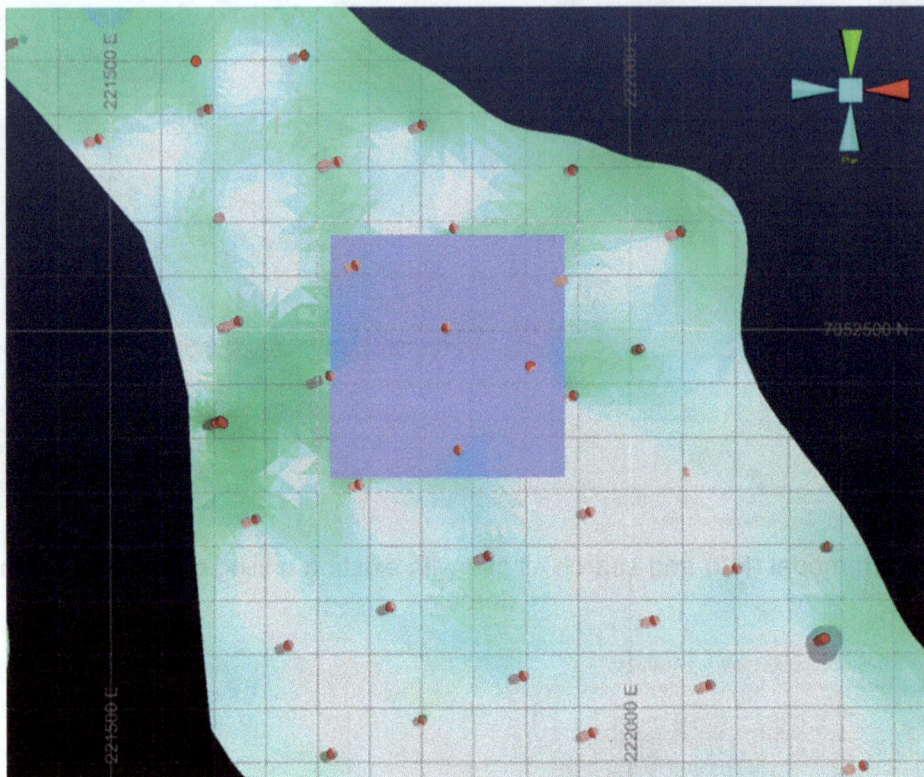

FIG 4 – Production block scale located within the mineralisation domain.

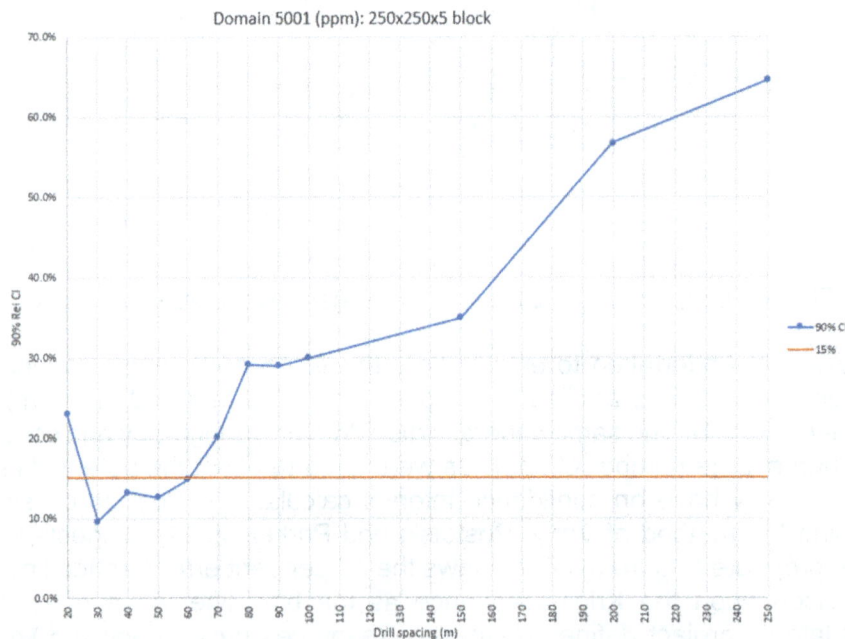

FIG 5 – Output of the 90:15 approach to variance-based drill hole spacing analysis.

It is important to reiterate that a key assumption of the KV methods is that the current variogram, based upon the existing drilling, will be preserved with infill drilling. A refinement of the KV-based approaches is to investigate the impact of a 'worse' and a 'better' variogram than the current evidence-based model, but these are not based upon real-world outcomes, just predictions.

SIMULATED REALITY APPROACHES

Overview

As mentioned above, the use of conditional simulation provides the DHSA practitioner with the potential to define a dense grid of data, using the parameters of the existing real-world data set, which can then be post-processed in a variety of ways, either to define a confidence interval as called for in the 90:15 pseudo-standard, or to look at the drilling required to generate a maximum error, or more sophisticated methods. The approach described here, and the associated case study, looks at the cost of misclassification and the drilling grid required to minimise the cost of misclassification, based upon some relevant cost and revenue assumptions. The key objective is to define an optimal grade control grid, but the same approach can be used in the 90:15 context to achieve a certain classification confidence level.

A typical workflow for this variant of the simulated approach would be as follows:

1. Using the real-world data, simulate on a dense grid over the area of interest.

2. From the suite of realisations, choose one which reflects the key parameters (histogram, variogram, multivariate statistics) of the existing drill holes.

3. Using the chosen realisation as the 'truth', resample to generate pseudo drill holes on the range of potential drilling grids.

4. Generate a grade estimate for the key variables for each potential drill grid.

5. Using a model of the misclassification and associated revenue and losses, calculate the total net revenue (or loss) associated with each drilling grid. The optimal drilling grid is that which gives the best cost-benefit.

This approach is similar to one of the workflows presented by Alfonseca and Silva (2022), but the emphasis is on defining the misclassification costs and benefits. As with the KV/CI methods, the outcome depends upon the assumption that the quality of the continuity model does not diminish as infill drilling takes place.

Cost optimisation

The theoretical relationship between estimated blocks and true blocks in a mineral deposit has been documented since the introduction of modern geostatistical techniques and first appeared in Journel and Huijbregts (1978). A version is presented in Figure 6.

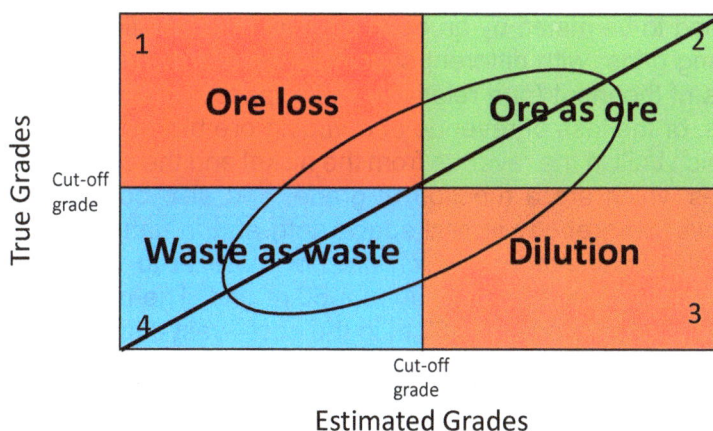

FIG 6 – Relationship between estimated and true grades above and below an applied cut-off.

In its simplest form, the relationship between estimated block grades and true block grades is portrayed with respect to an applied mining cut-off. This divides the blocks into four quadrants:

1. Blocks predicted to be waste which are in reality ore.

2. Blocks predicted to be ore which are truly ore.

3. Blocks predicted to be ore which are in reality waste.

4. Blocks which are predicted to be waste, and which actually are waste.

This rather simplistic formulation generates two quadrants, 2 and 4, where there is no misclassification, and two quadrants, 1 and 3, where potential misclassification exists. Figure 6 shows an ellipse reflecting the scatter of true versus estimated grades and a 1:1 regression line. Due to conditional bias the diagram typically exhibits a regression line having a slope less than unity; in other words, high-grade estimated blocks underestimate the true grade and low-grade estimated blocks overestimate the true grade.

The key application is that the 'true' blocks on the Y axis can be represented by a simulation which reflects the key parameters—histogram, variogram and inter-analyte correlation (if necessary)—of the sample drill holes. The estimated blocks on the X axis reflect ordinary kriged (or similar) models built from pseudo-drill holes selected from the 'reality' at various grid spacings. Thus any one block in an estimate will reside in one of the four quadrants.

Cost profiles can be assigned to the four quadrants. In the simplest possible formulation, this may be:

- quadrant 1 – the negative value of the material not processed ('ore loss')
- quadrant 2 – the revenue from the material sent for processing
- quadrant 3 – the cost of mining and processing waste material which has no value ('dilution')
- quadrant 4 – the cost of mining waste.

A simplified model (as in the case study below) is to assume an open pit scenario where all material (ie quadrants 1–4) has to be removed, whether to the waste dump or to the processing plant – thus the mining cost is not a factor in the cost as it is the same for each block in whatever quadrant. There are many refinements which may be made to this simple model, including a variable mining cost for ore and waste material.

The key point is to choose the drilling configuration which either minimises the loss or (hopefully) maximises the profit. For each proposed drilling grid, however, there is the cost of drilling; thus a very tight drilling grid should maximise the profit, but the cost of drilling may outweigh the revenue from the reduced misclassification.

Case study – nickel-copper project

The objective of this case study was to define the optimum drilling grid for grade control at a nickel-copper project scheduled to be mined by an open pit. The misclassification approach was adopted, and four potential drilling grids, with different spacings, were drawn from a 'truth' simulation which replicated the statistics of the input (and relatively wide-spaced) drilling on a fine grid. The revenue from mining ore blocks, or the loss of revenue of dumping ore blocks, was defined as a net smelter return (NSR) value, which built in the revenue from the nickel and the copper, adjusted for the relative metallurgical recoveries which are a function of grade, and also consider treatment and refining charges. Figure 7 shows the orebody in real space (left) and in flattened space (right). There are some artefacts of the flattening process which are not material to the outcome. The exploration drilling is on an approximate 25 m × 25 m to 50 m × 50 m grid. The mineralisation is around 750 m in the north–south dimension and around 950 m in the east–west dimension.

The reference simulations for the DHSA use an additive log ratio compositional transformation (ALR) method, followed by Gaussianisation to preserve the histogram, covariogram and correlation between the input nickel and copper grades. A detailed description of the multivariate simulation is beyond the scope of this paper, but the theory and practice are summarised in Cook et al (2023). Figure 8 shows the input composite nickel and copper data and the simulated values from one realisation, intended to represent reality. The selected realisation preserved the histogram of the nickel and copper values, the covariogram and the correlation between the samples.

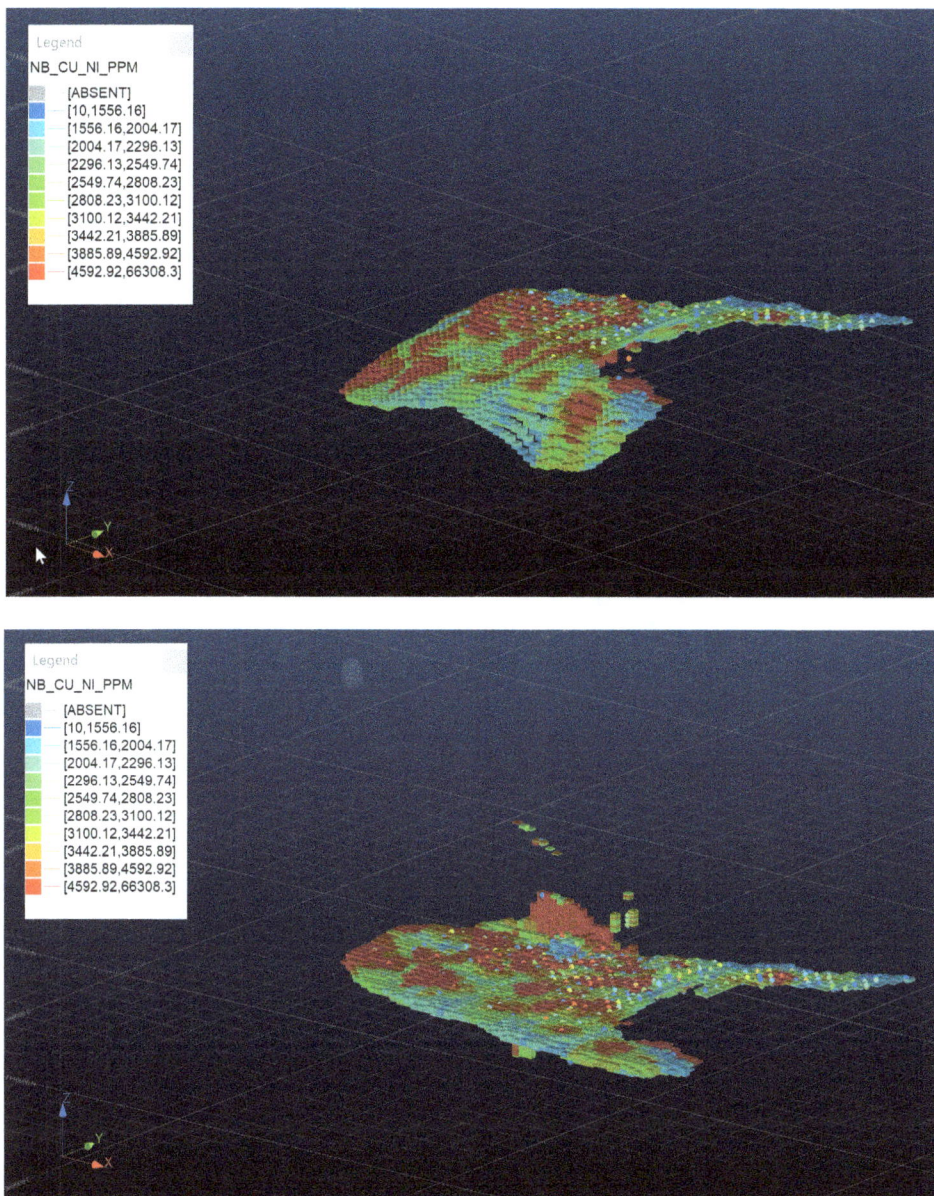

FIG 7 – 3D view of the mineralisation and drill holes, coloured on nickel grade in real-world space (left) and in flattened space (right).

Once a realisation of simulated nickel and copper grades was selected, this was used to generate pseudo drilling grids which were derived from the simulated samples. Four potential candidate grids were chosen; 10 m × 10 m, 15 m × 15 m, 20 m × 20 m and 25 m × 25 m, each with 2 m downhole samples from vertical holes. In each case holes were generated on an offset 'dice 5' pattern. Figure 9 shows two pseudo drilling grids selected from the realisation, 10 m × 10 m and 25 m × 25 m.

Once the four drilling grids had been extracted, these were used to generate an ordinary kriged block model at a 10 m × 10 m × 5 m SMU size. All four models used the same variograms, which were generated from the 10 m × 10 m pseudo-drilling, not from the original simulation. It would have also been appropriate to generate different variograms in each drill grid instance, but the differences in continuity were not considered material, given the generally good variograms of the nickel and copper mineralisation.

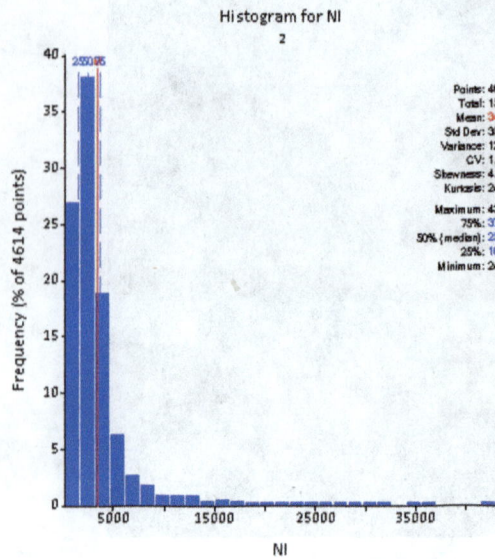

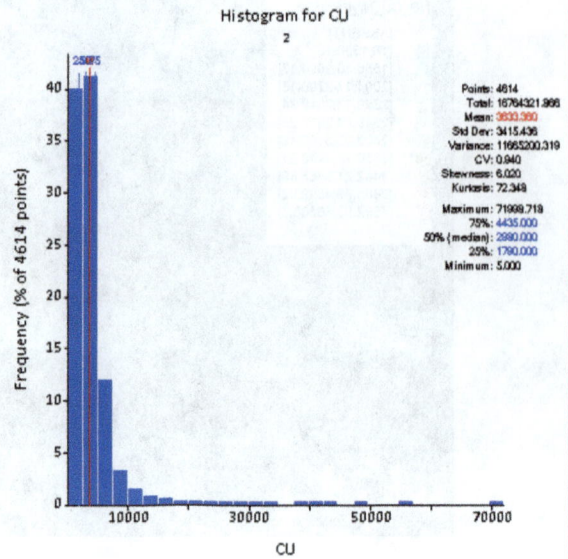

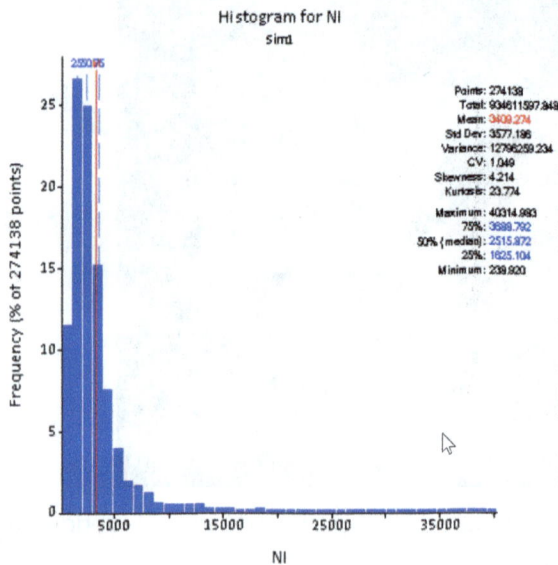

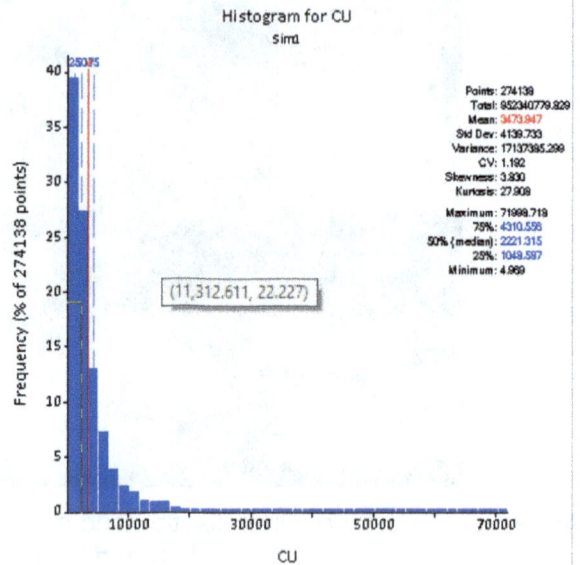

FIG 8 – Histogram of the sample nickel and copper composites (in ppm, top) and the simulated values from one realisation (bottom).

The OK models were validated against their own set of input pseudo-drill holes and then each were compared back to 'truth' simulation. Drilling and assaying costs were supplied and were based upon current contract prices, assuming that all samples were to be assayed for both nickel and copper. Figure 10 shows the scatter between estimated and 'true' grades for the 10 m × 10 m and 25 m × 25 m drilling cases, applying a cut-off of 0.15 per cent copper. The quadrants as per Figure 6 are identified.

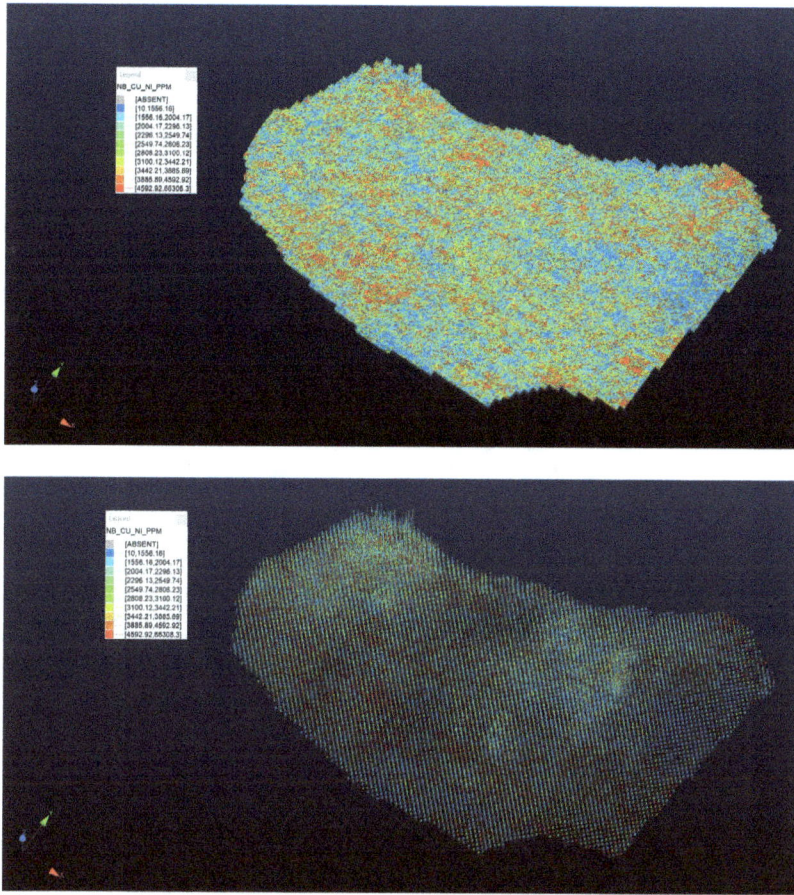

FIG 9 – Two instances of simulated drilling grids showing nickel grades, 10 m × 10 m (top) and 25 m × 25 m (bottom).

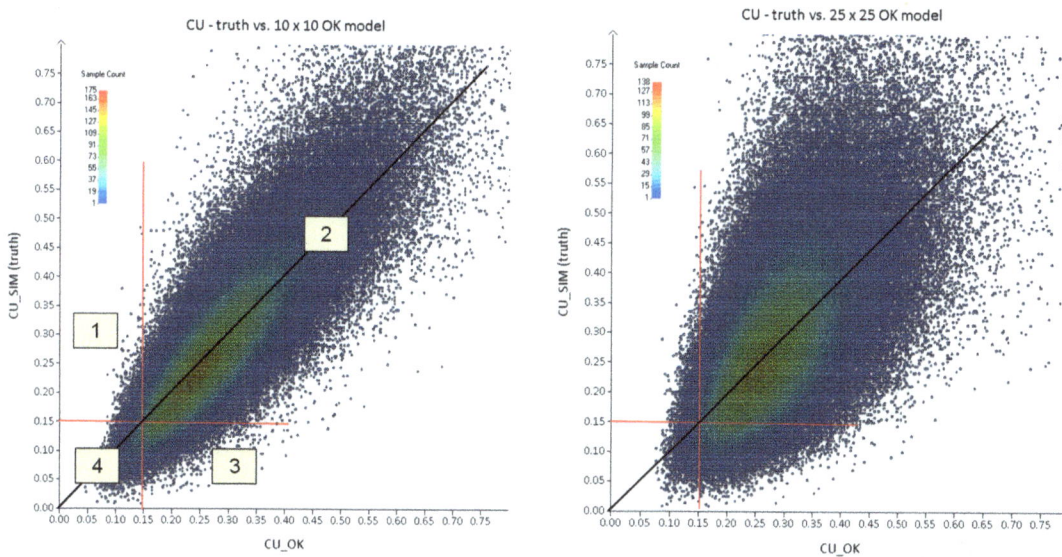

FIG 10 – Comparison of the estimated model versus the truth for copper grades in the 10 × 10 m drilling (left) and 25 × 25 m drilling (right) cases.

Figure 10 shows that quadrant 3 (dilution) is a greater contribution to misclassification than quadrant 1 (ore loss). It can be predicted in advance of the estimation-simulation comparison that a tighter drill spacing will result in less misclassification errors; however, this approach allows the quantification of the economic impact of the misclassification compared to the overall drilling cost.

For the misclassification algorithm, a simplified model was used, assuming that all tonnes had to be mined in an open pit and had the same mining cost; therefore, the following revenue/cost streams were allocated to the quadrants of Figure 6:

- quadrant 1 – the revenue from the lost ore sent to the waste dump
- quadrant 2 – the revenue minus the processing cost
- quadrant 3 – the processing cost of the waste material
- quadrant 4 – no cost or revenue.

The net revenue or cost per block is the sum of these four (three) items, accumulated over the entire deposit, minus the drilling cost for each scenario. Once the consolidated net revenue is available, several metrics can be generated to optimise the drilling grid; one is shown in Figure 11, which plots the net revenue relative to the 10 m × 10 m drilling case against the drilling grid.

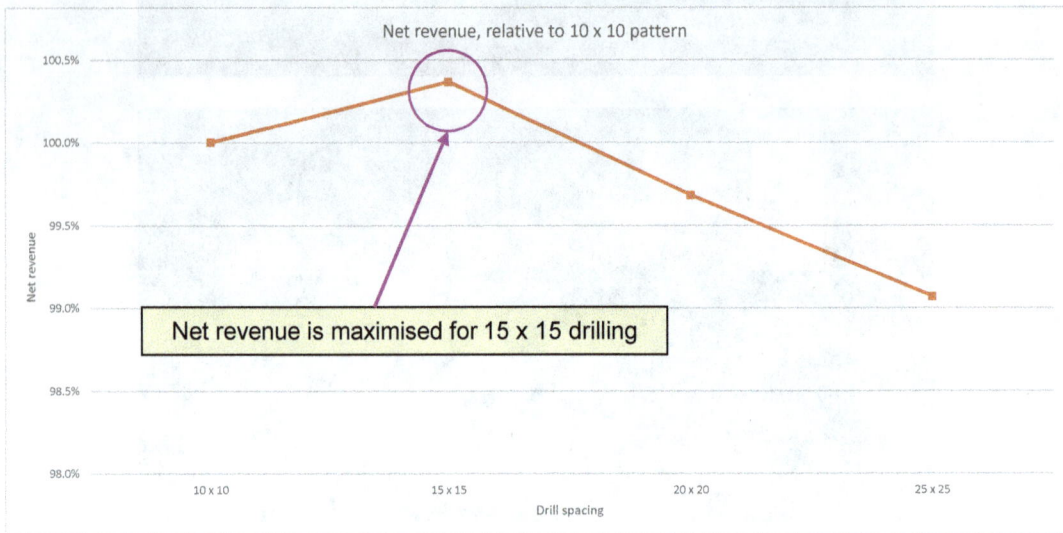

FIG 11 – Chart of net revenue relative to the 10 × 10 m drill pattern.

It is noted that the differences in revenue are small, but that the revenue is maximised for the 15 m × 15 m drilling case. Considering the cost of drilling alone, the 10 m × 10 m drilling grid costs 50 per cent more than the 15 m × 15 m grid.

The misclassification model used to generate the net revenue is robust but simplistic, and could be refined to reflect more realistic cost models, for instance:

- the relative and different costs of mining ore and waste can be used
- an increasing mining cost with depth could be applied
- the opportunity cost of displacing ore from the mill by processing waste can be invoked.

The sophistication of the misclassification revenue algorithm depends in part on the predicted mine-mill balance; if mine production outstrips mill capacity then quadrant 3 (dilution) incurs a higher penalty, while if the processing capability is greater than mine production, the cost of dumping ore is more than the cost of accepting dilution at the plant. This invokes the theories of risk profiles and loss functions (Glacken, 1996) which can be used to define a more realistic outcome.

CONCLUSIONS – THE FUTURE OF DHSA

The practice of drill hole spacing analysis is mature and the literature is replete with detailed case studies, especially over the past five years. DHSA is used either to define the spacing for a particular Mineral Resource classification, usually Indicated, based upon the industry 90:15 pseudo-standard, or to optimise the return from drilling by quantifying and minimising misclassification. There are a range of techniques available to the practitioner, which can be based upon relatively simple and straightforward treatments of Kriging variance for various production-related units, or upon more complex (and time-consuming) approaches which consider the risk inherent in grade (and sometimes thickness and material type) through conditional simulation. Because the simulation approaches have the ability to generate a dense grid of data which reflects the input sample data, multiple pseudo-drilling grids can be used, either for re-simulation (generating a confidence interval and associated mean error) or to compare estimates at the proposed scale of mining with the 'truth', allowing for some relatively sophisticated cost and revenue analysis.

Further developments will see the routine inclusion of these tools in either generalised or specialised mining software packages and the analysis of optimal drill spacing as part of estimation and classification workflows. The increased ability to quickly generate simulated data on a dense grid and then resample either for re-simulation or estimation will lead to more accurate DHSA studies which should result in cost and time optimisation. The introduction of risk profiles will provide a more realistic outcome which considers differential costing.

All techniques, however, whether simple or complex, do rely on the assumption of consistency of the spatial continuity model from the existing, sparse data to the future infill drilling, whether for resource classification or grade control purposes. Any scenario testing of alternative variograms can never be based upon reality but may allow the consideration of upside or downside scenarios.

Furthermore, while the DHSA methods described are complementary to the Mineral Resource Classification, the consideration of other classification criteria, such as data quality, geological uncertainty, and other factors *must* form part of the classification allocated by the Competent or Qualified Persons, and thus the DHSA results can only be a starting point, albeit a firm foundation for the ultimate outcome.

ACKNOWLEDGEMENTS

The permission of A-Cap Energy to publish results from the Wilconi Project is gratefully acknowledged. Permission from another minerals company to discuss the multivariate DHSA case study and findings is also appreciated. The paper has benefited from a review by Paul Blackney.

REFERENCES

Alfonseca, B V and Silva, V, 2022. Defining optimal drill-hole spacing; a novel integrated analysis from exploration to ore control, *Journal of the Southern African Institute of Mining and Metallurgy*, 122(6):1–12.

Cook, A, Rondon, O, Graindorge, J and Booth G, 2023. Iterative Gaussianisation for multivariate transformation, in *Geostatistics Toronto 2021 – Quantitative Geology and Geostatistics* (eds: S A Avalos Sotomayor, J M Ortiz and R Mohan Srivastava), pp 21–36 (Springer Verlag).

Glacken, I M, 1996. Change of support by direct conditional block simulation, Master's thesis, Stanford University, USA.

Journel, A G and Huijbregts, C J, 1978. *Mining Geostatistics*, 600 p (Academic Press: London).

Murphy, M, Parker, H M, Ross, A and Audet, M-A, 2004. Ore-thickness and grade resource confidence at the Koniambo Nickel Laterite (a conditional simulation voyage of discovery), *Geostatistics Banff 2004* (eds: O Leuangthong and C V Deutsch), pp 469–478 (Springer Verlag).

Neves, C M S, Costa, J F, Souza, L, Guimaraes, F and Dias, G, 2023. Methodology for defining the optimal drilling grid in a laterite nickel deposit based on a conditional simulation, *Geostatistics Toronto 2021 – Quantitative Geology and Geostatistics* (eds: S A Avalos Sotomayor, J M Ortiz and R Mohan Srivastava), pp 151–162 (Springer Verlag).

Nowak, M and Leuangthong, O, 2019. Optimal drill hole spacing for resource classification, *Mining Goes Digital* (ed: U Mueller), pp 115–124 (Taylor and Francis Group: London).

Owusu, S K A and Dagdelen, K, 2019. Critical review of mineral resource classification techniques in the gold mining industry, *Mining Goes Digital* (ed: U Mueller), pp 201–209 (Taylor and Francis Group: London).

Parker, H M and Dohm, C E, 2014. Evolution of Mineral Resource classification from 1980 to 2014 and current best practice, *Finex 2014 Julius Wernher Lecture,* 72 p.

Rivoirard, J, Renard, D, Celhay, F, Benado, D, Queiroz, C, Oliveira, L and Ribeiro, D, 2016. From the Spatial Sampling of a Deposit to Mineral Resources Classification, *Geostatistics Valencia 2016* (eds: J J Gómez-Hernández, J Rodrigo-Ilarri, M Elena Rodrigo-Clavero, E Cassiraga and J A Vargas-Guzmán), pp 329–344 (Springer-Verlag).

Usero, G, Misk, S and Saldanha, A, 2019. An approach for drilling pattern simulation, *Mining Goes Digital* (ed: U Mueller), pp 59–66 (Taylor and Francis Group: London).

Verly, G, Postolski, T and Parker, HM, 2014. Assessing uncertainty with drill hole spacing studies – applications to Mineral Resources, in *Orebody Modelling and Strategic Mine Planning Symposium 2014*, pp 109–118, AusIMM, Melbourne.

Wawruch, T M and Betzhold, J F, 2004. Mineral Resource classification through conditional simulation, *Geostatistics Banff 2004* (eds: O Leuangthong and C V Deutsch), pp 479–489 (Springer Verlag).

Resource and Reserve category inflation – known rewards, hidden risks

G W Booth[1], R R Hargreaves[2] and M Bond[3]

1. Associate Executive Consultant, Snowden Optiro, Perth WA 6000. Email: em64216@gmail.com
2. MAusIMM(CP), Principal Consultant, Snowden Optiro, Perth WA 6000.
   Email: rayleen.hargreaves@snowdenoptiro.com
3. Commodities Account Manager, S&P Global Market Intelligence, Perth WA 6000.
   Email: michael.bond@spglobal.com

ABSTRACT

There are numerous causes of speculative mineral resource and reserve inflation. Traditionally, these are linked to issues of data accuracy or misinterpretation, the use of overly optimistic estimation parameters or assumptions, an absence of independent expert verification, or the simple pressure to increase company value, covert or otherwise.

Owing to its subjective nature, classification of resources and reserves remains an opaque process. Of necessity, Competent Persons rely on personal heuristic interpretations, ideally underpinned by appropriate estimation and classification experience across commodity and deposit types.

Whereas independent verification of individual resource and reserve statements remains impractical, publicly available data can be interrogated for potential category inflation, particularly when viewed with exploration expenditure statistics included in annual company reports.

Abrupt changes in category tonnages and market capitalisation without concomitant expenditure can aid in outlier detection. Consequently, a comparison of common exploration and resource reporting parameters offers a basis for risk assessing classification robustness, over time.

In this introductory study, an analysis of globally listed companies was undertaken over a multi-year period, examining base and precious metal projects in both preproduction and operational asset classes. From these, a series of exploratory decision tree trigger points were generated against which target companies were anonymously identified and risk-ranked.

INTRODUCTION

Corporate fiduciary obligations stress the importance of an unbiased reporting of resource and reserve assets, with technical requirements typically sourced from stock exchanges and professional licencing and/or governing bodies. Bourses and professional institutes worldwide require listed entities and their officers to carefully consider economic viability and materiality when publicly presenting exploration and mining asset data.

Traditionally, such requirements are meant to protect the broader investment community and their advisors who rely on credible information on which to base their respective speculative strategies (eg Jowitt and McNulty, 2021). Unsurprisingly, at all stages of development, the size and quality of an exploration target or mineral asset have an unambiguous influence on investor enthusiasm.

However, as issues of resource sustainability have become increasingly important, companies are now required to plan for the long-term social and environmental impacts of their respective operations. In this regard, a biased estimate may impede or even preclude completion of key sustainability obligations, including mine rehabilitation and long-term asset closure monitoring (Jowitt, Mudd and Thompson, 2020). While frequently ignored, maintaining a sound business reputation is also crucial to corporate success. Too often, the short-term benefit of asset misrepresentation swiftly unravels as material reality sets in. The corporate reputational damage which inevitably results makes it all the more difficult to attract ongoing investment for any given target no matter how initially promising.

Depending on jurisdiction, any intentional or inadvertent misrepresentation of a mineral estimate constitutes a reporting breach, many of which attract a formal penalty, including fines. Regrettably, even when combined with all of the other stakeholder obligations, such punitive consequences may not moderate overzealous interpretations, with their misguided perception of long-term economic advantage. Too often, any reporting risks are outweighed by the considerable benefits of an embellished estimate. While there is clearly a need for exploration and mining companies to deliver value when competing for investor funding (Kreuzer, Etheridge and Guj, 2007), overstating an asset's potential is as perilous as it is proscribed.

Fortunately, for comparable projects within specific commodity classes, the rate of change in mineral resource and market capitalisation offers an opportunity to identify inflated resource and/or reserve estimations, particularly when viewed with exploration expenditures.

BACKGROUND

With its comprehensive database comprising global exploration budgets, reserves replacement analysis and other in-depth asset level metrics for mining properties, projects, companies and mines worldwide, S&P Global Metals and Mining platform was used to test for the presence of resource and reserve inflation. More specifically, the correlation between resource and/or reserve growth as well as category elevation against exploration expenditure was examined.

This investigation follows examination of global reconciliation reporting as to how well publicly reported resource and reserve depletions compare to metal production using nominal reconciliation factors (Hargreaves and Booth, 2019). Its focus is on discovery costs and how they can be used to test for resource and/or reserve inflation. Corporate and jurisdictional identities were purposefully concealed.

DATABASE

The S&P Global Metals and Mining database was interrogated for representative companies across two broad project classes including Developing (Target Outline, Advanced Exploration, Prefeasibility/Scoping, Reserves Development) and Operating (Construction Started, Operating, Expansion) categories.

Group formulation

From the more 1725 individual projects extracted worldwide, two commodity groups were created, including base (Cu, Pb, Zn, Ni, Li) and precious (Au, Ag, Pt, Pd) metals. From the 1604 ventures ultimately identified, a total of 381 multi- and single-project companies were selected for analysis. To create a more robust data set, company information was collected over a purposefully lengthy, seven-year (2015–2021) period.

Data selectivity

Where company information was questionable or total changes in corporate value were not attributable to a given project or series of projects against which expenditures were made, such data was automatically eliminated from consideration. This included asset changes associated with the release of a maiden resource and/or reserve, or the purchase/incorporation of an external mineral asset developed outside of the exclusive seven-year period. Accordingly, care was taken to ensure that projects were neither excluded or nor included which might otherwise introduce bias. However, of necessity, this study specifically sought and thereafter targeted outliers as prospective examples of both resource and/or reserve inflation.

Parameter generation

The following parameters were calculated and/or directly extracted from the S&P Global database:

1. Percentage Change in Resources and Reserves (tonnage – base metal/ounces – precious metal).

2. Percentage Change in Market Capitalisation.

3. Change in Market Capitalisation (USD) ÷ Exploration Spend (USD).

4. Percentage of Measured and Indicated Resource of Total Resource.

5. Percentage of Inferred Resource of Total Resource.

6. Total metres drilled (289 of 381 companies reporting).

7. Total Exploration Budget (USD).

8. Total Exploration Spend (USD).

ANALYSIS

Data evaluation commenced with univariate analysis and the creation of conventional boxplots and bar charts, with the determination of supporting group mean, median, minima and maxima group statistics, by development stage and metal class. Thereafter rank and percentile assessments were undertaken of four exploratory parameters from which trigger points were generated and anomalous projects individually risk-ranked over this multi-year period.

Resources and Reserves – percentage change

Percentage change in resource and reserve data (Figure 1) generate tight inter-quartile ranges linked to extreme maxima, particularly amongst developing base metal projects with their tonnes-based comparisons.

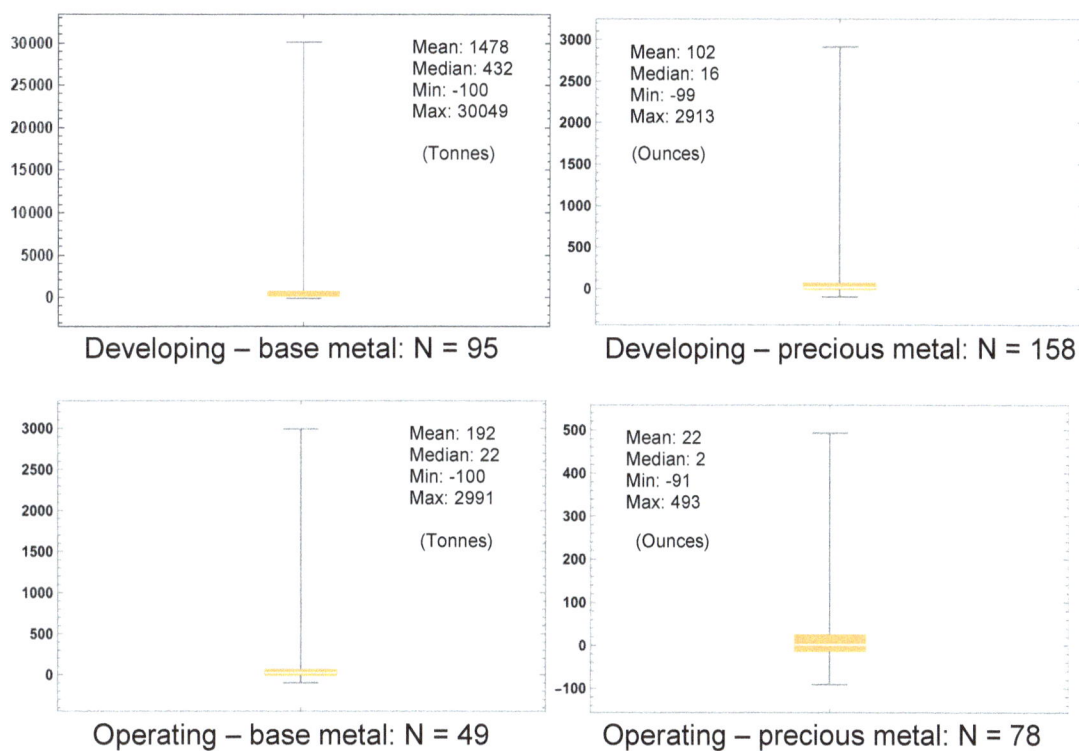

FIG 1 – Box and whisker plot – per cent change total Resources and Reserves by project class.

While this parameter is broadly sensitive to inflationary pressures, where projects commence with little to no resource, large percent changes need not uniquely reflect problematic resource or reserve growth. Irrespectively, contrasting mean with maximum percentage change (Figure 2), reinforces the comparative size of outliers observed in individual project classes.

Of note, the separation observed between the mean and median values of developing and operating projects of either commodity class are consistent with the relative maturity of the latter group. Unless highly successful reserve definition programs are in place, significant changes in this parameter at an operational level become less likely over time.

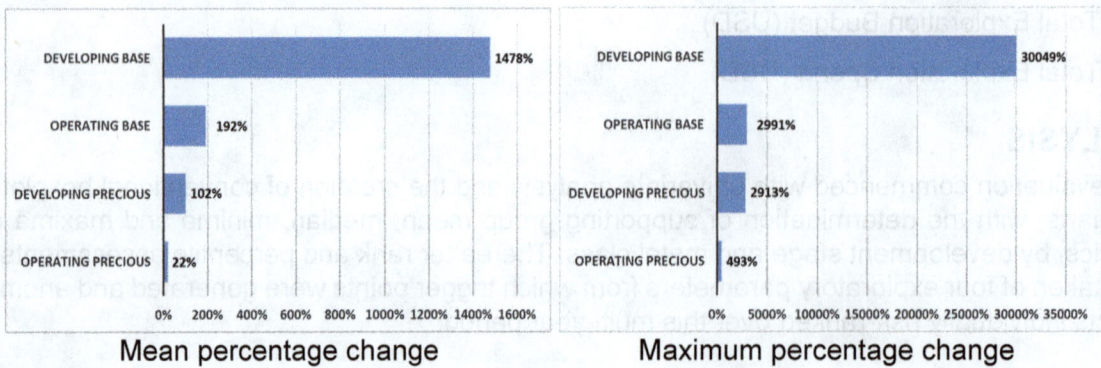

FIG 2 – Mean and maximum per cent change in total Resources and Reserves by project class.

Market capitalisation – percentage change

Percentage change in market capitalisation (Figure 3) is similarly influenced by large maxima (eg >50 000 per cent), particularly amongst developing base metal projects. Whilst sizeable changes are of greatest interest, taken individually, especially in early projects, supporting factors are clearly required for context definition.

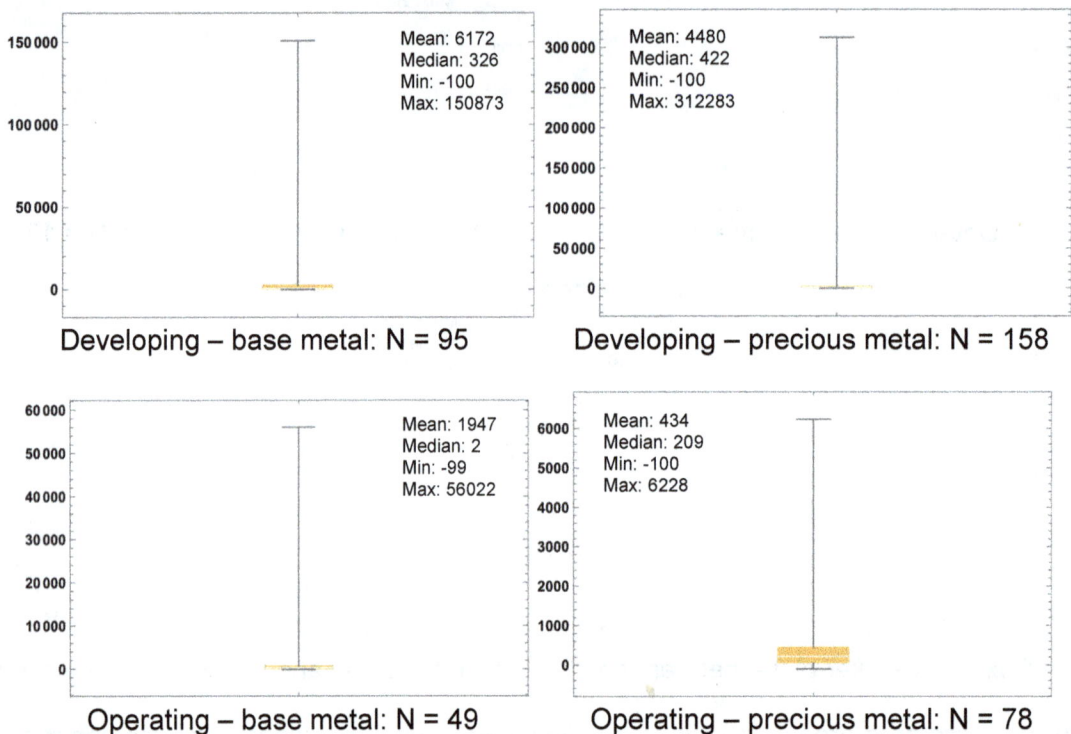

FIG 3 – Box and whisker plot – percentage change $market capitalisation (USD) by project class.

Nevertheless, the magnitude of mean and median values for both commodity groups and project classes underscores the benefits of a successful and discovery and resource definition program for any given company, especially when achieved over a short time frame. In this context, maximum percent change in developing project classes can be remarkably high (Figure 4), consistent with traditional Lassonde curve share price movement (eg Rijsdijk *et al*, 2022).

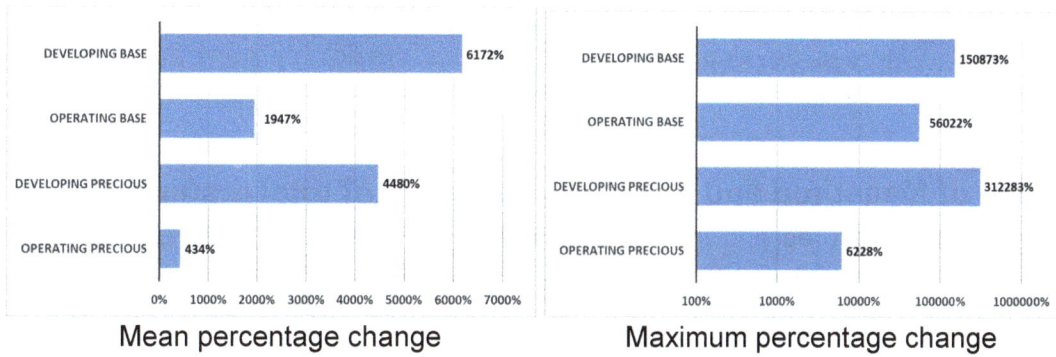

Mean percentage change

Maximum percentage change

FIG 4 – Mean and maximum per cent change $market capitalisation (USD) by project class.

Change in market capitalisation (USD) ÷ exploration expenditure (USD)

The market capitalisation and exploration expenditure ratio (Figures 5 and 6) was created to aid in the tracking of large changes in corporate value against exploration outlay, ideally linked to advanced deposit definition or physical discovery. In the presence of modest mean and median statistics, as well as any negative values, an excessively large ratio offers insight into a company's claimed success rate by way of exploration investment.

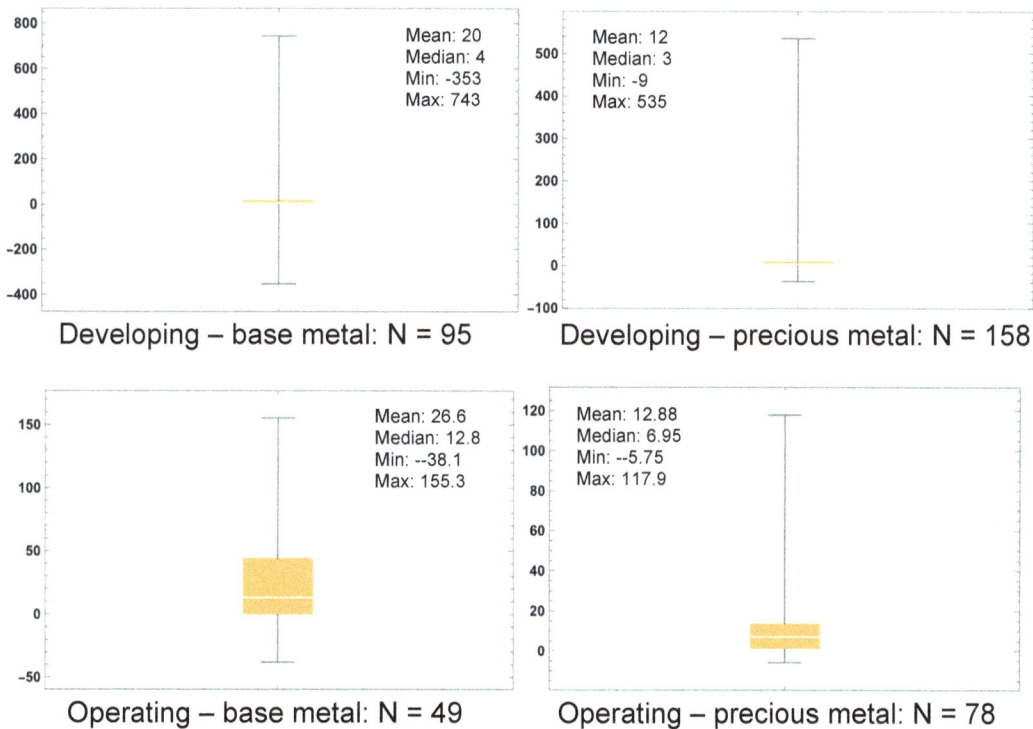

Developing – base metal: N = 95

Developing – precious metal: N = 158

Operating – base metal: N = 49

Operating – precious metal: N = 78

FIG 5 – Box and whisker plot – change in $market capitalisation ÷ $exploration expenditure.

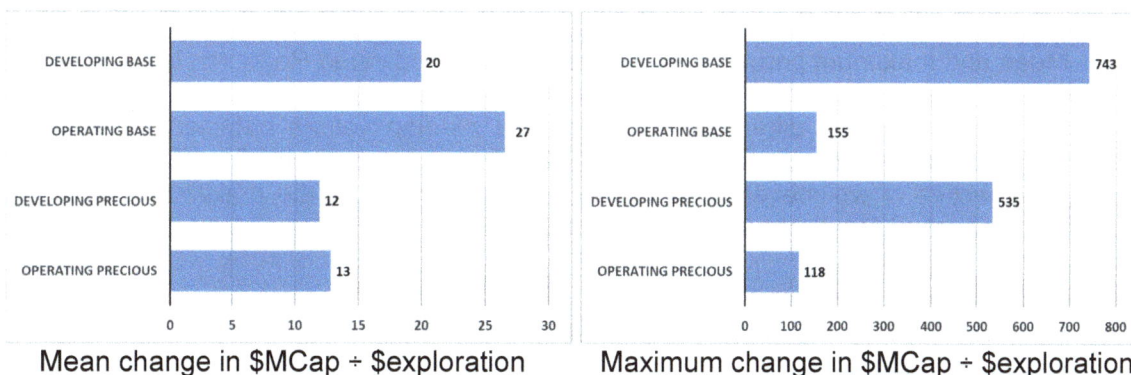

Mean change in $MCap ÷ $exploration

Maximum change in $MCap ÷ $exploration

FIG 6 – Mean and maximum percent change $market capitalisation (USD) by project class.

Accordingly, any elevated statistic represents an exceptionally high rate of return for exploration dollar expended. Whereas developing projects statistics predictably generated higher maxima, individual operating entities also offered significant outliers to test for the presence of resource and reserve inflation.

Percentage of Measured and Indicated Resource of total resource

The percentage of Measured and Indicated Resource of total resource (Figures 7 and 8) offers analysts a direct insight as to the likelihood of resource and/or reserve inflation. Where this parameter is elevated, so too are the proportions of these two upper resource classes and hence, the likelihood for estimation inflation to be the cause.

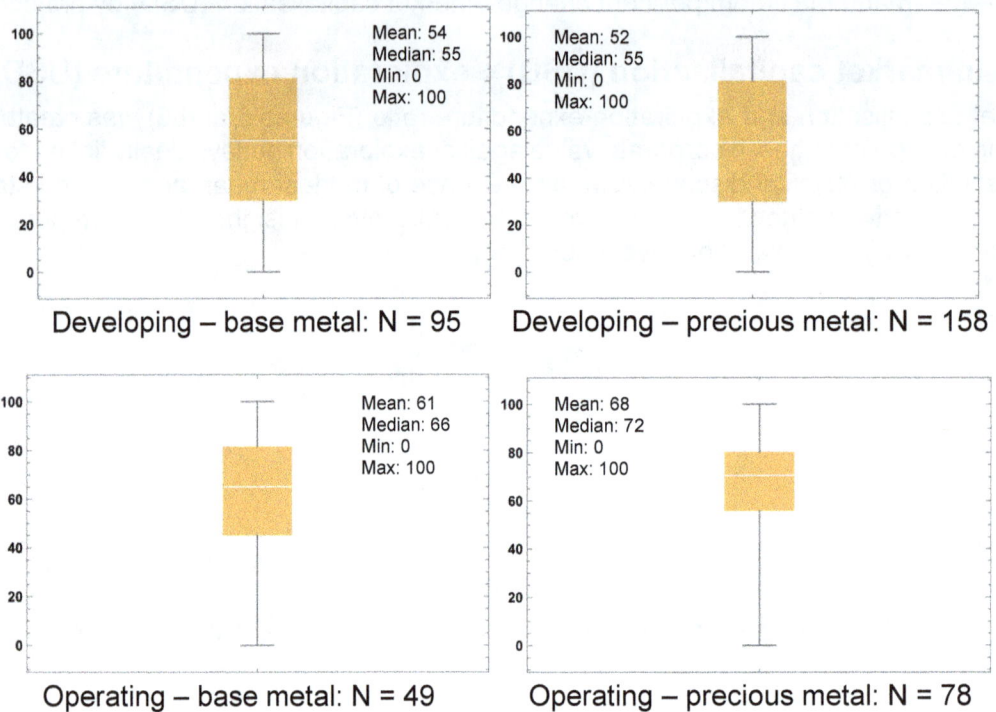

FIG 7 – Box and whisker plot – percentage of Measured and Indicated Resource of total resource.

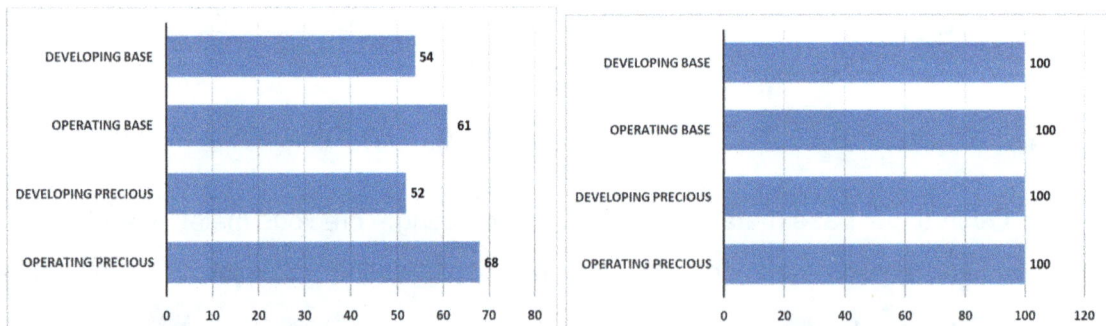

FIG 8 – Mean and maximum percentage of Measured and Indicated Resource by project class.

The uniform ~10–15 per cent difference in mean and median values between developing and operating project groups adds credibility to the sensitivity of this factor for its use in anomaly detection. Irrespective of whichever project group was interrogated, the maximum percentage of Measured and Indicated Resource was identical across all project classes, with multiple examples of 100 per cent observed.

Drilling and exploration budget/expenditure statistics

Owing to their partial reporting by contributing companies, cumulative drill metres could not be used in this study. The authors suggest that any future inclusion requires use of both category (ie exploration, grade control, geotechnical, sterilisation etc) and type (ie diamond, reverse circulation, aircore) statistics.

Such contextual information offers analysts a utilitarian benchmark for assessing resource classification integrity and even direct, per tonne discovery costs. In the absence of this level of granularity, the use of a partial data set was deemed problematic. However, a separate comparison of exploration budget and spend data found a relatively strong correlation between planned and actual expenditure across all commodity and project groups.

RISK RANKING

From the 381 individual and combined projects initially drawn in this study, the upper quartile of each data set was selected for risk ranking, using performance criteria derived by commodity group and project status. From this analysis, trigger points were applied and projects progressively assessed.

Percentile ranking

With their creation of a single scale, percentile-ranking techniques facilitate comparison of data with differing units. Accordingly, they enable parameter combination and ultimately the generation of sensible thresholds for variably skewed data sets. In this introductory analysis, these four foundation criteria underwent collective percentile ranking for which a range of options were considered.

In the presence of multiple prominent outliers, amongst many, interquartile range and Z scores were trialled and ultimately deemed unsuitable. In keeping with the broad goals of this introductory study, a straightforward upper quartile range was typically employed to generate period trigger/risk points. When examined against grouped data, this threshold ensured that all outliers, together with a reasonable underlying project buffer was available for inflationary mineral resource and reserve testing (Figure 9).

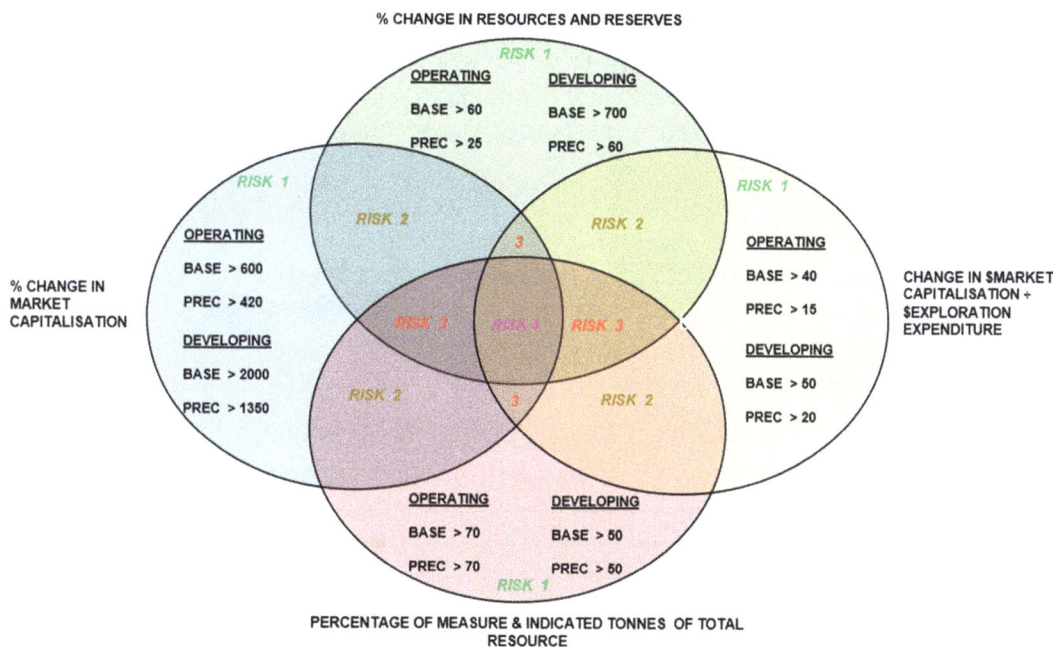

FIG 9 – Percentile rank risk criteria by parameter – period (seven year) thresholds.

Developing – base metal

From the ranking assessment, a total of 96 risk events were observed (Table 1) which underwent further classification by parameter attributes.

TABLE 1

Developing – base metal total risk events.

INFLATION   LIKELIHOOD ➡

DEV BASE	RISK 1	RISK 2	RISK 3	RISK 4
% CHANGE - RES & RER	12	7	4	1
% CHANGE - MKT CAP	6	8	9	1
$MKT CAP / $EXP SPEND	6	7	10	1
% M & I TONNES	12	4	7	1
TOTAL RISK EVENTS	36	26	30	4
% UPPER QRTL	38%	27%	31%	4%

A follow-up analysis targeted individual Level 3 and a single Level 4 project for changes in published resource tonnages over the interval period. Of immediate concern to the authors was the relative paucity of supporting data from which to make an informed decision regarding the reliability of published resource estimates, including the identity of supporting Competent Persons. Similarly, while company exploration expenditure claims appeared realistic, totals cited were invariably at odds with the size or scope of exploratory or infill drilling programs required for advanced resource category classification within normal developing projects.

Given that discovery costs are likely to have doubled over this seven-year interval and the likelihood of quickly converting any discovery into formal mineral resource is low, the number and size outliers identified was noteworthy. While the extent to which some resource inflation may have occurred remains uncertain, the authors were able to identify multiple projects with marked changes in market capitalisation, tied to sizable increases in resource, supported by apparently modest exploration investments.

Developing – precious metal

From the ranking assessment, a total of 156 at risk events were observed (Table 2) which underwent further classification by parameter attributes, with a focused assessment of risk 3 and 4 level projects.

TABLE 2

Developing – precious metal risk events.

INFLATION   LIKELIHOOD ➡

DEV PREC	RISK 1	RISK 2	RISK 3	RISK 4
% CHANGE - RES & RER	19	11	8	1
% CHANGE - MKT CAP	14	17	7	1
$MKT CAP / $EXP SPEND	11	19	8	1
% M & I TONNES	24	7	7	1
TOTAL RISK EVENTS	68	54	30	4
% UPPER QRTL	44%	34%	19%	3%

Of immediate concern amongst identified companies was the proportion of Inferred versus Measured and Indicated Resource, with the latter two categories often exceeding 85 per cent of the combined ounces/metal. Once again, such percentages appeared at odds with the cumulative exploration expenditure quoted for some of the junior companies identified, together with the age of the asset base.

The authors also noted that a project's location also appeared to play a part in predicting where the highest percentage of Measured and Indicated Resources were typically reported. In such instances, the need for a uniform implementation of reporting codes (eg United Nations Framework Classification for Resources (UNFC) or the Committee for Mineral Reserves International Reporting Standards (CRIRSCO) was reinforced. In this precious metal development group, multiple targets for follow-up resource and reserve inflation analysis were in evidence.

Operating – base metal

From the ranking assessment, a total of 52 risk events were observed (Table 3) which underwent subsequent classification by risk level attributes.

TABLE 3

Operating – base metal risk events.

INFLATION   LIKELIHOOD ➡

OPER BASE	RISK 1	RISK 2	RISK 3	RISK 4
% CHANGE - RES & RER	4	6	1	2
% CHANGE - MKT CAP	3	7	1	2
$MKT CAP / $EXP SPEND	4	5	2	2
% M & I TONNES	5	4	2	2
TOTAL RISK EVENTS	16	22	6	8
% UPPER QRTL	31%	42%	12%	15%

Over the seven-year review period, the authors observed repeated examples of projects with marked increases in posted resource and reserves which did not appear supported by proportionate exploration expenditures. Where an especially rapid increase in Measured and Indicated Resource was detected (ie over 1–2 years), care was taken to ensure that no purchases of pre-existing resources and/or reserves had occurred. Even where minor purchases may have been made, the authors observed that resource reclassification rarely if ever occurred, irrespective of the asset's probable quality. Multiple targets for resource inflation and/or reclassification were in evidence in this data set as well.

Operating – precious metal

From the ranking assessment, a total of 80 risk events were observed (Table 4) which underwent subsequent classification by attributes.

TABLE 4

Operating – precious metal total risk events.

INFLATION   LIKELIHOOD ➡

OPER PREC	RISK 1	RISK 2	RISK 3	RISK 4
% CHANGE - RES & RER	11	7	2	0
% CHANGE - MKT CAP	4	11	5	0
$MKT CAP / $EXP SPEND	7	7	6	0
% M & I TONNES	6	9	5	0
TOTAL RISK EVENTS	28	34	18	0
% UPPER QRTL	35%	43%	23%	0%

With a wide range of projects available for review and the comparative richness of associated data, the authors were offered examples of how a range of factors might have impacted discovery costs, including challenging geology and mining conditions for precious metal operators. In this context, differences in percent change in resource and market capitalisation appeared between projects located within established mining centres with easy access to infrastructure, or in countries with generally lower regulatory costs.

Irrespectively, evidence of inflation persisted in some of Level 3 designated projects wherein anomalously high proportions of Measured and Indicated Resource were reported, after comparatively elementary exploration programs. Similarly, amongst some of these same at-risk companies, examples of greenfield projects reporting little to no Inferred Resource were also observed. Invariably, such projects also offered high-level commentary on accelerated studies, with prospective links to near-term production.

SUMMARY

Inasmuch as both resource development and financing sectors are renowned for the comparative optimism of their expert classes, regrettably, resource overconfidence can lead to adverse outcomes for all mining stakeholders. The short-term benefits of inflated tonnages of increasing classification confidence are invariably outweighed by longer-term drawbacks, some of which may ultimately compromise project economics.

The use of extended time-based exploration and resource category ratios offers an elementary decision tree option for the identification of at-risk projects. With the advancement of machine learning techniques, future studies will undoubtedly benefit from supervised and unsupervised algorithms and neural network techniques to better understand anomalous data trends over progressively shorter time frames. Creation of a formal and reportable discovery cost KPI, ideally linked to metres drilled would undoubtedly benefit this investigation process as well.

At any given stage of project maturity, all prospective ore deposits should lie within a bounded range of exploratory growth and resource classification paradigms. Any evidence of a hasty conversion of Inferred Resource to either Indicated or Measured categories should raise investor and/or allied institutional concerns. The examination of publicly available data using straightforward statistical techniques suggests this practice and its associated risks need not remain hidden forever.

REFERENCES

Hargreaves, R R and Booth, G W, 2019. Transparency and standardisation in metal reconciliation reporting, in *Proceedings of the International Mining Geology Conference 2019*, pp 36–42 (The Australasian Institute of Mining and Metallurgy: Melbourne).

Jowitt, S M and McNulty, B A, 2021. Geology and Mining: Mineral Resources and Reserves: Their Estimation, Use, and Abuse, *SEG Discovery,* (125):27–36.

Jowitt, S M, Mudd, G M and Thompson, J F H, 2020. Future availability of non-renewable metal resources and the influence of environmental, social, and governance conflicts on metal production, *Communications Earth & Environment,* pp 1–8.

Kreuzer, O P, Etheridge, M A and Guj, P, 2007. Australian junior exploration floats, 2001–06, and their implications for IPOs, *Resources Policy,* 32:159–182.

Rijsdijk, T, Nehring, M, Mehmet, K and Roosta, F, 2022. Confirming the Lassonde Curve through life cycle analysis and its effect on share price: A case study of three ASX listed gold companies, *Resources Policy,* 77(Aug):102704.

Risk assessment of iron mineral resources using conditional simulations

W Patton[1]

1. MAusIMM, Principal of Resource Modelling, BHP Iron Ore, Perth WA 6000.
   Email: will.patton@bhp.com

ABSTRACT

Investment decisions in the mineral resources sector are made on the basis of an assessment of the economic potential for a mineral deposit. Given the supporting information for the location, scale, and quantity of the commodity of interest, miners and investors aim to make risk-informed decisions when considering further investment. The objective of resource classification is the synthesis of various types and quantities of geologic information used in resource model construction to assign confidence categories to the estimated mineral resources, which can then be used to support informed economic decisions. In this paper a risk-based classification scheme for iron ore mineral resources using conditional simulations is presented. The method involves categorisation of each block according to a confidence level, considering expected recoveries from a kriging model over mining production volumes. The realisations of the simulated model are used to assign risk-based confidence categories to zones of the ore deposit. One motivation for this work is to assess the sensitivity of the scheme to geologic uncertainties. Specifically, that which uncertainty in the layout of mineralising alterations and weathering overprints influence variabilities of associated rock properties and the likelihood of obtaining expected ore recoveries. It is shown these risks have material implications for a project's economic potential. How the scheme can be extended to the grade control context and the value subsequent grade control activities is also quantified.

INTRODUCTION

Throughout the various stages of development of a mineral resource, critical investment decisions, like: 'how much' and 'where' to drill, are made based on limited orebody knowledge from sparse observations such as surface mapping, geophysical surveys and borehole sampling. Traditionally, the industry has focused on producing single best estimates of categorical, continuous and compositional rock properties over the orebody in question for estimating ore tonnes, grade and mineral resources (Emery, Ortiz and Rodríguez, 2006; Snowden, 1996; Watson, Journel and Huijbregts, 1980). Deterministic modelling approaches are well suited for this objective, although in many cases these approaches do not fully capture the heterogeneity (Rossi and Deutsch, 2014) and uncertainties in the layout of rock types in orebody models (Madani and Emery, 2015). Increasingly, mining companies are considering risk as part of the financial evaluation of orebodies during scoping and (pre-)feasibility studies in order to rank and prioritise potential investments. This necessitates high-resolution stochastic orebody models that present the full range of possible scenarios for these properties for financial evaluation and the quantification of risk (Chilès and Delfiner, 2012; Rossi and Deutsch, 2014).

Geostatistical simulation is the practice of constructing a pool of equiprobable spatial models that characterise naturally occurring properties such as metal grades and geologic objects including sedimentary structures in aquifers, reservoirs or mineral deposits (Armstrong et al, 2011; Carr, 2003; Goovaerts et al, 2005; Journel, 1974; Rossi and Deutsch, 2014). These models aim to reproduce the statistical and spatial characteristics of samples taken from these features, such as the mean, variance and the semivariogram. In doing so, they can be upscaled to any volume of interest and used to evaluate the distribution of grades at any support using a probabilistic framework (Rossi and Deutsch, 2014). Gaussian geostatistical simulations are suitable for both categorical and continuous properties and require data to be normally distributed or transformed to such Gaussian variables prior to implementation (Chilès and Delfiner, 2012). If data are compositional, as is the case for geochemical assays used in evaluation of mineral resources, additional transformations are necessary to make data more amenable for (geo)statistical treatment (Aitchison, 1986; Pawlowsky-Glahn and Olea, 2004; Talebi et al, 2019; Tolosana-Delgado, Mueller and van den Boogaart, 2018; van den Boogaart and Tolosana-Delgado, 2013).

The objective of resource classification is the synthesis of various types and quantities of geologic information used in resource model construction to assign confidence categories to the estimated mineral resources. Quantitative classification approaches can be separated into geometric and geostatistical based techniques. Geometric approaches include confidence category assignment based on search neighbourhood, sample or borehole spacing criteria. Geostatistical approaches generally refer to the application of thresholds on block kriging variances using criteria derived from pre-defined borehole configurations (Rossi and Deutsch, 2014). Geostatistical based methods which utilise the realisations from a conditional simulation model have also been defined (Emery, Ortiz and Rodríguez, 2006; Silva and Boisvert, 2014; Snowden, 1996). These aim to characterise the local uncertainties in grade followed by assignment of confidence categories using confidence level tolerances. Each of these schemes each have their own advantages and disadvantages. For instance, geometric methods rely on few assumptions and are straightforward to apply, although do not consider grade variability. Kriging based geostatistical methods might only consider some aspects of resource confidence such as sample configuration and continuity of grades but omit others such as the actual sample grades themselves and mining selectivity. Simulation-based methods can take into account a broader range of factors influencing resource confidence, including heteroscedasticity near the mining cut-off and expected mining rates. This however relies on more modeller assumptions potentially making for less transparent disclosure to investors (Rossi and Deutsch, 2014).

Several simulation-based schemes for classification of mineral resources have been proposed. Local Selective Mining Unit (SMU) and long-term mining production (panel) scale uncertainties can be assessed by considering a series of confidence level tolerances over a uniform block volume (Emery, Ortiz and Rodríguez, 2006; Maleki Tehrani, Asghari and Emery, 2013; Snowden, 1996). Generally, these approaches require specification of the volume under consideration, precision, and the desired confidence interval. For example, the values of a quarterly production volume must fall within ±15 per cent of the mean 90 per cent of the time in order to be classified as low risk or Measured category. Lower-confidence category mineral resources can subsequently be assigned using wider precision and lower confidence thresholds. These approaches are flexible and can be tailored to objectives for the study, whether they be grade control or long-term mine planning in open cut or underground mining scenarios. A moving-window approach centring long-term mining production volumes (panels) over SMUs can yield improved results free of undesirable artefacts associated with the use of fixed grids (Silva and Boisvert, 2014).

The above approaches rely solely on the realisations of a conditional simulation model and their summary statistics, mean or expectation type (E-type) model. In some situations, a linear estimate of mineral resources is the desired product for further evaluations. This might be due to requirements such as high confidence local estimates or the reduced number of scenarios that require mining optimisations or stope designs. In these cases, it may be advantageous to evaluate the grade presented in linear model via a kriging with respect to the realisations of the stochastic model. This allows for the assessment of risks associated with averaging of grades over large volumes and deterministic geologic contacts given that the linear model will go forward for further financial analysis.

The remainder of the paper is organised as follows. In Methodology, an approach for construction of a conditional simulation model capturing uncertainties in layout of categorical rock properties and variabilities of multivariate compositional geochemical data is described. How this type of model can be used to quantify mineral resources and assign confidence in grade control ore/waste designations of an alternate deterministic model using a risk-based framework is also outlined. These techniques are then presented using a real data set called 'Orebody B', a complex strata-bound bedded iron deposit located in Western Australia's Pilbara region. The implications of incorporating uncertainties in the layout of geological contacts in the uncertainty model are also assessed. Finally, a summary of concluding remarks is then provided.

METHODOLGY

Compositional data

Compositions are vector representations of multivariate data with individual positive components z_i, $i = 1, ... D$, summing to a constant representing proportions of a whole, called the D-part Simplex (Aitchison, 1986; Pawlowsky-Glahn and Olea, 2004; van den Boogaart and Tolosana-Delgado, 2013). When the compositional vectors are considered dependent on their spatial location u, within a study area A, one refers to regionalised compositions (Pawlowsky-Glahn and Olea, 2004) defined as follows:

$$Z(u) = [z_1(u), z_2(u), ..., z_D(u)] \qquad (1)$$

Here, D is the number of components that sum to an arbitrary positive constant C for instance, proportions (1), percentages (100 percent) or parts per million (1 000 000 ppm) depending on the abundances of desired components.

Compositions encountered in mining data sets are usually subcompositions where unimportant or trace constituents are not reported. To ensure constituents are accounted for a complementary, or 'rest' variable is included, defined as:

$$z_R(u) = 100 - \sum_{i=100}^{D} z_i(u) \qquad (2)$$

and interpreted as the totality of elements not assayed. This approach has advantages for mining applications as it allows for the calculation of the mass of each component as well as the combined mass of components not assayed which contribute to the total haulage mass. Compositional data convey relative information of components and are appropriately represented by use of log ratio transformations. The log ratio transformations referenced in this contribution are the centred log ratio (*clr*) and additive log ratio (*alr*) (Aitchison, 1986). After geostatistical treatment, back transformation of log ratios is achieved using the *alr* back transform, known as the additive generalised logistic (*agl*) (Talebi *et al*, 2019; Tolosana-Delgado *et al*, 2015; Tolosana-Delgado, 2005).

Gaussian simulations

Geostatistical simulation of a set of Gaussian Random Functions (GRFs) is performed for mineralising alterations (ore types), weathering overprints and oxide grades. This is achieved using the Turning Bands method (Matheron, 1973) for Gaussian simulations and is conducted following a two-stage process. First, realisations of the layout of the ore types and weathering overprints are simulated at locations between boreholes to produce realisations of geological domains. This is followed by modelling of the continuous rock properties (ie oxide grade compositions) across each realisation of the geological domains. The Turning Bands Simulation (TBS) algorithm is used as the underlying geostatistical simulation algorithm for both purposes. It is based on the assumption that input data are multivariate Gaussian data (Chilès and Delfiner, 2012; Matheron, 1973). As this is rarely the case in real data sets normal scores transforms are routinely applied to obtain the corresponding Gaussian distribution. The manner in which the GRFs used in geostatistical simulations are related to the geological categories (lithotypes) and oxide grade compositions is dependent on their type. Lithotype properties are categorical in nature, consisting of a set of discrete levels. These are related to GRFs via a transformation to indicator variables followed by truncation of one or more GRFs using a truncated Gaussian (Matheron *et al*, 1987) modelling approach. Oxide grades are continuous multivariate data and are transformed to GRFs for modelling using the Minimum/Maximum Autocorrelation Factors (Desbarats and Dimitrakopoulos, 2000; Switzer, 1984) multivariate Gaussian transformation.

Truncated Gaussian and pluriGaussian simulations

The truncated Gaussian model is a geostatistical model originally developed for simulating sedimentary structures (lithofacies) in oil reservoirs. Typically, the aim is the reproduction a simple organisation of ordered lithotypes. Components of the model consist of the overall proportions of lithotypes across the study, their spatial distribution and contact relationships. Realisations of the individual lithotypes are obtained by applying thresholds on a single GRF. The thresholds used are calculated to match the proportions of each lithotype whilst the distribution and contact relationships are reproduced by relating the characteristics of the underlying GRF to that of the lithotype experimental indicator variograms. The pluriGaussian model is a generalisation of the truncated Gaussian model and allows for more complicated boundary configurations between lithotypes. It has previously been applied to complex mineral deposits (Madani and Emery, 2015; Maleki Tehrani, Asghari and Emery, 2013; Mery *et al*, 2017; Talebi *et al*, 2013, 2014, 2015).

The pluriGaussian model relies on several GRFs, upon which a truncation rule is formulated by defining a number thresholds in order to produce categories which represent geological domains. For example, two GRFs each with a single threshold could be used to produce a scenario where three lithotypes are in contact with one another. At locations where the first Gaussian field (Y_1) is less than the first threshold (t_1) lithotype one is found. Otherwise, we find lithotype two above this threshold when below the second threshold (t_2) on the second Gaussian (Y_2), or lithotype three when above threshold two. A truncation rule such as this can be represented by a two-dimensional image or flag, where each axis represents one Gaussian random field and thresholds by colour change boundaries such as that presented in Figure 1a. The truncation rule is multivariate and any number of Gaussian variables can be accommodated. However, to simplify parameter inference and implementation, the number of GRFs is typically restricted to two or three.

The flag controls the allowed and forbidden contacts between geological domains whilst the magnitude of thresholds control the domain proportions. Where the stationary case is assumed, truncation thresholds are fixed and proportions of various domains will be constant across the modelled space. Alternately, thresholds can be allowed to vary spatially to account for the non-stationary case where domain proportions change across the modelled space. The spatial behaviour of domains also depends on that of the underlying GRFs as can be observed in Figures 1b–1d.

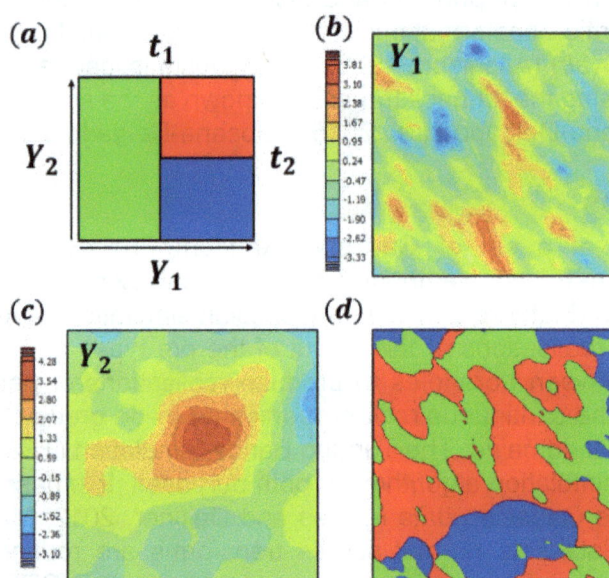

FIG 1 – Flag representing the truncation rule between three rock types (a), first Gaussian random field Y_1 (b), second Gaussian random field Y_2 (c) and three rock types obtained applying the truncation rule (d).

Risk based mineral resource classification

Mineral resource classification is conducted on a model discretised at the scale of the SMU. This volume represents the minimum parcel of rock that can be extracted and dispatched to the mill, stockpile, other ore handling facilities or the waste dump. The SMU is also used to account for the inevitable ore-loss and dilution associated with open cut mining (Rossi and Deutsch, 2014). The three types of dilution described there-in include; external geologic contact dilution, internal geologic dilution related to the *volume variance effect*, and operational dilution that occurs at the time of mining. The classification scheme used here aims to account for at least the first two of these types of dilution. A regular model discretised to scale of the SMU using a volume weighted averaging procedure accounts for external and potentially some aspects of operational mining dilution that will occur. Classification of mineral resources presented in a linear estimate from a kriging with respect to the realisations of the stochastic model accounts for internal geologic dilution and uncertainties relating to local scale geologic and grade variability.

At each SMU u_i the summed quantities of ore (tonnes), metal or mean grade above a mining cut-off grade over a larger panel representing a mining production volume (quarterly or annual) are computed for the kriging Q_c^K and each realisation $Q_c^{r=1,...,R}$ of the stochastic model. The mineral resource category at any SMU is then given by:

$$\text{Mineral resource category at } u_i = \begin{cases} \textbf{Measured } if\ f(\boldsymbol{Q_{diff}}(u_i) \leq \boldsymbol{p_1}) \geq \textbf{c}_1 \\ \textbf{Indicated } if\ f(\boldsymbol{Q_{diff}}(u_i) \leq \boldsymbol{p_1}) < \textbf{c}_1\ and\ f(\boldsymbol{Q_{diff}}(u_i) \leq \boldsymbol{p_2}) \geq \textbf{c}_2 \\ \textbf{Inferred } if\ f(\boldsymbol{Q_{diff}}(u_i) \leq \boldsymbol{p_1}) < \textbf{c}_1\ and\ f(\boldsymbol{Q_{diff}}(u_i) \leq \boldsymbol{p_2}) < \textbf{c}_2 \end{cases} \quad (3)$$

where $f(\boldsymbol{Q_{diff}}(u_i) \leq p_x)$ is the frequency at which the absolute relative difference in the quantities Q_c^K and $Q_c^{r=1,...,R}$ are lower than the desired precision tolerance for measured p_1 and indicated p_2 category resources and the desired confidence interval for measured and indicated are c_1 and c_2, respectively.

A critical decision is how to define the production volume. This in principle relates to ore reserves and unless access to this information is available after a (pre)feasibility study, some assumption on the size of the operation and the production rate must be made.

Assessing risk in a grade control setting

During mining production focus shifts from understanding the risk associated with the expected quantities and quality of ore over large volumes to assessing risk at the scale of mining extraction processes. A critical objective is the designation of each individual SMU as either ore or waste based its average grade relative to the mining cut-off (Verly, 2005). Here, errors in the estimation of a blocks true grade may result in misclassification of the blocks designation as ore or waste. For example, where a blocks estimated grade is above the mining cut-off and its true grade is below, a dilution type misclassification error will be incurred. This will result in lower than the expected actual mining production grades observed at the mil or stockpile. Conversely, where a blocks estimated grade is lower than the mining cut-off and its true grade is above, an ore-loss type misclassification error has occurred representing lost revenue.

The risk of incurring such misclassification errors can be estimated by comparing the designation for each SMU in the grade control estimate from a kriging against that of the realisations of a conditional simulation model. Where the misclassification frequency of this ore/waste designation is lower than that of a desired probability threshold, the SMU is assigned to the low-risk category. Otherwise, it is assigned to the high-risk category. This binary scheme aims to identify zones where the objectives of the grade control model have been satisfied and where they have not.

CASE STUDY

The data set for this contribution was supplied by BHP PTY LTD Western Australia Iron Ore. A database containing borehole data collected through Reverse Circulation (RC) and Diamond Core (DC) drilling and a block model discretised on 10 mE × 10 mN × 4 mRL centres was provided. Grades were interpolated using block Ordinary co-Kriging and Inverse Distance Weighting for Fe, P, SiO_2, Al_2O_3, LOI, Mn and a *Rest* variable. In addition to this surface mapping of mineralising alterations collected by geologists and aerial photogrammetry was supplied in the form of two-dimensional polygons and a hypothetical mining sequence in the form annual mining period wire frame solids were also supplied. A fictitious name, 'Orebody B' was used in the interest of protecting the company's intellectual property.

Deposit geology

The Hamersley Province covers the southern third of the Pilbara Craton, south of and including the Chichester Range, to the margins of the various Mesoproterozoic sedimentary basins to the south (Kepert, 2001). The province contains sediments from late Archaean to Lower Proterozoic age (2800–2300 Ma) of the Mount Bruce Supergroup which unconformably overly the mid Archaean granite-greenstone terrain of the Pilbara Craton (Figure 2a). Within the Mount Bruce Supergroup are three subgroups; the Fortescue, Hamersley and Turee Creek Groups (Harmsworth et al, 1990). The Hamersley Group hosts all strata-bound Banded Iron Formation (BIF) iron ore deposits in the

Hamersley Iron Province, with enrichment generally although not exclusively occurring within the Brockman and Marra Mamba Iron Formations (Figure 2b).

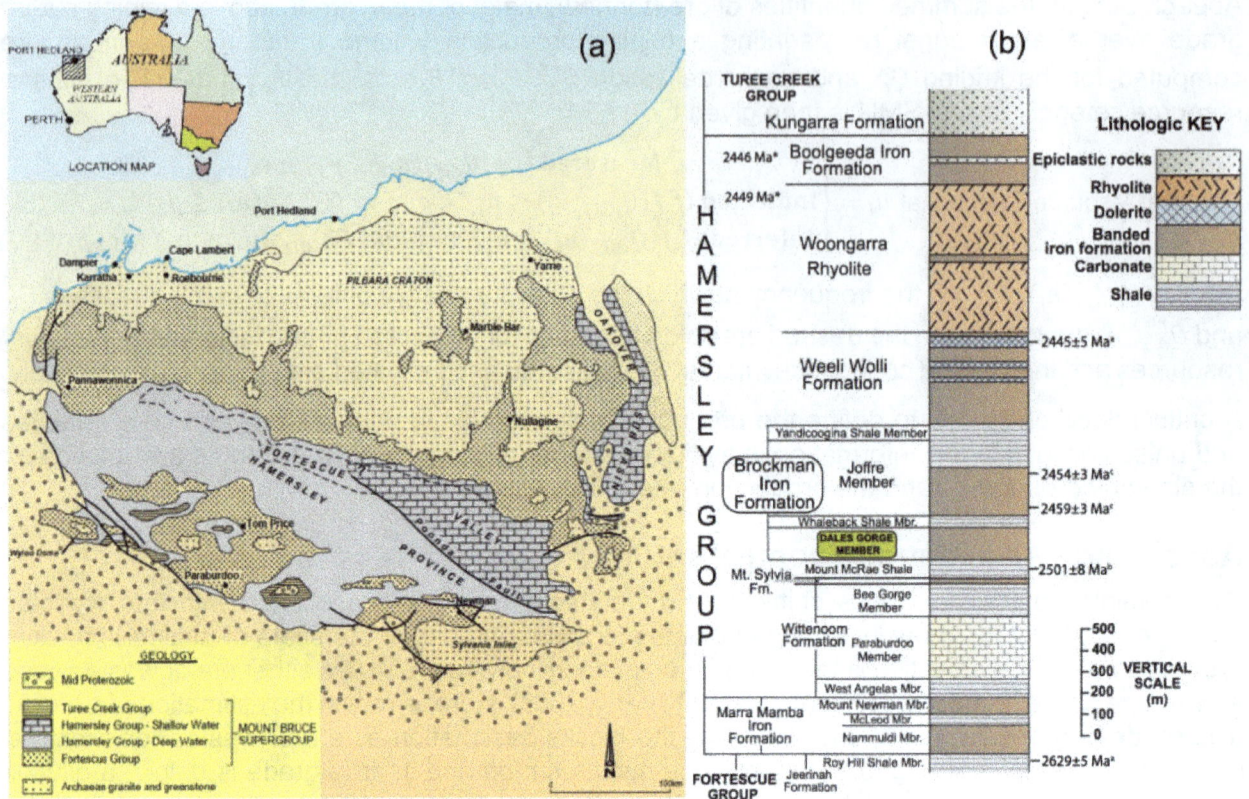

FIG 2 – Pilbara area geological sketch and location plan (a) modified from Harmsworth *et al* (1990); and Hamersley area stratigraphic column (b) from de Oliveira Carvalho Rodrigues, Hinnov and Franco, 2019.

Orebody B is a 4.0 km long zone of semi-continuous bedded iron enrichment of the Brockman Iron Formation (PHb) ranging from 0.2–1.0 km in width at surface (Figure 3a). It extends from surface vertically down dip to a maximum depth of 200 m. The bulk of the prospective mineralisation is located in the hinge zone of the syncline structure forming the eastern end of the low range at the site. This structure plunges to the west (10°–15°) and is terminated at its western end by the abutting volcanics of the Jeerinah Formation (AFjb) at the Mt Whaleback Fault to the north-west. Mineralisation is predominantly hosted in an interbedded sequence of shales, jaspilites and cherts of the Joffre (PHbj), Whaleback Shale (PHbw) and Dales Gorge (PHbd) Members of the Brockman Iron Formation (Figure 3b). These are bounded by a conformable sequence of interbedded shales and cherts of the Mount McRae shale (AHr) and Mount Sylvia Formation (AHs) to the south and east, and by the Weeli Wolli Formation (PHj) to the north in the core of the syncline structure. On the southern and eastern flanks of the range this sequence is overlain by unconformable Quaternary re-worked semi-consolidated scree, colluvial and alluvial sediments (Qa) associated with erosional processes.

Mineralisation is a combination of typical high phosphorous (>0.1%P) martite-goethite (M-G) supergene ores and low phosphorous (<0.1%P) martite-microplatey hematite (M-mplH) hypogene ores (Morris, 1980, 2011, 2012). Hypogene M-mplH ores are associated with fault structures located in the hinge zone of the syncline structure at the eastern end of the ore deposit. Supergene M-G ores occur throughout the PHbd and PHbj members whilst hypogene M-mplH ores are limited to the PHbd member. These ore types are similar to Brockman Formation type stratiform ores found throughout the Pilbara. Both ore types are present in the near surface hydrated and oxidised zones extending to depths of the ore deposit and can be distinguished by their contrasting geochemical and physical properties. Hydrated zone ores are an overprint feature across the orebody from surface to a maximum depth of approximately 40–100 m and represent a leaching process related to interaction with meteoric fluids. Iron enriched scree (Qa) deposits also referred to as 'Canga' are situated on the flanks of the range in limited quantities and represent erosion of the outcropping bedded iron deposit.

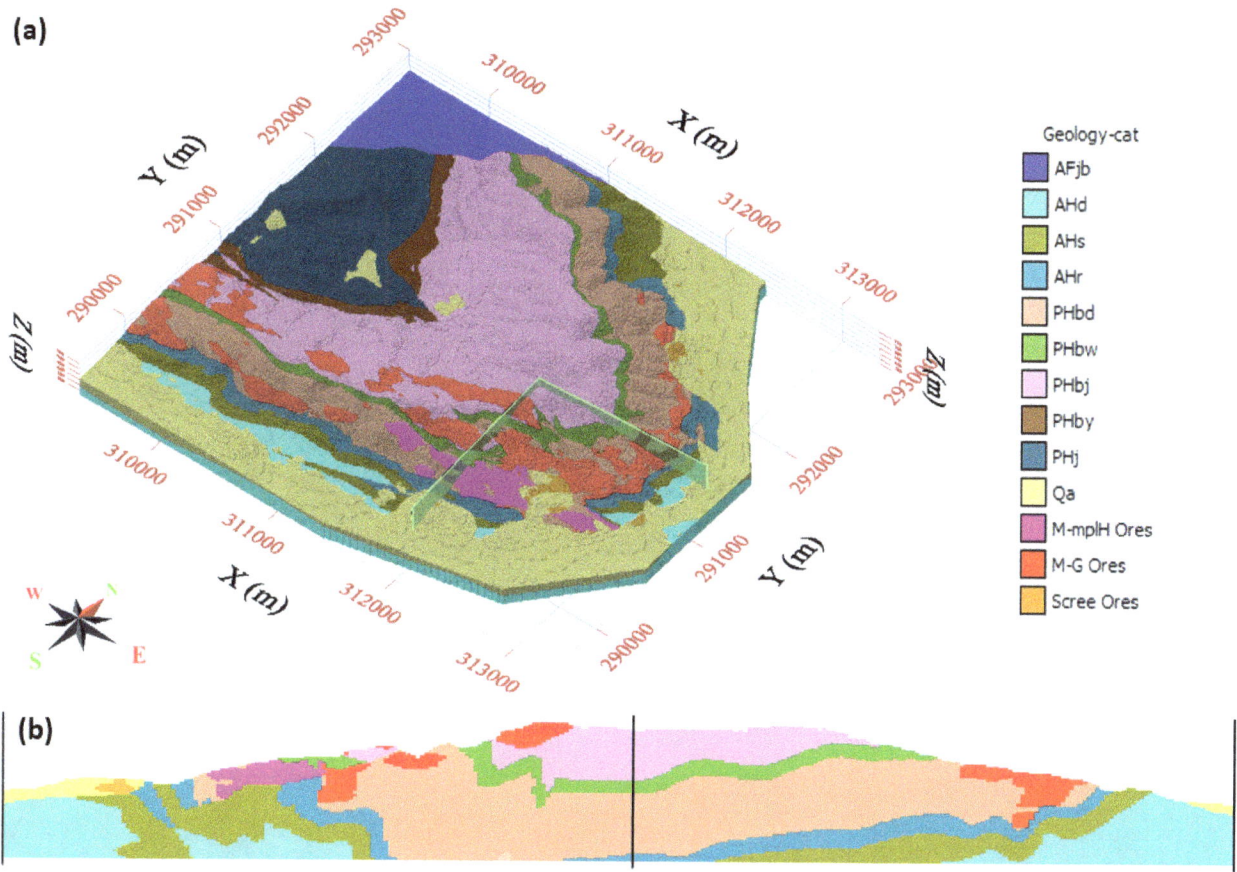

(a)

(b)

FIG 3 – Oblique view of Orebody B geology (a) and fence section looking west and north, left to right respectively (b).

Data

The borehole data consist of a set of 421 RC and 3 DC boreholes spaced nominally on 50 mE × 50 mN separations across the ore deposit. Many irregularities and gaps in this nominal drilling pattern are observed as a result of steeply inclined topography unable to be accessed safely by drilling contractors. RC boreholes were sampled exclusively on 3 m intervals whilst DC are cores of rock extracted intact from boreholes on intervals between 0.1 m and 7 m in length. Drill cuttings from RC and DC boreholes were assayed using X-ray Fluorescence (XRF) to determine oxide compositions Fe, P, SiO_2, Al_2O_3, Loss on Ignition (LOI) and Mn. A rest variable (REST) is calculated for each analysed interval to close this composition defined using Equation 2. Other rock properties interpreted by geologists including lithofacies, ore-type and weathering overprint are also reported for all samples.

The raw borehole sample data were composited to 13 605 uniform 3 m intervals using a length weighted averaging procedure. Descriptive statistics for 13 534 samples (composites) with oxide grades are presented in Table 1. Histograms for Fe, P, Al_2O_3 (Figure 4) reveal multi-modal distributions representing the differing rock (ore) types across the study area. High Fe grades and P concentration can distinguish the several types of iron enrichment. Lower Fe and elevated SiO_2 and Al_2O_3 represent unmineralised rock types away from the orebody.

TABLE 1

TABLE 1

Descriptive statistics for all samples.

Continuous Variables	Count	Minimum	Maximum	Mean	Standard Deviation
FE [%]	13,534	1.10	68.31	40.84	15.87
P [%]	13,534	0.005	0.714	0.093	0.062
SIO_2 [%]	13,534	0.49	89.28	29.47	20.43
AL_2O_3 [%]	13,534	0.01	46.17	5.24	6.32
LOI [%]	13,534	0.16	45.34	5.23	3.63
MN [%]	13,534	0.001	28.880	0.215	1.215
REST [%]	13,534	1.30	50.60	18.93	6.26

Categorical Variables									

Lithofacies	AFjb	AHd	AHs	AHr	PHbd	PHbw	PHbj	PHby	PHj	Qa
	2100	4000	5300	5400	5600	5700	5800	5900	6100	8100
Count	2	450	837	1,167	7,542	623	2,588	26	107	448
Proportion [%]	0.01	3.26	6.07	8.46	54.69	4.52	18.77	0.19	0.78	3.24

Ore type	Unmin.	M-G (incl. Canga)	M-mpIH	Weathering overprint	Fresh rock	Oxidised zone	Hydrated
	0	1	2		0	1	2
Count	9,447	3,381	962	Count	1,164	10,745	1,881
Proportion [%]	68.51	24.52	6.98	Proportion [%]	8.44	77.92	13.64

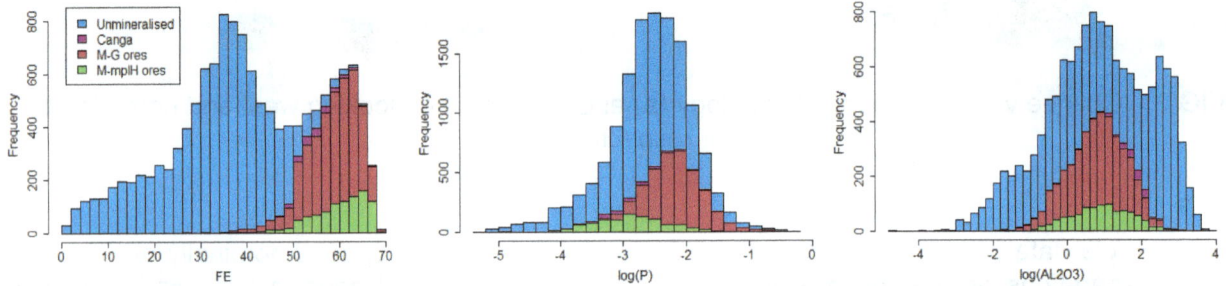

FIG 4 – Stacked histograms for Fe, log(P) and log(Al_2O_3) for all samples coloured by ore type.

Multivariate statistical relationships between components of a subcomposition such as oxide grades can be inspected using compositional Principal Component Analysis (PCA) (Aitchison and Greenacre, 2002). PCA is an interpretation of the Singular Value Decomposition (SVD) of a data matrix (Hotelling, 1933; Pearson, 1901) allowing for the optimal representation of multivariate data at the lower dimensions of the biplot. In compositional PCA, singular value decomposition is applied to the data matrix after a *clr* transformation of data (van den Boogaart and Tolosana-Delgado, 2013). The result is a data object that contains loadings, a number of Principal Component (PC) variables equal to that of the number of input variables and a set of scores equal to the number of samples in the data set.

The scree plot for samples from iron enrichment (Figure 5a) allows for inspection of the portion of variance in the data set explained by each PC. In this case the total variance explained by the first two PCs is high at close to 77 per cent suggesting conclusions may be drawn from interpretation of biplots for the first two PCs (Figure 5b). Loadings for the first two PCs are presented on the biplot as rays or arrows. The links between arrow heads represent the pairwise log ratios between variables. Proportional variables have near constant log ratios with very short links and arrow heads that lie together such as Fe and the Rest variable. The pairs Fe and P, P and LOI, and SiO_2 and Al_2O_3 also have short links suggesting proportionality. Conversely, long links and angles between rays close to 120° indicate highly variant log ratios such as those between Mn and SiO_2, or Mn and P and will have ternary diagrams with a widespread. The angle between links approximates the correlation coefficient of the two log ratios. Links at orthogonal angles have log ratios that are likely uncorrelated such as Fe (or REST) and Mn or Al_2O_3 and LOI. Links for variables lying on a common line forming either 0° or 180° may be perfectly correlated, either directly or inversely as for LOI and Mn and Fe and the REST variable and will produce one-dimensional pattern ternary diagrams. Groups of three or more variables with links that form a straight line are possibly collinear subcompositions, for example the trio SiO_2, Al_2O_3 and Mn or SiO_2, LOI and P.

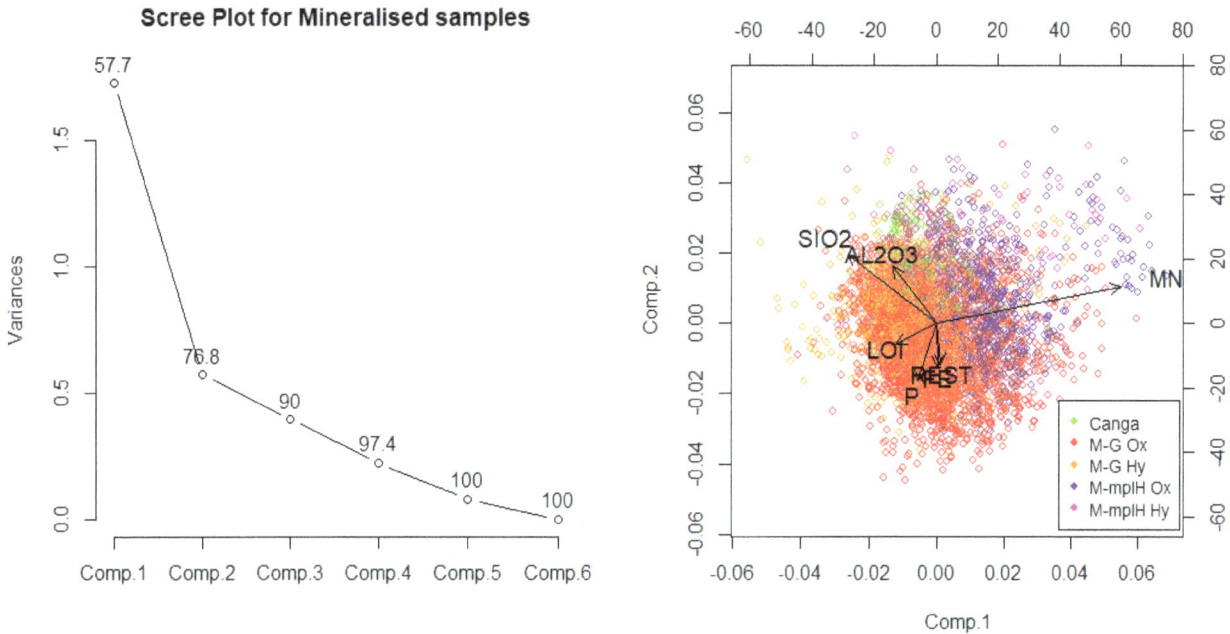

FIG 5 – Robust (clr) PCA scree plot and compositional covariance biplot for PC1/2 for the oxide grade compositions for mineralised samples.

Colouring of PCA scores by ore type and weathering overprint combinations allows for interpretation of the relationships between these categories and oxide grade composition. We can see hydrothermally altered Mn-rich ores explain the largest portion of variability in the data set and are separated from LOI-rich supergene ores along PC1. Increased proportions of Fe (Rest) and P are prevalent in M-G oxidised zone ores and decreased Fe (Rest) and P are associated with M-mplH and Canga (scree) ores along PC2. M-G and M-mplH hydrated ores show decreased P and increased SiO_2, Al_2O_3 and LOI compared to their oxidised zone equivalents. Intermingling of scores for the various ore types suggesting ores have similar mineral compositions.

Uncertainty model construction

A geologic block model containing information on the prevailing lithofacies at each location was constructed in a grid discretised on 5 mE × 5 mN × 2 mRL spaced centres. The lithofacies model is deterministic and does not consider uncertainties in the prevailing lithofacies unit at each location of the deposit model. Models that characterise uncertainties in the spatial layout of the various categories of materials that constitute the ore types and weathering profile are constructed independently of one another following similar truncated (pluri)Gaussian simulation workflows.

The weathering profile model consists of three categories; the near-surface hydrated zone (2), oxidised zone (1) and fresh rock below the base of oxidation (0). Borehole samples and geologic model blocks from the reworked weathering and redeposition products of the tertiary detrital (Qa) units are omitted from the weathering profile model as the model is not applicable for these materials. Inspection of transition probabilities between these categories (Table 2) reveal a 'layer-cake' arrangement where the transitional oxidised zone is in contact with both, the near-surface hydrated and fresh rock at depth whilst no contact between hard cap and fresh rock is observed. This might support an interpretation that the prevailing weathering overprint category at any location is dependent on the proximity to surface landforms. As such both input boreholes and the geologic model grid were transformed to unfolded coordinates using the topography as a reference surface (Figure 6a).

The mineralising alteration model consists of three categories that also must be predicted across the grid; Unmineralised 'waste' rock (0), supergene M-G ores (1) and hypogene M-mplH ores (2). Outcrop mapping of iron mineralisation in the form of polygon data subsampled on 50 mE × 50 mN centres (Figure 6b) was used as additional conditioning data alongside borehole samples (Figure 6c). Given transition statistics observed in Table 2, all ore type categories are assumed to be in contact with one another.

TABLE 2

Transition probability matrices for input data pairs for the weathering overprint and ore type models calculated in the downhole direction. Direct contacts refer to situations where row category is situated immediately above column category in the downhole direction whilst indirect contacts refer to the opposite situation.

Transition Probabilities		Fresh rock	Oxidised	Hydrated	Pairs
Direct contacts	Fresh rock	100	0	0	692
	Oxidised	0.43	99.57	0	10 347
	Hydrated	0	22.22	77.78	1881
Indirect contacts	Fresh rock	93.89	6.11	0	737
	Oxidised	0	96.07	3.9	10 723
	Hydrated	0	0	96.44	1517
Transition Probabilities		Unmineralised	M-G ores	M-mplH ores	Pairs
Direct contacts	Unmineralised	98.36	1.39	0.25	9024
	M-G ores	8.55	91.21	0.24	3380
	M-mplH ores	8.73	0	91.27	962
Indirect contacts	Unmineralised	95.97	3.12	0.91	9249
	M-G ores	3.9	96.1	0	3208
	M-mplH ores	2.53	0.88	96.59	909

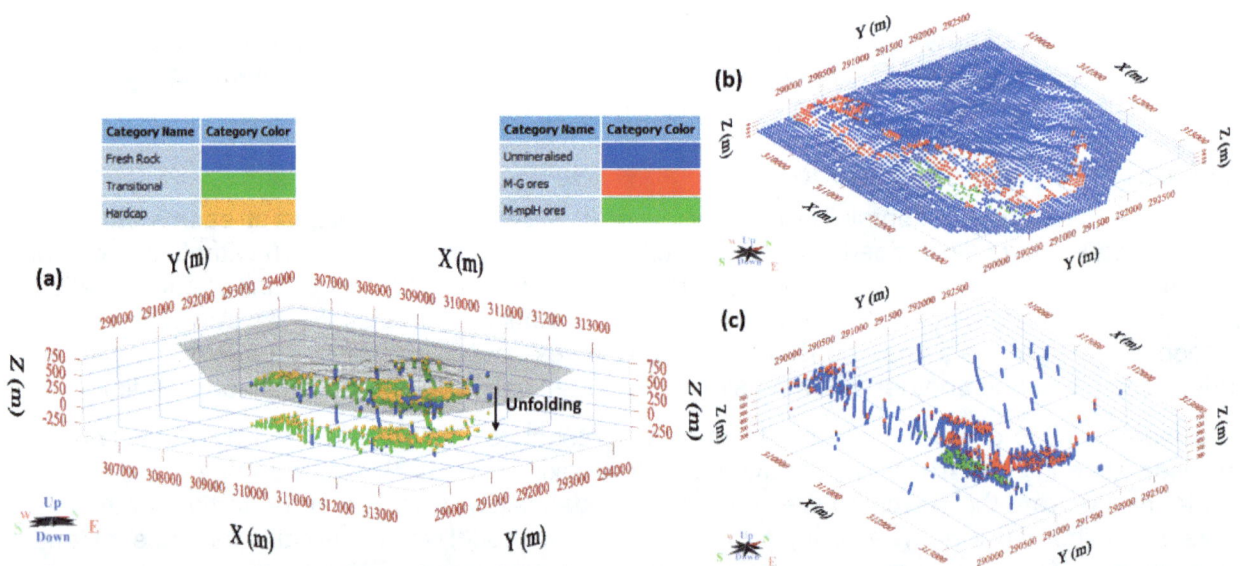

FIG 6 – Unfolding reference surface (grey) and borehole sample locations coloured by weathering overprint category in structural (upper) and unfolded (lower) coordinates (a). Point data set after subsampling outcrop mapping (b) and boreholes data coloured by ore type (c).

Non-stationarity in the layout of weathering overprint and ore type categories across the model is accounted for by allowing the thresholds for their truncation rules to vary in line with local proportions for each weathering overprint (Figure 7a–7d) and ore type (Figure 7e–7f) category. This 'soft' data for conditioning are computed at each location of the grid using a moving window search for borehole samples. When computing the proportions for each category, samples are weighted by the volume of rock each sample is the closest sample to using nearest neighbour declustering. The moving window dimensions used are 1000 mE × 1000 mN × 600 mRL for the weathering overprint and 800 mE × 600 mE × 400 mRL for the ore type models and were determined experimentally.

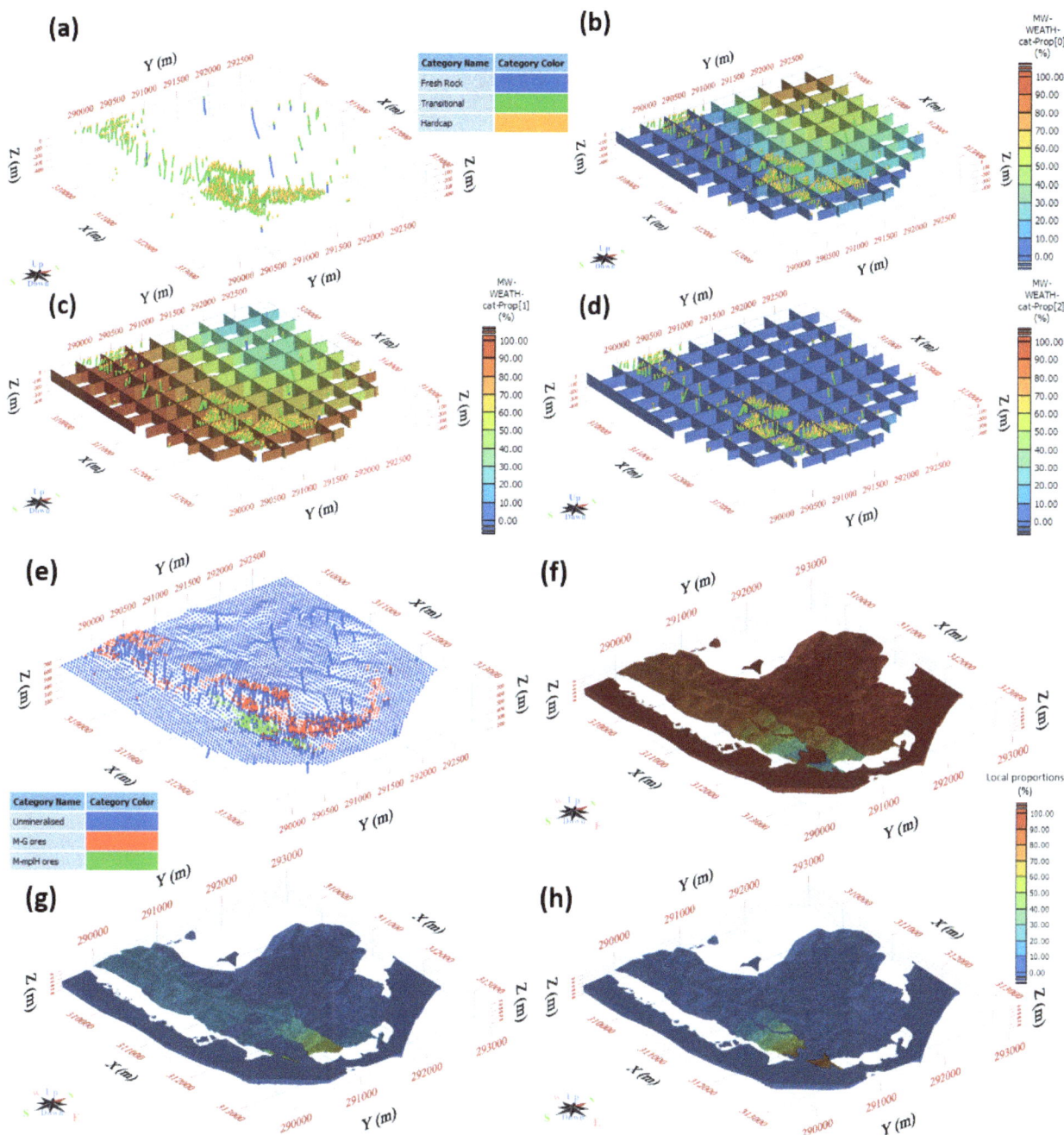

FIG 7 – Unfolded borehole sample locations used as hard conditioning data coloured by weathering over-print category (a). Unfolded grid coloured by moving window local proportions (soft data) for each weathering category; fresh rock (b), oxidised (c) and hydrated (d) along 10 easting and northing sections at 400 m spacing. Hard data from which local proportions for ore type categories are computed (e). Local proportions of unmineralised (f), M-G ore (g) and M-mplH ore (h) categories computed using moving windows.

The layer-cake style contact relationship of the weathering overprint is a simple arrangement and is modelled using truncated Gaussian model with thresholds for each contact between each weathering overprint categories (t_1, t_2) applied to a single GRF (Y_1) as depicted in Figure 8a, following:

$$\text{Weathering overprint category at location } x = \begin{cases} \textbf{Fresh rock } \textit{if } Y_1(x) \leq t_1 \\ \textbf{Oxidised } \textit{if } Y_1(x) > t_1 \leq t_2 \\ \textbf{Hydrated } \textit{if } Y_1(x) > t_1 > t_2 \end{cases} \quad (4)$$

Ore type contact relationships are more complex requiring a pluriGaussian model with a truncation rule involving two GRFs (Y_1, Y_2) with two thresholds (t_1, t_2) as depicted in Figure 8c, following:

$$\text{Ore type at location } x = \begin{cases} \textbf{\textit{Unmineralised if } } Y_1(x) \leq t_1 \\ \textbf{\textit{M} } - \textbf{\textit{G Ore if } } Y_1(x) > t_1 \textit{ and } Y_2(x) \leq t_2 \\ \textbf{\textit{M} } - \textbf{\textit{mplH Ore if } } Y_1(x) > t_1 \textit{ and } Y_2(x) > t_2 \end{cases} \quad (5)$$

The variogram models for the underlying GRFs for weathering overprint and ore type models were fitted to experimental indicator variograms computed using the method by Desassis *et al* (2015). Strong zonal anisotropy is evident between horizontal and vertical directions for the GRF associated with the weathering overprint model (Figure 8b). This is in agreement with the understanding of geologic processes resulting in the layer cake arrangement of categories. Variograms for GRFs associated with the ore type model (Figure 8d and 8e) exhibit geometric anisotropies in keeping with the complex structure of bedded lithofacies host to these ore types. Models for all GRFs were fitted in accordance with the basic nested structures implied by the experimental data and are given in Table 3.

TABLE 3

Variogram model specification for weathering overprint and ore type models.

Model	Gaussian Random Field (GRF)	Basic nested structure	Geologist direction (Dip/Dip Az./Pitch)	Sill	Range (m) U	V	W
Weathering overprint		Exponential	0°/80°/90°	0.13	300	300	40
	1	Spherical	0°/80°/90°	0.27	∞	∞	60
		Spherical	0°/80°/90°	0.6	∞	∞	300
Ore type	1	Cubic	0°/90°/90°	0.4	300	100	85
	1	Spherical	0°/90°/90°	0.6	650	400	85
	2	Gaussian	0°/90°/90°	1.0	1200	300	200

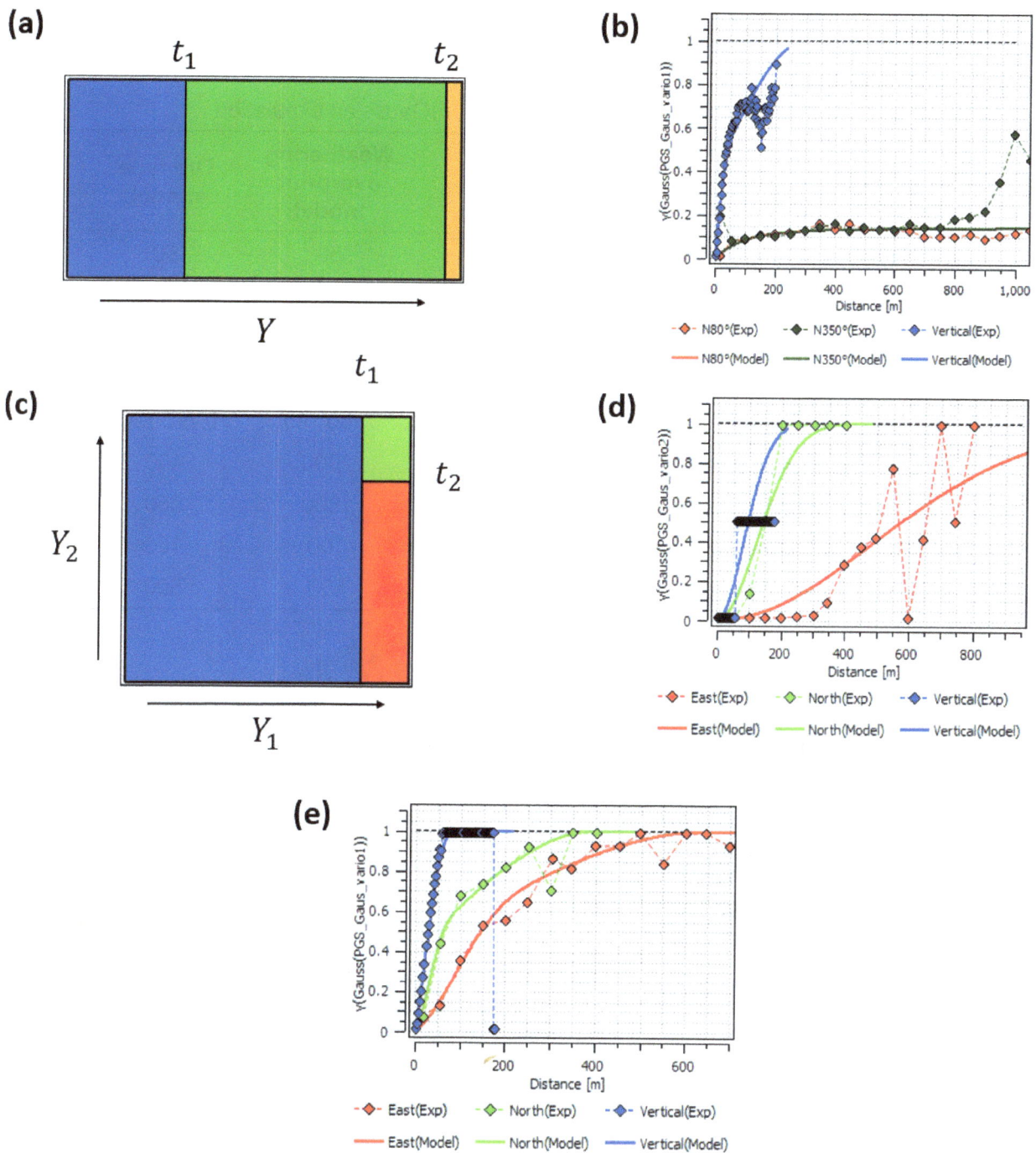

FIG 8 – Graphical representation of the weathering overprint model truncation rule (a) and model (solid lines) fitted to the experimental indicator variogram (dot-dash) for the GRF (b). Graphical representation of the ore type model truncation rule (c), experimental and fitted variograms for underlying GRF 2 (d), GRF 1 (e).

With specification of the conditioning data (hard and soft), the truncation rule and the variogram model for the underlying GRFs complete, a total of 30 realisations were constructed for the unknown weathering overprint and ore type categories at locations between the boreholes. Implementation parameters for the truncated (pluri)Gaussian models are tabulated in Table 4. Artefacts manifesting in the form of disordered, 'grainy' contact transitions between simulated categories observed in Figure 9 are related to inconsistencies between local neighbourhood conditioning data (hard data) and varying local proportions (soft data). These artefacts can have material implications for reserve calculations and were addressed in post-processing of the realisations using a moving window based smoothing algorithm. An image processing filter is used to assign the highest probability category computed from a local moving window defined at each grid node. Moving window dimensions used was 1 × 1 × 2 grid cells in the X, Y and Z directions respectively.

TABLE 4

Implementation of the truncated (pluri)Gaussian simulation.

Parameter	Weathering overprint model	Ore type model
Gibbs sample iterations	50	50
Number of angular sectors	4	4
Optimum number of conditioning data per sector	6	4
Split ellipse vertically	No	Yes
Geologist Rotation (Dip/Dip Azimuth/Pitch)	0°/80°/90°	0°/90°/90°
Radius (m) of search along U direction	2000	400
Radius (m) of search along direction	1800	250
Radius (m) of search along vertical/normal direction	100	150
Number of turning bands	500	500

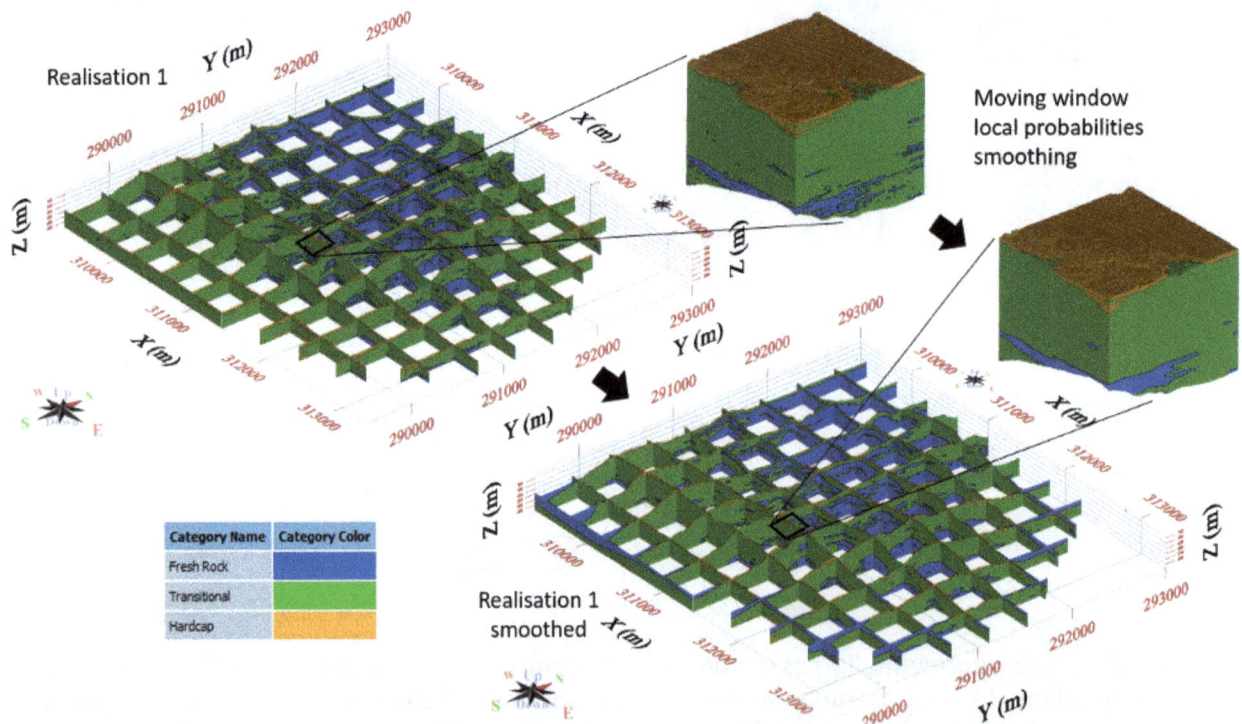

FIG 9 – Realisation 1 of truncated Gaussian model for weathering profile (top left) and smoothed output using moving windows local probabilities for realisation 1 (bottom right) in structural coordinates of the original grid.

Realisations of the weathering overprint and ore type models are in agreement with input data at borehole locations, in global proportions (Figure 10) and in so far as contact relationships between domains are honoured as controlled by the truncation rule. The absolute global proportions for fresh rock and oxidised categories are overestimated by approximately 2 per cent and 5 per cent (respectively) in the simulated realisations in comparison with declustered sample proportions (Figure 10a). Absolute global proportions for the hydrated category are underestimated by less than 1 per cent. In the case of the ore type model absolute global differences for the unmineralised and M-mplH ores are overestimated by approximately 0.5 per cent and 0.02 per cent (respectively) whilst

M-G ores are underestimated by approximately 0.6 per cent (Figure 10c). Inspection of absolute global proportions might under represent the magnitude of differences when quantities of each category vary greatly. Relative differences in the global proportions of input data and realisation categories range considerably from plus or minus 5–15 per cent for the weathering over print model (Figure 10b) and from plus or minus 1–20 per cent for that of the ore type model (Figure 10d). These biases might be resolved with further parameterisation of the model, improved methodologies for calculation of local proportions or realisation post-processing techniques that aim to preserve global proportions.

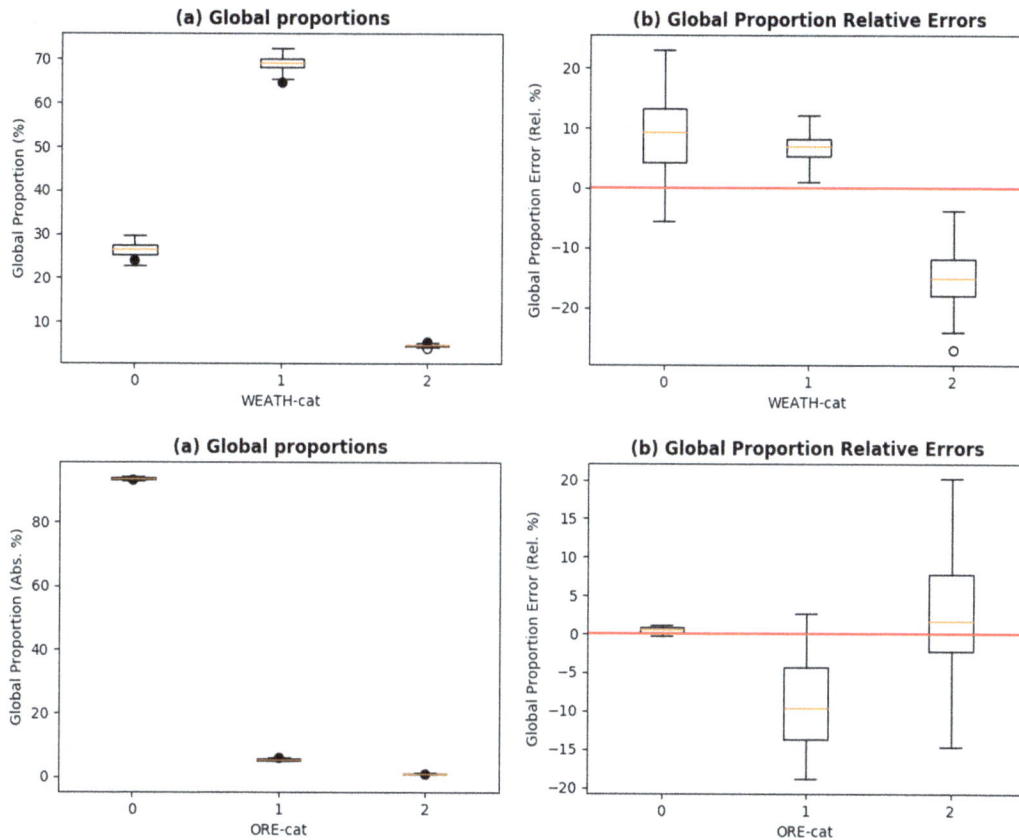

FIG 10 – Comparison category global (decluster weighted) sample and realisation proportions for the weathering overprint (a) and ore type (c) models. Relative differences (errors) between global sample and realisation category proportions for the weathering overprint (b) and ore type (d) models.

Realisations are also in agreement with understanding of geological processes. In the weathering overprint model (recall Figure 9), hydrated category blocks are located at near-surface locations forming a thin blanket. This zone is occasionally recessed in topographic gullies and adjacent steeply inclined terrain placing oxidised zone category blocks at surface which might be explained by increased erosion rates in these areas. The oxidised zone category constitutes the bulk of intermediate elevations of the deposit model. Whilst fresh rock is situated at depth or at higher elevations in zones of higher relief towards the core of the syncline structure to the north-west of the deposit model. Fresh rock exhibits a lensing or inter-fingered boundary with transitional category blocks of the oxidised zone as exhibited in transition probability matrices (Table 2) following changes in topographic relief. The ore type model presented in Figure 11a shows M-G ores constitute the bulk for the prospective mineralisation and outcrop extensively at surface in keeping with their supergene origin. Whilst M-mplH (2) ores are constrained by structures to the south which are presumably sources of the hypogene fluids leading to their formation.

The probability of occurrence of each category at each block is calculated over the 30 realisations to produce probability maps for each model. Probability maps for the M-G and M-mplH ores are presented in Figure 11b and 11c (respectively) and complement maps of simulated domains allowing an assessment of the risk of encountering a specific domain. These are further supported by the most-probable ore type at each location map (Figure 11d).

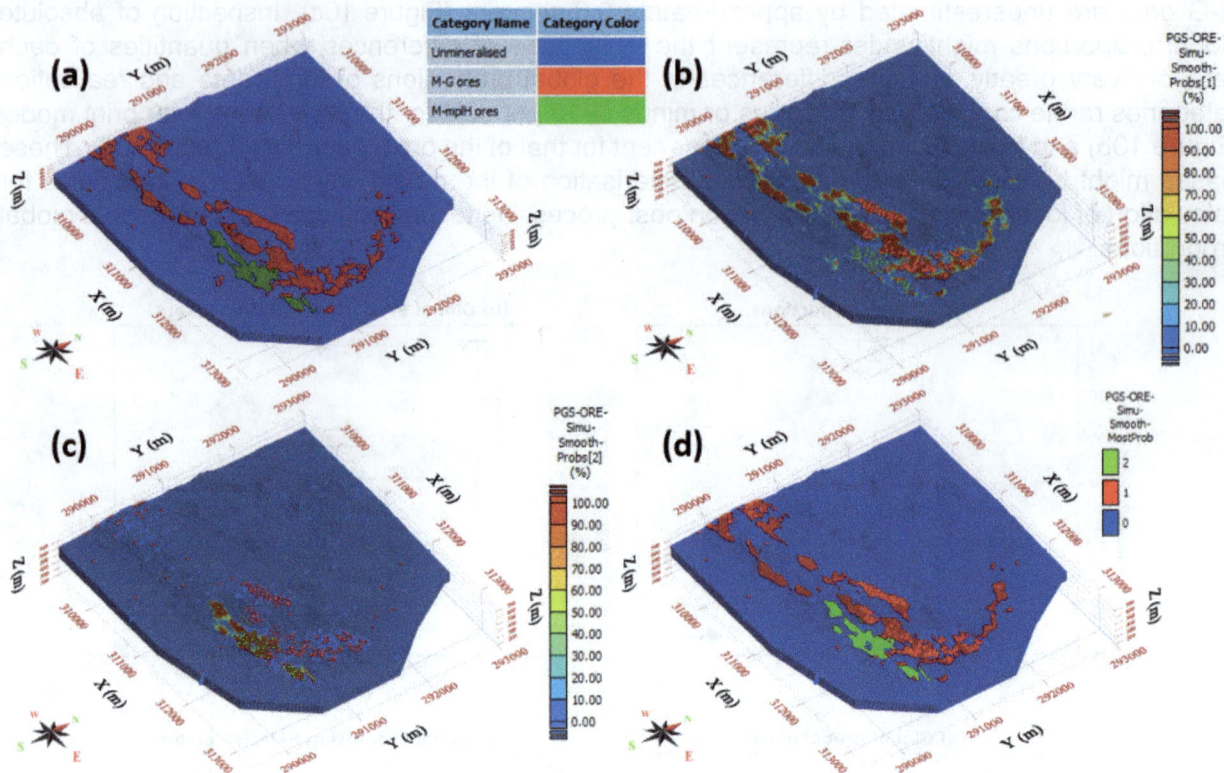

FIG 11 – Realisation 1 and of the ore type model depicting the layout of M-mplH (2), M-G (1) ores and unmineralised rock (0) (a). Probabilities of encountering M-G (b), M-mplH (c) ores and the most-probable ore type category at each location (d).

Oxide grades were modelled in a hierarchical fashion with respect to lithofacies and realisations of the ore type and weathering overprint categories previously shown to control their spatial distributions (recall Figure 5). This was done using Gaussian simulation following a chained transformation to obtain a set of multivariate normal variables after taking into account the compositional nature of multivariate geochemical data. The subcomposition of raw oxide grades Fe, P, SiO_2, Al_2O_3, LOI, Mn and the Rest variable are presented in Figure 12a. This vector of amounts was then transformed to a new $D - 1$ length vector of ratios using the *alr* transformation (Figure 12b). The *alr* scores are transformed to univariate normal scores (Figure 12c) using Gaussian anamorphosis and finally on to multivariate normal factors (Figure 12d) using the MAF transform. The absence of structure observed in cross-variograms (Figure 13) indicate that the MAF factors are also spatially decorrelated which is a desirable property in multivariate Gaussian simulations.

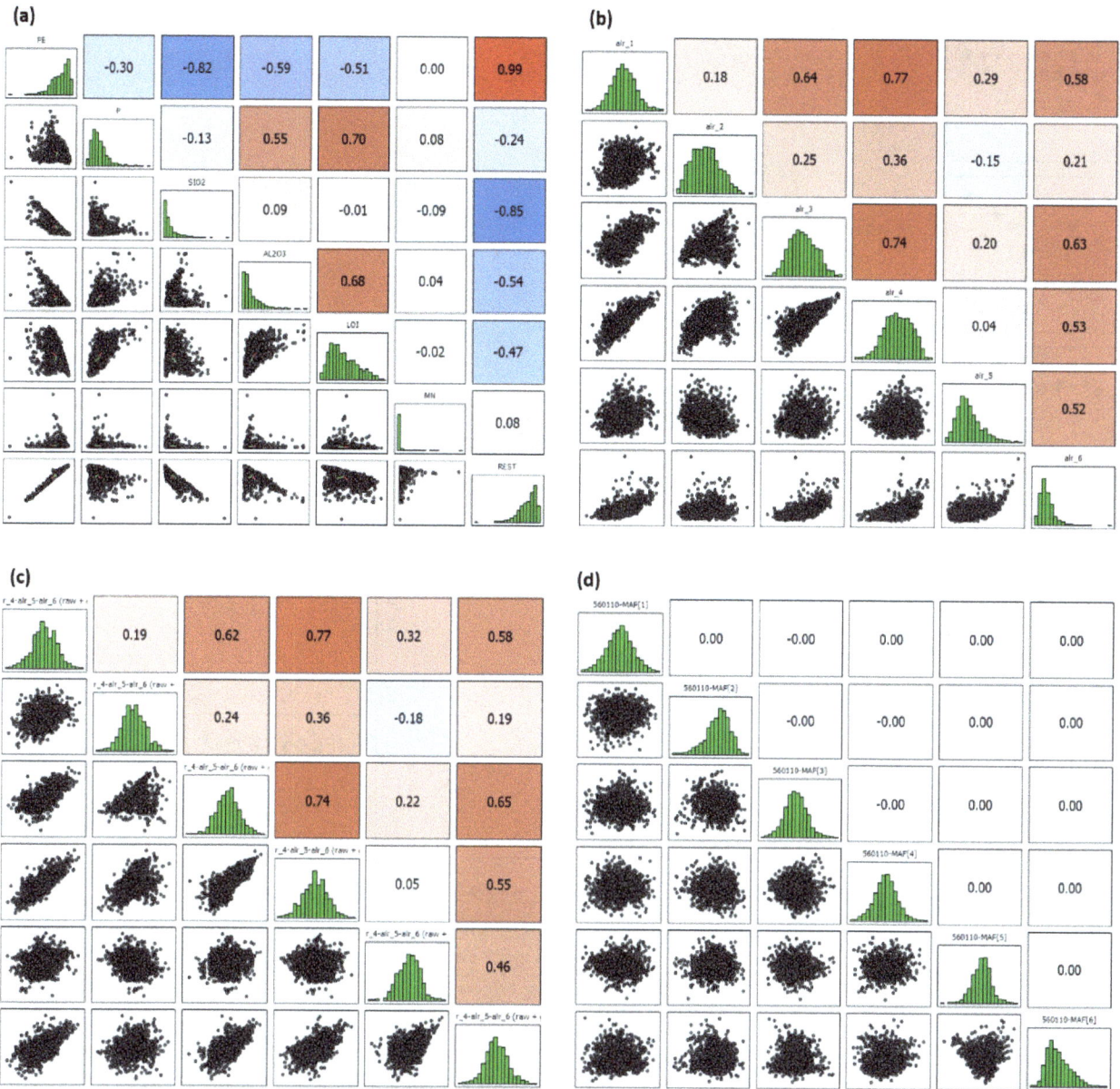

FIG 12 – Scatter plots (low off diagonal), histograms (on diagonal) and Pearson's correlation coefficients (upper off diagonal) for raw oxide grades (a) *alr* transformed log ratios (b), Normal scores of log ratios (c) and MAC factors (d) for the PHbd (5600) oxidised zone (1) M-G ores (1) – ie 560110.

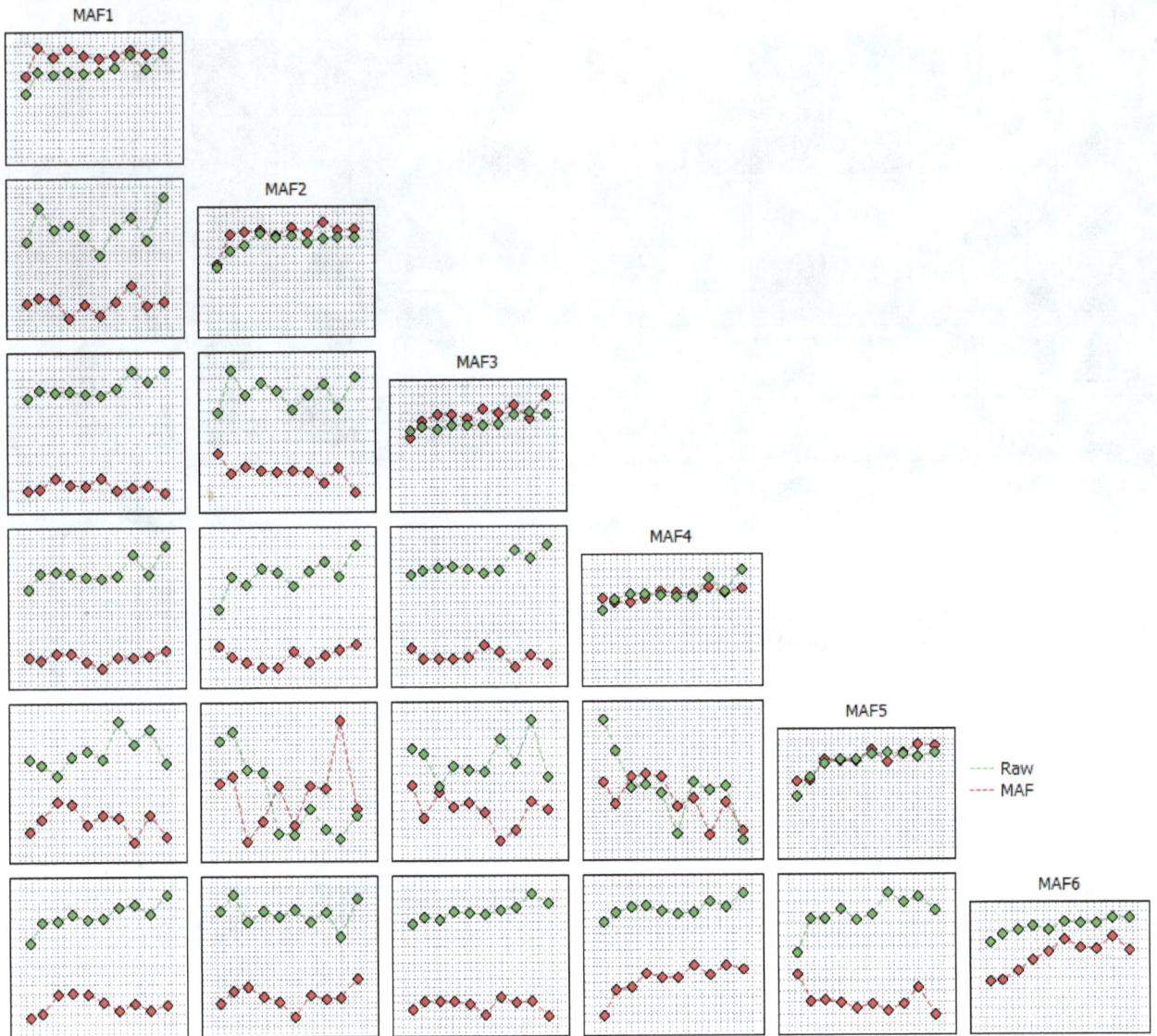

FIG 13 – Direct (on diagonal) and cross (off diagonal) omni-directional variograms for the spatially decorrelated MAF factors in the PHbd oxidised zone M-G ores (560110) plotted using 10, 50 m lag separations.

With multivariate normal scores (MAF Factors) at input data locations and variogram models for each domain (omitted to conserve space), simulations were conducted over the grid using the TBS algorithm. A total of 30 realisations of MAF factors were obtained, one for each realisation of the hierarchical truncated (pluri)Gaussian model of ore type and weathering overprints. Parameters used for TB simulation of MAF factors for domains differ slightly between domains and are provided in Table 5.

TABLE 5

Implementation of turning bands simulations for mineralised domains.

Parameter	Domain								
	560210	560220	560110	560120	560111	560121	580110	580120	810100
Dip	0°	0°	15°	15°	10°	0°	10°	0°	0°
Dip azimuth	80°	80°	0°	0°	270°	270°	345°	90°	90°
Pitch	90°	90°	90°	90°	90°	90°	90°	90°	90°
Radius (m) of search along U direction	300	250	300	150	300	300	200	200	150
Radius (m) of search along V direction	100	100	100	150	300	300	150	200	150
Radius (m) of search along W direction	30	24	30	24	30	21	24	24	30
Number of angular sectors	4	4	4	4	4	4	4	4	4
Optimum number of conditioning data per sector	4	4	4	4	4	4	4	4	4
Maximum number of conditioning data from a single borehole	2	3	2	3	2	3	2	3	2
Minimum number of conditioning data per block/node	4	4	4	4	4	4	4	4	4
Number of bands	400	400	400	400	400	400	400	400	400

Simulated MAF factors were carefully checked prior to back-transformation to compositional data. Images in Figure 14 show that variogram reproduction for simulated MAF factors was satisfactory for PHbd oxidised zone M-G ores (560110) with similar results obtained other simulated domains. Moreover, realisations back-transformed to oxide grade compositions are in agreement with conditioning data statistics. Grade-tonnage curves presented in Figure 15 exhibit satisfactory reproduction of conditioning data oxide grade means at all cut-offs. Scatter plots presented in Figure 16 confirm compositional nature of data including heteroscedasticities, closure features and presence of outliers are reproduced exceptionally well. Reproduction of spatial trends in realisations is shown using swath plots of Figure 17 and exhibit satisfactory results for all domains. Despite this, some bias can be observed in grade-tonnage curves (Figure 15) and in swath means (Figure 17) for several oxides including Fe. These inconsistencies are attributed to the assignment of weathering overprint categories by geologists which may not adequately capture non-stationarity in conditioning data.

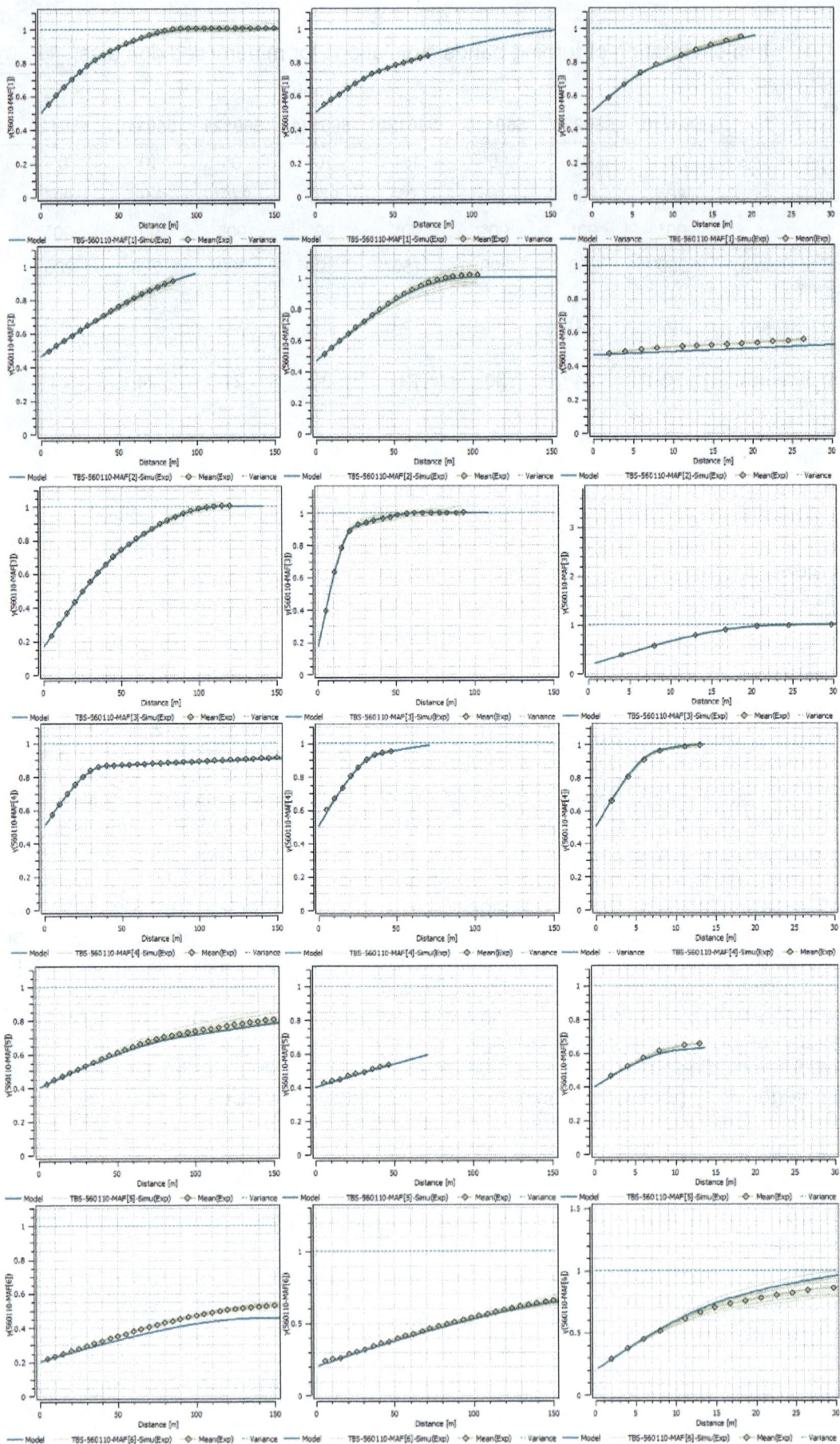

FIG 14 – Variogram reproduction for MAF factor variables 1–6 (rows 1–6) along the west, north (dip 15° from horizontal) and normal directions (left, central and right columns) for first ten realisations of the simulated model.

Mineral Resource Estimation Conference 2023 | Perth, Australia | 24–25 May 2023

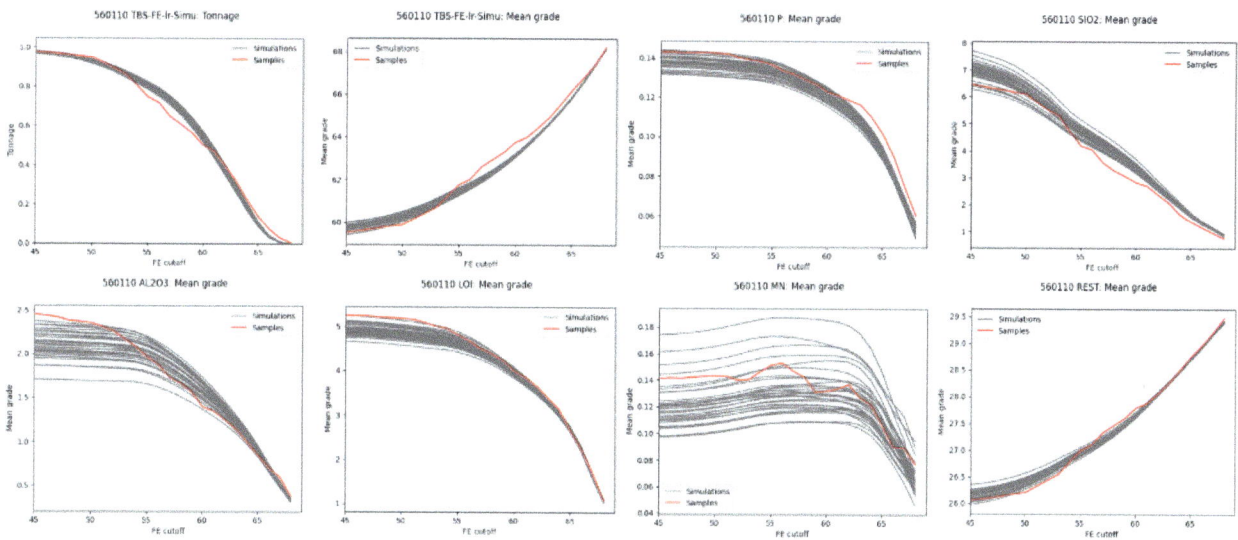

FIG 15 – Tonnage and mean grade by Fe cut-off plots for Fe, P, SiO$_2$, Al$_2$O$_3$, LOI, Mn and the Rest variable (top left to bottom right) comparing reproduction of conditioning data (red) in realisations (grey) for the PHbd oxidised zone M-G ores (560110).

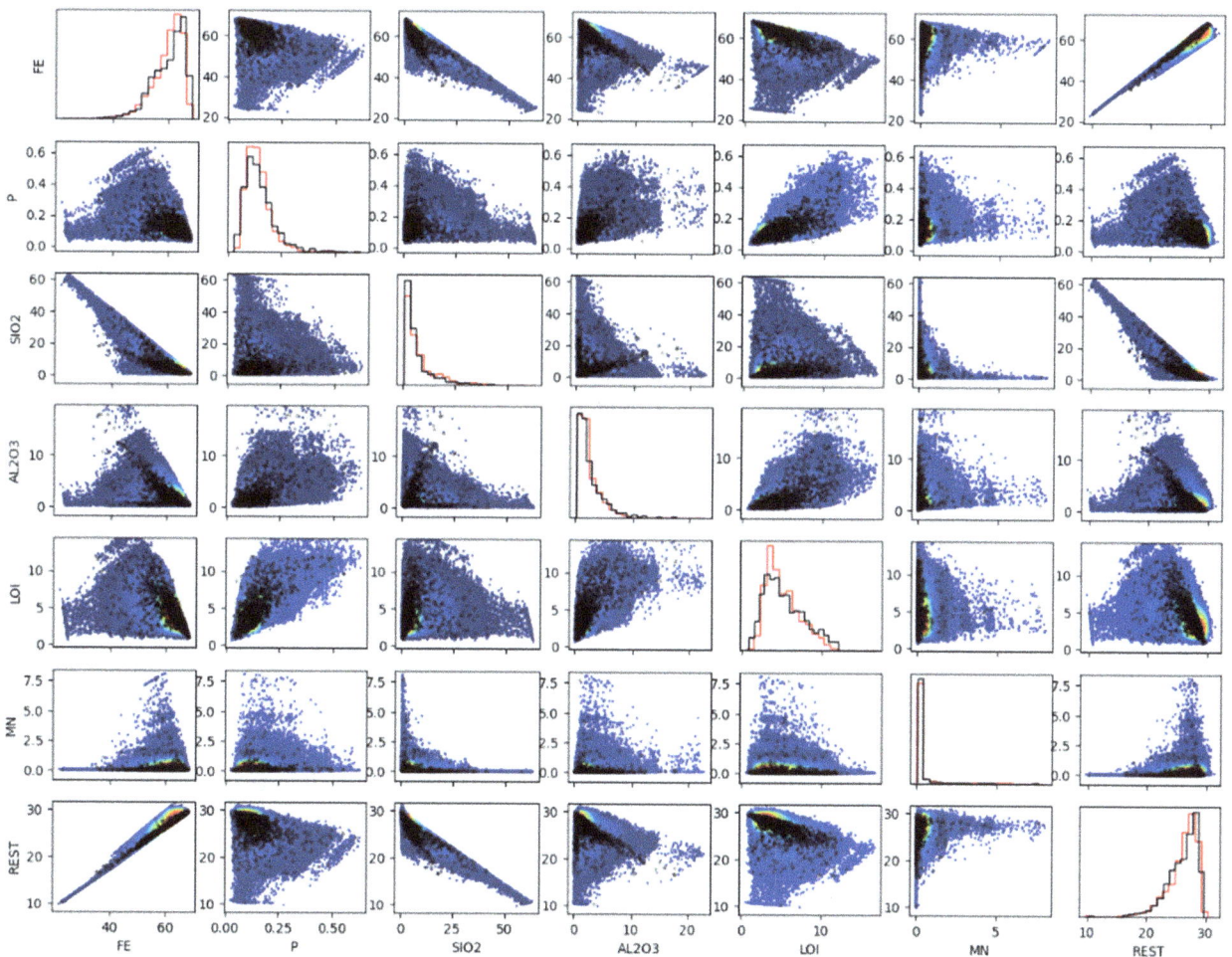

FIG 16 – Histograms (on diagonal) and scatter plots (off diagonal) comparing reproduction of univariate and bivariate statistics of conditioning data (black/points) with realisation 1 (red/density image) for the PHbd oxidised zone M-G ores (560110).

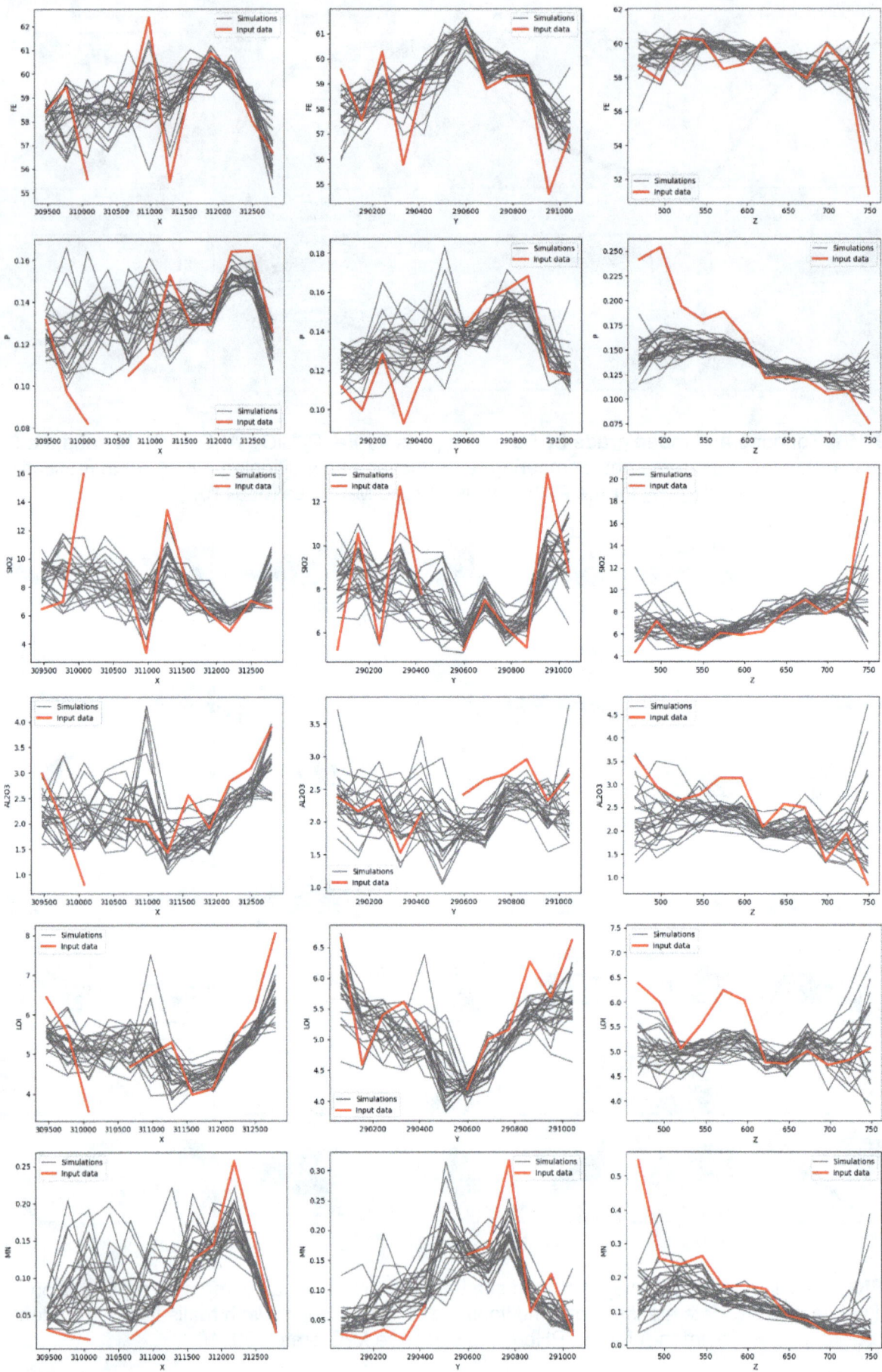

FIG 17 – Swath plots comparing conditioning data and simulation local means along 300 m easting (X), 100 m northing (Y) and 24 m elevation (Z) swaths for Fe in the PHbd oxidised zone M-G ores (560110). Rest is variable omitted due to similarity with Fe.

Classification of mineral resources

Realisations of the simulated model were compiled in a manner to produce two scenarios. Construction of the first scenario, *stochastic domains* is described in the previous section and captures the uncertainties in both the categorical properties that define the layout of various ore types and variabilities in oxide grade concentrations across each. The second scenario represents a situation where only variability of grades is considered and uncertainties in the layout of the weathering overprint ore type categories are not. To produce this scenario probabilities of occurrence of each category are computed with the highest probability category assigned to each block (recall Figure 11d). These *Most-probable domain* categories are then used to define geologic domain limits across the grid and grades from each realisation are assigned to volumes defined by these categories.

Each scenario is re-blocked using volume weighted averaging to the selectivity of mining extraction assumed to be 10 mE × 10 mN × 4 mRL. Local moving windows centred over each SMU representing the volume of ore expected to be extracted from the ore deposit over an annual mining period are used to sum the tonnages above a hypothetical mining ore grade cut-off greater than or equal to 58 per cent Fe in both the kriging model and the realisations. This annual volume is assumed to be a panel of dimensions 200 mE × 150 mN × 36 mRL. At each SMU the frequency which the relative difference in panel ore tonnage for the realisations is within the interval -0.15 to 0.15 of the kriging model tonnage is computed. Selective Mining Units for which panels achieve within tolerance tonnages at frequencies equal to or greater than 90 per cent (27 of 30 realisations) are considered measured category. Those for which differences in ore tonnage are within the interval -0.30 to 0.30 of the kriging model at a frequency equal to or greater than 50 per cent (15 of 30 realisations) are considered indicated category. Remaining SMUs with mining volumes failing to meet these criteria are considered inferred category.

Results for each scenario are tabulated in Table 6 and are reported from volumes representing the model as whole, as well as the volume of the prospective zone of the ore deposit as defined by the ore type categories of the deterministic model provided. The quantities of high confidence measured category mineral resources are significantly (>20 per cent) lower in the stochastic domains scenario when proportions are calculated from the whole model volume. Differences in proportions of measured category mineral resources between scenarios are close to negligible (~1 per cent) when the volume from they are reported from is restricted to that of prospective zone or through increasing Fe cut-off grade used for reporting from the kriging model. Intuitively this reflects the uncertainty in ore tonnages related to the unknown layout of both ore type and weathering overprint categories in less well-informed marginal zones of the ore deposit. Inspection of global tonnage by cut-off curves for the two scenarios (Figure 18b and 18d) reveal the increased proportion of higher risk categories (indicated and inferred) are related to additional resource potential over that of the kriging model only identified in the realisations of the stochastic models.

TABLE 6

Risk-based mineral resource classification for the most-probable and stochastic domains scenarios presented as percentages of total tonnage at a zero cut-off grade. The prospective zone is defined by the ore type categories M-G (1) and M-mpIH (2) ores of the deterministic model.

Zone	Category	Most-probable domains	Stochastic domains	Difference
Whole model	Measured	90.1%	67.9%	-22.2%
	Indicated	2.0%	17.4%	15.4%
	Inferred	7.9%	14.7%	6.8%
Prospective zone	Measured	14.7%	13.7%	-0.9%
	Indicated	35.0%	35.7%	0.7%
	Inferred	50.3%	50.6%	0.2%

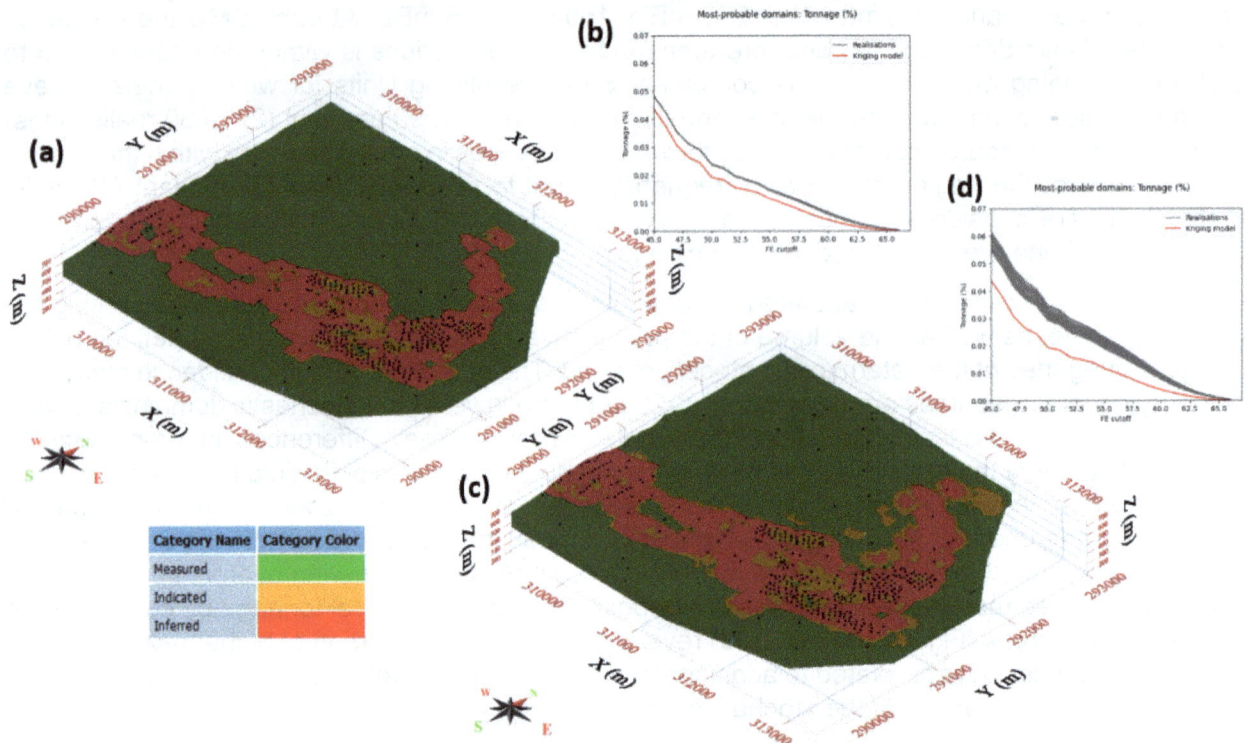

FIG 18 – Risk based mineral resource classification for the most-probable (a) and stochastic (b) domains scenarios. Borehole collar locations are indicated by black points.

The stochastic domains scenario is selected to demonstrate the method for assessing confidence in the grade control setting. Selective Mining Units are designated as either ore or waste determined by their estimated Fe grade relative to the hypothetical mining-cut-off. The ore/waste designation for each SMU as given by the kriging model is now compared to that of realisations and the frequency of misclassifications computed. The SMUs achieving misclassification frequencies equal to or below the threshold of 10 per cent are assigned to the low-risk category whilst remaining SMUs are assigned to the high-risk category. The grade control misclassification risk map produced by this procedure is presented in Figure 19a. It can be seen that quantities of materials within the prospective zone at low risk of misclassification are limited to small volumes proximal to boreholes or where average grades are consistently higher or lower than that of the hypothetical mining cut-off. This is unsurprising at this stage where models are informed by sparse and irregular drilling for the purposes of resource definition as opposed to the close spaced sampling required to obtain high confidence local estimates desired for grade control.

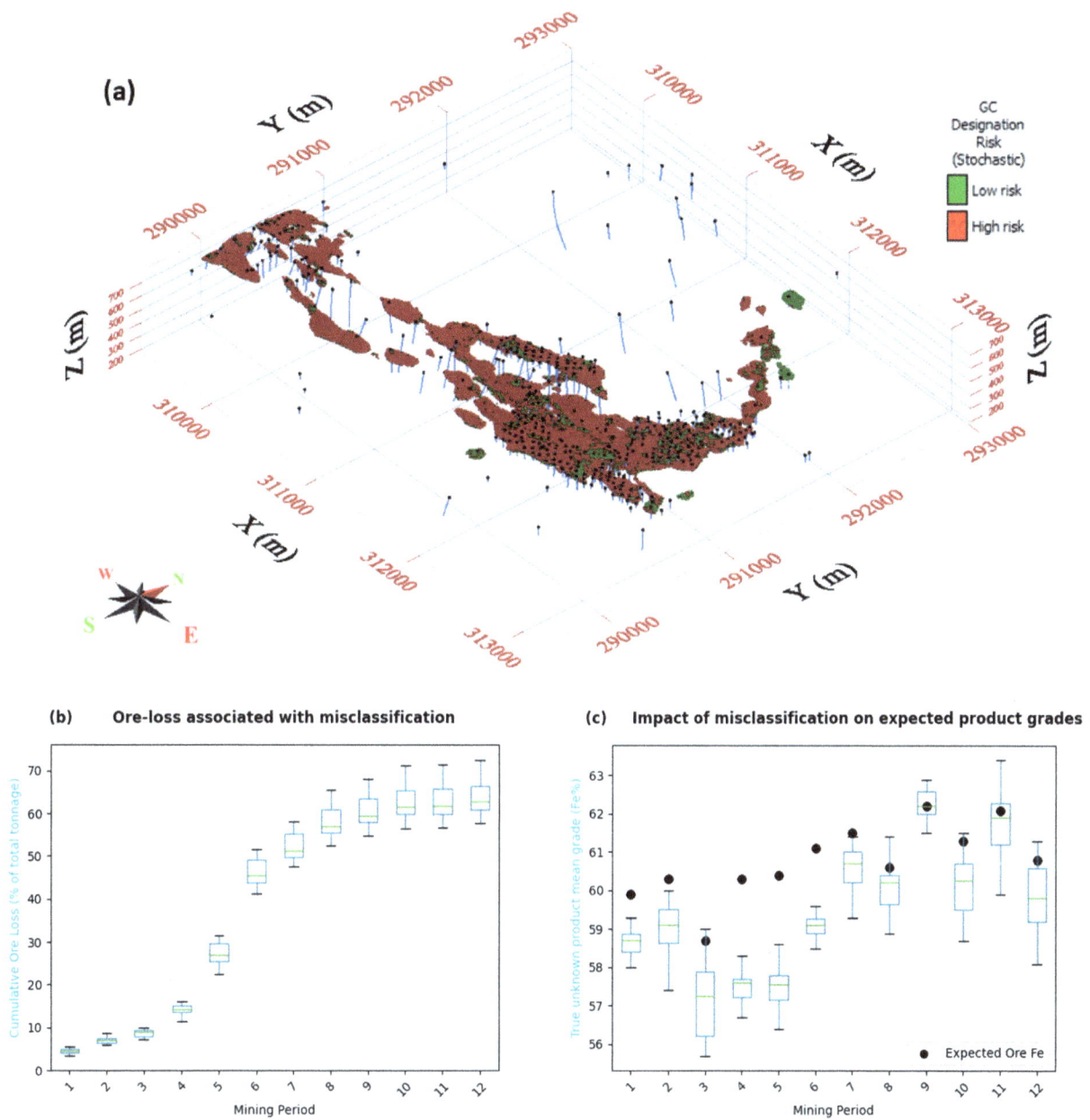

FIG 19 – Ore/waste designation misclassification risk map for the prospective zone (a), cumulative ore-loss (b) and expected verses recovered product grades (c) resulting from misclassifications of ore/waste designation by year given a hypothetical mining sequence.

The mining sequence supplied is used to assess risk were one to attempt to use grades predicted in the kriging model to make grade control designation decisions. The hypothetical mining sequence represents a series of annual mining volumes planned for extraction sequentially over the life of the mine. It is a realistic representation of an open cut mining scenario which progresses from surface in early periods to depths of the ore deposit during later periods and fluctuating mining rates representing production ramp up and eventual decommissioning. Here, each realisation is treated as one possible version of the true but unknown layout of ore within the ore deposit. Ore/waste designations given by the kriging model are compared to that of each realisation of the stochastic model and quantities of ore-loss and dilution misclassification errors are computed for each mining period. The cumulative ore-loss as a percentage of the total ore tonnage given by the kriging model for each realisation is plotted using boxplots presented in Figure 19b. It can be seen over the life of the mine a quantity equivalent to between 55–70 per cent of the total ore tonnage will be lost. In addition to this, the expected product grades for a majority periods will not be realised in actual mining production results (Figure 19b).

CONCLUSIONS

Iron ores at Orebody B are characterised by the relationships between lithofacies, mineralising alterations (ore types) and weathering overprints resulting in material types with distinct geochemical compositions. Compositional data analysis techniques involving log ratio transformations allow for statistical analysis revealing important relationships between these categories and oxide grade compositions as well as the geostatistical modelling of such constrained multivariate data.

The realisations of ore type and weathering overprint categories constructed using the truncated (pluri)Gaussian model reproduced the global proportions, contact arrangements and presented uncertainties in their layout at unknown locations between boreholes. How additional information from field mapping by geologists can be incorporated into model constraints was also demonstrated. Realisations of the oxide grade composition conditional to these categories reproduce the histogram, multivariate relationships and variogram. This allows for a more holistic assessment of uncertainties in the spatial distribution of grades at a scale relevant to mining extraction processes (the SMU).

The mineral resource classification scheme presented aims to categorise grades presented in traditional deterministic model constructed using kriging following a risk-based framework considering an uncertainty model. The classification criteria are flexible and are defined with the objective of quantifying risks relevant for the purposes of long-term mine planning for consideration in (pre)feasibility studies. When additional sources of uncertainty such as the layout of materials controlling the distribution of grades are incorporated in the uncertainty model, associated risk to forecast ore recoveries are shown to increase. Indeed, understanding the risks associated with the complex geology informed by sparse and irregularly spaced boreholes was one motivation for selecting Orebody B for this study. The spatial risk maps shown here are a useful tool for identifying zones where a deterministic model might materially under or overestimate mineral resource potential. An advantage of this scheme is that zones of sparse drilling are not unnecessarily classified as high-risk when economic potential is low, unlike what might result from geometric or other geostatistical mineral resource classification schemes. This knowledge could be used to direct cost-effective collection of additional data in high-risk zones improving forecast production performances.

The flexibility of this risk-based framework was demonstrated by also presenting an example of the methodology in the grade control setting. Here, one key objective is to minimise the occurrence of ore-loss and dilution misclassification errors maximising the quantity of correctly designated blocks. Assessing risk using these misclassification metrics is advantageous as they are intuitive and the consequences in terms of lost revenue and increased processing costs (or product penalties) are straightforward to evaluate.

ACKNOWLEDGEMENTS

Access to the Orebody B data set and permission to publish was provided by BHP PTY LTD Western Australia Iron Ore. The author would like to acknowledge the BHP PTY LTD Western Australia Iron Ore Geoscience department and its leadership team for supporting the publication of this article.

REFERENCES

Aitchison, J and Greenacre, M, 2002. Biplots of compositional data, *Journal of the Royal Statistical Society: Series C (Applied Statistics)*, 51(4):375–392. https://doi.org/https://doi.org/10.1111/1467-9876.00275

Aitchison, J, 1986. The statistical analysis of compositional data, *Monographs on statistics and applied probability* 25, 44(2) (Chapman and Hall). https://doi.org/10.1111/j.2517-6161.1982.tb01195.x

Armstrong, M, Galli, A, Beucher, H, Loc'h, G, Renard, D, Doligez, B, Eschard, R and Geffroy, F, 2011. *Plurigaussian Simulations in Geosciences* (Springer). https://doi.org/10.1007/978-3-642-19607-2

Carr, J R, 2003. Geostatistical reservoir modeling, *Computers and Geosciences,* 29(1) (OUP USA). https://doi.org/10.1016/s0098-3004(02)00101-2

Chilès, J-P and Delfiner, P, 2012. Geostatistics, in *Geostatistics: Modeling Spatial Uncertainty*, second edition (John Wiley and Sons, Inc). https://doi.org/10.1002/9781118136188

de Oliveira Carvalho Rodrigues, P, Hinnov, L A and Franco, D R, 2019. A new appraisal of depositional cyclicity in the Neoarchean-Paleoproterozoic Dales Gorge Member (Brockman Iron Formation, Hamersley Basin, Australia, *Precambrian Research*, vol 328. https://doi.org/10.1016/j.precamres.2019.04.007

Desassis, N, Renard, D, Beucher, H, Petiteau, S and Freulon, X, 2015. *A pairwise likelihood approach for the empirical estimation of the underlying variograms in the plurigaussian models.* http://arxiv.org/abs/1510.02668

Desbarats, A J and Dimitrakopoulos, R, 2000. Geostatistical simulation of regionalized pore-size distributions using min/max autocorrelation factors, *Mathematical Geology.* https://doi.org/10.1023/A:1007570402430

Emery, X, Ortiz, J M and Rodríguez, J, 2006. Quantifying Uncertainty in Mineral Resources by Use of Classification Schemes and Conditional Simulations, *Mathematical Geology*, 38(4):445–464. https://doi.org/10.1007/s11004-005-9021-9

Goovaerts, P, Avruskin, G, Meliker, J, Slotnick, M, Jacquez, G and Nriagu, J, 2005. *Geostatistical modeling of the spatial variability of arsenic in groundwater of southeast Michigan*, 41(April). https://doi.org/10.1029/2004WR003705

Harmsworth, R A, Kneeshaw, M, Morris, R, Robinson, C J and Shrivastava, P K, 1990. BIF derived iron ores of the Hamersley Province, *Geology of the Mineral Deposits of Australia and Papua New Guinea*, pp 617–642 (The Australasian Institute of Mining and Metallurgy: Melbourne).

Hotelling, H, 1933. Analysis of a complex of statistical variables into principal components, *Journal of Educational Psychology*, 24(6):417–441. https://doi.org/10.1037/h0071325

Journel, A G, 1974. Geostatistics for Conditional Simulation of Ore Bodies, *Economic Geology*, 69(5):673–687. https://doi.org/10.2113/gsecongeo.69.5.673

Kepert, D A, 2001. The mapped stratigraphy and structure of the Mining area C region (the Black Monolith), An eclectic synthesis of geological mapping by BHPBIO Exploration 1994–2001.

Madani, N and Emery, X, 2015. Simulation of geo-domains accounting for chronology and contact relationships: application to the Río Blanco copper deposit, *Stochastic Environmental Research and Risk Assessment*, 29(8):2173–2191.

Maleki Tehrani, M A, Asghari, O and Emery, X, 2013. Simulation of mineral grades and classification of mineral resources by using hard and soft conditioning data: Application to Sungun porphyry copper deposit, *Arabian Journal of Geosciences.* https://doi.org/10.1007/s12517-012-0638-y

Matheron, G, 1973. The Intrinsic Random Functions and Their Applications LK, *Advances in Applied Probability TA–TT*, 5(3):439–468. https://ecu.on.worldcat.org/oclc/5546215114

Matheron, G, Beucher, H, de Fouquet, C, Galli, A, Guerillot, D and Ravenne, C, 1987, September 27. Conditional Simulation of the Geometry of Fluvio-Deltaic Reservoirs, *All Days.* https://doi.org/10.2118/16753-MS

Mery, N, Emery, X, Cáceres, A, Ribeiro, D and Cunha, E, 2017. Geostatistical modeling of the geological uncertainty in an iron ore deposit, *Ore Geology Reviews*, vol 88. https://doi.org/10.1016/j.oregeorev.2017.05.011

Morris, R C, 1980. A textural and mineralogical study of the relationship of iron ore to banded iron-formation in the Hamersley iron province of Western Australia, *Economic Geology*, 75(2):184–209. https://doi.org/10.2113/gsecongeo.75.2.184

Morris, R C, 2011. Microplaty haematite of the high-grade iron ores - Its nature and genesis, in *Proceedings of the Iron Ore 2011 Conference*, pp 117–124 (The Australasian Institute of Mining and Metallurgy: Melbourne).

Morris, R C, 2012. Microplaty hematite—its varied nature and genesis, *Australian Journal of Earth Sciences*, 59(3):411–434. https://doi.org/10.1080/08120099.2011.626453

Pawlowsky-Glahn, V and Olea, R A, 2004. *Geostatistical Analysis of Compositional Data* (Oxford University Press: USA).

Pearson, K, 1901. LIII, On lines and planes of closest fit to systems of points in space, *The London, Edinburgh and Dublin Philosophical Magazine and Journal of Science*, 2(11):559–572.

Rossi, M E and Deutsch, C V, 2014. *Mineral Resource Estimation* (Springer: Dordrecht). https://doi.org/10.1007/978-1-4020-5717-5

Silva, D S F and Boisvert, J B, 2014. Mineral resource classification: A comparison of new and existing techniques, *Journal of the Southern African Institute of Mining and Metallurgy*, 114(3):265–273.

Snowden, D V, 1996. Practical interpretation of resource classification guidelines, *AusIMM Annual Conference*, Perth, vol 68 (The Australasian Institute of Mining and Metallurgy: Melbourne).

Switzer, P, 1984. Inference for Spatial Autocorrelation Functions, *Geostatistics for Natural Resources Characterization.* https://doi.org/10.1007/978-94-009-3699-7_8

Talebi, H, Asghari, O and Emery, X, 2013. Application of plurigaussian simulation to delineate the layout of alteration domains in Sungun copper deposit, *Central European Journal of Geosciences.* https://doi.org/10.2478/s13533-012-0146-3

Talebi, H, Asghari, O and Emery, X, 2014. Simulation of the lately injected dykes in an Iranian porphyry copper deposit using the plurigaussian model, *Arabian Journal of Geosciences.* https://doi.org/10.1007/s12517-013-0911-8

Talebi, H, Asghari, O and Emery, X, 2015. Stochastic rock type modeling in a porphyry copper deposit and its application to copper grade evaluation, *Journal of Geochemical Exploration.* https://doi.org/10.1016/j.gexplo.2015.06.010

Talebi, H, Mueller, U, Tolosana-Delgado, R and van den Boogaart, K G, 2019. Geostatistical Simulation of Geochemical Compositions in the Presence of Multiple Geological Units: Application to Mineral Resource Evaluation, *Mathematical Geosciences*, 51(2):129–153. https://doi.org/10.1007/s11004-018-9763-9

Tolosana-Delgado, R, 2005. *Geostatistics for Constrained Variables : Positive Data, Compositions and Probabilities.*

Tolosana-Delgado, R, Mueller, U and van den Boogaart, K G, 2018. Geostatistics for Compositional Data: An Overview, *Mathematical Geosciences*. https://doi.org/10.1007/s11004-018-9769-3

Tolosana-Delgado, R, Mueller, U, Van Boogaart, K G D, Ward, C and Gutzmer, J, 2015. Improving processing by adaption to conditional geostatistical simulation of block compositions, *Journal of the Southern African Institute of Mining and Metallurgy*, 115(1):13–26.

van den Boogaart, K G and Tolosana-Delgado, R, 2013. Analyzing Compositional Data with R, In *Analyzing Compositional Data with R* (Springer: Berlin). https://doi.org/10.1007/978-3-642-36809-7

Verly, G, 2005. Grade Control Classification of Ore and Waste: A Critical Review of Estimation and Simulation Based Procedures, *Mathematical Geology*, 37(5):451–475. https://doi.org/10.1007/s11004-005-6660-9

Watson, G S, Journel, A G and Huijbregts, C J, 1980. Mining Geostatistics, *Journal of the American Statistical Association*, 75(369):245. https://doi.org/10.2307/2287429

Estimation methods

Modelling metal recovery by co-kriging the feed and concentrate masses of metal

A Adeli[1], P Dowd[2], C Xu[3] and X Emery[4]

1. Postdoctoral Research Fellow, University of Adelaide, Adelaide SA 5005.
   Email: am.adeli@gmail.com
2. Professor, University of Adelaide, Adelaide SA 5005. Email: peter.dowd@adelaide.edu.au
3. Associate Professor, University of Adelaide, Adelaide SA 5005.
   Email: chaoshui.xu@adelaide.edu.au
4. Professor, University of Chile, Santiago, Chile. Email: xemery@ing.uchile.cl

ABSTRACT

Geometallurgical modelling is increasingly being incorporated into mineral resource modelling and estimation as a means of increasing efficiency, decreasing operating costs and reducing risk in mining operations. Traditional geostatistical workflows for geometallurgical modelling face several difficulties, in particular, related to the non-additivity of process response variables, to highly heterotopic data sets and/or to preferential sampling designs, all of which may explain why most attempts to estimate geometallurgical response variables from assay data and mineralogy are still based on statistics and machine learning.

The aim of the work presented here is to demonstrate the applicability of geostatistics to estimate copper recovery from copper sulfide ores at the Prominent Hill Iron Oxide Copper-Gold (IOCG) deposit in Australia. As recovery is non-additive, the preferred workflow is the joint estimation of the masses of metal in the feed and in the concentrate, both of which are additive variables. While the mass of metal in the feed can be estimated from abundant online assay analyses, the mass of metal in the concentrate can only be estimated from a very limited number of laboratory-scale batch flotation tests.

Traditional approaches to estimating the masses of metal in the feed and in the concentrate, such as simple and ordinary co-kriging, are described and discussed. A modified version of co-kriging is introduced, which incorporates linear relationships between the mean values of the input variables, in addition to a spatial correlation model. Co-kriging with related means is shown to outperform traditional simple and ordinary co-kriging in terms of precision, accuracy and consistency of the estimates, and offers a practical and efficient tool to deal with highly heterotopic and preferential sampling designs.

INTRODUCTION

There is a need for the mining sector to improve its efficiency while overcoming several obstacles such as deeper and lower grade mineral deposits, more complex geology, volatility of commodity prices, high energy consumption, and difficulties related to global warming. These challenges require geoscientists and engineers to seek new approaches to produce more accurate predictive models in all phases of the mining process. Geometallurgical modelling has shown promise as a major opportunity for mineral resource estimation, mine planning and project evaluation (Dunham and Vann, 2007; Navarra, Grammatikopoulos and Waters, 2018).

Geometallurgy is a multi-disciplinary approach that combines geology, geostatistics and extractive metallurgy, it emphasises the geological controls on mineral processing and the prediction of process performance from rock properties and it offers a route to more effective energy-efficient implementation, enhanced forecasting, increased certainty, technical risk reduction, improved economic optimisation of mineral production and sustainable mine development (Dobby *et al*, 2004; Williams and Richardson, 2004; Powell, 2013; Coward and Dowd, 2015; Dominy *et al*, 2018; Sepúlveda, Dowd and Xu, 2018).

Geometallurgical modelling enables incorporating ore characterisation data into 3D block models that complement the traditional geological and grade modelling. This requires the quantification of mineral deposits in terms of process parameters such as hardness work indices, comminution energy, size reduction, acid consumption, liberation potential, mineral flotation kinetics, plant throughput and product recoveries.

Most studies in the field of geometallurgical modelling are statistics-based and do not explicitly account for the spatial correlations between available data such as assay or mineralogical data. In this respect, machine learning techniques such as classification and clustering are widely used to discover relationships between process response variables and assays, grain sizes and mineralogical compositions (Hunt *et al*, 2014; Sepúlveda, Dowd and Xu, 2018; van den Boogaart and Tolosana-Delgado, 2018). Other techniques include reducing the number of variables by merging or removing them (Boisvert *et al*, 2013).

Geostatistical approaches provide an opportunity to improve the estimation of response variables, as they can account for the spatial correlation of the geometallurgical data, as well as for the information provided by secondary variables (either process variables or geological variables that are intrinsic to the processed ore). At the same time, several challenges often arise with geometallurgical data sets. The first is the limited sampling of the target variables, which motivates the search for covariates to compensate for the lack of direct information. A second challenge relates to the sampling design, which may be preferentially oriented and non-representative of the entire mineral deposit. A third challenge is the lack of additivity of many geometallurgical variables, such as metal recoveries, which prevents upscaling these variables by a simple linear averaging (Carrasco, Chilès and Séguret, 2008).

To circumvent non-additivity issues, the recovery can be calculated from Equation 1 and the estimation of the components:

$$Rec = \frac{w_c}{w_f} = \frac{Cc}{Ff} \tag{1}$$

where F is the feed weight (assumed to be 100 g), f is the metal assay in the feed, C is the concentrate weight and c is the metal assay in concentrate, while w_f and w_c are the masses of metal in the feed and concentrate respectively, both of which are additive variables. Two additive variables w_f and w_c can be estimated (separately or jointly) and recovery can then be estimated from Equation 1 (Emery and Séguret, 2020).

The content of this paper is outlined as follows. Section 2 reviews several co-kriging variants together with their drawbacks in situations where the target variable is highly under-sampled with respect to covariates. Section 3 compares the results of the mentioned co-kriging approaches for an Iron Oxide Copper-Gold (IOCG) deposit in Australia where assay data are abundant, but a very limited number of flotation tests are available. Conclusions follow in Section 4.

GEOSTATISTICAL JOINT ESTIMATION BY CO-KRIGING

Fewer samples may have been collected for a target variable (primary) due to technical difficulties or costs, a common situation in geometallurgical modelling (Figure 1). Co-kriging can then produce more accurate results compared to kriging the variables separately if there is a strong dependence between the primary and secondary coregionalised variables (Wackernagel, 2003). Simple co-kriging (SCK) and ordinary co-kriging (OCK) are the two most common co-kriging variations, depending on whether the mean values of the variables under study are assumed to be known or unknown.

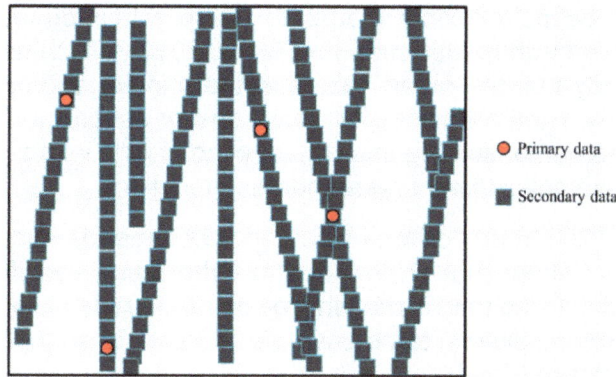

FIG 1 – A heterotopic sampling design where a large number of secondary data will be used to estimate a primary variable with a limited number of samples.

Simple co-kriging

The mean values of the primary and secondary variables are assumed to be known and the sum of the co-kriging weights applied to the primary or secondary data is unrestricted. Although theoretically optimal, the estimation can deteriorate considerably if the mean values are miss-specified. This issue is likely to happen when (1) the primary variable is under-sampled compared with the secondary variables, and/or (2) the samples for the primary or secondary variables have a preferential design, eg they are mostly taken from high-grade zones. Regrettably, both situations are frequent in geometallurgical modelling, with the preferential sampling entailing not only a loss of precision in the estimates, but also systematic errors (biases).

Ordinary co-kriging

As the means of the primary and secondary variables are assumed to be unknown, co-kriging should not be affected by a miss-specification due to a lack of data or a preferential sampling design. The weights of the primary data should sum to one and the weights of each secondary data should sum to zero in order to meet the unbiasedness condition in ordinary co-kriging. However, due to the constraint that the weights for each secondary variable must sum to zero, these weights are often small, which leads to little influence of the secondary variables on the estimation, a feature that has been recognised as a weakness of ordinary co-kriging (Goovaerts, 1998). In highly heterotopic cases, there is another issue related to the constraint on the weights for the primary data: as there are many cases with no primary data in the neighbourhood to meet the condition, ordinary co-kriging causes numerous fails in the estimation of the primary variable.

Standardised ordinary co-kriging

To avoid the excessive underweighting of the secondary data in OCK, a variant of co-kriging has been developed, in which a single constraint is used to force all primary and secondary data weights to sum to one (Goovaerts, 1998). This strict condition for generating unbiased estimates is the main difficulty of this variant, as the mean values for all the primary and secondary variables should be the same. This condition occurs in very few specific situations, eg when the same quantity is measured by different devices or on different support sizes. On the positive side, by giving the secondary data higher weights, this variant of co-kriging increases the secondary data contribution to the estimation.

Co-kriging with related means

The previous statements illustrate the fact that a 'good' estimator should take into account the dependence relationships between primary and secondary variables. However, these relationships are not limited to spatial cross-correlations. The mean values may also play a central role and ignoring them (ordinary co-kriging approach) makes the covariates less informative and less useful for estimation purposes. In real situations, neither the simple nor the standardised ordinary co-kriging are satisfactory as they assume known mean values or unknown, but equal, mean values.

Can co-kriging be made to be more flexible in order to incorporate information from covariates? One approach is to use *co-kriging with related means* (Emery, 2012), which uses standardised ordinary co-kriging but assumes only that the mean values of the primary and/or secondary variables are linearly related, ie there are known weights for which one or several weighted sums of the mean values are constant. Before demonstrating this variant of co-kriging with a geometallurgical data set, we provide the following examples in mineral resources modelling:

- The specific gravity (SG) in iron or lead-zinc deposits depends strongly on the metal grades, eg: $m_{SG} = a + b \, m_{Fe}$, where m_{SG} and m_{Fe} denote the mean specific gravity and iron grade, respectively. Instead of using this relationship on the iron grade block estimates, it can be used internally in the co-kriging system to co-estimate SG and Fe so that, on average, the desired relationship is reproduced.

- Linear relationships between total iron (Fe), recoverable iron (rFe) and silica (SiO_2) grades are also widely recognised in such deposits, which can translate into formulae of the form $m_{rFe} = c \, m_{Fe}$ and $m_{Fe} = d - e \, m_{SiO_2}$.

- Quite often, the metal grade also depends on the prevailing rock type, which can be codified through one or several indicator (binary 0/1) variables. For the sake of simplicity, considering two rock types with approximate mean iron grades m_1 and m_2, the relationship between mean values can be written as $m_{Fe} = m_1 + (m_2 - m_1) \, m_I$, with m_I denoting the mean indicator; the latter varies locally and is close to 0 in the central part of rock type 1 and close to 1 in the central part of rock type 2.

CASE STUDY – PROMINENT HILL IOCG DEPOSIT

Presentation of the data

The co-kriging methods described earlier have been tested and evaluated at OZ Minerals' Prominent Hill IOCG deposit located in South Australia. Although the amount of uranium in this copper-gold-silver deposit is significant, it is not economically feasible to recover it, and there is a penalty associated with high contents of uranium in the final products. The mineralisation occurs mainly in a fault zone and is hosted by hematite-matrix breccia containing fragments of sandstone, siltstone, dolostone, and mafic to intermediate volcanic rocks. The ore is mined from the Malu open pit and two (Malu and Ankata) underground mines, where the predominant form of copper mineralisation is sulfide disseminations of chalcocite, bornite and chalcopyrite in the breccia matrix (Belperio, Flint and Freeman, 2007; Schlegel and Heinrich, 2015).

More than 2600 boreholes have been drilled at Prominent Hill and around 600 000 samples have been taken from cores and analysed for the grades of more than 40 elements, including Cu, Au, Ag, U, Fe, S, Si, Co, Al, Ca, K and Mg. Assay data are composited into 2 m lengths and, after removing duplicates and erroneous data, there were 282 359 samples of metal assay in feed (f). In this study, we focus on the mass of copper in the feed (w_f), calculated by multiplying the copper grade by 100 (as F is 100 g). Figure 2 shows its distribution in the deposit, from which it can be seen that w_f data are abundant, although on an irregular sampling design.

In contrast, there are 96 laboratory-scale batch flotation tests for copper recovery, but 44 of them are not considered in this study because they are taken from a conveyor belt, stopes, or are a combination of more than one borehole, leading to unknown spatial coordinates. The remaining 52 copper recovery samples are used for estimating the mass of metal in the concentrate (w_c). The spatial distribution of these samples in the deposit is shown in Figure 3 aligned with w_c for copper (known from flotation tests). The number of flotation test data makes any geological domaining unfeasible for block modelling. Table 1 shows the general statistics of the data available for this study.

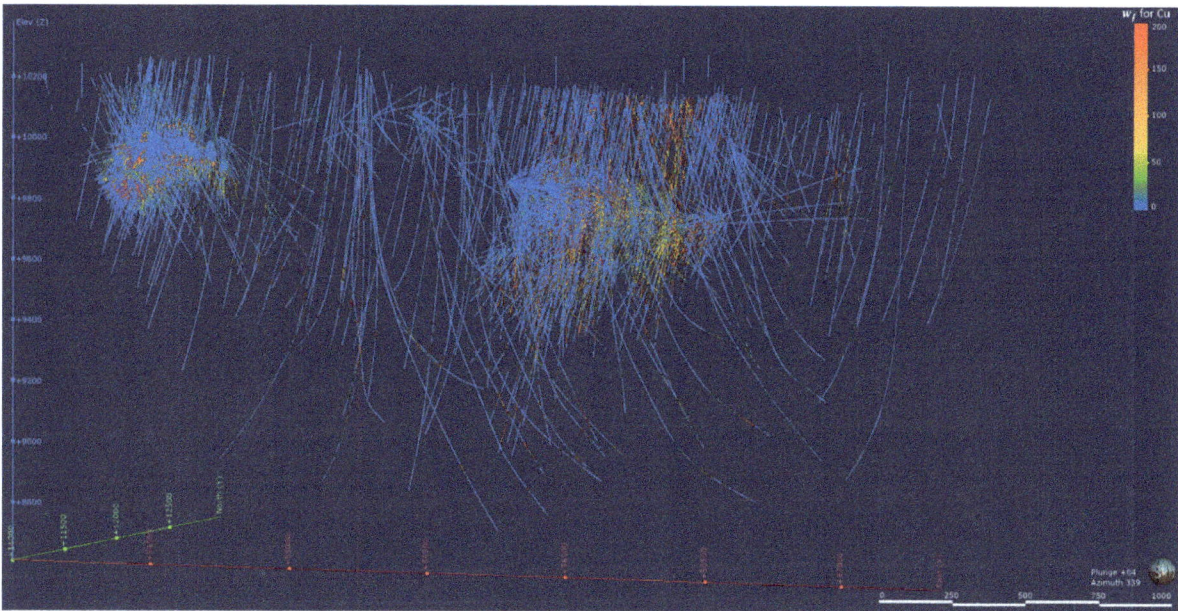

FIG 2 – 3D view of the distribution of assay samples in the deposit with w_f values for copper.

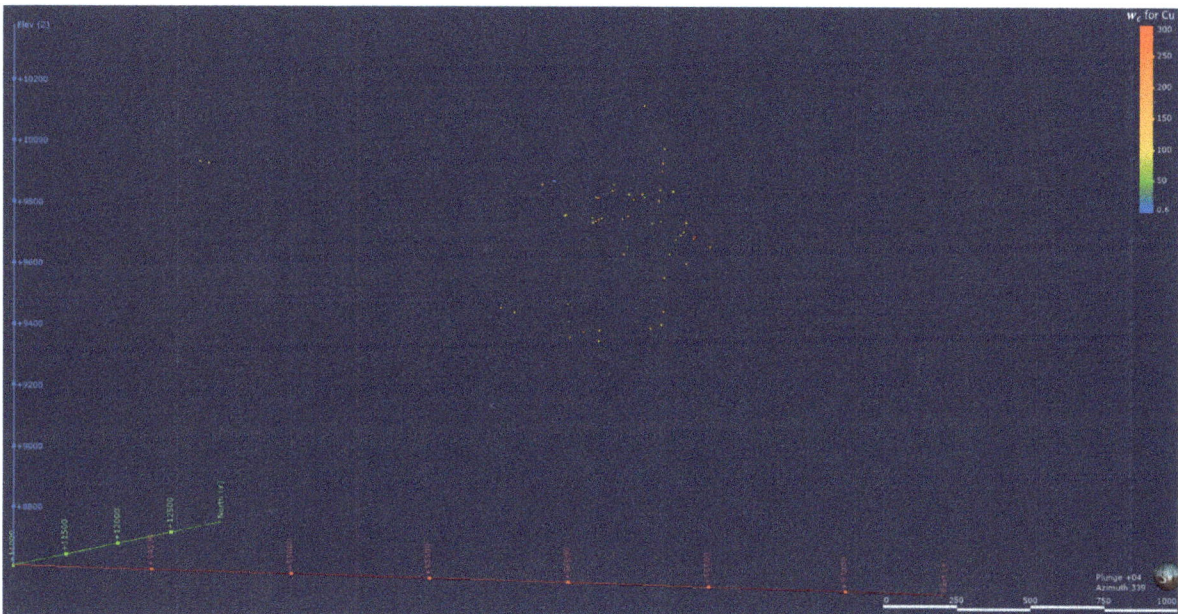

FIG 3 – The distribution of batch flotation tests in the deposit with w_c values for copper.

TABLE 1

Statistics of the mass of copper in the feed (w_f) and in the concentrate (w_c).

Variable	Unit	# of data	Minimum	Maximum	Mean	Variance
w_f	gr × %	282 359	0.00	2061.20	37.37	9442.62
w_c	gr × %	52	35.80	322.10	137.47	3717.62

Modelling the relationship between mean values

The relationships between different metals assays in the feed (f) and w_c for copper (Figure 4), reveal a strong linear relation between the masses of copper in the feed (w_f) and in the concentrate (w_c), which makes co-kriging a suitable option. This linear relationship is displayed in the fitted linear regression models (Figure 5), with a correlation coefficient of 0.988. As $w_f = 0$ results in $w_c = 0$, the intercept for the regression lines is assumed to be zero and $w_c = 0.854\, w_f$ shows the relationship between the mean values based on a regression of w_c as a function of w_f, or $w_f = 1.169\,w_c$ when fitting the opposite regression of w_f as a function of w_c.

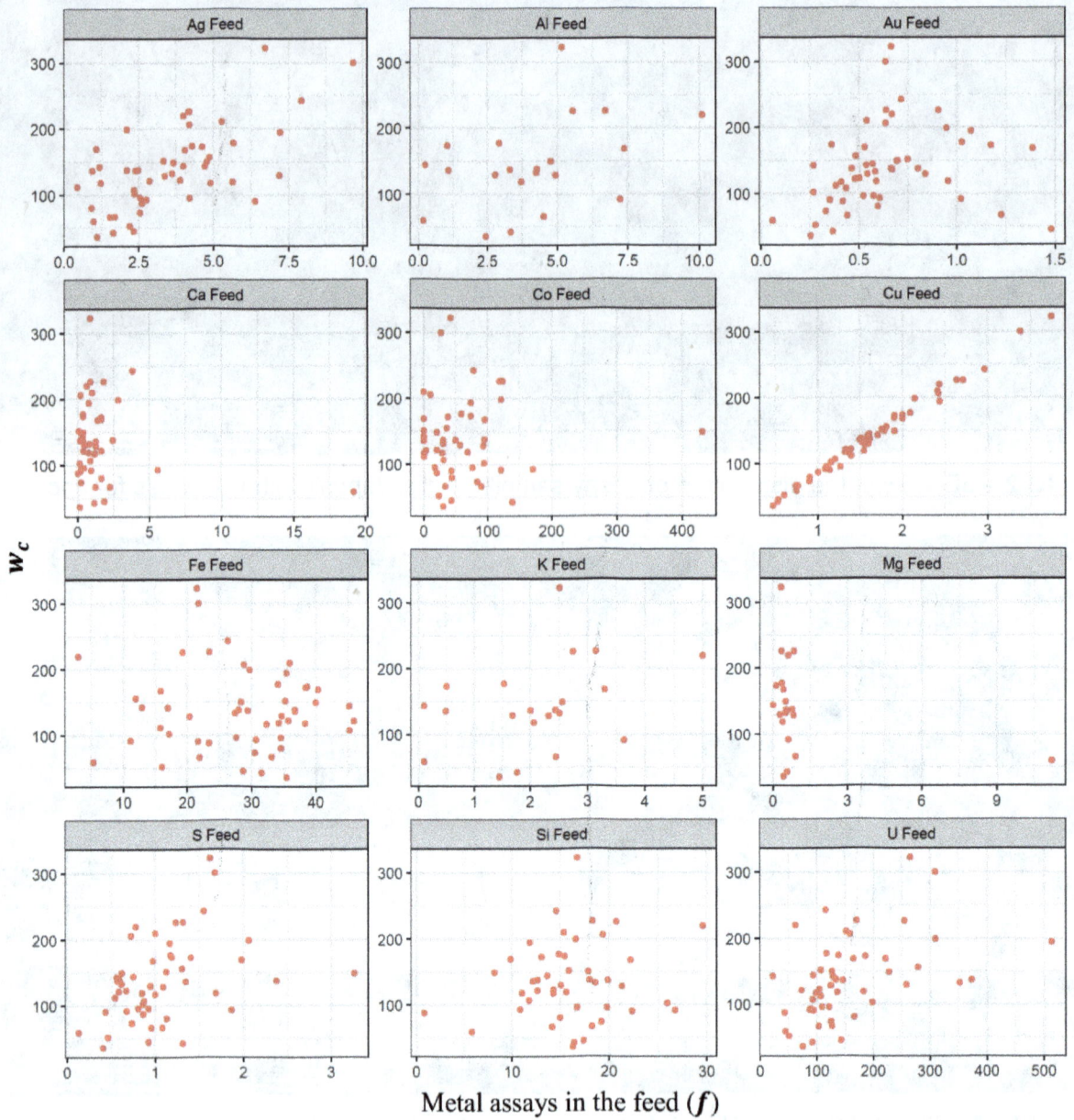

FIG 4 – Scatter plots of different metal assays in the feed (f) against w_c for copper in 52 batch flotation samples.

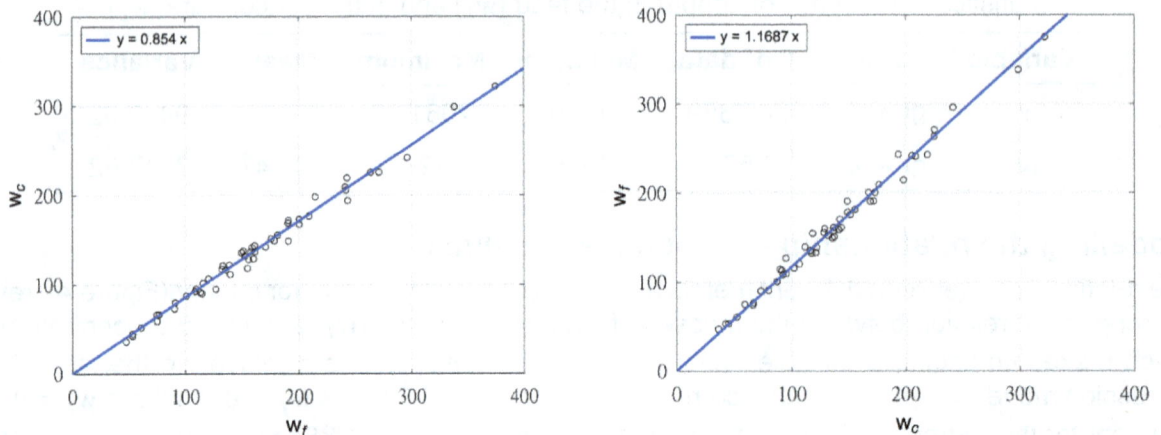

FIG 5 – Scatter plots between w_f and w_c for copper with fitted linear trendlines (intercept = 0).

To model the relationship between the mean values, possible biases due to preferential sampling should be identified. In particular, because a very limited number of flotation tests for copper sulfide ores were taken and all are from high-grade copper zones, the experimental mean for w_c is significantly higher than the experimental mean for w_f (Table 1) which is an inaccurate representor of the true mean of w_c. However, the ratio of the mean values of w_c to w_f for the 52 samples for which the values of both variables are available is 0.852, which is close to the slope (0.854) or its reciprocal (1/1.169 = 0.855) calculated from the regression lines shown in Figure 5. As a result, the estimated ratio of 0.852 will be considered as an accurate and unbiased representation of the relationship between the mean values of w_c and w_f and will be applied in the co-kriging system with related means.

Modelling spatial correlations

The next stage for implementing co-kriging is the identification of the spatial correlation structures of the target (w_c) and secondary (w_f) variables. This stage is difficult because of the significant difference in the amount of information available for both variables. Ignoring the data for which w_c is unavailable leads to unrobust experimental variograms or covariances. For this reason, our choice is to consider all the available data. In addition, as the experimental cross-variograms can be calculated using only the pairs of collocated data, they are not robust. Instead, we have used the non-ergodic experimental covariances (Isaaks and Srivastava, 1988).

The direct and cross covariances are calculated on the basis of different (and highly unbalanced) data sets, which makes the experimental curves subject to biases. This is particularly so for the direct covariance of w_c and the cross-covariance, for which the expected theoretical relationships, such as the Cauchy-Schwarz inequality, are violated. This requires a post-processing step to ensure that the covariance model is (1) mathematically valid and (2) is as representative as possible of the true spatial correlation between the two variables. The proposed strategy is the following:

- Calculate the experimental direct and cross-covariances of w_f and w_f.

- The experimental direct covariance of w_f (abundant variable) is assumed to be accurate.

- The experimental cross-covariance is assumed inaccurate and is rescaled in order to match the collocated covariance between w_f and w_f. Because of the preferential location of the 52 w_c data, the experimental covariance calculated from these data is likely to be biased. A better option is to trust the slope of the regression of w_c on w_f (0.854), which is theoretically equal to the covariance divided by the variance of w_f. Using the variance of w_c equal to 9942.62 (Table 1), the collocated covariance is 8064.

- Similarly, the experimental direct covariance of w_c is inaccurate and is rescaled in order to match the 'true' variance of this variable. While the variance shown in Table 1 is not reliable due to the preferential sampling design, the regression slope of w_f on w_c (1.169) and the previously calculated collocated covariance (8064) provide a variance of w_c equal to 6900.

- The rescaled experimental covariances are fitted with a linear model of coregionalisation (Wackernagel, 2003) consisting of a nugget effect and two isotropic exponential structures with practical ranges of 30 m and 100 m (Figure 6).

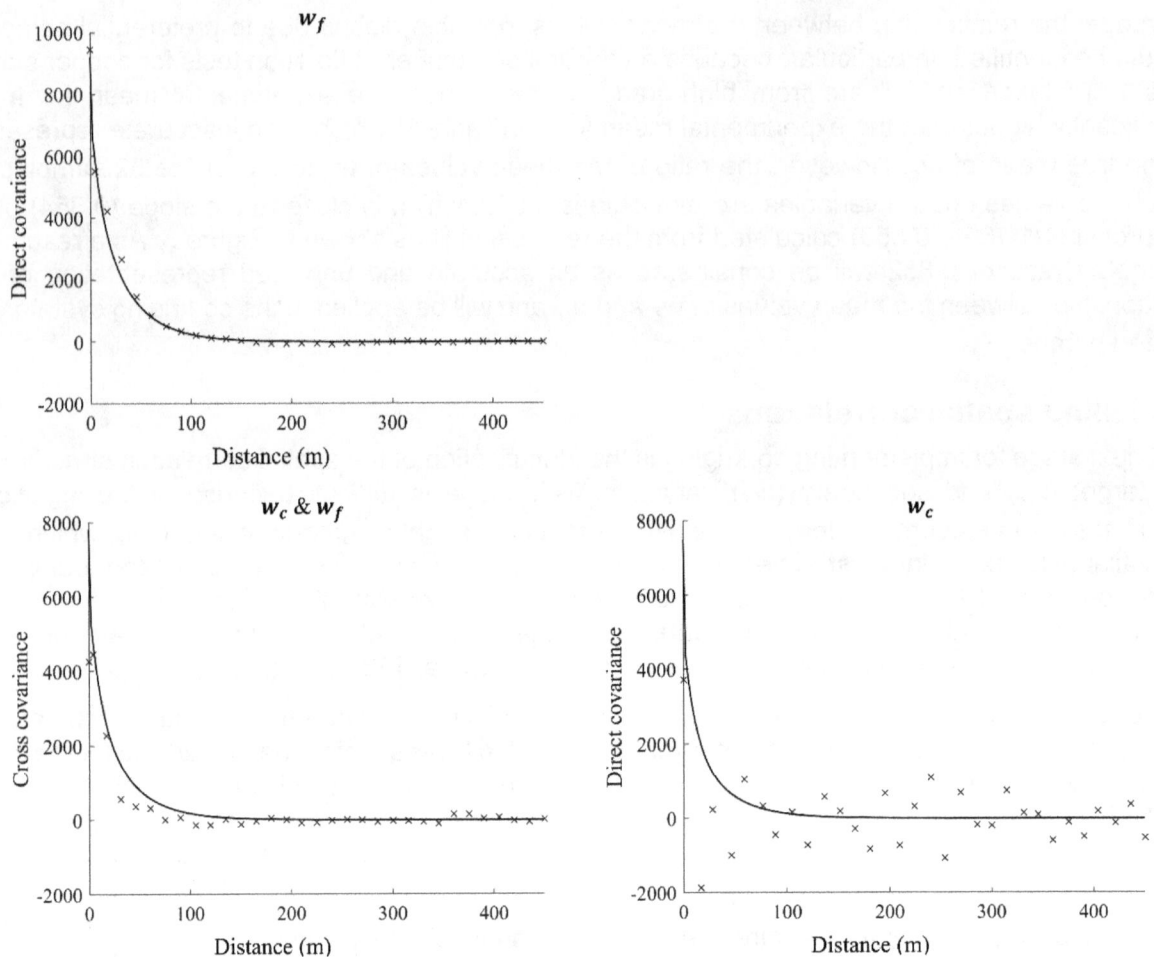

FIG 6 – Rescaled experimental (crosses) and fitted (solid lines) direct and cross-covariances for w_c and w_f of copper.

Cross-validation

A legitimate question is how much improvement can be expected by using co-kriging with related means instead of kriging or another variant of co-kriging? To address this question, we performed cross-validation, using a local search neighbourhood containing the 32 nearest data to the target (four data per octant), and compared the following kriging and co-kriging approaches:

- Simple kriging to separately estimate w_f (mean value assumed to be 37.37, as per Table 1) and w_c (mean value assumed to be 0.852 × 37.37 = 31.91).

- Simple co-kriging to jointly estimate w_f and w_c.

- Ordinary kriging to separately estimate w_f and w_c, without assuming any mean value.

- Ordinary co-kriging to jointly estimate w_f and w_c.

- Co-kriging with related means to jointly estimate w_f and w_c, assuming that the mean of the latter variable is 0.852 times the mean of the former.

Although the estimates for w_f do not vary much between the various approaches, the reverse is true for w_c and, consequently, for the copper recovery *Rec* calculated as the ratio between w_c and w_f (Equation 1). An examination of the cross-validation scores (Table 2) and the scatter plots of estimated versus true values (Figure 7) indicates that:

1. Separate kriging leads to inaccurate estimates of w_c and *Rec*, with a root mean square error (RMSE) for *Rec* about one order of magnitude greater than that for co-kriging.

2. Ordinary co-kriging fails to estimate w_c, because of the lack of primary data in the search neighbourhood.

3. Co-kriging with related means is the most accurate approach, with a RMSE half of its nearest competitor, simple co-kriging, both co-kriging variants being able to reproduce the linear dependence relationship between w_c and w_f, as observed on the data (Figure 5).

TABLE 2

Cross-validation statistics for the estimated mass of copper in the feed (w_f), mass of copper in the concentrate (w_c), and copper recovery (Rec). Ordinary co-kriging did not provide any estimate due to the lack of primary data in the search neighbourhood.

Approach	Average value for w_f (282 359 data)	Average value for w_c (52 data)	Slope of regression of w_c upon w_f	Root mean square error for Rec
Simple kriging	37.64	52.60	0.252	0.5151
Simple co-kriging	37.64	130.8	0.810	0.0727
Ordinary kriging	37.37	135.6	0.657	0.6643
Ordinary co-kriging	-	-	-	-
Co-kriging with related means	37.37	136.3	0.845	0.0356

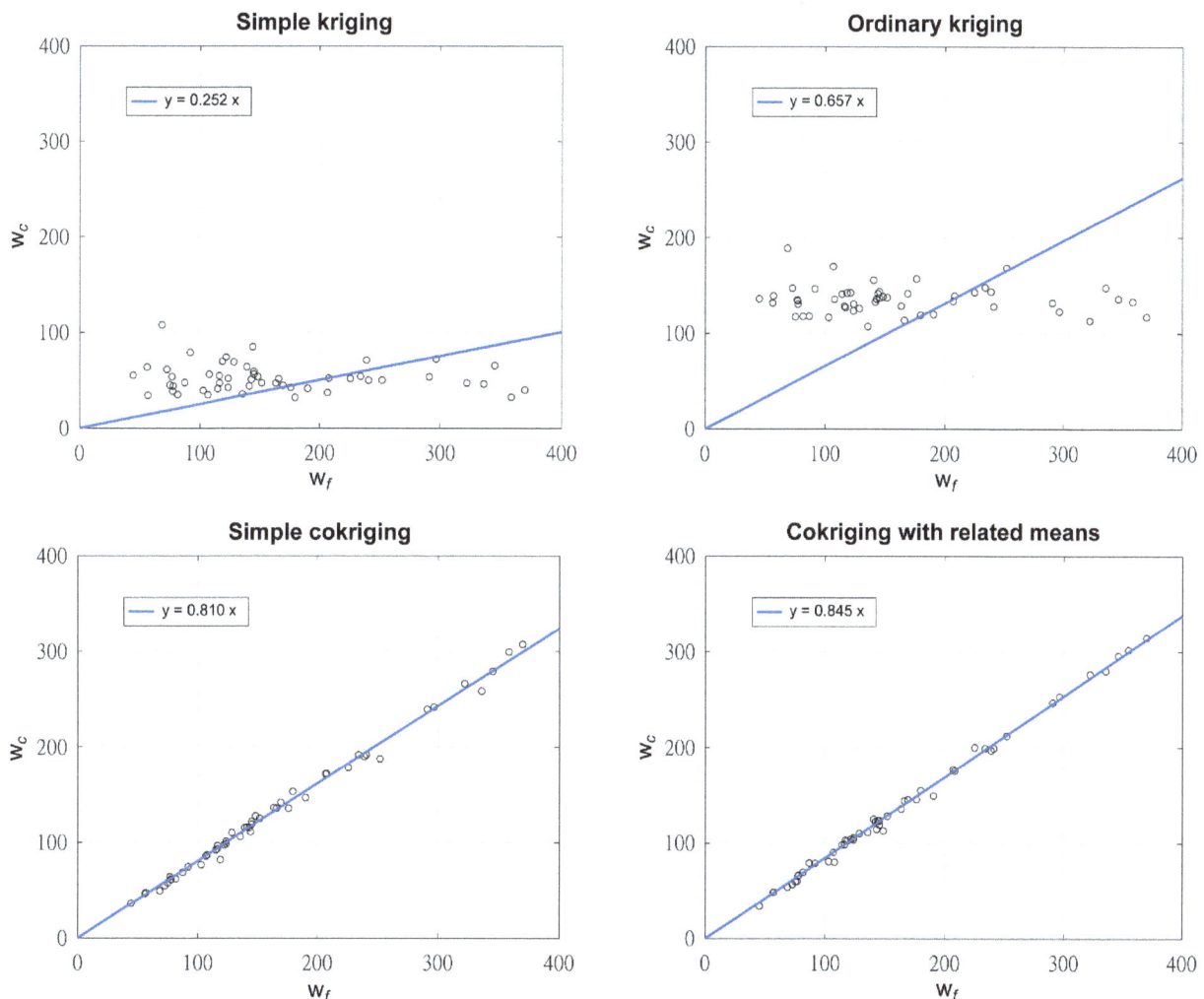

FIG 7 – Scatter plots between cross-validation estimates of w_c and w_f of copper for simple and ordinary kriging, simple co-kriging and co-kriging with related means. The upper scatter plots show the regression lines without intercepts, none of which are close to the regression observed on the data.

Block modelling

A relevant question whether the cross-validation results give a full representation of the accuracy of the estimations at the block support and at the scale of the entire deposit. To elucidate this question, the three co-kriging approaches are applied to construct a block model, with $101 \times 41 \times 96 = 397\,536$ blocks of size $10\,\text{m} \times 10\,\text{m} \times 10\,\text{m}$, each of which are discretised into $5 \times 5 \times 2 = 50$ core samples for block co-kriging. As the two variables w_f and w_c are additive, block co-kriging provides meaningful estimates of these variables at the block support, and so is their ratio as an estimate of the copper recovery Rec. An illustration is given in Figure 8.

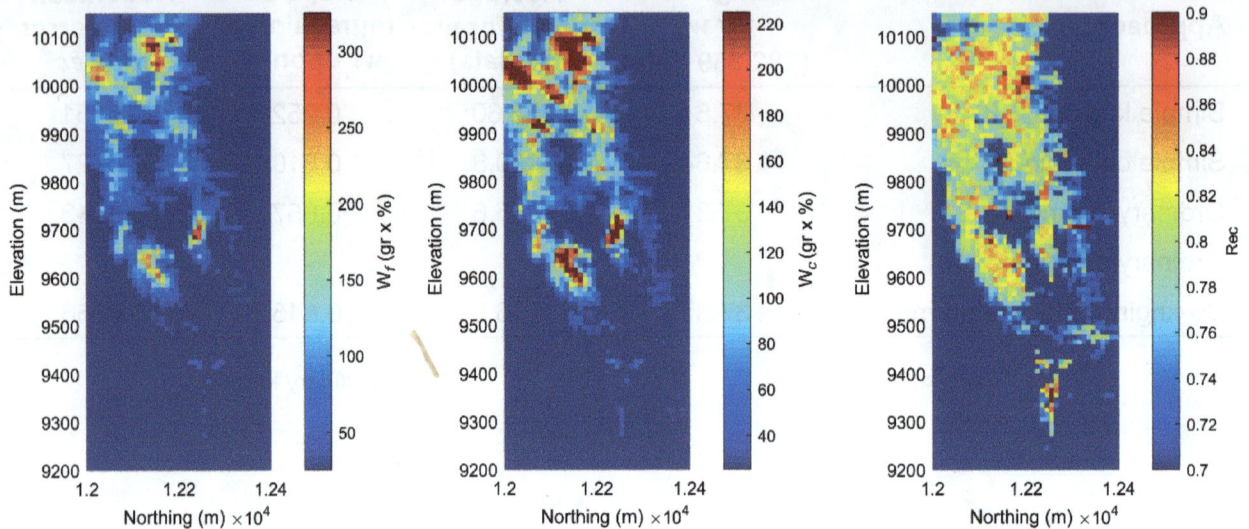

FIG 8 – Colour maps showing a cross-section with easting coordinate 56 000 m, with the estimates of w_f, w_c and Rec obtained by co-kriging with related means at a block support.

The following comments arise from the resulting block model statistics (Table 3):

- Ordinary co-kriging fails to estimate most blocks, for the same reason as given in the previous subsection (lack of primary data in the search neighbourhood). The estimated blocks are mostly located in high-grade areas, which is why the mean estimates are much higher than those of the other co-kriging approaches.

- The outputs from simple co-kriging and co-kriging with related means do not have the same mean values of the primary and secondary variables. While simple co-kriging provides mean estimates of 32.03 gr × % and 28.09 gr × % for w_f and w_c, the mean values obtained by using co-kriging with related means are 23.90 gr × % and 20.36 gr × %. This shows that, where there is under-sampling and preferential sampling designs, a statistic as simple as the global mean value remains largely subject to estimation errors. In simple co-kriging, the input mean values were 37.37 gr × % and 31.91 gr × % for w_f and w_c, which are much larger than the output mean values, which suggests that these input means were miss-specified. Spatial declustering techniques may help to reduce the miss-specification error, but their application is subject to subjective choices. In this respect, the benefits of co-kriging with related means are two-fold: on the one hand, the method does not require specifying global mean values for the input variables, and on the other hand it allows the mean values to vary locally (at the scale of the co-kriging neighbourhood), thus it can account for a locally stationary behaviour for both the primary and secondary variables.

- In addition to these issues on the specification of the mean values, co-kriging with related means provides estimates that better reproduce the linear relationships between the variables (Figure 9), in terms of slope and intercept. This suggests an improved consistency of the block models and a better estimation of the copper recovery (ratio between the primary and secondary variables).

TABLE 3

Block model statistics for the estimated mass of copper in the feed (w_f), mass of copper in the concentrate (w_c), and copper recovery (Rec). The total number of blocks is 397 536 blocks, many of which remain un-estimated when using ordinary co-kriging.

Approach	Average value for w_f	Average value for w_c	Slope of regression of w_c upon w_f	Average value for Rec
Simple co-kriging	32.03	28.09	0.809	1.24
Ordinary co-kriging	74.91	62.60	0.844	0.82
Co-kriging with related means	23.90	20.36	0.843	0.87

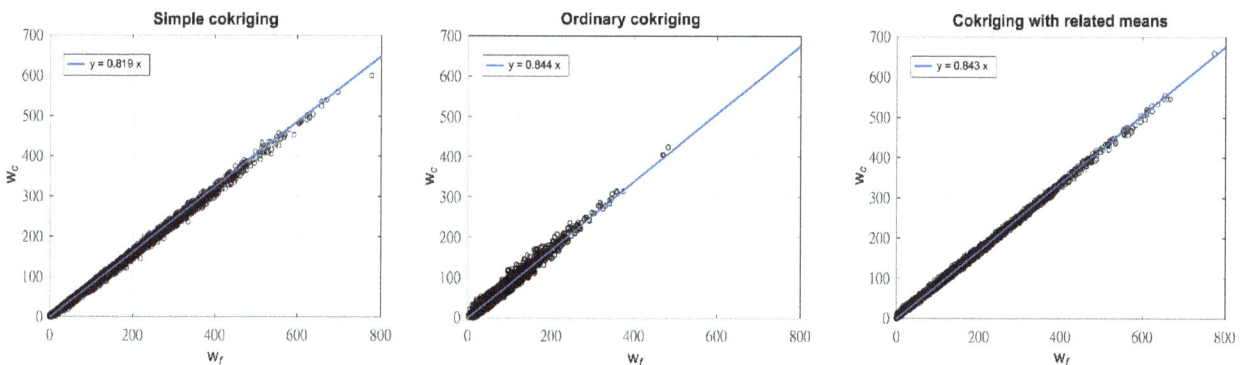

FIG 9 – Scatter plots between the block estimates of w_f and w_c obtained by simple co-kriging, ordinary co-kriging and co-kriging with related means. Regression lines without intercept are superimposed.

CONCLUSIONS

Multivariate modelling techniques provide an opportunity to improve predictive models in the fields of mineral resource assessments and geometallurgy, in particular when a target (primary) variable is under-sampled with respect to secondary variables that provide more data. However, there are significant difficulties in the practical implementation of such techniques in the presence of preferential sampling designs. In this paper, these difficulties have been illustrated in a case study of geometallurgical modelling for an IOCG deposit located in South Australia.

The chosen case study is characterised by a preferential sampling design for both the primary (mass of copper in the concentrate) and the secondary (mass of copper in the feed) variables, and by a strong imbalance in the sampling data (52 versus 282 359 data). Even the estimation of the global mean values of the primary and secondary variables is challenging, let alone the local estimation of these variables on a block-by-block basis. The solution presented in this work is to use co-kriging with related means, which is a variant of co-kriging that takes advantage not only of the spatial correlation between the primary and secondary variables, but also of the linear relationships between their mean values. The latter are assumed unknown and can vary locally, which is the advantage of ordinary kriging over simple kriging. Care must be taken when modelling the spatial correlation structure of the primary and secondary variables, insofar as the modelling uses covariances rather than variograms and the experimental curves are subject to a rescaling before fitting a model to ensure mathematical consistency and to rectify distortions caused by the preferential sampling design.

The effort required in the work presented here is worth the reward, as can be seen in the cross-validation scores and the properties of the estimates. Co-kriging with related means outperforms all its competitors (simple and ordinary kriging and co-kriging) in terms of precision and accuracy (root mean square error) and in terms of consistency (reproduction of the linear relation between the variables), while its implementation is simple and accessible to practitioners.

ACKNOWLEDGEMENTS

This research was supported by funding from the South Australia Government through the Premier's Research and Infrastructure Fund (PRIF) Research Consortia Programme via the Integrated Mining Consortium at the University of Adelaide (https://www.adelaide.edu.au/imer/integrated-mining-consortium/). OZ Minerals also provided funding as a partner organisation in the Integrated Mining Consortium and provided the data used in this study and guidance in their use. The participation of Xavier Emery was supported by the National Agency for Research and Development of Chile, through grants ANID PIA AFB220002 and ANID Fondecyt 1210050.

REFERENCES

Belperio, A, Flint, R and Freeman, H, 2007. Prominent Hill: A hematite-dominated, iron oxide copper-gold system, *Econ Geol*, 102(8):1499–1510.

Boisvert, J B, Rossi, M E, Ehrig, K and Deutsch, C V, 2013. Geometallurgical modeling at Olympic Dam Mine, South Australia, *Math Geosci*, 45:901–925.

Carrasco, P, Chilès, J P and Séguret, S A, 2008. Additivity, metallurgical recovery and grade, in *Proceedings of the Eighth International Geostatistics Congress* (eds: J M Ortiz and X Emery), pp 237–246 (Gecamin Ltda: Santiago).

Coward, S and Dowd, P A, 2015. Geometallurgical models for the quantification of uncertainty in mining project value chains, in *Proceedings of the 37th Symposium on Application of Computers and Operations Research in the Mineral Industry* (ed: S Bandopadhyay), pp 360–369 (Society of Mining, Metallurgy and Exploration: Englewood).

Dobby, G, Bennett, C, Bulled, D and Kosick, G, 2004. Geometallurgical modelling – the new approach to plant design and production forecasting/planning, and mine/mill optimization, in *Proceedings of the 36th Annual Meeting of the Canadian Mineral Processors* (ed: J Abols), pp 227–240 (The Canadian Institute of Mining, Metallurgy and Petroleum: Ottawa).

Dominy, S, Connor, L, Parbhakar, A, Glass, H and Purevgerel, S, 2018. Geometallurgy: A route to more resilient mine operations, *Minerals*, 8:560.

Dunham, S and Vann, J, 2007. Geometallurgy, geostatistics and project value – Does your block model tell you what you need to know?, in *Proceedings of the AUSIMM Project Evaluation 2007 Conference*, pp 1–8 (The Australasian Institute of Mining and Metallurgy: Melbourne).

Emery, X and Séguret, S A, 2020. *Geostatistics for the Mining Industry – Applications to Porphyry Copper Deposits*, 247 p (CRC Press: Boca Raton).

Emery, X, 2012. Cokriging random fields with means related by known linear combinations, *Comput Geosci*, 38:136–144.

Goovaerts, P, 1998. Ordinary cokriging revisited, *Math Geol*, 30(1):21–42.

Hunt, J, Berry, R, Bradshaw, D, Triffett, B and Walters, S, 2014. Development of recovery domains: Examples from the Prominent Hill IOCG deposit, Australia, *Miner Eng*, 64:7–14.

Isaaks, E H and Srivastava, R M, 1988. Spatial continuity measures for probabilistic and deterministic geostatistics, *Math Geol*, 20(4):313–341.

Navarra, A, Grammatikopoulos, T and Waters, K E, 2018. Incorporation of geometallurgical modelling into long-term production planning, *Miner Eng*, 120:118–126.

Powell, M S, 2013. Utilising orebody knowledge to improve comminution circuit design and energy utilisation, in *GeoMet 2013: The Second AusIMM International Geometallurgy Conference*, pp 27–35 (The Australasian Institute of Mining and Metallurgy: Carlton).

Schlegel, T U and Heinrich, C A, 2015. Lithology and hydrothermal alteration control the distribution of copper grade in the prominent hill iron oxide-copper-gold deposit (Gawler Craton, South Australia), *Econ Geol*, 110(8):1953–1994.

Sepúlveda, E, Dowd, P A and Xu, C, 2018. Fuzzy clustering with spatial correction and its application to geometallurgical domaining, *Math Geosci*, 50(8):895–928.

van den Boogaart, K G and Tolosana-Delgado, R, 2018. Predictive geometallurgy: An interdisciplinary key challenge for mathematical geosciences, in *Handbook of Mathematical Geosciences* (eds: B Daya Sagar, Q Cheng and F Agterberg), pp 673–686 (Springer: Cham).

Wackernagel, H, 2003. *Multivariate Geostatistics: An Introduction with Applications*, 387 p (Springer: Berlin).

Williams, S R and Richardson, J M, 2004. Geometallurgical mapping: a new approach that reduces technical risk, in *Proceedings of the 36th Annual Meeting of the Canadian Mineral Processors* (ed: J Abols), pp 241–268 (The Canadian Institute of Mining, Metallurgy and Petroleum: Ottawa).

Failures and disasters

An estimation error

D A Sims[1]

1. FAusIMM(CP), Director, Dale Sims Consulting, Bolwarra NSW 2320.
   Email: dalesims@tpg.com.au

ABSTRACT

This paper outlines the contributing factors to a series of erroneous Mineral Resource estimates which led to considerable loss, including significant reduction of mine life and consequently a large write-down of asset value. The error prompted divestment of the mineral asset in the short-term by the owners, a major mining company.

There were multiple and complex contributions to the result:

- The mine was undergoing a transition in the deposit geology with increasing depth that was not accommodated in the models. Early indications of the geological transition in initial drilling were generally unrecognised in resource modelling. The shortcomings of the models became apparent with increasing infill drilling and subsequent production reconciliation.

- There was a lack of recognition of, and confidence in, site-based input and skills by corporate technical management. The site had a reputation internally for over-delivery to plan early in its life and that supported the anticipation of ongoing metal delivery into the mine plan by management.

- A corporate lack of confidence in site being able to adequately domain the mineralisation led to the use of a CIK domaining process, rather than one which was geologically based. This domaining process led to gross over-statement of mineralisation trends reflected in the subsequent OK estimate.

- Timelines on the outsourced estimation process were so short that validation was not possible. Obvious errors in the estimate versus data were not able to be used to correct the issues and the site's Competent Person sign-off occurred without adequate checks.

This case study is presented to reinforce the concept that Mineral Resource estimates are not solely based on the ability of the Competent Person to satisfactorily guide the process. There are usually multiple complex issues in the estimation process which need to be recognised and considered in judging reconciliations of model to actual.

INTRODUCTION

This case study is a true story, but some details are omitted as in reality they are not important. The name of the company, the name of the site, the commodity being mined, and the names of the people involved will not be discussed. These details are not important because, in the author's experience, such circumstances can and do occur in any company, at any site, in any commodity and with any people. They don't occur all the time, but they can occur when sets of recurring circumstances arise and generally when issues compound and timelines shorten.

In this example the author's involvement in the circumstances or knowledge of the occurrences was either firsthand or through discussion with people directly involved at stages working on the project in various duties.

This is not being put forward as a fully comprehensive or definitive review, yet it contains enough detail of the main causative issues for the reader to appreciate the circumstances as the author sees them. There will be many sides to any story. The presentation of this paper is not intended to vilify any participants nor companies, but rather contribute to the conversation around Mineral Resource estimation (MRE) and public reporting. The issues of competency, judgement and the responsibility of the Competent Person need to be assessed in relation to corporate activity. There are significant responsibilities attached to 'signing off' as a Competent Person, yet the people involved are often in varying circumstances and may not have a clear mandate to manage the work under their direction as intended in the JORC Code.

Case studies like these are rarely presented, mainly because company permission is not granted for presentation. In this instance the owners of the asset have approved the publication of this paper.

THE SITE

The site had been in development and production for around a decade. Ownership had changed in that time and the mine had a reputation for being a 'strong performer' with a 'high cash margin'. The company which owned it had several operating sites, a sound reputation in the market, and an established internal technical team based in their head office overseeing all their operations.

Extraction had been dominantly by open cut methods but was progressing to include underground production on an extension of the pit orebodies at depth. A portal had been established in the pit and an underground contractor engaged to mine a decline and stope the ore. In the pit, mineralisation was hosted in thin and steep subparallel structures spaced throughout the mineralised sections of the host rock in a broadly sub horizontal zone. As depth increased, the steep structures were less widespread yet still present in favourable structural zones. Several mineralised zones or domains had been identified in the deposit.

The resource had been defined with a mix of diamond core and reverse circulation (RC) drilling, while pit grade control (GC) was undertaken with RC. The extensions below the pit were also defined initially by diamond drilling then infilled with RC holes either drilled from overlying pit benches or adjacent ex-pit areas. The deposit was relatively well drilled, and data collection/assay methodology were at industry standard or better. Multielement assay data was being regularly collected in both resource development and GC drilling.

Mineral Resource estimation had been undertaken dominantly in house for much of the operation's early history, with external reviews by consultant groups undertaken every few years. Internal site reviews by corporate technical services professionals were undertaken annually or six monthly; any subsequent recommendations were the responsibility of the site to address.

The operation usually employed a resource estimation geologist, but that position was not always able to be filled and the MRE process had moved to external estimation modelling. Reserve estimates were usually undertaken in house with contractor/consultant support. The most senior site geoscientist was at Superintendent level, and they reported to the site's Technical Services Manager.

Within the company, the site was not a large component of their overall asset base but had been a reliable performer. It had a reputation for over-delivery in comparison with annual plans, either through scheduling conservatism or overperformance of the resource model compared to expectation.

Site technical team members had been stable for much of the operations early life but more recently a number of major roles had changed incumbents or been vacated and unable to be filled. In some instances, incoming technical personnel did not stay for extended periods, potentially reflecting the level of operational stress or technical dissatisfaction at the site.

The operation was transitioning from being a relatively steady and reliable producer to having increased operational pressure. This was due to mining complexity in the pit as it deepened into the structurally controlled mineralisation, as well as the move to incorporate higher grade underground production.

Short-term reconciliation issues plagued the operation in its need to meet monthly production schedules, which were based on regularly updated GC models incorporating all available drilling. This circumstance led to daily senior site management engagement with the issue and subsequent dynamic mine planning and ore delivery practices. Fresh ore was relatively hard, and the site had operational challenges around blast heave, and the subsequent displacement of GC dig polygons and ore dispatch.

WHAT HAPPENED?

Figure 1 outlines the approximate value profile of the project through time in current metal price and exchange rate terms. Early in the project life there had been a significant adjustment to the Mineral Resource dropping around 30 per cent of gross value over the first year of production (Figure 1 –

Year B). This was reported to be due to infill drilling finding increased geological complexity coupled with the application of more rigorous mine planning processes after acquisition.

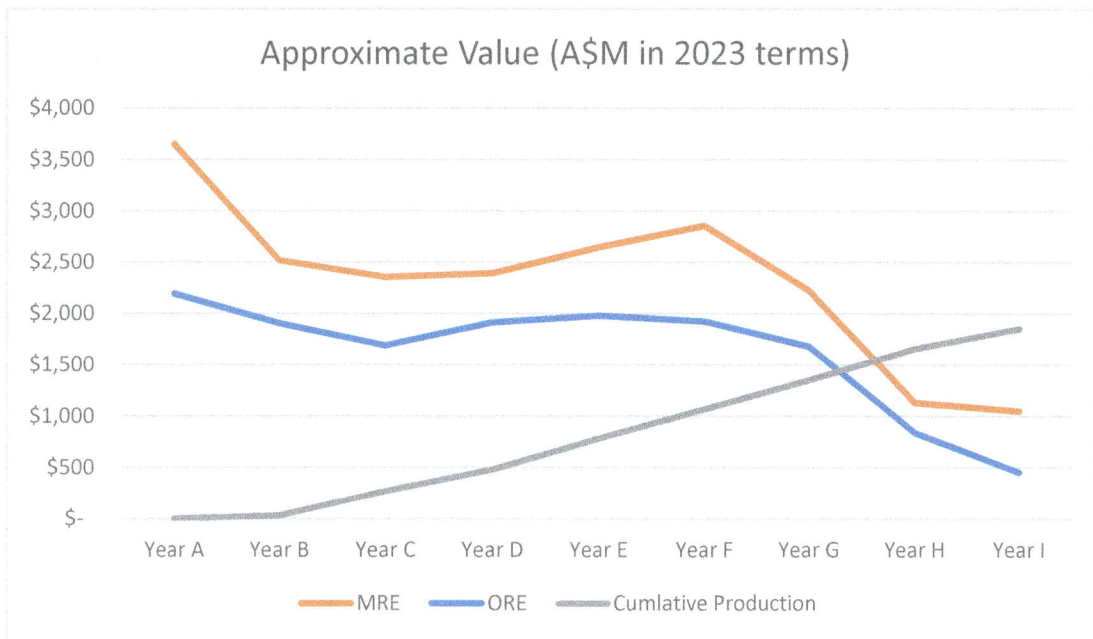

FIG 1 – Approximate value through time.

Following this adjustment, the resource value had remained relatively 'static' for a few years despite production depletion, then increased with significant addition of Inferred Resource in Year F (Figure 1). The Ore Reserve estimate (ORE) value mirrors the MRE trends although at a lower value given the non-conversion of Inferred Resource and design-based exclusions in remaining categories.

Between Year F and Year H, the MRE value dropped by around 60 per cent while the ORE value decreased by 75 per cent. These major reductions in estimated metal availability for the project shortened the mine life by around 5 years and ultimately prompted divestment of the asset by the owner. Cumulative production value to year I approximated the projects initial ORE value.

HOW DID IT HAPPEN?

The major reduction in asset value resulted primarily from the impacts of infill drilling prompting the reassessment of the resource models and estimation methodology. It was clear the MRE was overestimated, and short-term mining reconciliation was increasingly variable.

Several factors contributed to the overestimation.

Corporate-site relationship

Over time, there had been a reduction of the corporate technical service team's confidence in the site's ability to undertake resource estimation duties. Perceptions of modelling and estimation capability shortfalls, site turnover/non-replacement of resource estimators in the team structure, and the focus of the site geology team on the daily tasks of GC, blend management and reconciliation eroded the independence of the site to undertake Mineral Resource and GC estimation.

With responsibility for the site's MRE and GC modelling moving off-site, external consultants were engaged to undertake the estimate under the guidance of the technical services team and specific key personnel. Site then became responsible for providing data, a level of guidance into the process and then reviewing the outcomes for final approval, yet with little ownership of the results. Significant infill drilling programs were underway, including on the recently approved underground operation.

Loss of the geological model underpinning the resource estimate

When the project was initially discovered and developed, the owners put significant effort into developing the geological understanding of the deposit and producing a robust geological model for the mineralisation and its controls. The main geological architecture required, for successful modelling later in the project life, was established relatively early and documented at the time. Between that initial work, and the MRE undertaken in approximately Year F, the modelling process had largely lost its understanding of the geological controls on the mineralisation.

Although the broad controls of favourable stratigraphy and alteration, along with the geometry of sporadic steep veins hosted in narrow structures, was well understood by GC geologists, it was not being translated into the MRE or GC models at depth. Initial deeper drilling held indications of the system's transition to a more structurally controlled/discrete mineralisation style, yet these observations were largely not being included into the modelling process.

This is potentially in part to the loss of on-site estimation functionality, but it was also coupled with a corporate belief that the mineralisation could not be adequately domained by site geologists – hence the need for a more geostatistical domaining approach. The aim was to 'avoid poor quality geological interpretations'.

Ironically, past site modellers had been able to produce geologically domained MREs, yet their contributions had been over-ridden by corporate guidance once site resources or skills were not deemed adequate to produce site models in the required time frames. The site MRE model included available GC data so the process of updating the MRE as well as the GC model were identical, as the same model was used for both levels of planning and reporting.

The application of Categorical Indicator Kriging

The modelling approach applied by the consultants to address the lack of geological domains was to utilise Categorical Indicator Kriging (CIK).

The methodology involved the following steps:

1. A very low-grade outer envelope was developed to separate background from mineralised waste/ore material. This very broad domain was subdivided into a few orientation subdomains to allow the application of varying anisotropy in the CIK and subsequent estimation.

2. Within that, and the subdomains, all data was converted to an indicator based on the modeller's assessment of the cut-off point of low-grade to higher grade material – nominally the 80th percentile. This process was also guided by site and corporate advice.

3. Variography was undertaken on the indicators per subdomain, and other estimation parameters determined for the CIK run such as search distances, sample numbers, and anisotropy.

4. An indicator estimate was run into the block model and the composite grades (not indicators) are compared to the CIK model.

5. A low probability threshold is chosen to discriminate low-grade from high-grade indicator blocks.

6. Once an indicator block was classified as high-grade, it was estimated with Ordinary Kriging (OK) using the actual values for the composites, but only using composites flagged as high-grade by the indicator threshold (ie a value of 1). In this way, no low-grade composites inform the model within the high-grade domain and vice versa. Variogram and estimation parameters were determined for each subdomain per high-grade or low-grade composite set. Top-cutting was also applied to both high-grade and low-grade composites.

7. The resultant OK models were validated and exported for the client.

A stated advantage of the approach is the rapid time in which a model can be produced, compared to manual wireframing domain development. Hence, it is suitable for short time frame estimates, allowing modelling updates to be produced quickly following data updates.

The major problem with this approach was that the broadly flat-lying anisotropy from the higher levels of the pit could apply into zones where the geology was transitioning to more discrete and steeply dipping structural mineralisation at depth.

Broad envelopes of the CIK high-grade domains were developed at depth and laterally extensively away from the closer drilled areas. With these areas estimated with only the high-grade composites, the extent and intensity of the mineralisation was significantly overstated.

The value increase in Year F (Figure 1) and the decision to include underground mining are thought to be directly linked to this approach. As infill drilling was undertaken, the subsequent revisions to the models indicated major overstatement of mineralisation extent and continuity. Detailed examination of the modelling approach led to a removal of CIK from the MRE workflow and mineralisation domains were produced by site for direct OK estimation in Year H.

Compressed timelines

The site was undertaking considerable amounts of infill drilling in advance of staged pit expansion and underground mining. The significant reduction in the asset value in Year G (Figure 1) led to an increased corporate focus on the drilling programs, and regular model updates were needed to manage both short-term and long-term planning since the same model was used for both processes. Poor local reconciliation was activating site and corporate management on the issue of model updates.

Mining always works in overdrive; hence this is not an unusual circumstance. Yet, the need for rapid model turnaround as data sets were updated potentially drove continuance of CIK-based estimation, by the external consultants, as alternatives were not deemed possible within the available timelines.

The site-based Competent Person was involved in the oversight of drilling and data management, as well as liaison with the external consultants as required; although, the consultants were also instructed by the corporate technical services team. Often timelines did not adequately allow inspection or validation of the model by the Competent Person.

CONCLUSIONS

Engaging and empowering site professionals to be closely involved in the MRE process is a key feature of ensuring model reliability as well as ownership of results from reconciliation.

This case study identifies several contributing factors leading to a significant reduction in asset value. Although, the real issue was how the overstated value was put into the estimate(s) in the first place. Mining, like so many other human endeavours, is driven by good news. Ultimately our responsibility as resource estimation professionals is to ensure that good news is real news.

ACKNOWLEDGEMENTS

The author thanks the owners of the operation who reviewed and approved this paper for publication. He is also thankful for reviews and verification by prior workers from the project, and reviews and comments from several professional colleagues and the two reviewers from the conference committee.

Geological modelling and estimation software – development, opportunities and limitations

Overcoming implicit modelling software limitations using Python scripting – an innovative geological modelling workflow for George Fisher Mine, Queensland, Australia

L Bertossi[1] and D Carvalho[2]

1. MAusIMM(CP), Senior Resource Geologist, Glencore, Mount Isa Qld 4825.
   Email: laercio.bertossi@glencore.com.au
2. Principal Geologist, Glencore, Brisbane Qld 4000. Email: dhaniel.carvalho@glencore.com.au

ABSTRACT

In the last decade, the use of advanced implicit modelling software/algorithms in mineral resource workflows has become a best practice in the mining industry. In most cases, these workflows can easily replicate the level of detail of explicitly modelled geometries, even for very complex deposits. These techniques are flexible and can produce fast results since software is responsible for generating 3D surfaces and solids based on geological and interpretation data.

Nevertheless, depending on how unique and complex some deposits are, intense manual editing and input are required to honour the geological context and produce the desired outcome. The current geological modelling workflow at George Fisher Mine (GFM) is based on the semi-automatic built-in tool that generates stratigraphic sequences (stacked lithology layers) based on logging and mapping data. This tool (*stratigraphic sequence* from Leapfrog Geo™) works for the majority of areas in the mine but fails to capture observed mineralisation pinch-outs in operational areas and at extrapolated boundaries of the deposit.

These pinch-outs have the potential to negatively impact grade forecasts if not resolved manually. Using built-in software tools only, the alternative for a more accurate model is to generate each individual layer and adding tens of thousands of manual interpretation points to honour pinch-outs (based on the amount of existing drilling in the mine and number of layers). This approach would require an estimated minimum of two to four years of a full-time employee (FTE) to be exhaustively completed.

This study presents an automated approach to generate pinch-out points using a set of Python scripts that only take a few minutes to run. The newly generated points are then combined into the commercial software workflow as interpretation points, honouring observed pinch-outs. The remarkable time difference between approaches removes chance of manual errors and allows mine geologists to spend valuable time in validation and testing of scenarios, ensuring they are more representative of the deposit's geology.

INTRODUCTION

Mineral resource models at any stage of mineral exploration, whether in project phase or in an operating mine, must be robust and reliable. Data collection methods, quality of information, geologist knowledge of the type of mineralisation and the tools used to construct the mineral resource model are directly reflected on the quality of work (Bertossi *et al*, 2013). The foundation for a robust and accurate mineral resource estimation is a reliable geological model. Unsupported assumptions and erroneous modelling criteria, that generate questionable models, are one of the main sources for unsuccessful resource estimates.

Traditional methods of 2D geological interpretation through geological cross-sections and subsequent extrusion lines for constructing solids or surfaces are still widely used in the mining industry (Sinclair and Blackwell, 2002). They are also known as explicit modelling techniques. However, given the large quantities of drill holes increased yearly to either handle the mine production requirements or analyse the feasibility of a mineral exploration project, this methodology is becoming less popular. Another important aspect of the method is the time spent on updating geological models. Such action can be very time consuming in huge deposits as new geological data set is generated, ranging from many weeks to several months.

To overcome speed constraints, heavy data sets, and pressure for faster models and scenarios, implicit modelling (Bloomenthal, 1997) became the mainstream method. This method is based on the use of Radial Basis Functions (Broomhead and Lowe, 1988) from which a set of lithological contacts, structural and chemical data are described by a single mathematical function. Iso-surfaces are defined by mathematical functions and represented as wireframes. These iso-surfaces can be created for any desired value and resolution, and this process is only limited by the processing capacity of the hardware. There are many advantages to using implicit modelling over traditional modelling. One example is that drawing hundreds or thousands of cross-sections is not required, as the algorithm already considers all the available information and interpolate contacts points automatically. Furthermore, processing time is substantially reduced, flexibility is increased significantly, as well as the dynamism of the modelling workflow.

Implicit modelling tools have recently been increasingly applied to mineral deposit studies, expanding conventional analyses of geological data with the visualisation of continuities and spatial trends, geological units or orebody geometries, and structural geological frameworks. Some examples include 3D geological modelling applications in lateritic nickel deposits (Bertossi *et al*, 2013), phosphate-bearing carbonatite deposits (de Oliveira and Sant'Agostino, 2020), base metal deposits (Kampmann, Stephens and Weihed, 2016; Schetselaar *et al*, 2016; de Oliveira *et al*, 2019, 2020; de Oliveira, Torresi and Rossi, 2022), and gold deposits (Vollgger *et al*, 2015; Naranjo *et al*, 2018; Cowan, 2020). Regardless, even with implicit modelling, some unique and complex deposits demand a strong level of manual interaction and editing to honour geological complexities. In these cases, implicit modelling can become as time consuming as explicit modelling.

This work presents a hybrid methodology for the George Fisher Mine (GFM), in which geological surfaces are generated using implicit modelling software, but the majority of control points are produced with a set of Python programming language scripts. These scripts are responsible for automating the creation of hundreds of thousands of control points to honour stratigraphical units pinch-outs. Scripts outputs (control points) are then incorporated to the 3D implicit modelling workflow.

REGIONAL AND LOCAL GEOLOGY

The L72 and P49 deposits/mines (also known as George Fisher North and South, respectively) are located within the north–south trending Western fold belt of the Mt Isa inlier which is a remnant of an intracontinental extensional basin system (active 1800–1600 Ma). The chronostratigraphic framework of the Western Fold Belt includes four unconformable cover sequences that represent a series of rift, volcanic, and sag events (Chapman, 2004).

Cover Sequence Four is of significance here which includes mid-Proterozoic basement rocks that are unconformably overlain by rift-fill sedimentary and volcanic sequences and intrusives. In the Mt Isa district, the Mt Isa Group hosts the Zn-Pb-Ag mineralisation and forms part of Cover Four Sequence. The Mt Isa Group consists of basal quartzite, siltstones and carbonates deposited in a shallow marine environment that occurred from 1710 to 1595 Ma. The L72 and P49 deposits are entirely hosted within the Urquhart Shale Member from the Upper Mt Isa Group which comprises interbedded dolomitic, shales, siltstones and mudstones (Davis, 2004).

The Mt Isa Group unconformably overlies the Haslingden Group of Cover Sequence Two. This major unconformity is represented by the north–south trending Mt Isa fault zone where Mt Isa Group sediments ramp onto basement rocks of the Eastern Creek Volcanics (ECV). The deposition of Cover Sequence Four (Mt Isa group and Urquhart shale) and basin development terminated at the onset of the Isan orogeny which culminated with several episodes of deformation (Chapman, 2004).

Three major deformation events (D1 to D3) during the polyphase Isan Orogeny (1620–1500 Ma) are described in the region. D1 deformation, produced minor ENE–WSW tending open folds and 'S' shaped flexures with moderate plunges from 20 to 50° south–south-west in the George Fisher deposit (Murphy, 2004). D2 deformation characterised with W–E directed shortening that produced district-scale upright folds with N–S trending axis (Bell, 1983; Williams, 1998; Hand and Rubatto, 2002). D3 deformation resulted in NE–SE and NE–SW trending chevron-like folds associated with shear zones at the Hilton deposit (Valenta, 1994), and minor symmetrical N–NNE trending upright folds at George Fisher deposit (Chapman, 1999; Murphy, 2004). Post to late-D3 NE-striking normal faults offset Pb-Zn-Ag mineralisation from 2 to 50 m at the George Fisher and Hilton deposits (Valenta, 1994; Chapman, 1999; Murphy, 2004). The L72 and P49 deposits occur east of the Mt Isa Fault within the western limb of a regional D2 north trending anticline.

The geological units in the GFM area and surroundings from oldest to youngest are as follows: Judenan Beds, Myally Subgroup, Eastern Creek Volcanics, (Haslingden Group), Warrina Park Quartzite, Moondarra Siltstone, Breakaway Shale, Native Bee Siltstone, Urquhart Shale (with Zn-Pb-Ag ore horizon), Spears Siltstone, Kennedy Siltstone and Magazine Shale (Mount Isa Group). Primary host rocks in GFM are interpreted as being part of sediments deposited in intrabasinal or lacustrine environments (Lilly, Taylor and Spanswick, 2017).

Similarly to Mt Isa deposit, both L72 and P49 stratiform Zn-Pb-Ag orebodies at GFM are within the Urquhart Shale. In addition to dolomitic shales, siltstone, and mudstones, bedding commonly contains carbonaceous fine-grained pyrite. Other minerals that occur in both deposits include coarse grained pyrite, chalcopyrite, and pyrrhotite. Both deposits contain mainly pyritic siltstones for each of their stratigraphic intervals. These deposits are spatially separated by approximately 2 km of barren to sub economic pyritic shales, carbonaceous shales and siltstone as a result of faulting (Gidyea Creek Fault). Both have similarities in terms of marker beds, ore zones, waste geometry and stratigraphy. Nevertheless, the deposits exhibit differences around metal distribution, lens sizes and fault geometries. In general terms, P49 contains narrower high-grade orebodies, with high proportions of Pb and Ag, while L72 (Figure 1) contains wider mineralised zones at lower Pb grades but higher proportions of Zn (Grenfell and Haydon, 2006).

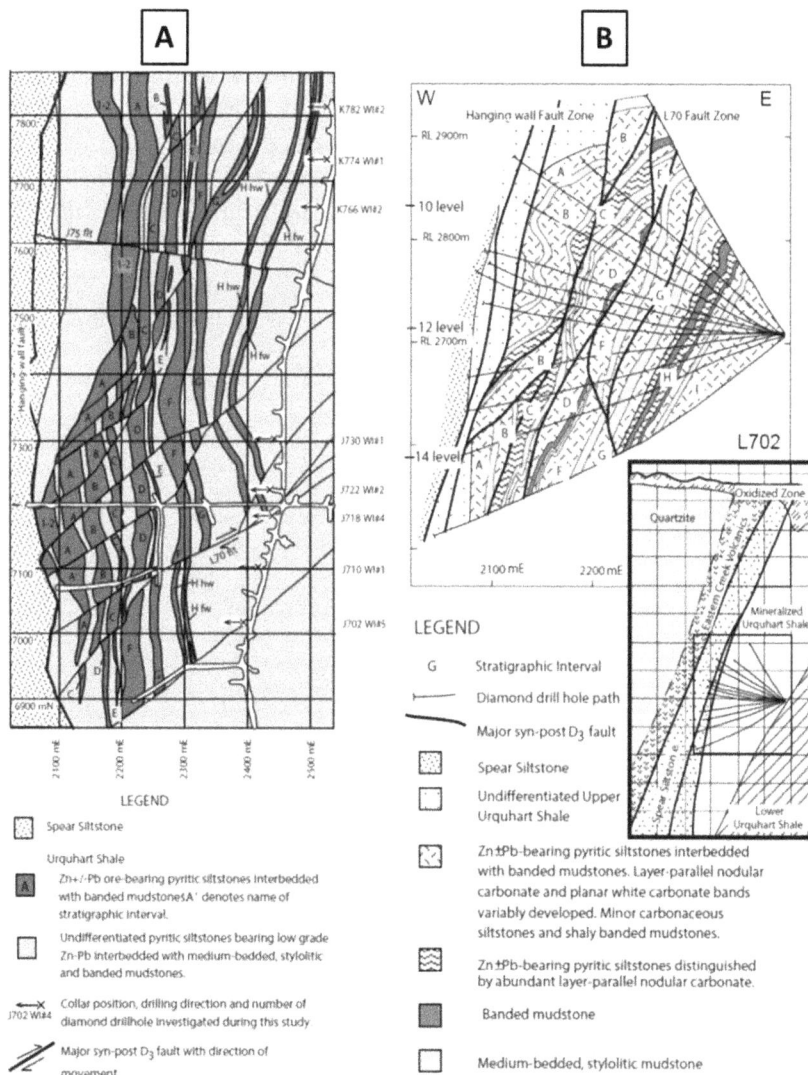

FIG 1 – Schematic plan view (A) and section (B) from L72 deposit (from Valenta, 2018, modified from Murphy, 2004).

MINERALISATION

From Chapman (1999, 2004) description of the mineralisation, the earliest event post-dates the formation of early bedding parallel veins and pre-dates regional deformational events. The main stage of sphalerite mineralisation (Stage IV) consists of layer parallel disseminated sphalerite and layer parallel vein-hosted sphalerite. Deformation-related mineralisation includes breccia hosted sphalerite (Stage VI) and vein- and breccia-related galena (Stage VII), which is the major stage of galena mineralisation.

Based on geochronology, other studies suggest that the L72 and P49 deposits are an epigenetic syn-deformational model for Zn-Pb-Ag mineralisation (Cave *et al*, 2023). The age of pre-mineralisation calcite at the George Fisher deposit and the age of quartz ± albite ± K-feldspar alteration/veining in the dolerite dyke, constraining the maximum age of overprinting Zn-Pb-Ag mineralisation. Furthermore, textural relationships suggests that Zn-Pb-Ag mineralisation overprints of dolerite dykes along their margins with occasionally mineralisation within dykes. This would evidence that the main mineralisation is not cross-cut or intruded into by the dolerite dykes.

The mineralisation generally occurs in north–south striking, west-dipping stratigraphic sequence (Figure 2) with approximately 350 m thickness, strike length up to 1.2 km and vertical extent of at least 1 km (Chapman, 2004). The L72 deposit comprises 13 mineralised stratigraphic intervals that occur as stacked stratabound lenses, hosted by pyritic siltstone, interbedded pyritic siltstone and banded mudstone units of the Urquhart Shale. These units are separated by thicker bands of barren Banded mudstone. The mineralised intervals are named from the west to east as units 1 and 2 and sequentially from A to I. Internal subdivisions are denoted by numerical subscripts. Detailed stratigraphic correlations throughout the ore-bearing sequence have been made based on observations of the bedding characteristics and location of tuffaceous marker beds in the mineralised intervals (Johnston *et al*, 1998; Tolman, Shaw and Shannon, 2002). The P49 deposit comprises 13 stacked ore lenses within a 250 m thick stratigraphic interval. A barren dolomitic siltstone, 40 to 50 m in thickness divides the orebodies into hanging wall orebodies (1 to 3) and footwall orebodies (4 to 7). The hanging wall orebodies are relatively sulfide-rich and tend to be thicker than the footwall orebodies (Valenta, 1994).

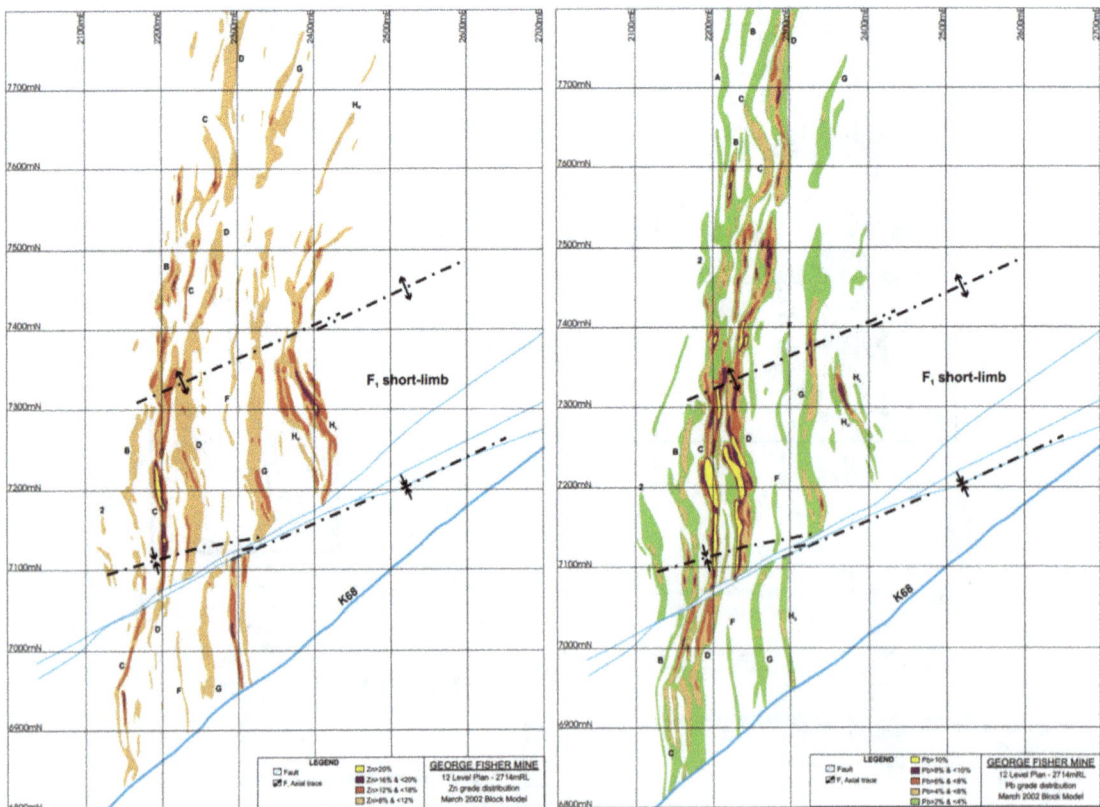

FIG 2 – Mineralisation plan maps from L72 deposit showing Zn and Pb grades distribution. Note that higher grades occur at the short limb F1, I which mineralisation strikes NNW (from Valenta, 2018, modified from Murphy, 2004).

GEOLOGICAL MODELLING

As emphasized by Grenfell and Haydon (2006), the stratigraphy at George Fisher is of paramount importance when strategizing the methodology used for geological modelling and subsequent accuracy of resource estimation and mine planning. There are multiple stratigraphic markers used within the Urquhart shale mineralised zone. Specific identifiers are present at both deposits, the tuffaceous marker beds (TMBs). The TMBs are thin layers of potassic material, which have distinguishing geological features and are present within the stratigraphy. These identifiers can be recognised over kilometres of strike length and are critical to locate sections of stratigraphy and interpretation of mineralised orebodies and estimation domains.

Currently, there are 83 different stratigraphic domains modelled at GFM, being 35 at P49 and 48 at L72. For the geological model, more than 700 individual solids are expected from the result of the number of stratigraphic domains repeated in each separate fault block (within Urquhart shale). Comprehensive stratigraphic columns are present at the site, with several characteristics and consistent features for each domain. These materials and their detailed information are critical for the logging procedure and have direct impact at the generation of the geological model.

The main challenge for geological modelling at GFM is to honour the structural complexity of the faulting blocks, the stratigraphical domains and the mineralisation pinch-outs and discontinuities. Both past and current modelling methodologies aimed to generate the best possible geological domains from the drill hole, mapping and interpretation data. Past methodologies proved to be satisfactory to capture geometrical complexities but with the downside of being remarkably time consuming. Some of the current methodologies solved the issue with amount of time required for modelling but failed to capture the pinch-out and swell from mineralised domains. Such pinch-outs have direct impact in resources estimation, reporting, mine planning and scheduling.

Past modelling methodology

The previous methodology applied at George Fisher Mines for building geological models used in resource estimation were mainly explicit modelling (Sinclair and Blackwell, 2002), also known as sectional modelling. The mining software package used at the time was MineSight 3D™ (by Hexagon Mining). Grenfell and Haydon (2006) describe the workflow at time, the required data management, number and types of files. The described workflow from the authors focuses at L72 geological modelling. The geological data set for L72 consisted of a large number of hard points that represent actual data and soft points that are based on interpretation, wireframes and solids, being a total of 1423 files to be managed for each resource or grade-control update. The number of files for P49 was not reported in the mentioned study but is assumed to be equivalent, given dimensions and complexity similarity with L72. The number of files and data reflected the modelling complexity for GFM (number of stratigraphic domains and fault blocks). Three distinct sets of data were generated at the time: interpretation data, geological planning data, and modelling data (each set with its own purpose).

The *interpretation data set* consisted of drilling points, mapping points and interpretation points. They were used to generate extrapolated stratigraphic, mineralised domains and fault wireframes. These wireframes were all extended past fault blocks boundaries for estimation and display purposes. A set of each of these files existed per stratigraphic domains per fault block. The *geological planning data* was a simplified version and subset of interpretation data. Stratigraphic domains, faults and wireframes were all cut by fault blocks (not extrapolated past them). Faults were terminated against each other, representing chronological relationship between them. This set of files provided simple and accurate geological information for daily planning requirements and designs. The *modelling data set* was a combination of the interpretation data, wireframes and solids created for the block modelling process. In order to execute the unfolding process, interpretation wireframes were extrapolated to the East. Fault blocks and domains were them coded to drill hole data, displaying density of drilling. This subset was used for resource classification.

All of mentioned procedures of extrapolation, fault truncation, solids creation and cutting were essentially manual by means of triangulation, Boolean processes, string linking in MineSight 3D™. Figure 3 illustrates the complexity for updating stratigraphic and mineralisation wireframes when faults and fault blocks were updated, since there were multiple domains multiplied by each fault block. Geological modelling works and estimation runs were responsibility of mine geologists and could take up to three months for each resource update, being the majority of time from stratigraphic domains and wireframe modelling.

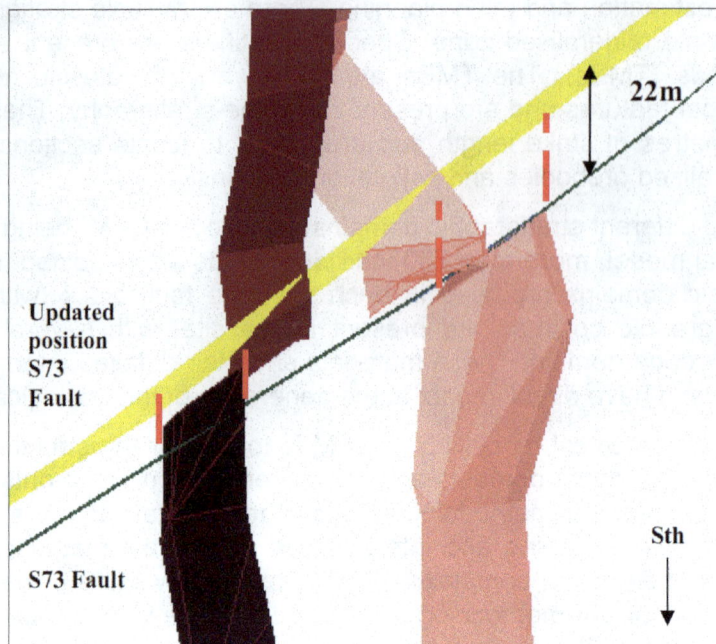

FIG 3 – Plan view displaying wireframes generated with past methodology, highlighting changes in fault block boundaries, complexity in updating and the impact on mineralised wireframes (Grenfell and Haydon, 2006).

Current modelling methodology

The current geological modelling can be characterised as general implicit modelling, using the software package Leapfrog Geo™ (Seequent, 2022). The software uses radial basis function (RBF) interpolation (Hardy, 1971) to generate the geological models (Cowan *et al*, 2003).

The workflow is described by some internal documents, procedures and manuals at the GFM site. There are two different stages of the implicit modelling implementation: starting in 2016, the transition from explicit modelling using the *vein tool* and the current methodology, since 2017, with the *stratigraphic sequence tool*. A brief description of both tools is presented below, followed by some of their shortcomings related to pinch-out modelling.

At the initial implementation (2016), previous modelled wireframes were used to back code drill holes with interpreted domains. The back coded domains were then validated with logging information about GFM stratigraphic domains. The selected tool for modelling the domains was the *vein tool*. Based on lithological codes and contacts at drill holes, this method automatically creates hanging wall/footwall contact points. With created points, surfaces are then interpolated with a user-specified resolution and a final watertight solid is generated in between hanging wall/footwall surfaces. This tool results in significantly fast processing of surfaces and solids with less than 25 minutes per deposit for an entire update. This allows the mine or resource geologist to spend critical time on validation and, if required, reinterpretation of logging and data input. The shortcoming of this technique is that it requires tens of thousands of manual inserted control points to avoid solids overlapping each other and follow the stratigraphy. If there is a low number of interpolation points per fault block, the stratigraphic domains can sometimes be generated at the wrong strike and dip direction. Since this method automatically creates hanging wall and footwall surfaces, there would be also a duplication of surfaces for subsequent domains. This would result in volumetric gaps and overlaps of different stratigraphic domains.

The current implicit modelling technique utilised at GFM (since 2017) is based on the *stratigraphic sequence tool*. For this phase of the implementation, similarly to the original explicit technique, the same type of data was used for the interpretation of stratigraphic domains: drill hole data, mapping points, and interpretation points (the latter for faults and manual local adjustments). This resulted in much more adherent geological domains compared to available data. As this tool was designed for sequences uniform in thickness with a consistent stacking order, the simplicity to set-up the model was unprecedent. The initial step for building the geological model was to inform the software the expected sequence of lithological codes from drill holes. Contact points for each unit are then extracted and surfaces automatically generated for the stacked sequence. The stratigraphic domains solids are created based on hanging wall and footwall surfaces from previous step. After all solids are built, faults are interpolated based on mapping and interpretation points and activated. Once activated, all fault blocks are generated and stacked units honour fault displacements, drill hole data and expected strike and dip directions. The current processing time for building wireframes is only a couple of minutes for an entire deposit update. An example of the generated stratigraphic domains can be seen on Figure 4.

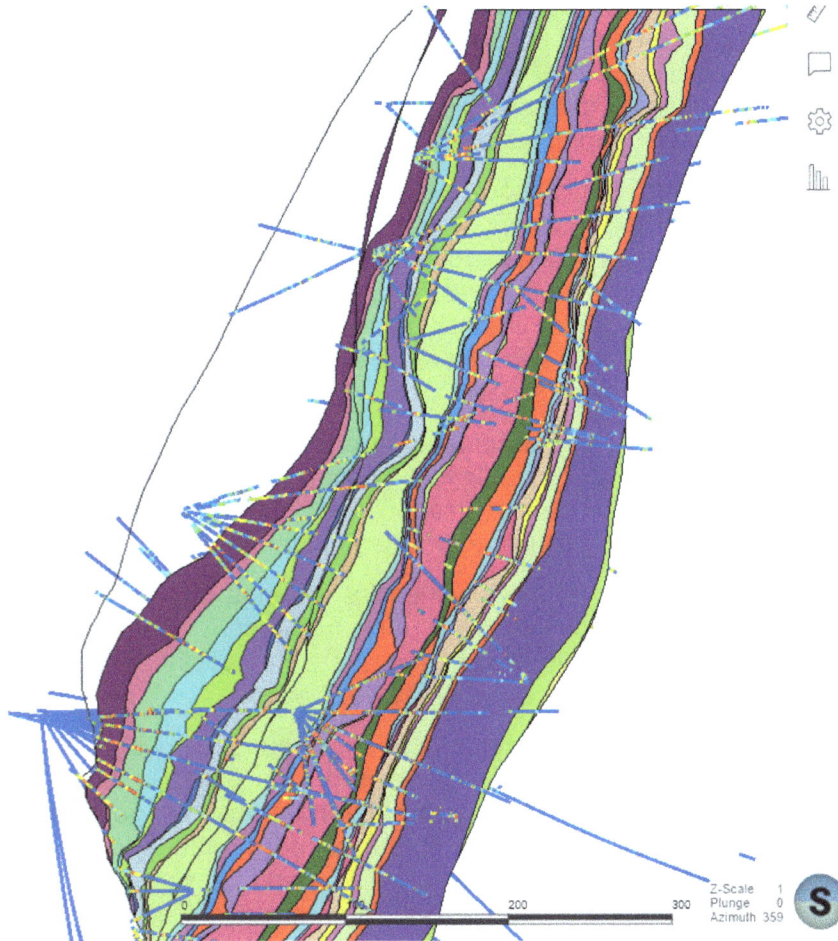

FIG 4 – Cross-section in P49 displaying wireframes generated with the current methodology, using stratigraphic sequence tool from Leapfrog Geo™.

Disadvantages of current methodology

Even though the *stratigraphic sequence tool* is a successful implementation of implicit modelling in GFM given its agility and easiness for set-up, processing speed and robustness on wireframe generation, there are shortcomings. The main disadvantage of the current methodology is that there is no guarantee that pinch-outs of stratigraphic domains and manual interpretation will be honoured. If a layer varies in thickness, the contact surface may not match the contact points in some places. The reason for this behaviour is in the nature of the algorithm of the *stratigraphic sequence tool*. When creating the stacked surfaces, the software tries to match the overall strike and dip direction of the entire sequence while trying to maintain the thickness of all domains. The result is that, for the majority of cases, the algorithm forces a singular solution in which all stratigraphic domains are present, even if drill hole data displays the pinching-out. This characteristic also results in a wireframe 'stiffness' in which manual polyline editing (interpretation lines) and mapping points would not always be honoured. In areas where drill hole spacing is denser, the pinch-out issues are better controlled since thicknesses can lower significantly but, nevertheless, are still present. Where drill hole spacing is sparser, extrapolated solids may have more significant impact in resources. In fairness, the most recommended tools for honouring the pinch-outs are the *offset surface* or *deposit/erosion tools*. Both methods have similar mechanisms, in which they automatically detect contact points and interpolate surfaces between data points (drill holes, mapping and interpretation points). Even though previous alternative tools (offset, deposit/erosion) are suitable for pinch-outs, they require the creation up to hundreds of thousands of control points per mine in order to honour the geometry complexities of GFM.

Discussion and challenge summary

As stated previously, the current major challenge to model stratiform mineralisation geometries such as GFM is that the individual orebodies pinch and swell, and whilst they have a broadly stratiform shape, they are irregular and discontinuous due to brittle and ductile deformation (Forrestal, 1990) with four known distinct (D_1, D_2, D_3 and D_4) deformational events (Chapman, 1999). Given the large number of stratigraphic domains present in GFM (83 in total summing L72 and P49 deposits), and structural complexity of fault blocks it is possible to state that current commercial software tools do not provide a straightforward semi-automated solution.

Using traditional explicit 2D interpretation could potentially solve the issues with honouring pinch-out geometries. However, drawing thousands of polylines and digitising pre-interpreted geological cross-sections are highly time-consuming activities (can take months for each mine without all pinch-out points. Moreover, there is a strong dependency in the interpretation and experience of the geo-modeller with the software. There is also the risk of creating gaps and overlaps between modelled sections unless many additional control sections are created (which would definitely add much more time for the completion of the modelling process).

The *vein tool* (Seequent, 2022), which automatically creates hanging wall and footwall surfaces constructed from selected input data is suitable for pinching-out horizons, when a specific lithofacies is not present in the drill hole. Nevertheless, this tool is not recommended for stacked units since this can produce uncontrolled gaps and overlaps between domains.

The *stratigraphic sequence tool*, which is designed for a series of continuous layers (Seequent, 2022), is able to honour the stratigraphy shape of units, however, it fails to represent the pinch and swell in individual units. Given its stiffness, surfaces are not easily controlled by interpretation polylines and points, so pinch-outs geometries are not achieved.

The *deposit tool* (Seequent, 2022) generates as sheet-like surfaces, with the primary lithology on the young side and the contacting/avoided lithologies on the older, in which contact points are created based on different lithological codes at the drill hole. As they generate surfaces that are independent from each other stratigraphic unit, it is required manual addition of control points to avoid gaps and overlaps between lithological units. Similarly, to reproduce pinch-outs and the stratiform style of the mineralisation it is required intense manual interaction. For simpler geological models and with much lower quantity of holes, manual insertion of control points is feasible and could potentially be implemented. However, in very large and complex deposits such as GFM, the amount of data (5000–6000 drill holes per mine), this manual approach is inconceivable. As mentioned before, it would be necessary to add up to 100 000 points manually per mine to account for all pinch-outs and discontinuities. A full-time employee (FTE) would take roughly up to four years per mine to manually insert all points (considering 100 000 points and 5 min each point).

Consequently, given the uniqueness of the George Fisher deposits and its geometric complexities, it is possible to state that the there is no straightforward solution or workaround present at the software in use. Any non-automated solution would require large amounts of exclusive time to reach desired outcomes.

PROPOSED MODELLING METHODOLOGY

The new proposed methodology is a hybrid workflow, in which a set Python scripts, developed in-house, are used to automatically generate hundreds of thousands of pinch-out/control points in conjunction with Leapfrog Geo™ for geological surface interpolation (hanging wall modelling).

The set of scripts was developed to overcome two major weaknesses of *deposit tool* when used for modelling a stacked/stratiform deposit such as GFM: (1) mimic pinch-outs and discontinuities at the drill hole location, in which data indicates that the lithological unit is not present (2) to avoid surfaces crossing other units of the stratigraphic sequence where a drill hole starts or finishes in the middle of the sequence (in other words, when the drill holes does not cross the entire sequence). Respectively, these points will be called (1) pinch-out points, and (2) stacking control points. In general terms, the scripts read the drill hole database and generate 3D control points to guide all geological surfaces in the correct spatial position, creating the desired geometries and honouring stratigraphy and geological controls.

Pinch-out points generation

A schematic logic for generation of pinch-out points is shown at Figure 5. In summary, the code executes the following described tasks (1) to (6), being (2) to (5) looped through all drill holes:

1. Reads the given complete stratigraphic sequence in a stacking/chronological order (There is a different list of sequential units for each mine, L72 and P49).

2. Stores the sequence of units present at the drill hole.

3. Checks the sequence present at the drill hole against the complete stratigraphic sequence list.

4. Stores the expected missing units for the drill hole.

5. Calculates spatial coordinates for missing units. Coordinates X, Y and Z are retrieved from the next sample midpoint (older/below contacting unit).

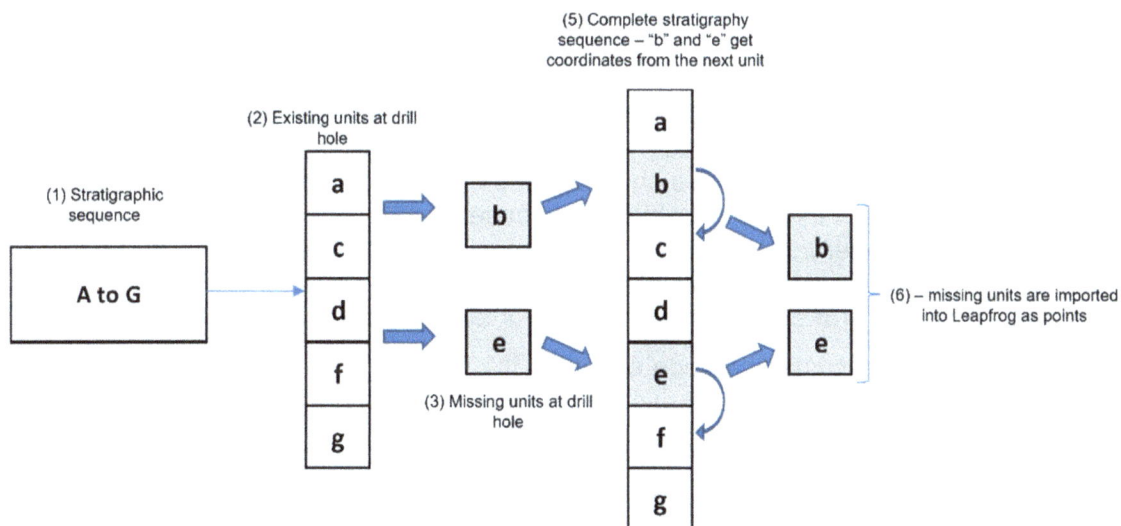

FIG 5 – Schematic logic for generation of pinch-out points based on missing stratigraphic units per drill hole. Control points are generated in Python programming language.

Finally, after looping through all drill holes, these points are imported into Leapfrog Geo™ (6) and used in the final interpolation as contact points in the *deposit tool*. Recalling that hanging wall is modelled, the created points in the space (at drill hole sample location) allow the younger/above surface to truncate underneath an older/below unit in the stratigraphic domain, creating the pinch-out shape. It is important to state the loop described above runs for each drill hole at each fault block. Drill holes are separated in different data sets by each fault block. The importance of this procedure is to avoid generating erroneous pinch-out points at drill holes cross-cutting fault blocks with repeating stratigraphic units.

Figure 6 shows on the left side a schematic representation of how the *deposit tool* interpolation would generate contact surfaces if there was no contact point for unit 'b' in the central drill hole. It is possible to visualised that, after generation of solids between hanging wall and footwall surfaces, the unit 'b' will be present at the central drill hole (even though data shows the absence). This behaviour is also present at the *stratigraphic sequence tool*. On the same schematic figure, the right side shows how the unit 'b' would be generated if a pinch-out/control point is added into the *deposit tool* interpolation, forcing the surface 'b' to go underneath the surface unit 'c'. Solids are then generated between hanging wall and footwall surfaces, mimicking a pinching-out like for unit 'b'.

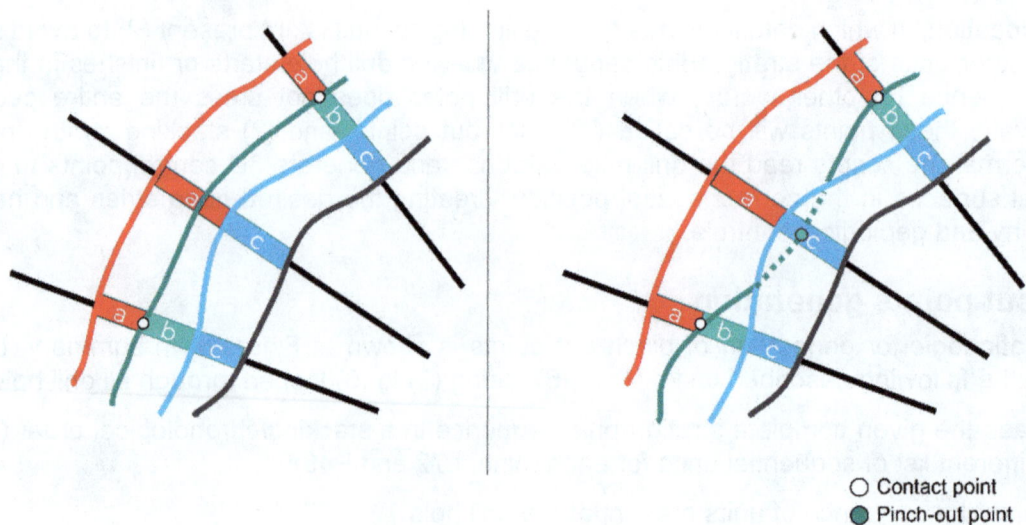

○ Contact point
● Pinch-out point

FIG 6 – Schematic representation of original surface/solids interpolation using the deposit tool in a case where unit 'b' pinches-out (left); and surfaces when adding the generated pinch-out points for unit 'b' (right).

Stacking control points generation

In order to control all surfaces in the correct spatial stacking sequence throughout the deposit, an additional step was coded auxiliary to the pinch-out workflow, as is shown in Figure 7. This step is required to guarantee all surfaces in the stratigraphic sequence to be correlated or dependant on each other, since the *deposit tool* (Seequent, 2022) works with independent surfaces. As surfaces are originally independent between units, it is not always possible to ensure that the stacking sequence will be honoured properly. This characteristic issue usually occurs when drill holes do not cross the entire stratigraphic sequence, which is the case for most drill holes at GFM. Drill holes are usually executed from underground, and so, are in most cases already starting and ending inside the Urquhart shales. There are three possible drill hole situations: start and end inside the sequence; start inside sequence but crossing top/base; starting outside sequence but crossing top/base.

Similarly to the pinch-out scripts, the code executes the following described tasks (1) to (6), being (2) to (5) looped through all drill holes:

1. the top and the base temporary units are included in the complete stratigraphic sequence list.

2. Stores the sequence of units present at the drill hole.

3. Checks the sequence present at the drill hole against the complete stratigraphic sequence list with the additional temporary top/base.

4. Given the drill hole does not cross the entire sequence, it stores the expected missing units for the drill hole at start or end of hole.

5. Calculate control points coordinates based on next or previous unit sample midpoint, depending on whether it is in the top or in the base of the deposit. A defined offset following the same azimuth and dip projection from previous intervals at the drill hole. The offset ensures the control point is going to be distanced from the contact point and the pinch-out points.

Likewise the pinch-out points, the stacking control points are exported to Leapfrog Geo™ (6) and used in the surface interpolation, ensuring all drill holes have control points for every unit in the stratigraphic sequence, even if the stratigraphic domain is not present in the given drill hole. This step guarantees that surfaces are not cross-cutting other stacked younger/older units.

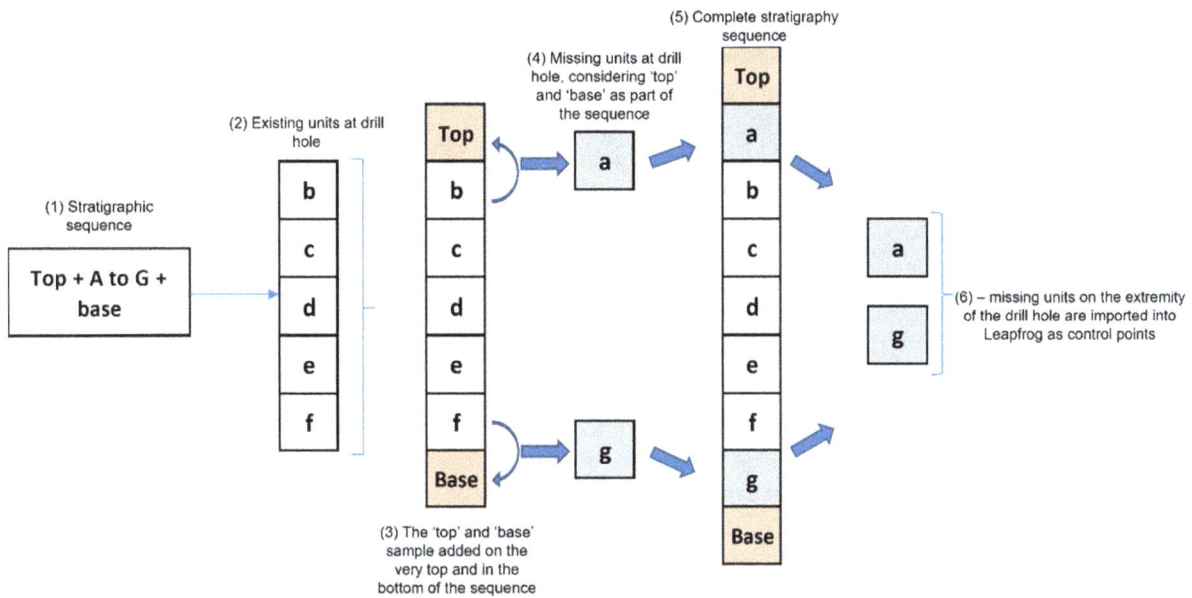

FIG 7 – Schematic logic for generation of stacking control points based on top and bottom of the stratigraphic units per drill hole. Control points are generated in Python programming language.

On left at Figure 8, an example is displayed where one drill hole does not intercept the oldest unit on the bottom of sequence, unit 'c'. As there is no point to control the hanging wall surface of this unit 'c' in that last drill hole, the interpolated surface cross-cut over a youngest unit creating a void in the model interpretation (or pinch-out of unit 'b'). On the right size the same figure, another example is displayed using a control point generated with the Python scripts. This time, no void is generated and the hanging wall of unit 'c' does not cross-cut unit 'b', honouring drill hole data.

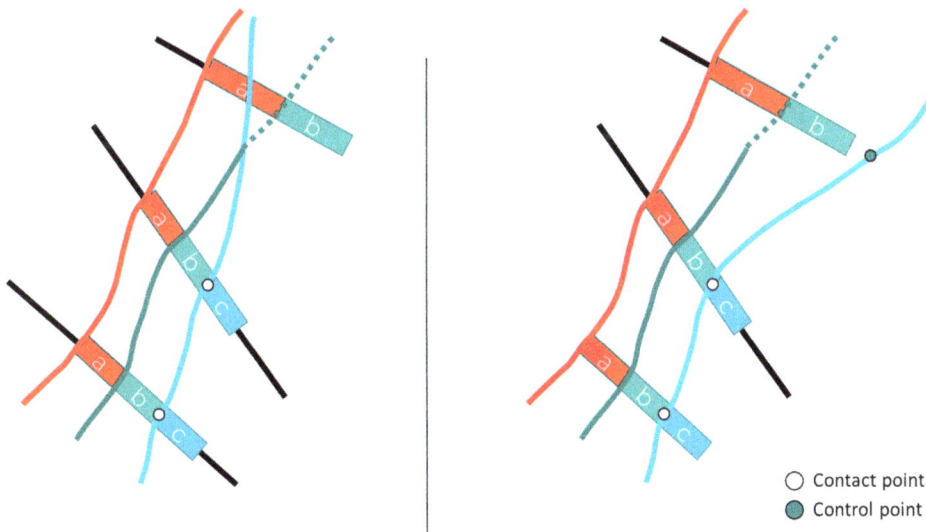

FIG 8 – Schematic representation of original surface/solids interpolation using the *deposit tool* in a case which unit 'c' cross-cut younger units (left); and surfaces when adding the stacking control points for unit 'c' (right).

CASE STUDY

This new methodology was first implemented and validated for P49 mine. The drill hole database was downloaded late 2022 from the local database located at GFM, including collar, survey, assay and lithological (field geological log) data. The database has a total of 13 620 drill holes over 356 886 m. The drill holes are in a semi regular spacing with distances ranging from 100 to 20 m in some infill areas. Most of the drilling was executed from underground. As the GFM stratigraphic units generally strikes within North–south direction, drill holes have directions pointing towards East or West direction with significant variable dips depending on the structures and goal of the drilling.

The P49 layers of Zn-Pb-Ag orebodies vary in thickness and can present discontinuities laterally. The lithological model considers 35 stratigraphical domains from the base to the top. The detailed stratigraphy lithological codes, base to top, can be described as: native bee dark shale (*n*), the dark banded shale '8' (*8d, 8cm, 8b* and *8am*), siltstone/shale '7' (*7c, 7bm* and *7a*), pale calcitic shale/siltstone '6' (*6f, 6el, 6dm, 6c, 6bl* and *6am*), shaley siltstone '5' (*5cl, 5bm* and *5bl*), pale siltstone/shale '4' (*4g, 4fm, 4e, 4dl, 4cm, 4b* and *4al*), siltstone/dark shale '3' (*3em, 3dl, 3cm, 3b* and *3a*), chloritic siltstone '2' (*2dm, 2c, 2bm* and *2al*), chloritic shale rib '1' (*1m*), hanging wall upper shale and the spears siltstone (*hwl*).

These codes are logged at the lithological column *doma* of the database, which is the input data used in the Python scripts to create pinch-out and control point. Based on the stratigraphic sequence, an approximate total of 130 000 were generated in about five minutes of processing time. The newly created points are then imported into Leapfrog Geo™ in order to interpolate all geological surfaces (*deposit tool*).

Both the structural and geological implicit models are built simultaneously, ensuring that dynamic updates are possible in between different fault blocks. As mentioned, the *deposit tool* is utilised since it is the most suitable tool for stratiform mineralisation geometries, in which thicknesses are not constant and frequent swells are present. The output points from the script are utilised to guide modelled hanging wall surfaces.

Initially, the hanging wall surface of each unit is based on automatically extracted contact points between adjacent within Leapfrog Geo™. All generated pinch-out points are imported as a single data file, filtered by each unit and added to the surfaces as edit points. Surfaces are snapped to automatically extracted contact points, pinch-out points and stacking control points. Final solids are then generated in between modelled hanging wall. Both models were compared with the data to check whether pinch-outs and discontinuities were honoured (Figure 9).

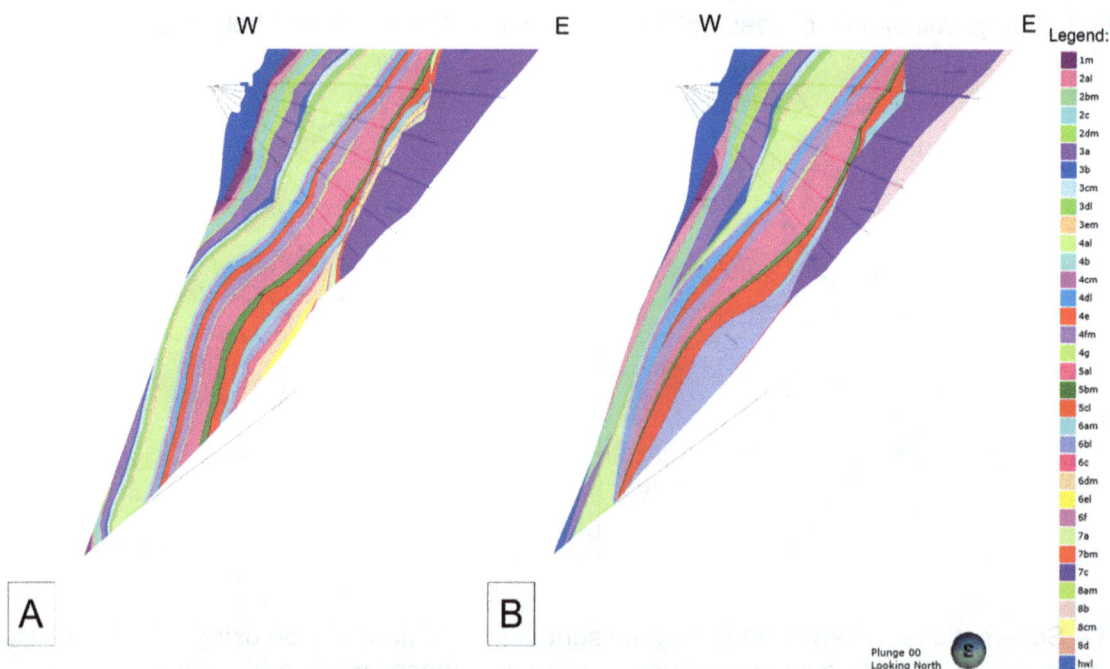

FIG 9 – W-E cross-section on the 3D implicit model for P49: (a) current model using stratigraphy sequence; (b) the new methodology combining implicit modelling with Python script.

Additionally, a volumetric comparison was performed on current methodology against the proposed one. This comparison focused on a specific non-operational area of the P49 deposit. As the proposed approach is able to reproduce pinch-outs and discontinuities observed in GFM (Figure 8), a decrease in the volume is expected. The total observed global volume decrease for the tested area, without cut-off or resource reporting criteria, is approximately -17 per cent for mineralised domains and -3 per cent in the waste domains. Since domaining is more refined, the inclusion of lower grades and dilution is decreased, resulting in higher grades at mineralised and ore domains.

The final impact in terms of metal and using resource reporting criteria is negligible. This is in agreement with metal reconciliation results from the mine that corroborate production forecasts. This shows the robustness of the current and proposed modelling methodologies. Exact grade, tonnages and metal figures comparison can't be published since this is considered confidential and sensitive information.

It is observed that the new approach improves excessive extrapolation of solids on extremities of the deposit (Figure 10). The original solids using *stratigraphic sequence tool* enforces continuity and sometimes thickness of units, even when the unit is not present at the drill hole location. It is important to state that comparison shown in this figure does not include any resource classification and reporting criteria, but only global comparison at zero cut-off. The current resource estimation workflow has additional steps to ensure minimum extrapolation is included in reported mineral resources.

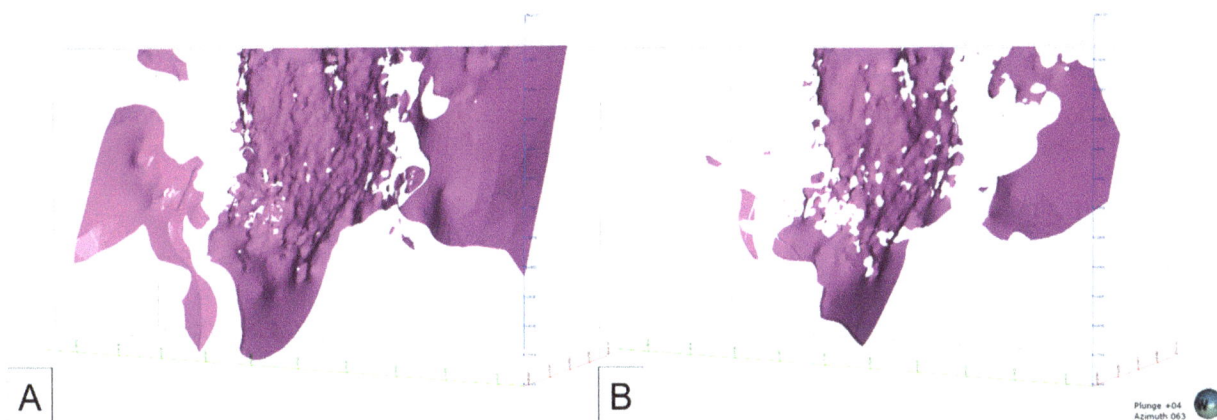

FIG 10 – Volume solid for domain 6f: (a) current model using *stratigraphic sequence tool* only; (b) the new methodology combining implicit modelling with Python scripting for generation of pinch-out and control points.

CONCLUSIONS AND CONSIDERATIONS

The new hybrid workflow proposed, using deposit tool (Seequent, 2022) simultaneously with Python scripts that generate pinch-out/control points, allowed a better definition and representation of the irregular geometry and discontinuous stratigraphic units due to brittle and ductile deformation observed in Urquhart Shale. The same results could not be accomplished using explicit or implicit modelling techniques in an acceptable time frame using manual insertion of control points. As previously stated, due to the complexities of the deposit and the amount of drill hole data, the completion could take up to four years of an FTE. Given the monotonous and repetitive nature of the task, there is a high probability of human error and less time assigned for validation. Also, the extremely repetitive work is an unacceptable misspend of the geologist or modellers time; even more when it comes to a problem that can be solved using technology. There are many crucial tasks with higher priority at the mine that requires critical thinking and solving skills from a motivated geologist.

No other commercial software packages were tested for this study. Nevertheless, the majority of available tools utilises similar algorithms and workflows from Leapfrog Geo™, which pioneered and popularised implicit modelling techniques in the industry. As the GFM deposit complexities and modelling requirements are specific, a tailored solution needed to be developed in-house. Furthermore, given all personnel is highly trained with current software workflows, a disruptive implementation is not recommended.

A considerable amount of time is required for validation, which is now enabled given speed for generation of pinch-out points. Mine and resource geologists need to check control points insertion and are able to reject any misplaced points. From the case study, it is possible to state that most problems are easily detected when checking wireframe geometries in cross, plan and long sections. Points are not individually checked, but rather entire sections. If surfaces are displaying an incorrect behaviour, control points can be evaluated. Another advantage of the method is that it highlights many logging mistakes by generating 'incorrect geometries' (driven from incorrect/outdated/missing logging). Geological and structural logging are then revised as well as stratigraphical domains and structures.

In general, the methodology is not necessarily of exclusive implementation with Python scripting. This workflow can be generated using any other programming language. Nevertheless, Python was selected given personnel knowledge, its easiness to read and learn, the extensive number of available libraries, one of the strongest communities in the world and its free to use. To run the scripts built for GFM it is not required vast expertise with the programming language. However, to achieve consistent results, the user needs to follow the exact steps recorded in the internal procedure (created for the purpose of updating the model) and validate the output thoroughly.

The automation of repetitive tasks is of utmost importance in present days. Automation allows for quick generation of different scenarios, with minimal risk of human error due to fatigue or inattentive employees. The majority of time is spent on validation of results rather than at the repeating task. Reproduction is also achieved, given the workflow is based on a sequence of scripts, allowing auditability. Software limitation can be overcome with the intelligent design of such tasks using in-house solutions. Automation in geological modelling, allied with strong understanding of the mineralisation, the deposit and the structural relationship between faults and stratigraphical units can now result in more realistic geological models that honours field observations.

ACKNOWLEDGEMENTS

The authors would like to thank Glencore and George Fisher Mines for the use of data. We also would like to thank Chantelle Lower, Ian Fahey and Matt Blennerhassett, from GFM, for their constructive comments, validations and suggestions, which helped us to improve the manuscript.

REFERENCES

Bell, T H, 1983. Thrusting and duplex formation at Mount Isa, Queensland, Australia, *Nature*, 304:493–497.

Bertossi, L G, Usero, G, Verde, R L V and Lopes, J A, 2013. Implicit geological modelling versus traditional modelling, A Case Study Applied to a Lateritic Nickel Deposit, in *Proceedings of the 36th APCOM Symposium Applications of Computers and Operations Research in the Mineral Industry*, pp 1–8.

Bloomenthal, J (ed.), 1997. *Introduction to Implicit Surfaces* (Morgan Kaufmann Publishers Inc: San Francisco).

Broomhead, D S and Lowe, D, 1988. Multivariable functional interpolation and adaptive networks, *Complex Systems*, 2:321–355.

Cave, B, Lilly, R, Simpson, A and McGee, L, 2023. A revised model for the George Fisher and Hilton Zn-Pb-Ag deposits, NW Queensland: Insights from the geology, age and alteration of the local dolerite dykes, *Ore Geology Reviews*, vol 154.

Chapman, L H, 1999. Geology and genesis of the George Fisher Zn-Pb-Ag deposit, Mount Isa, Australia, PhD, James Cook University.

Chapman, L, 2004. Geology and Mineralisation Styles of the George Fisher Zn-Pb-Ag Deposit, Mount Isa, Australia, *Economic Geology,* 99:233–255.

Cowan, E J, 2020. Deposit-scale structural architecture of the Sigma-Lamaque gold deposit, Canada—insights from a newly proposed 3D method for assessing structural controls from drill hole data, *Miner Deposita*, 55:217–240. Available from: <https://doi.org/ 10.1007/s00126-019-00949-6>

Cowan, E J, Beatson, R K, Ross, H J, Fright, W R, McLennan, T J, Evans, T R, Carr, J C, Lane, R G, Bright, D V and Gillman, A J, 2003. Practical implicit geological modelling, in *Proceedings of the Fifth International Mining Geology Conference 2003*, pp 89–99 (The Australasian Institute of Mining and Metallurgy: Melbourne).

Davis, T, 2004. Mine-scale structural controls on the Mount Isa Zn-Pb-Ag and Cu ore bodies, *Economic Geology and the Bulletin of the Society of Economic Geologists*, 99:543–559.

de Oliveira, S B and Sant'Agostino, L M, 2020. Lithogeochemistry and 3D geological modeling of the apatite-bearing Mesquita Sampaio beforsite, Jacupiranga alkaline complex, Brazil, *Braz J Genet*, 50(1–11):2317-4889202020190071.

de Oliveira, S B, Juliani, C, Monteiro, L V S and Tassinari, C C G, 2020. Structural control and timing of evaporite-related Mississippi valley-type Zn–Pb deposits in Pucara group, northern Central Peru, *J S Am Earth Sci*, 103:102736. Available from: <https://doi.org/ 10.1016/j.jsames.2020.102736>

de Oliveira, S B, Leach, D L, Juliani, C, Monteiro, L V S and Johnson, C A, 2019. The Zn–Pb mineralisation of Florida Canyon, an evaporite-related Mississippi valley-type deposit in the Bongara district, northern Peru, *Econ Geol*, 114:1621–1647.

de Oliveira, S B, Torresi, I and Rossi, D A L, 2022. 3D structural control and spatial distribution of Zn-Pb-Cu grades in the Palmeiropolis, VMS deposit, Brazil, *B Appl Earth Sci*, 131:69–85.

Forrestal, P J, 1990. Mount Isa and Hilton Silver-Lead-Zinc Deposits, *Geology of the Mineral Deposits of Australia and Papua New Guinea,* pp 927–934 (The Australasian Institute of Mining and Metallurgy: Melbourne).

Grenfell, K and Haydon, M, 2006. Challenges in modelling large complex orebodies at the George Fisher North Mine, in *Proceedings Sixth International Mining Geology Conference*, pp 143–152 (The Australasian Institute of Mining and Metallurgy: Melbourne).

Hand, M and Rubatto, D, 2002. The scale of the thermal problem in the Mt Isa Inlier, in *Proceedings of the Geological Society of Australia*, p 173.

Hardy, R L, 1971. Multiquadric Equations of Topography and other Irregular Surfaces, *Journal of Geophysical Research*, 176:1905.

Johnston, M, Versace, S, Shaw, A, Arnott, D, Bojcevski, D and Sims, D, 1998. Data collection and analysis for the George Fisher project, Mount Isa, Qld, *Australasian Institute of Mining and Metallurgy Conference Proceedings,* vol 4/98, pp 171–176.

Kampmann, T C, Stephens, M B and Weihed, P, 2016. 3D Modelling and Sheath Folding at the Falun Pyritic Zn-Pb-Cu-(Au-Ag) Sulphide Deposit and Implications for Exploration in a 1.9 Ga Ore District, Fennoscandian Shield, *Mineralium Deposita, Sweden*, 51:665–680. Available from: <https://doi.org/10.1007/s00126-016-0638-z>

Lilly, R, Taylor, D and Spanswick, N, 2017. Mount Isa Cu-Pb-Zn deposit including George Fisher, *Australian Ore Deposits,* pp 473–478 (The Australasian Institute of Mining and Metallurgy: Melbourne).

Murphy, T E, 2004. Structural and stratigraphic controls on mineralisation at the George Fisher Zn-Pb-Ag deposit, northwest Queensland, Australia PhD, James Cook University.

Naranjo, A, Horner, J, Jahoda, R, Diamond, L W, Castro, A, Uribe, A, Perez, C, Paz, H, Mejia, C and Weil, J, 2018. La colosa Au porphyry deposit, Colombia: mineralisation styles, structural controls and age constraints, *Econ Geol,* 113:553–578.

Schetselaar, E, Pehrsson, S, Devine, C, Lafrance, B, White, D and Malinowski, M, 2016. 3D geologic modeling in the Flin Flon mining district, Trans-Hudson orogen, Canada: evidence for polyphase imbrication of the Flin Flon-777-Callinan volcanogenic massive sulfide ore system, *Econ Geol*, 111:877–901.

Seequent, 2022. User Manual for Leapfrog. https://help.seequent.com.

Sinclair, A J and Blackwell, G H, 2002. *Applied mineral inventory estimation* (Cambridge University Press).

Tolman, J, Shaw, A and Shannon, K, 2002. Stratigraphic relationships between stratiform polymetallic base metal deposits within the Mount Isa Inlier, NW Queensland, Australia, *Australasian Institute of Mining and Metallurgy Conference Proceedings*, vol 6/00 (The Australasian Institute of Mining and Metallurgy: Melbourne).

Valenta, R, 1994. Deformation of host rocks and stratiform mineralisation in the Hilton Mine area, Mt Isa, *Australian Journal of Earth Sciences*, 41(5):429–443.

Valenta, R, 2018. NW Queensland Mineral Province Deposit Atlas Prototype Report – the Mount Isa and Ernest Henry Deposits, DNRME-GSQ Commissioned study and report.

Vollgger, S A, Cruden, A R, Ailleres, L and Cowan, E J, 2015. Regional dome evolution and its control on ore-grade distribution: insights from 3D implicit modelling of the Navachab gold deposit, Namibia, *Ore Geol Rev*, 69:268–284. Available from: <https://doi.org/ 10.1016/j.oregeorev.2015.02.020>

Williams, P J, 1998. An introduction to the metallogeny of the McArthur-Mount Isa Cloncurry Minerals Province, *Economic Geology*, 93:1120–1131.

SBRE framework – application to Olympic Dam deposit

I Minniakhmetov[1] and D Clarke[2]

1. Principal Global Modelling and Data, BHP, Perth WA 6010.
   Email: ilnur.minniakhmetov@bhp.com
2. MAusIMM(CP), Principal Resource Geologist, BHP Olympic Dam, Adelaide SA 5000.
   Email: david.clarke@bhp.com

ABSTRACT

This paper describes the application of the SBRE framework to model the Olympic Dam deposit, one of the largest copper and uranium deposits in the world. Geostatistical simulations are the best practice for quantifying uncertainty in the mineral value chain and estimating resources. However, these simulations are currently undertaken on an *ad hoc* basis in the mining industry due to the high requirements in computing resources and capabilities in advanced geostatistics. The SBRE project provides a standard framework to test, build, and run conditional simulation workflows, which has been applied to the Olympic Dam deposit.

Conventional geostatistical methods face challenges when modelling the Olympic Dam deposit, such as non-stationarity, non-linear correlations, and a large number of grid blocks to simulate. To address these challenges, a combination of methods was used, including projection pursuit multivariate transform, trend-residual decorrelation, pluri-Gaussian simulations, and turning-bands simulations. Various tools and applications were used to build the model, including in-house developed applications, Autovariogram software (Varify), and the RMSP python library. All computations were accomplished on the AWS cloud using the EC2-Batch service.

The final set of simulations was validated using various methods, such as grade-tonnage curves, histograms, swath plots, and variogram reproductions, as well as validation using historical samples back-testing. The study highlights the importance of an automated, modular, and scalable framework in the development of conditional simulation workflows and describes the learning gained from trials of different combinations of methods and software

INTRODUCTION

The mineral value chain is a complex system that involves multiple parameters and decisions across different process streams and time horizons, ranging from daily operations to the life of the mine, which could span up to 100 years. Due to the presence of variable bottlenecks, uncertainty and variability play a critical role in this complex system. One of the major sources of uncertainty is the resource model, which is a representation of *in situ* resources constructed from a limited number of sparse drill hole samples. The resource model is the most upstream process of the value chain and serves as an input for all the downstream processes and decisions. However, due to the complexity of the mineral value chain, the uncertainty of *in situ* resources expressed as a block model, such as the standard deviation per block of an attribute of interest or resource category model, cannot be transferred through the mine planning, processing, and closure processes.

Monte-Carlo based geostatistical simulations or conditional simulations are the only feasible solutions to address this challenge. These simulations can quantify the uncertainty in the mineral value chain and provide an unbiased estimation of resources. By considering the spatial distribution of the mineral deposit, geostatistical simulations can capture the variability and uncertainty of the resource model, making it possible to transfer the uncertainty through the mine planning, processing, and closure processes. Geostatistical simulations are therefore essential for accurate decision-making in the mining industry, as they provide a better understanding of the risks and uncertainties associated with the mineral value chain.

Conditional simulations have been a well-known and established approach for producing equi-probable realisations of possible values of *in situ* resources (Journel and Alabert, 1990) consistent with observations, such as geophysical data and spatial statistics of data, including grades continuity. Despite their efficacy, the application of conditional simulations in the mining industry has been limited due to the complexity of geostatistical tools and computational requirements. Moreover, the difficulty in trialling different combinations of geostatistical tools has further hindered their adoption within the industry, leading to their infrequent use.

In order to address the challenges associated with geostatistical simulations, BHP has implemented the Simulation Based Risk Evaluation (SBRE) framework, which offers a simple, flexible, and scalable solution for creating conditional simulation workflows (Minnniakhmetov and Ashford, 2022). This framework allows for the combination of various applications, including Isatis Neo, RMSP, Varify, in-house python scripts, and Vulcan, through a plug-in modularised capability. The SBRE framework is similar in concept to Google's KuberFlow technology, but is tailored specifically for geostatistical applications. SBRE also documents key considerations and parameters during the execution of a workflow, which facilitates internal or external audits. The framework can run various steps on laptops or cloud services, such as AWS and Azure, to improve computational time and facilitate the design of the optimal combination of geostatistical tools. Additionally, SBRE provides multi-realisation capability for any data utilised in a workflow, enabling full uncertainty quantification and sensitivity analysis, such as accounting for uncertainty in variogram parameters or data quality.

The SBRE framework has been successfully employed in a range of deposits, including coal, iron ore, and IOCG. This paper, however, concentrates on the application of the framework specifically to the Olympic Dam deposit. The reason for focusing on this deposit is due to its substantial complexity, which creates unique modelling challenges. Specifically, the Olympic Dam deposit is characterised by its vast size, intricate multivariate relationships between variables of interest, and a high degree of non-stationarity and non-Gaussianity (Minniakhmetov and Clarke, 2022).

The structure of this article is as follows. Firstly, a brief overview of the geology of the Olympic Dam deposit is presented. Next, modelling challenges are outlined. Following this, the experiments conducted to determine the optimal combination of methods and software for modelling the deposit are described and the resulting workflow is proposed with a details on cloud computing. The quality of the simulations generated by the workflow is evaluated using a variety of validation using original data samples and historical samples back-testing, and potential applications of the simulations are discussed. The article concludes with a summary of the findings and their implications.

OLYMPIC DAM GEOLOGY

The Olympic Dam Deposit is IOCG deposit found by Western Mining Corp in 1975. It formed 1.59 Ga in the upper 2 km of the earth's crust via tectonic-magmatic-hydrothermal processes, and is primarily located in the breccia complex of Roxby Downs granite. It is covered by flat, unmineralised sedimentary units, 300–400 m in-depth (Figure 1).

The Roxby Downs granite (RDG) has been extensively altered by brecciation and hydrothermal alteration to form the Olympic Dam breccia complex (ODBC). The primary minerals in the granite have been altered and replaced by hematite-magnetite, quartz, sericite, siderite, chlorite, fluorite, barite, and sulfides. To a lesser extent, the deposit also contains various other lithologies such as bedded sedimentary facies, and mafic/ultramafic dykes and lavas, which are pre-, syn- and post-mineralisation.

This package of differing lithologies within the ODBC have all experienced multiple brittle structural deformation events, intense brecciation, and/or hydrothermal alteration to varying degrees of intensity. The range of brecciation can scale from relatively coherent 'facies' to irregular, polymictic and up to a chaotic polyphase breccia. The hematite alteration has variably replaced the primary lithology by iron oxides—forming a compositional continuum from where the original protoliths remain—up to total iron oxide breccias, where the original lithology texture is completely replaced by iron oxide alteration. Lithological contacts are typically obscured by faulting and brecciation.

Typically, metals exhibit strong spatial correlation, although there exist isolated occurrences of gold. The distribution of metals is distinguished by fluctuations on a small scale. The Olympic Dam deposit displays a prominent hematite-quartz breccia zone at its centre that is devoid of substantial amounts of copper, uranium, and silver.

The mineralisation in the deposit is primarily influenced by a set of key factors including the type and abundance of sulfide minerals, the intensity of hematite alteration, the protolith or lithology, and the structure or geometry. These factors work in tandem to directly control both the occurrence and abundance of mineral grades throughout the deposit. Research conducted by Ehrig *et al* (2017) indicates a spatial correlation between hematite, copper, uranium, gold (associated with sulfide), and silver mineralisation due to the co-precipitation of these elements. As the controls on copper mineralisation are correlated, the same factors are associated with uranium, gold (associated with sulfide), and silver mineralisation.

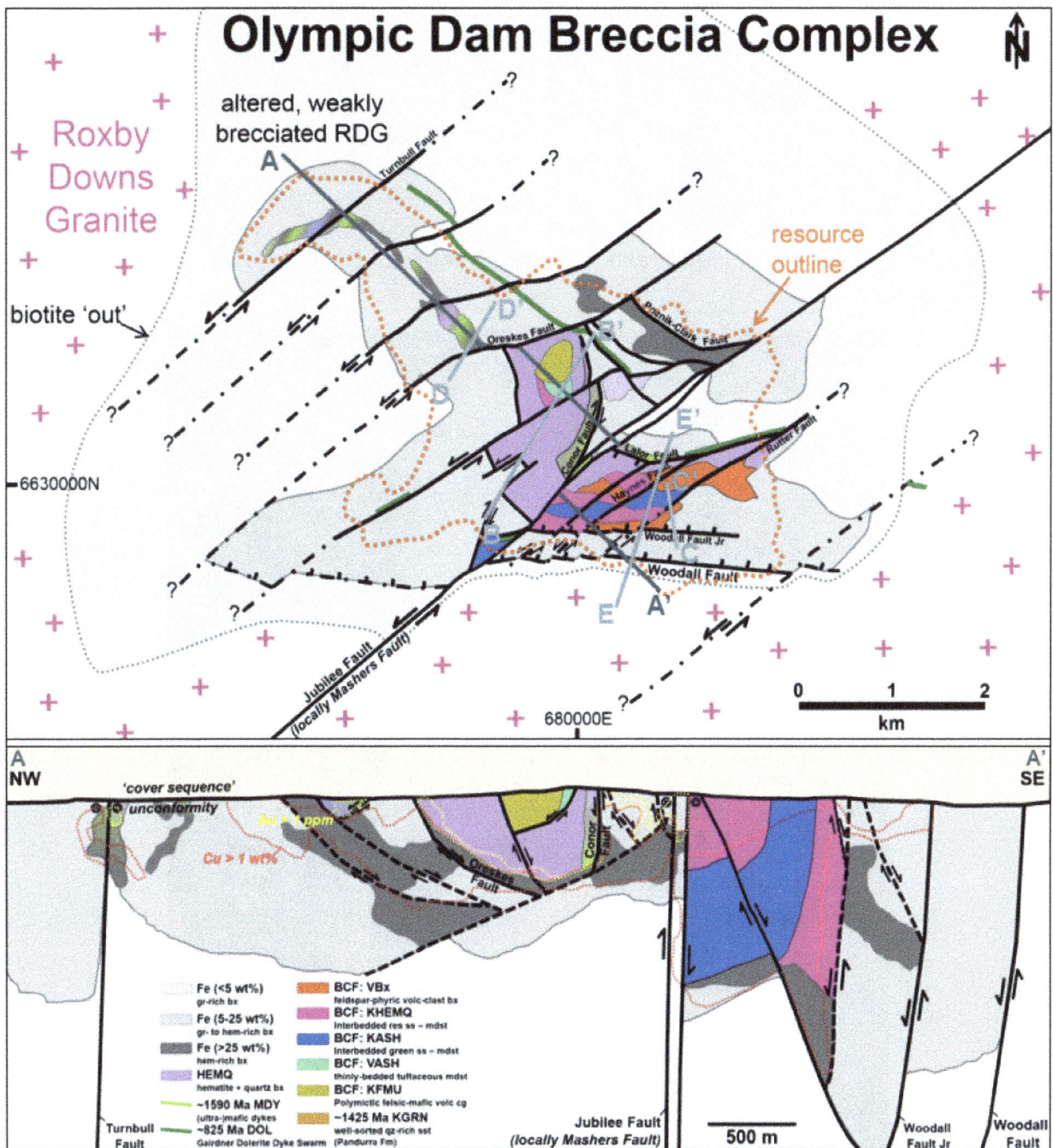

FIG 1 – (a) Simplified basement geological map (at 450 m depth) of the Olympic Dam deposit. Location of long- and cross-sections in this paper are displayed by blue lines. The biotite-out contour depicts the transition from relatively unaltered Roxby Downs granite (RDG) in the Olympic Dam breccia complex (ODBC) and is defined by the last occurrence of magmatic biotite. (b) Composite long section (A–A') showing the significant post-mineral structural disruption of major lithology and alteration domains (Clark and Ehrig, 2019).

The dominant sulfide assemblages responsible for copper mineralisation at the Olympic Dam deposit are chalcopyrite, bornite, and chalcocite. These minerals are characterised by strong zonation and dissemination throughout the mineralised system. The relative distribution of these three primary sulfide minerals is a critical factor in understanding the distribution of copper grades at Olympic Dam.

MODELLING SPECIFICATIONS

In this study, the modelling process is confined to the potential mineralisation wireframe, which is obtained by creating a 200 m-buffered envelope around the existing drilling. A regular grid with dimensions of 3625 × 5860 × 1695 m is employed to discretise the modelling extents into 66 million 5 × 5 × 5 m grid blocks, with variables of interest including Au, barium-adjusted sulfur (Ba-S), density (SG), and U_3O_8.

The block size is determined based on the highly selective mining methods, specifically underground sublevel open stoping with a minimum stope size of 30 × 30 × 40 m. A 5 × 5 × 5 m block size is chosen to provide sufficient discretisation (6 × 6 × 8) for determining an optimal stope location and performing change-of-support calculations without relying on additional assumptions such as Gaussianity, two-point spatial relations, or linear dependency of variables.

The primary data used for modelling is derived from approximately 700 000 assay samples obtained from around 15 000 diamond core drill holes. These samples are used to define 38 structural domains, where data in each domain exhibit similar direction of Cu-sulfur mineralisation continuity (Figure 2).

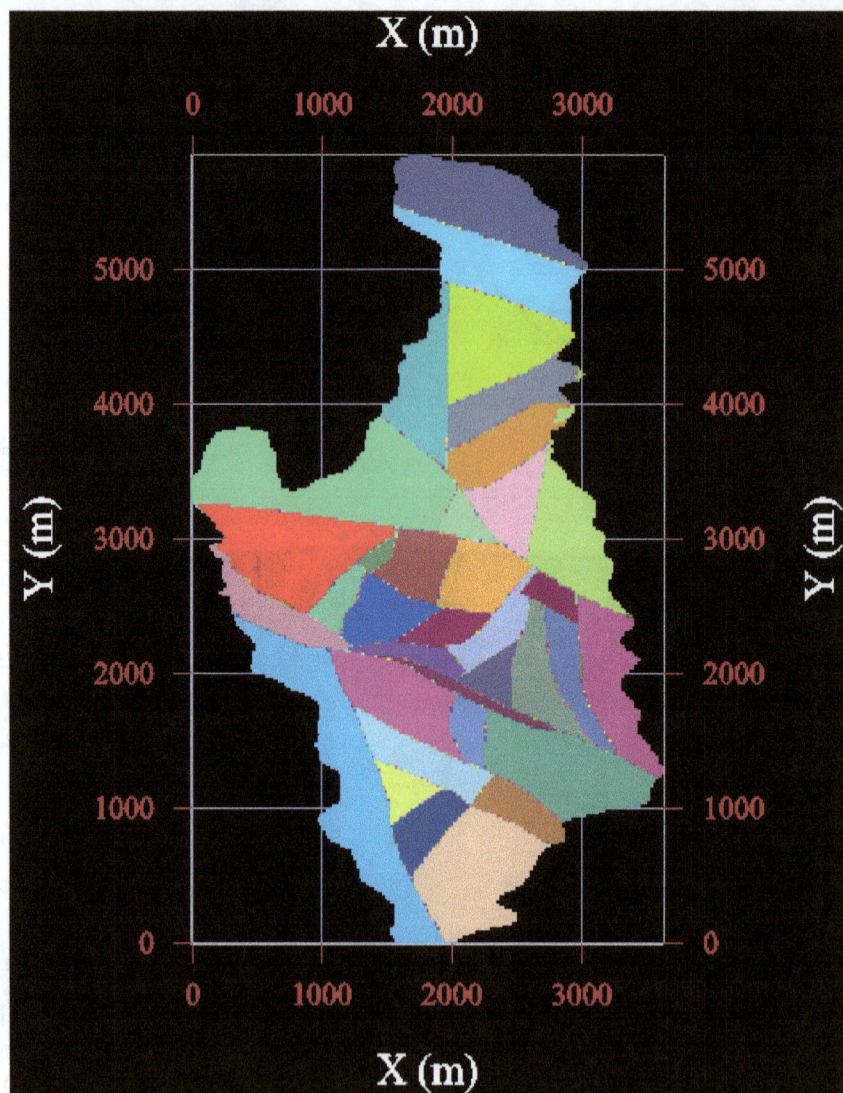

FIG 2 – Plan view of colour-coded structural domains.

CHALLENGES IN CONVENTIONAL GEOSTATISTICAL METHODS

In the presented study, Figure 3 depicts an east–west vertical section that displays the distribution of copper grades (ranging from 0 per cent in blue to 4 per cent in red) and copper mineralisation (depicted in red for chalcopyrite-bornite, yellow for chalcocite, and blue for non-sulfides/pyrite), respectively.

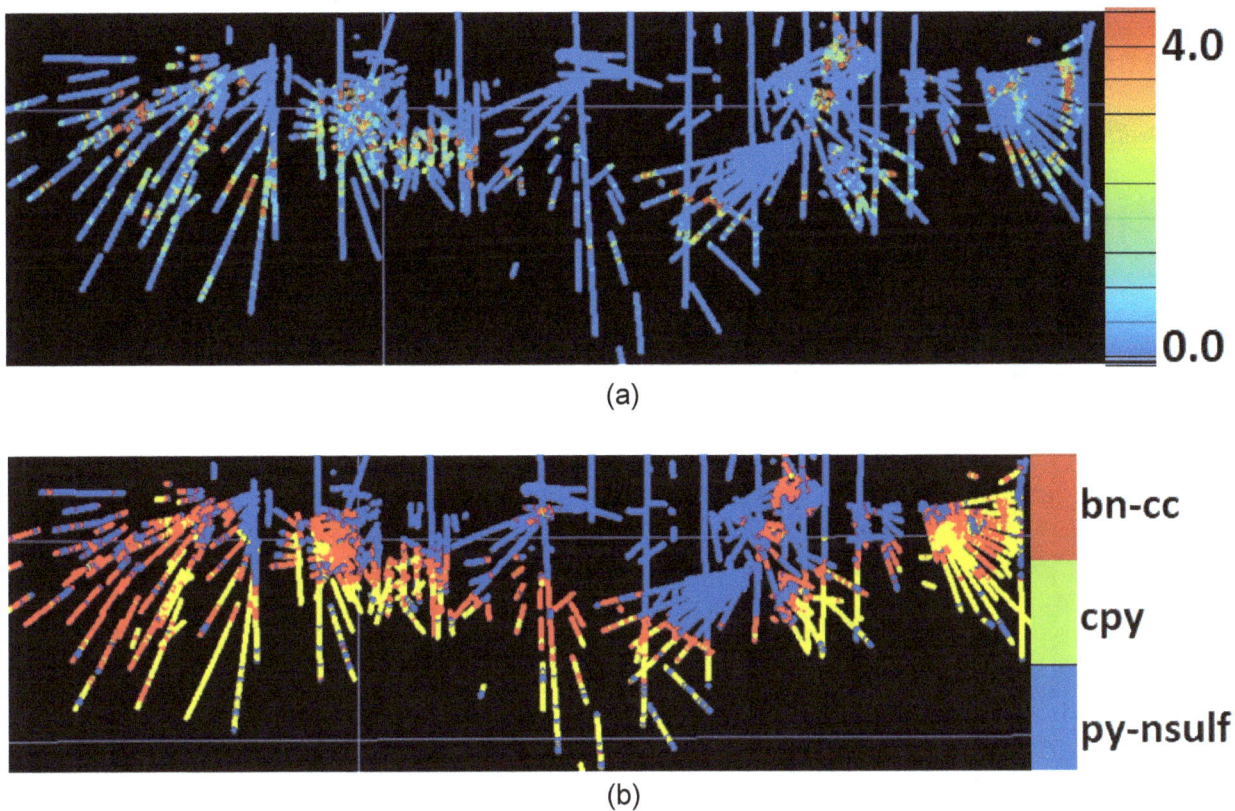

FIG 3 – Vertical east–west section: (a) copper grades (blue 0 per cent; red 4 per cent), (b) copper mineralisation, depicted in red for chalcopyrite-bornite, yellow for chalcocite, and blue for non-sulfides/pyrite.

The data analysis reveals that both grades and mineralisation exhibit a high degree of non-stationarity, indicating that they are not constant throughout the deposit. Figure 3b specifically shows how the proportions of mineralisation type categories change spatially, with chalcopyrite-bornite, chalcocite, and non-sulfides being the three main categories observed. Additionally, the spatial relationships between the different mineralisation types are variable in space, eg red and blue categories are in contact with small yellow incorporations on the left of Figure 3b and yellow being mostly in contact with red with minor incorporations of blue on the right. The grades of copper also vary throughout the deposit, changing gradually in copper bearing sulfides (red and yellow domains) from high-grade core areas to low-grades.

Cross-variate relations are shown to be non-linear and multi-modal in Figure 4. Furthermore, the distribution of U_3O_8 and Au also exhibits a high degree of non-Gaussianity and skewness, highlighting the complexity of the deposit's mineralisation patterns.

The majority of the data used in this study was obtained through underground fan drilling, which results in an uneven distribution of samples. While declustering methods can be employed to address this issue, the problem of data representativeness remains unresolved. In some cases, the copper grades can vary significantly from 4 per cent in the core area to 0.5 per cent in the peripheral area of the same structural domain. Unfortunately, the peripheral domains are poorly sampled due to limitations imposed by the fan geometry and cannot be treated as a separate domain. As a result, conventional variography will be mainly driven by the densely drilled areas and may fail represent continuity in the peripheral regions.

Lastly, the simulation of hundreds of realisations involving categorical (with three categories) and continuous (with five variables) attributes, using around 700 000 data points and approximately 60 million modelling blocks, can take several weeks to complete if run without parallel computing.

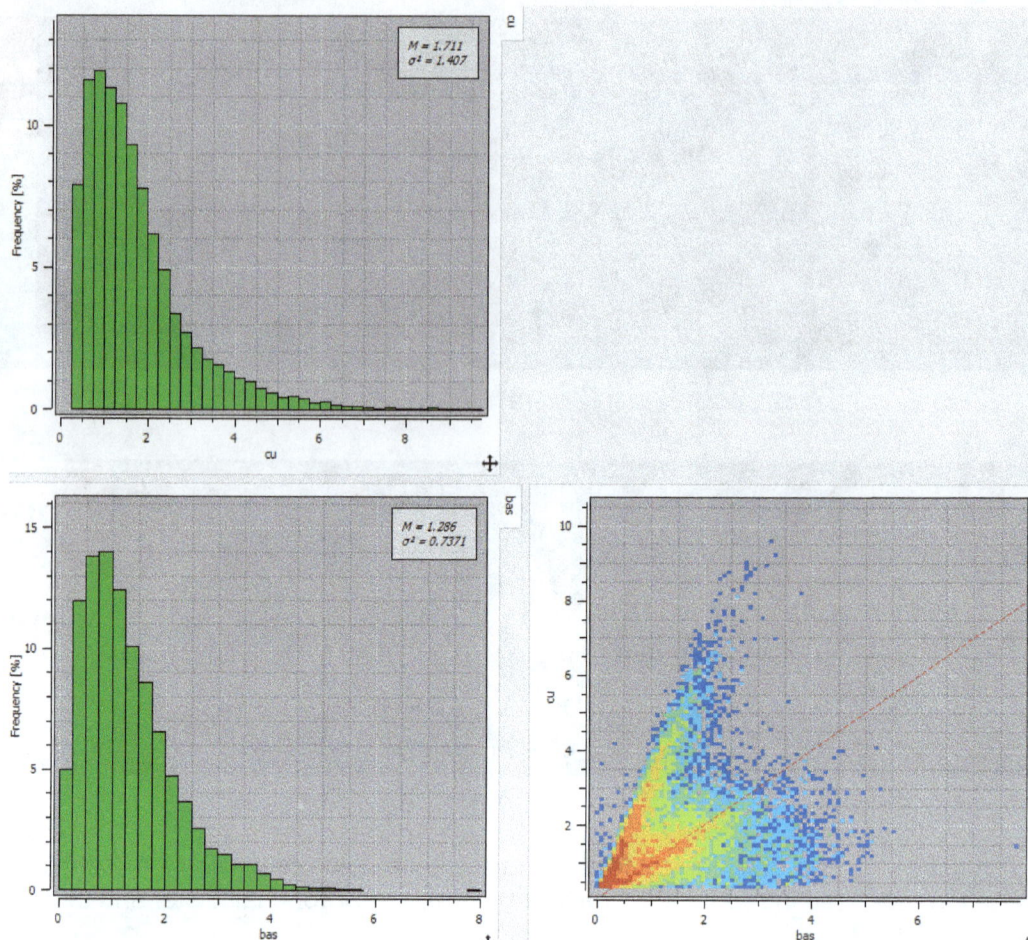

FIG 4 – Histograms and scatter-plot for copper and barium adjusted sulfur (BaS).

CONDITIONAL SIMULATIONS

Workflow development

The initial workflow for multivariate simulation utilised a conventional approach involving the following steps: normal score (Deutsch and Journel, 1998), trend calculation, minimum/maximum autocorrelation factors (MAF) (Desbarats and Dimitrakopoulos, 2000), variography, turning bands Gaussian simulation (Mantoglou and Wilson, 1982) and back-transformations, which was completed in Isatis 2017 (Geovariance), as illustrated in Figure 5a. However, it was discovered that the linear correlation assumption of MAF was inadequate in reproducing multivariate relationships. Furthermore, uncertainties regarding the sulfide/non-sulfide boundary became apparent, necessitating categorical simulations.

Since the non-linear multivariate decorrelation methods, namely projection pursuit multivariate transform (PPMT by Barnett, Manchuk and Deutsch, 2014) or Flow Anamorphosis (van den Boogaart, Tolosana-Delgado and Mueller, 2015), were unavailable in Isatis software, the workflow was updated and shifted to an internal python script. The revised workflow, as shown in Figure 5b, incorporated additional steps, namely, categorical simulation (Silva and Deutsch, 2019) with two categories (sulfides and non-sulfides) and PPMT decorrelation technique, rather than MAF. Moreover, local variance scaling was introduced to account for non-stationarity of variance, such that in low-grade peripheral areas, the variance of copper (or underlying factor after PPMT) was considerably smaller than that in high-grade core areas.

FIG 5 – Evolution of workflows: (a) Isatis workflow, (b) internal py-scripts workflow, (c) RMSP based workflow, (d) final workflow. Steps highlighted by red are changes with respect to the previous version of the workflow.

Prior to the availability of SBRE platform, changes in the workflow were time-consuming and required calculations to be performed on four separate Z840 desktop machines, with numerous manual file handling and manual runs of the script on different machines. Given the complexity of the deposit under consideration, a trial-and-error approach was utilised to tailor the workflow to achieve the desired results. That required tens and hundreds of iterations.

However, with the emergence of the SBRE framework and the maturation of the RMSP python library, the workflow was migrated to the SBRE platform, and most of the steps were moved to the RMSP python library (Figure 5c). Methodologically, only the trend modelling step was updated from conventional trend modelling (local mean trend was subtracted from sample values for residuals calculation) to trend-residual decorrelation technique (Qu and Deutsch, 2018). The former approach generated unrealistic values that were not observed in the data, such as negative grades and large area of 2 per cent copper grades in areas where grades changed sharply from 1 per cent to 3 per cent.

It should be noted that the SBRE platform provided cloud computing capabilities, allowing most of the workflow steps to run in parallel for each domain. It is important to have parallel computing capabilities not based on parallel runs for each realisation but generating a single realisation efficiently with parallel computing. The quality of simulations can be adequately assessed from few realisations, as all simulations share similar univariate and spatial statistics, and there are multiple trail runs required before desirable quality of simulations is achieved.

Although the latest workflow provided all standard validation graphs consistent with observed statistics, the grade-tonnage curves with a cut-off on copper and average grade of sulfur did not match the observed data statistics (Figure 6), which is a critical drawback as Cu:S ratios are the most important characteristic for geometallurgy/processing. Additionally, the turning bands method did not allow for a smooth transition of grades from one domain to another, creating artificial sharp contacts on the boundary of domains.

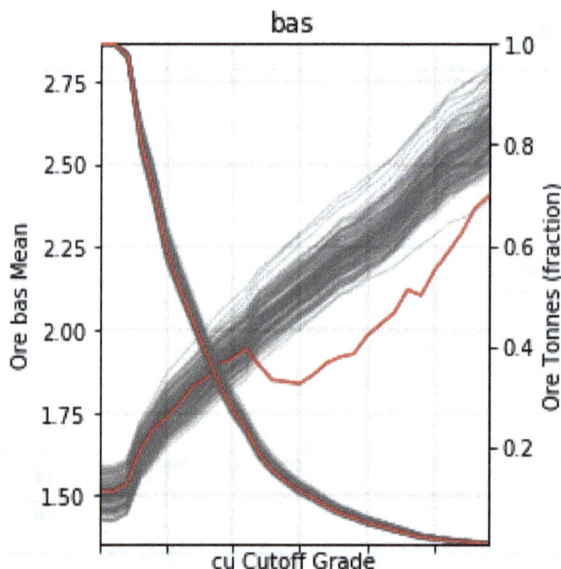

FIG 6 – Misrepresentation of Cu:BaS relations in simulations (grey lines) versus samples statistics (red line).

To address the Cu:S ratio issue, several solutions were tested, including creating an additional PPMT step before trend modelling, adding Cu:S ratio as an additional variable, or implementing categorical simulation with three categories: chalcopyrite-bornite, chalcocite, and non-sulfides/pyrite. The latter was proven to provide best outcomes. The SBRE framework allowed for efficient testing of each approach without changing any downstream steps, thanks to its modularisation capability. Moreover, the testing was performed on the most representative domain, which significantly reduced the calculation turnover time. The changes required to switch from a full-deposit simulation to a domain simulation involved only changes in the 'Data loading' step parameters.

The issue of artificial sharp contacts was resolved by internally implementing the spectral method, similar to that described in Pardo-Igúzquiza and Chica-Olmo (1993). The workflow update only required substituting the 'Turning bands simulation' step without any additional changes in code or data handling.

It is noteworthy that the SBRE framework possesses robust documentation capabilities. Specifically, the version of the code, python modules, data, and parameters have been documented and internally stored. By executing a simple command from the command line interface, the workflow can be rerun to reproduce the same outputs. Additionally, any imported or generated data during the calculations can be exported in various formats, including CSV, Datamine, pandas Dataframe, and GSLIB.

To further improve the latest workflow, a final modification was made to validate the models with actual data. This involved separating all data into a training set, prior to 2020, and samples from 2022 onwards for historical back-testing. All statistics have been calculated at both sample and stope scales. The details and results of the historical back-testing process are presented in Clarke and Minniakhmetov (2023). To implement this modification, two new steps were added to the workflow: cross_validation_setup and cross_validation, while no other steps were modified (Figure 5d).

The SBRE framework's library of data converters and geostatistical operations is expanding, offering the potential for further workflow updates, extensions, experimentation, and improvement in the future.

Computational environment

In the workflow, different modes of computation can be selected for each step, such as running locally on a single machine or in parallel on AWS EC2-Batch Service. Depending on the mode chosen, data is automatically uploaded or downloaded from S3 buckets. For example, load_data, domaining, and variography can be done on a local machine, while computationally intensive steps like trend modelling, PPMT, Gaussian simulations, and back transformation are performed on the cloud. Results are then automatically downloaded back.

For OD simulation, due to the large amount of input data (~3 GB) and output data (~500 GB), all steps are run on the cloud and stored on S3 buckets. Validations and upscaling are also performed on the cloud. Users can inspect statistical graphs and download upscaled models during multiple trial runs until the desired quality is achieved, at which point 100 realisations can be downloaded from the cloud in CSV format.

The computation of 100 realisations, including 100 categorical and 100 grade simulations, took two days using 20 extra-large cloud instances (32 CPUs and 64 GB of RAM). It should be noted, that not all workflow steps can be efficiently parallelised, and data transfer from worker cloud instances to the master cloud instance takes up to 30 per cent of the calculation time. Timing can be further improved by increasing the number of cloud instances to 40, but this is currently not possible due to RMSP licensing constraints.

Validation of simulations

Various univariate, multivariate, and spatial validation techniques have been utilised in the study, such as grade-tonnage curves, histograms, swath plots, and variogram reproductions. While there are various techniques available to correct proportions and histograms, the most reliable validation technique is cross-validation, including back-testing with historical samples. Figure 7 shows accuracy plot (Goovaerts, 2001), classical cross-validation plot, and variogram reproduction at validation drill holes. It is important to note that the uncertainty presented by the resulting realisation is slightly underestimated, as indicated by the steep slope of the accuracy plot, and there is a slight bias in the average Cu values. This is attributed to the use of fixed trend, variogram, global histograms, and mineralisation type categories. The incorporation of uncertainty in these parameters is expected to improve the quantification of uncertainty.

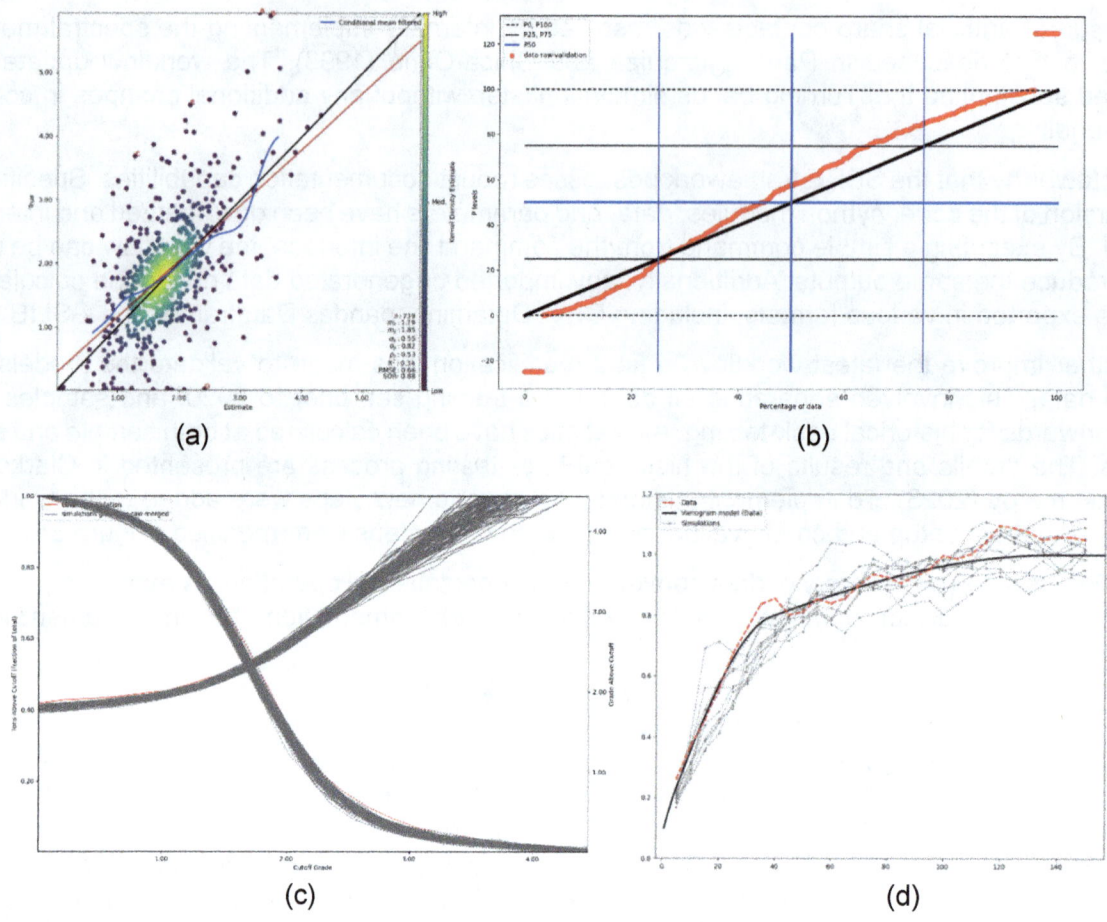

FIG 7 – Cross-validation graphs at stope-scale. (a) Classical cross-validation graph comparing average of simulations with actual values. Each point represent estimated and true value of average grade within a stope. (b) accuracy plot. X-axis is for proportion of data, Y-axis for corresponding P-value in simulations. (c) Copper grade-tonnage curve created from average grade in each stope. Red line represents actual values, whereas grey lines show average grade in stopes for each realisation. (d) Variogram for copper created from validation samples (red line) and from each realisation (grey lines) at validation samples locations.

An additional limitation of the approach is the tendency to underestimate average copper grades, as indicated by the validation data statistics (the red line) falling outside of the range of simulated values (represented by the grey lines) and the p50 value not aligning with the 50 per cent mark on the accuracy plot. This issue arises from the use of geostatistical techniques that rely on Gaussian simulations, which are unable to capture the spatial connectivity of high values, as noted in previous studies by Minniakhmetov, Dimitrakopoulos and Godoy (2018) and de Carvalho, Dimitrakopoulos and Minniakhmetov (2019).

CONCLUSIONS

This study demonstrates the successful application of the SBRE framework to model the Olympic Dam deposit, one of the world's largest copper and uranium deposits. The challenges associated with modelling such a complex deposit were overcome by utilising a combination of geostatistical methods, including PPMT, trend-residual decorrelation, pluri-Gaussian simulations, and turning-bands simulations. The resulting workflow was built using a range of tools and applications, including in-house developed software, Autovariogram software (Varify), and the RMSP python library, and was run on the AWS cloud using the EC2-Batch service.

The study highlights the benefits and the importance of a standardised and automated framework in the development of conditional simulation workflows and describes the learning gained from trials of different combinations of methods and software. The proposed workflow can provide a more accurate estimation of resources and better understanding of the risks and uncertainties associated with the mineral value chain.

Mineral Resource Estimation Conference 2023 | Perth, Australia | 24–25 May 2023

The SBRE framework offers a simple, flexible, and scalable solution for creating conditional simulation workflows, which can be applied to a range of deposits. The framework provides a standard approach for testing, building, and running conditional simulation workflows, allowing for the combination of various applications and documentation of key considerations and parameters. It is also suitable for cloud-based computations, improving computational time and facilitating the design of the optimal combination of geostatistical tools.

The forthcoming research endeavours will concentrate on utilising the framework's ability to taking into account various sources of uncertainty, including global distribution uncertainty and parameter uncertainty, enabling comprehensive assessment of uncertainties and sensitivities. This will render the framework an indispensable instrument for making well-informed decisions amidst uncertainty within the mining industry.

ACKNOWLEDGEMENTS

The authors would like to thank BHP for the data and collaboration and Resource Modeling Solutions for developing the back-end of the SBRE platform.

REFERENCES

Barnett, R M, Manchuk, J G and Deutsch, C V, 2014. Projection pursuit multivariate transformation, *Mathematical Geosciences*, 46:337–359.

Clark, J M and Ehrig, K J, 2019. What controls high-grade copper mineralisation at Olympic Dam?, in *Proceedings of the International Mining Geology Conference 2019*, pp 222–236 (The Australasian Institute of Mining and Metallurgy: Melbourne).

Clarke, D and Minniakhmetov, I, 2023. Benchmarking and cross validation of the multivariate conditional simulation model of the Olympic Dam deposit, in *Proceedings of the Mineral Resource Estimation Conference 2023*, pp 299–314 (The Australasian Institute of Mining and Metallurgy: Melbourne).

de Carvalho, J P, Dimitrakopoulos, R and Minniakhmetov, I, 2019. High-Order Block Support Spatial Simulation Method and Its Application at a Gold Deposit, *Math Geosci*, 51:793–810. <https://doi.org/10.1007/s11004-019-09784-x>

Desbarats, A J and Dimitrakopoulos, R, 2000. Geostatistical Simulation of Regionalized Pore-Size Distributions Using Min/Max Autocorrelation Factors, *Mathematical Geology*, 32:919–942. doi: 10.1023/A:1007570402430

Deutsch, C V and Journel, A G, 1998. *GSLIB: Geostatistical software library and user guide* (Oxford University Press).

Ehrig, K, Kamenetsky, V S, McPhie, J, Cook, N J and Ciobanu, C L, 2017. Olympic Dam iron oxide Cu-U-Au-Ag deposit, in *Australian Ore Deposits* (ed: G N Phillips) pp 601–610 (The Australasian Institute of Mining and Metallurgy: Melbourne).

Goovaerts, P, 2001. Geostatistical modelling of uncertainty in soil science, *Geoderma*, 103(1–2):3–26. Available from: <https://doi.org/10.1016/S0016-7061(01)00067-2>

Journel, A G and Alabert, F, 1990. New method for reservoir mapping, *Journal of Petroleum Technology*, 42(2):212–218.

Mantoglou, A and Wilson, J W, 1982. The Turning Bands Methods for Simulation of Random Fields Using Line Generation by a Spectral Method, *Water Research*, 18(5):1379.

Minniakhmetov, I and Ashford, J, 2022. SBRE – a simple, flexible and scalable framework to perform conditional simulations for uncertainty quantification, in *Proceedings of the International Mining Geology Conference 2022*, pp 462–469 (The Australasian Institute of Mining and Metallurgy: Melbourne).

Minniakhmetov, I and Clarke, D, 2022. Utilising local varying anisotropy in the multivariate conditional simulation model of Olympic Dam deposit, 21st Annual IAMG Conference 2022, International Association for Mathematical Geosciences.

Minniakhmetov, I, Dimitrakopoulos, R and Godoy, M, 2018. High-Order Spatial Simulation Using Legendre-Like Orthogonal Splines, *Math Geosci*, 50(7):753–780. <doi: 10.1007/s11004-018-9741-2>

Pardo-Igúzquiza, E and Chica-Olmo, M, 1993. The Fourier Integral Method: An efficient spectral method for simulation of random fields, *Mathematical Geology*, 25:77–217. Available from: <https://doi.org/10.1007/BF00893272>

Qu, J and Deutsch, C V, 2018. Geostatistical simulation with a trend using gaussian mixture models, *Natural Resources Research*, 27(3):347–363.

Silva, D and Deutsch, C V, 2019. Multivariate Categorical Modeling with Hierarchical Truncated Pluri-Gaussian Simulation, *Math Geosci*, 51:527–552, doi:10.1007/s11004-018-09782-5

van den Boogaart, K G, Tolosana-Delgado, R and Mueller, U, 2015. An affine equivariant anamorphosis for compositional data, in *Proceedings of IAMG 2015 – 17th Annual Conference of the International Association for Mathematical Geosciences*, pp 1302–1311.

Open to anything

Benchmarking and cross validation of the multivariate conditional simulation model of the Olympic Dam deposit

D Clarke[1] and I Minniakhmetov[2]

1. MAusIMM(CP), Principal Resource Geologist, BHP Olympic Dam, Adelaide SA 5000. Email: david.clarke@bhp.com
2. MAusIMM, Principal Geoscientist Data and Numeric Modelling, BHP Geoscience Centre of Excellence, Perth WA 6000. Email: ilnur.minniakhmetov@bhp.com

ABSTRACT

The Olympic Dam deposit is truly a world-class IOCG-Ag deposit, the world's fourth largest deposit of copper and the world's largest known single uranium deposit. Traditional and non-traditional methods have been used to accurately model the geology and mineralisation of the Olympic Dam deposit. Conditional simulation models have long been used as the most robust method for estimating risk and uncertainty in mine planning decisions at other mines around Australia and the world.

There are many challenges associated with simulating the Olympic Dam deposit at a selective mining unit size sufficient for underground mine planning purposes. These include the size of the deposit, resulting in hundreds of millions of grid blocks; the disseminated nature of the zoned sulfide mineralisation. And lastly the direct relationship of the relative abundances of the sulfide mineral species that are crucial to the Olympic Dam smelter performance.

Several geostatistical techniques have been utilised in the workflow for the Olympic Dam Conditional Simulation model to address these challenges. The key techniques are:

- Projection Pursuit Multivariate Transform to de-correlate the main elements before simulation.
- Pluri-Gaussian implicit boundary simulation method to simulate the mineralisation domains.
- Spectral simulation method of de-trended and de-correlated attributes with account of local varying variance.

The resulting conditional simulation models have been validated against histograms, grade-tonnage curves, cross-plots and assessed for variogram reproduction. This paper presents how the final conditional simulation models compare to benchmarking models created using alternate estimation techniques of ordinary kriging and local uniform conditioning. It also presents a validation process using historical samples to test final estimate accuracy of the simulation models.

INTRODUCTION

The Olympic Dam deposit is truly a world-class iron oxide-copper-gold deposit (IOCG) and was the first identified of its kind. The breccia-hosted deposit is one of the largest in the world and contains copper, uranium, gold and silver in economic quantities. The controls on mineralisation are well understood and have been studied in numerous publications, with the predominant controls on mineralisation being hematite alteration followed by structural and lithological controls (Clark and Ehrig, 2019).

Essentially, the most prospective zones for Cu-U-Au-Ag in the deposit are preserved within the shallower lateral parts of the mineralised system. Therefore, understanding and mapping out the deformation history is important for identifying the relevant high-grade mineralisation. These mappable zones can show complex geometries and are fundamental in honouring the high-grade mineralisation.

Mining at Olympic Dam is by means of underground sublevel open stoping. Therefore, the overall feasibility of this deposit is dependent on accurately defining the higher-grade portions of the deposit for extraction (Badenhorst, O'Connell and Rossi, 2016). The primary objective of the resource estimation in this context is to produce block estimates at a high enough resolution that mining engineers can develop accurate stope designs and associated mining schedules.

The Olympic Dam deposit poses many challenges to conventional estimation methods, such as complex alteration, structural architecture, mixed populations and non-stationarity. The motivation for taking this approach is to address these complexities and the following aspects and limitations of traditional linear estimation techniques:

- estimation smoothing effect
- element cross-correlation representativeness
- grade-tonnage curve representativeness
- conditional bias.

Advancements in geostatistical applications and computer hardware allow conditional simulations to be run at a global resource scale, hence this approach is being used as demonstrated in this study.

Conditional simulation is a method of producing images or realisations using stochastic modelling techniques that do not involve local averaging as per common grade interpolation methods, such as kriging. Geostatistical simulation is a spatial extension of the concept of Monte Carlo simulation and provides a method to quantify uncertainty and to minimise risk. Most geostatistical simulation methods generate grade estimates on a dense grid pattern conditional to known sample data at sampled locations that honour the input sample statistics.

The simulation realisations aim to reproduce the global histogram and variogram, and their short-scale variability, at a scale below or close to the drill hole sample spacing. This provides additional flexibility for modelling the actual in situ grade variability which can then be simulated and modelled at a scale suitable to the mining operation and planning. Conditional simulations can also offer greater flexibility for selective mining unit (SMU) emulation for resource modelling of grades at the SMU scale and can be directly modelled by upscaling simulations to SMU blocks, bypassing change of support models and assumptions.

In conjunction with the BHP Resource Centre of Excellence, the development and implementation of a conditional simulation model for the Olympic Dam deposit was first completed in March 2020 to facilitate risk-based analysis of the resource. The original conditional simulation model used a combination of decorrelation technique projection pursuit multivariate transform (PPMT) (Barnett, Manchuk and Deutsch, 2014) and univariate conditional simulation utilising spectral simulation method from the SPECSIM program (Fourier Integral Method, FFT-MA) (Pardo-Igúzquiza and Chica-Olmo, 1993). The updated conditional simulation constructed during 2022 has been further expanded to include pluri-Gaussian domain simulation (Silva and Deutsch, 2019) for the boundaries between sulfide and non-sulfide as well as the bornite and chalcopyrite interface. The complex geometries of the hematite alteration that often underpins high-grade copper mineralisation is better represented using these three simulated boundaries.

The main objective of the Olympic Dam condition simulation model is to gain an understanding of the spatial grade changes expected from infill drilling through to Measured Resource category (nominally 20 m × 20 m) spacing, for evaluation of risk and uncertainty in strategic underground mine planning and financial valuation purposes. Stochastic modelling (conditional simulation), is the only method technically suited for application in underground mining that can provide this grade resolution. Stochastic modelling can give a view of the global resource (grade/tonnage and a spatial representation) which is independent of resource confidence.

To measure the performance of the conditional simulation model it is compared to conventional deterministic approaches such as ordinary kriging (OK) (Journel and Huijbregts, 1978) and localised uniform conditioning (LUC) (Abzalov, 2006). Apart from, standard validation tools such as grade-tonnage curves, histograms, swath plots, and variogram reproductions, the models are compared against samples from new drillings, ie historical back-testing/reconciliation.

DEPOSIT GEOLOGY

The Olympic Dam deposit is a typical example of an IOCG and was the first of its type discovered in 1975 by Western Mining Corporation Ltd (Reeve *et al*, 1990). The deposit formed 1.59 Ga in the upper 1–2 km of the earth's crust by tectonic-magmatic-hydrothermal processes. It is primarily hosted within the Olympic Dam breccia complex that occurs entirely within the Roxby Downs granite

(1.59 Ga). The deposit is overlain by some 300 to 400 m of flat-lying, unaltered and unmineralised sedimentary units. Figure 1 displays a simplified basement geological map showing the biotite-out contour which depicts the transition from relatively unaltered Roxby Downs granite (RDG) in the Olympic Dam breccia complex (ODBC) and is defined by the last occurrence of magmatic biotite (Clark and Ehrig, 2019).

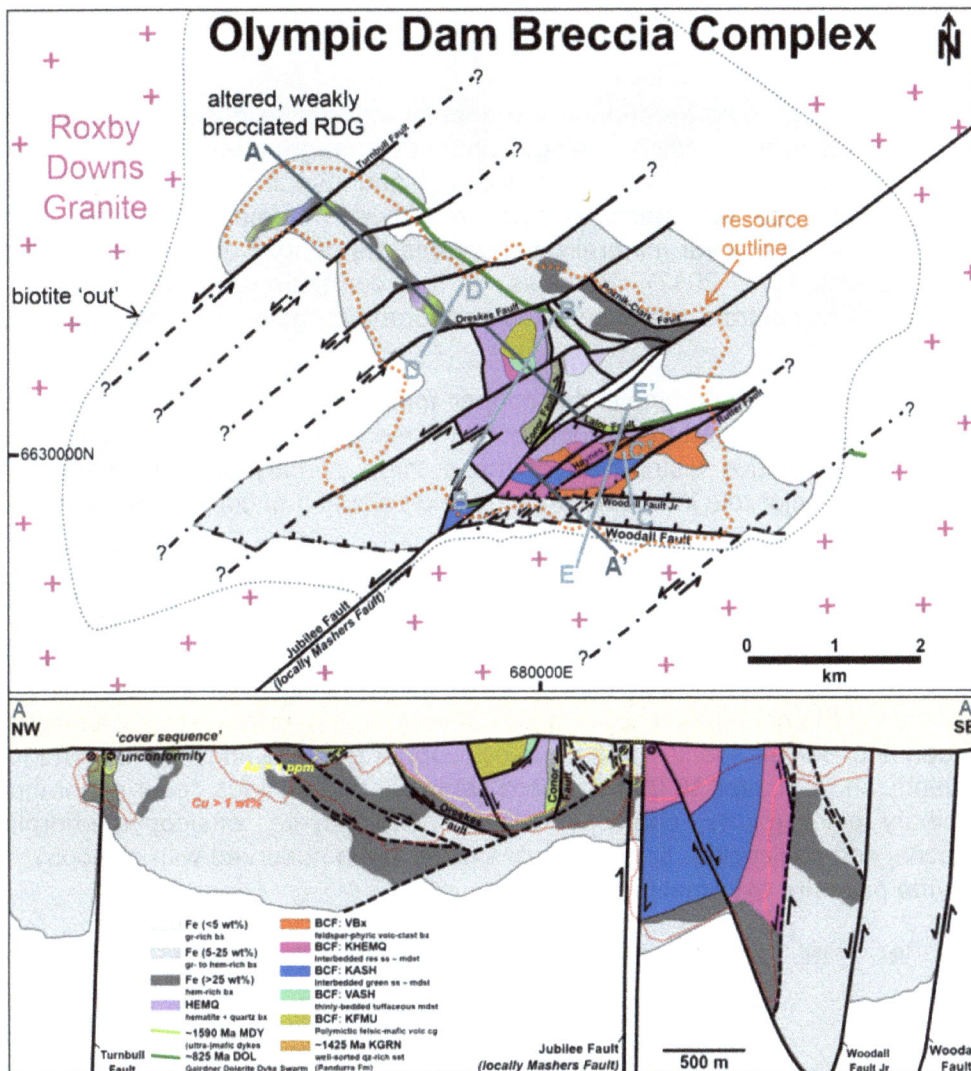

FIG 1 – Simplified basement geological map (at 450 m depth) of the Olympic Dam deposit.

The Roxby Downs granite (RDG) has been extensively altered by brecciation and hydrothermal alteration to form the Olympic Dam breccia complex (ODBC). The primary minerals in the granite have been altered and replaced by hematite-magnetite, quartz, sericite, siderite, chlorite, fluorite, barite, and sulfides. To a lesser extent the deposit also contains various other lithologies such as bedded sedimentary facies, and mafic/ultramafic dykes and lavas which are pre-, syn- and post-mineralisation.

This package of differing lithologies within the ODBC have all experienced multiple brittle structural deformation events, intense brecciation, and/or hydrothermal alteration to varying degrees of intensity. The range of brecciation can scale from relatively coherent 'facies' to irregular, polymictic and up to a chaotic polyphase breccia. The hematite alteration has variably replaced the primary lithology by iron oxides – forming a compositional continuum from where the original protoliths remain – up to total iron oxide breccias, where the original lithology texture is completely replaced by iron oxide alteration. Lithological contacts are typically obscured by faulting and brecciation. Where boundaries are sharp, there is either a fault contact or a dyke or vein that has intruded the breccia. The ODBC is predominantly characterised by a granite-hematite breccia continuum recognised to represent a RDG protolith. The alteration minerals show a zonation laterally and vertically across the deposit.

In general, all metals are spatially well correlated, although there are some gold-only occurrences. The metal distribution is characterised by short-scale variability. The Olympic Dam deposit also features a central large barren hematite-quartz breccia zone with very low concentrations of copper, uranium, and silver.

Mineralisation controls

The key factors controlling mineralisation in the deposit are sulfide mineral type and abundance, hematite alteration intensity, protolith/lithology, and structure/geometry. These processes work together and directly control both the occurrence and abundance of mineral grades across the deposit. There is evidence for spatial correlation between hematite, copper, uranium, gold (associated with sulfide) and silver mineralisation across the deposit due to the co-precipitation of these elements (Ehrig *et al*, 2017). Because of the correlation of the controls on copper mineralisation the same controls are associated with uranium, gold (associated with sulfide) and silver mineralisation.

The sulfide assemblages that dominate the copper mineralisation are chalcopyrite, bornite and chalcocite. They are characterised by strongly zoned dissemination across the mineralised system at Olympic Dam. The relative distribution of these three primary sulfide minerals have been established to play the major role in understanding the distribution of copper grade at Olympic Dam.

The four main dominant copper-bearing sulfide minerals display a noticeable upwards and inwards zonation from pyrite → chalcopyrite → bornite → chalcocite. Iron displays a positive correlation with increasing Cu, U_3O_8; and Au with increasing Fe. However, this is only true up to 40–45 wt. per cent or approximately 75 wt. per cent hematite, after which there is a shape decline (Clark and Ehrig, 2019) (Figure 2). This shows two fundamental relationships: (i) sulfide mineral species fundamentally control the contained copper grade, and (ii) hematite alteration intensity controls the relative abundance of sulfide species and therefore copper grade. Ehrig, McPhie and Kamenetsky (2012) also clearly show in over 10 000 drill core samples, that sulfides occur either individually or in discrete binary pairs (pyrite, chalcopyrite-pyrite, chalcopyrite, chalcopyrite-bornite, bornite, bornite-chalcocite, and chalcocite). Chalcocite has never been observed with chalcopyrite or pyrite, and bornite-pyrite pairs are extremely rare.

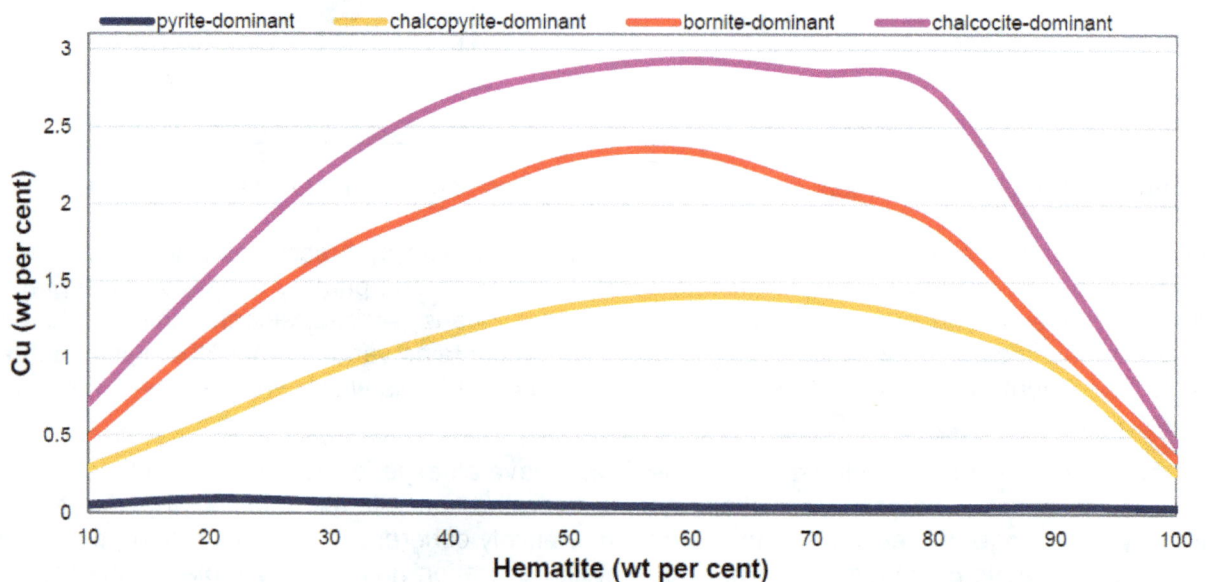

FIG 2 – Global conditional means plot showing the distribution of dominant sulfide species against hematite and copper.

Structure/geometry

There is an obvious and underlying structural control to the mineralisation, as observed from the dislocation of the high-grade hematite alteration and associated Cu-Au mineralisation (Clark,

Passmore and Poznik, 2017). Notwithstanding that early faults are typically isolated, discontinuous, fragmented or annealed, the relatively continuous nature (hundreds of metres of strike) of orebody geometries, sharp geochemical fronts and observation of coherent ca. 1590 Ma mafic-ultramafic dykes intruding along planar discontinuities, are strong indicators of the presence of pre- and syn-mineralisation structures (Clark *et al,* 2018). Some significant differences in continuity, geometry and extent of mineralisation are observed at the local scale. The deposit can thus be divided into localisations of several unique geometrical areas based on overall continuity and orientation of observed mineralisation.

Protolith/lithology

The original Roxby Downs granite host plays a relatively small and insignificant role in controlling mineralisation. Its main function is essentially that of a host rock for brecciation, which in turn provides the conduits for the mineralising fluid along structural highways and sites for metal precipitation.

CONDITIONAL SIMULATION APPROACH

Overview

The conditional simulation model has been completed using a combination of current best practices for simulating complex deposits, such as PPMT and spectral grade simulation. The general workflow is shown in Figure 3 and is summarised as such:

- Data preparation and loading: dividing the deposit into spatial domains for modelling based on continuity and anisotropy of Cu; loading composites, creating simulation grid, spatial domains flagged onto both samples and simulation grid.

- Spatial modelling of sulfide mineralisation zones (chalcopyrite, bornite-chalcocite, and non-sulfides) using truncated pluri-Gaussian simulation: declustering, trend modelling, varyography, pluri-Gaussian simulations.

- Spatial modelling of grades for Au, Barium-adjusted sulfur (BaS), Cu, bulk density, U_3O_8: declustering (Deutsch and Journel, 1998), trend-residual transformation (Qu and Deutsch, 2018), variography, PPMT, spectral simulation, and all back-transformations.

- Validation and reconciliation with new drillings, and benchmarking with OK and LUC.

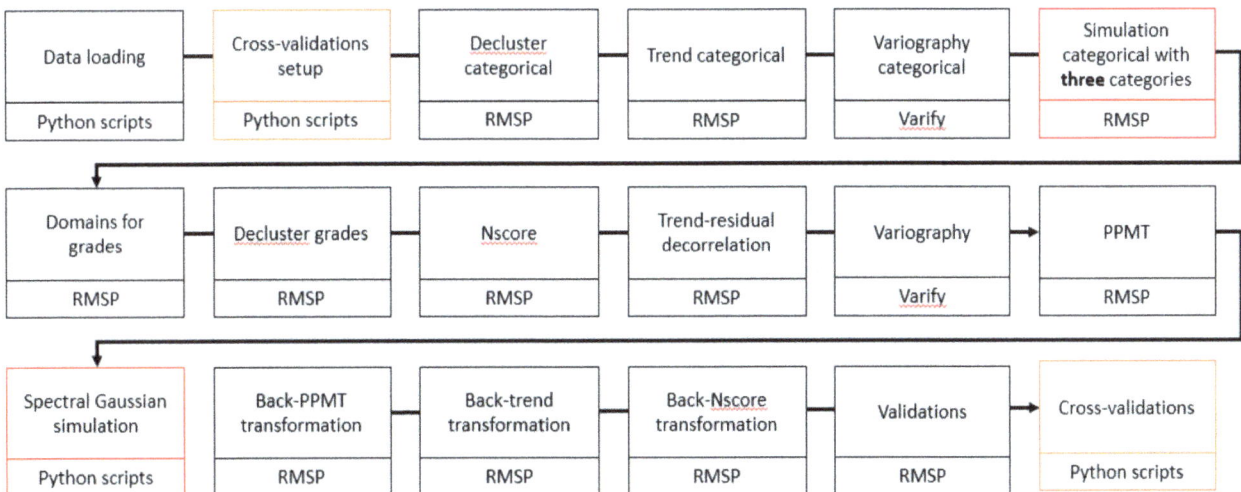

Data loading	Cross-validations setup	Decluster categorical	Trend categorical	Variography categorical	Simulation categorical with **three** categories
Python scripts	Python scripts	RMSP	RMSP	Varify	RMSP

Domains for grades	Decluster grades	Nscore	Trend-residual decorrelation	Variography	PPMT
RMSP	RMSP	RMSP	RMSP	Varify	RMSP

Spectral Gaussian simulation	Back-PPMT transformation	Back-trend transformation	Back-Nscore transformation	Validations	Cross-validations
Python scripts	RMSP	RMSP	RMSP	RMSP	Python scripts

FIG 3 – General workflow for conditional simulation methodology.

The workflow is designed and implemented using internal BHP SBRE framework (Minniakhmetov and Clarke, 2023) that provides a standard approach for testing, building, and running conditional simulation workflows, allowing for the combination of various applications and documentation of key considerations and a record of all input parameters.

Overall, 100 independent realisations are produced that reflects the uncertainty in both the mineralisation boundaries and grades. Using local uncertainty test, or accuracy plots (Goovaerts, 2001), of the simulations it has been confirmed that 100 realisations are enough to reasonably capture full extent of uncertainty evident in both the mineralisation boundaries and grade uncertainty, however other sources of uncertainty should be integrated in future work to capture the full extent of the uncertainty not currently measured, such as variogram parameters, trend, data quality, and others.

Domaining

Domaining the key zones in the deposit utilises grade continuity and anisotropy to best define stationary domains for simulation. There were 38 structural domains created based on major deposit-scale faults and fault zones. Some of the smaller fault bounded domains with minimal sampling were added to an adjacent domain to meet a minimum number of samples required for simulation. Figure 4 shows the plan view of the modelling area and coloured by fault block domains.

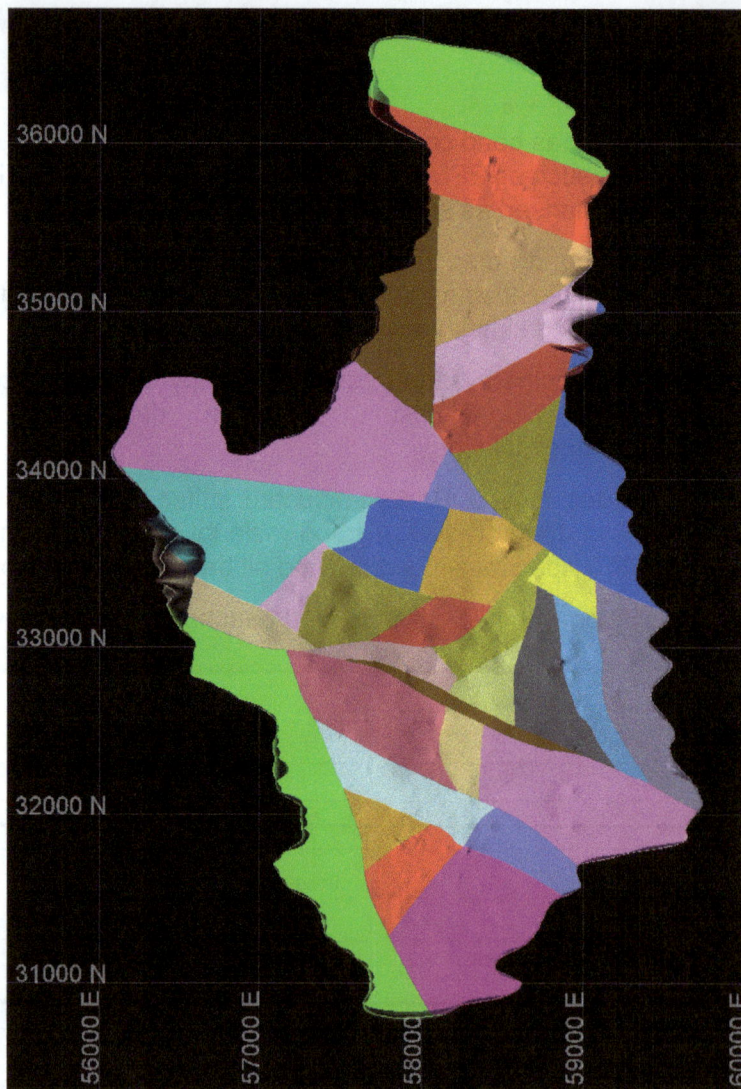

FIG 4 – Plan view of domains in mine plan grid. Section at Z = -450 m.

Mineralisation zone simulation

Three mineralisation zones, chalcopyrite, bornite-chalcocite, and non-sulfides/pyrite are modelled using a hierarchical truncated pluri-Gaussian simulation approach described in detail by Silva and Deutsch (2019).

The input data for the mineralisation zones is based on indicator data defined from the 5 m composites drilling data for the three sulfide zones. The samples Cu:S ratio is used to calculate the proportion of each of the sulfide species and to code by the dominant species.

Resulting realisations are validated using visual inspection of various sections of the deposit and comparing with geological wireframes, local knowledge, vertical proportions, and global proportions (Figure 5). Bornite-chalcocite zones are generally found up against the core non-sulfide zone as is geological consistent with the alteration profile and is surrounded by the boarder chalcopyrite zones.

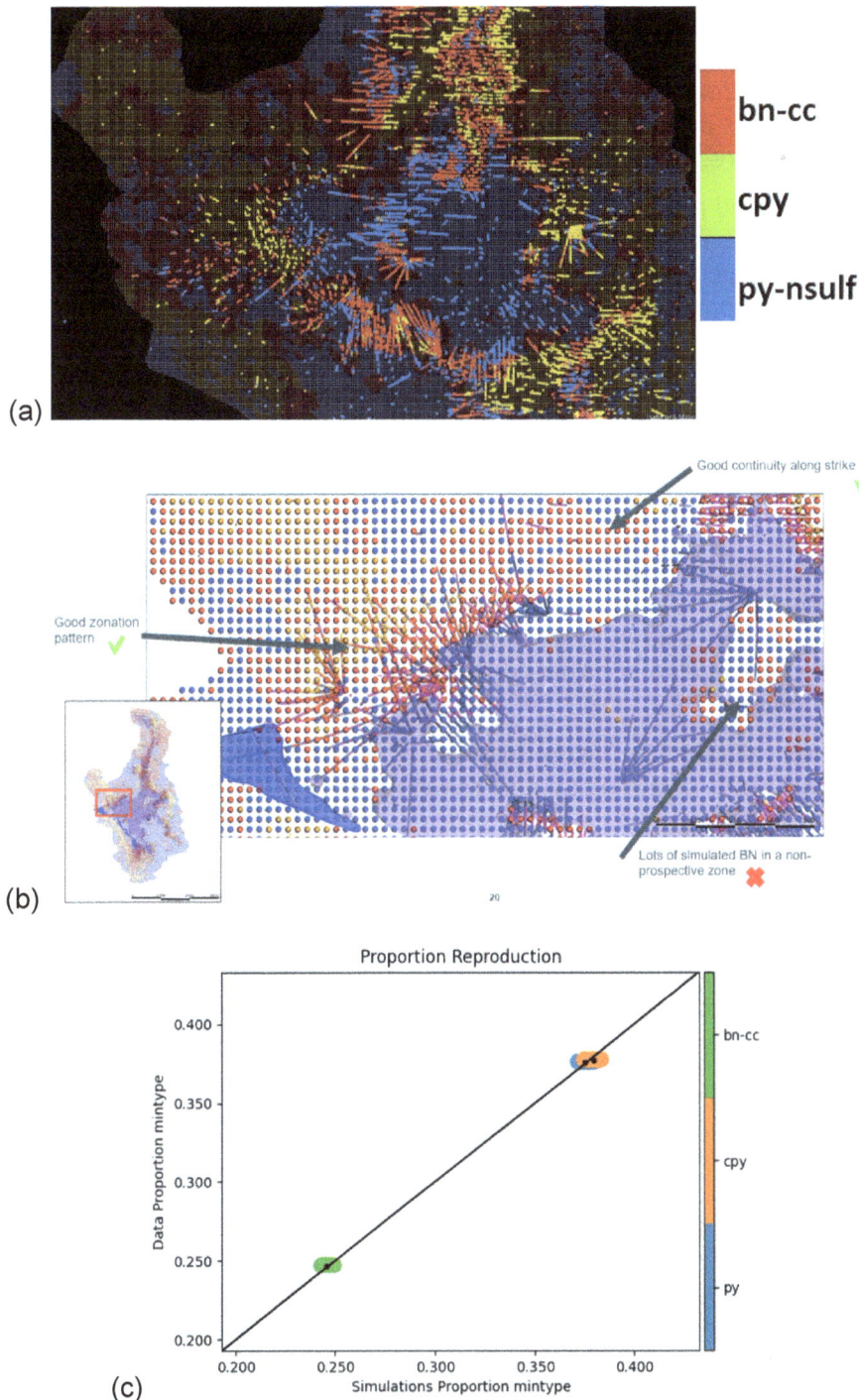

FIG 5 – Categorical simulation: (a) example of realisation; (b) example of visual validation using local knowledge; (c) global proportion reproduction.

Grade simulation

In this study, 100 scenarios of domains for grade simulations are produced by combining each realisation of categorical simulation with spatial domains definition (Figure 4). For each simulation domain, grade simulation is independently conducted using declustering, normal score (Deutsch and Journel, 1998), trend-residual transformation (Qu and Deutsch, 2018), variography, PPMT, spectral simulation, and all back-transformations (Figure 3).

To address non-stationarity, the trend-residual transformation approach is deemed critical. This approach differs from the classical method of subtracting local mean from the samples (Isaaks and Srivastava, 1989), which has a main limitation of not ensuring positivity at all locations or other important constraints of residuals (Leuangthong and Deutsch, 2004). The level of variability represented by the trend must be selected adequately. Over-fitting or under-fitting the trend is neither desirable. Over-fitting the deterministic component may result in too little variability in the residuals, leading to stationary models that do not reproduce the observed geologic trend. In this work, the optimal smoothing of the trend is chosen based on a p-value accuracy metric by selecting the size of the moving window that provides a slope of the p-value versus data-proportion accuracy graph close to 1, ie neither underestimating nor overestimating variability.

To transform the correlated residuals into multivariate Gaussian space with independent variables, or factors, the PPMT approach is utilised. Then, independent simulations can be carried out, which yields realisations that match the distribution of the transformed data, and the back transformation reintroduces the original complexity and cross-correlation.

Spectral simulation (Emery and Arroyo, 2018) is employed to conduct Gaussian simulations of decorrelated factors, allowing for a smooth transition of grades from one domain to another, without creating artificial sharp contacts on the boundary of domains as using the turning bands simulation method.

BENCHMARKING

The conditional simulation model is compared with OK and LUC models on a representable spatial domain, for this study the comparison will focus on spatial domain 32. This domain was chosen as it has a mix of well-informed area by drill samples and a more sparsely drilled area. The geological complexity in this domain is also considered representative of the deposit as well as the alteration profile.

All three models are validated by visual inspection of the sections, univariate statistics, grade-tonnage curves, cross-plots, swath-plots, and assessed for variogram reproduction.

Visual validation

Figures 6a and 6b show two realisations of conditional simulation model. They exhibit high contrast in grade texture and replicates distribution of high and low values of evident in the data. The spatial continuity varies among simulations with continuous horizontal high-grades in first figure, to rather isotropic shapes in second figure. Conditioning is adequate and vertical trending of high copper in shallow parts transitioning to low copper grades in deeper parts is preserved.

The ordinary kriging model in Figure 6c has high degree of smoothing and misrepresents variability of grades. However, it provides best guess of copper grades in each location and does not exhibit conditional bias.

LUC model (Figure 6d) has high contrast of grade texture similarly to realisations, however, those high-grade areas are highly connected (highlighted by yellow circle) and do not reflect spatial continuity of copper observed in data. Unlike to conditional simulation model where location of high-grade zones varies among simulations, those high values in LUC model are located next to the samples with high-grade values, hence creating conditional bias. Besides that, LUC model exhibits edge artefacts (white circle) due to the panel size selected for conditioning, which are not supported by surrounding data.

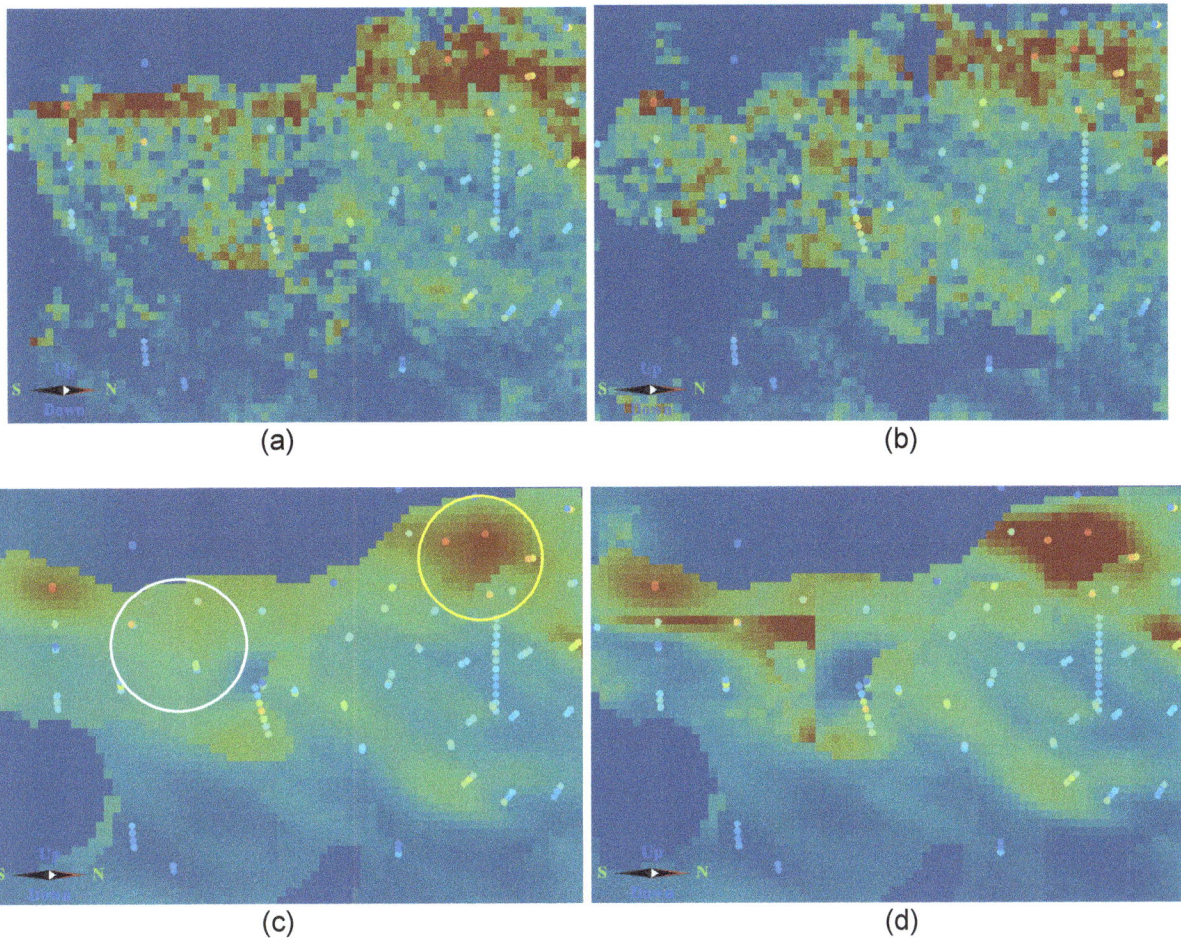

FIG 6 – Vertical sections for copper: (a) Realisation #1; (b) Realisation #2; (c) ordinary kriging; (d) LUC.

Statistical validation

A statistical validation process was carried out for the different methods to compare against the conditional simulation models. Statistics for the models and declustered data samples are shown in Table 1. All models are built on 5 × 5 × 5 m blocks which is considered as sample support given that minimum stope size at Olympic Dam is 30 × 30 × 40 m. For realisations all statistics are calculated for each realisation and then minimum, mean, and maximum of those statistics are reported in the table.

TABLE 1

Statistics of the data samples and models for Cu grades.

	Data	Simulations minimum	Simulations average	Simulations maximum	OK	LUC
mean	0.51	0.45	0.48	0.51	0.47	0.47
stdev	0.72	0.67	0.70	0.74	0.55	0.64
cv	1.42	1.42	1.48	1.55	1.17	1.36
min	0	0	0	0	0	0
P10	0.01	0.01	0.01	0.02	0.03	0.02
P50	0.22	0.14	0.17	0.2	0.28	0.21
P90	1.36	1.26	1.32	1.42	1.22	1.21
max	9.35	8.6	9.33	9.35	5.91	8.94

All data samples statistics are within minimum and maximum values from simulations statistics.

Each of the model methods reproduce well the average grade. However, OK misrepresents the total variance by a large amount due to the smoothing effect. Although LUC is built to mitigate smoothing limitations of linear estimation such as OK, it still produces lower variance than exhibited in the original sample data, as is noticeable in P90 and maximum values reported by OK and LUC.

Table 2 shows the statistics at Cu cut-off 1.5 per cent which is the representable cut-off used in stope design decisions. Above a cut-off the same issues are present in both the OK and the LUC where the simulation preforms better. Simulations reproduce well all the statistics including tonnages above cut-off which is reflected by proportion of blocks above the cut-off in the first row of Table 2. OK and LUC are shown to vastly underestimate tonnages above cut-off, ~30 per cent and ~20 per cent of tonnes mismatch relative to data statistics, respectively. Average grade is also much lower in the OK model (~12 per cent mismatch relative to data statistics) whereas LUC model predicted well the average grade above the cut-off.

TABLE 2

Statistics of the data samples and models for Cu>1.5 per cent.

	Data	Simulations minimum	Simulations average	Simulations maximum	OK	LUC
proportion	0.09	0.07	0.08	0.09	0.06	0.07
mean	2.34	2.28	2.32	2.43	2.04	2.31
stdev	0.93	0.82	0.88	1.04	0.53	0.84
cv	0.40	0.36	0.38	0.43	0.26	0.36
min	1.50	1.50	1.50	1.50	1.50	1.50
P10	1.57	1.57	1.58	1.60	1.56	1.58
P50	2.07	2.02	2.06	2.11	1.89	2.03
P90	3.52	3.25	3.39	3.61	2.70	3.42
max	9.35	8.60	9.33	9.35	5.91	8.94

The statistics above is further reinforced by the validation graphs in Figure 7 which shows the tonnes and grade curves for both Cu and BAS as well as the swath plot and variogram reproduction plot.

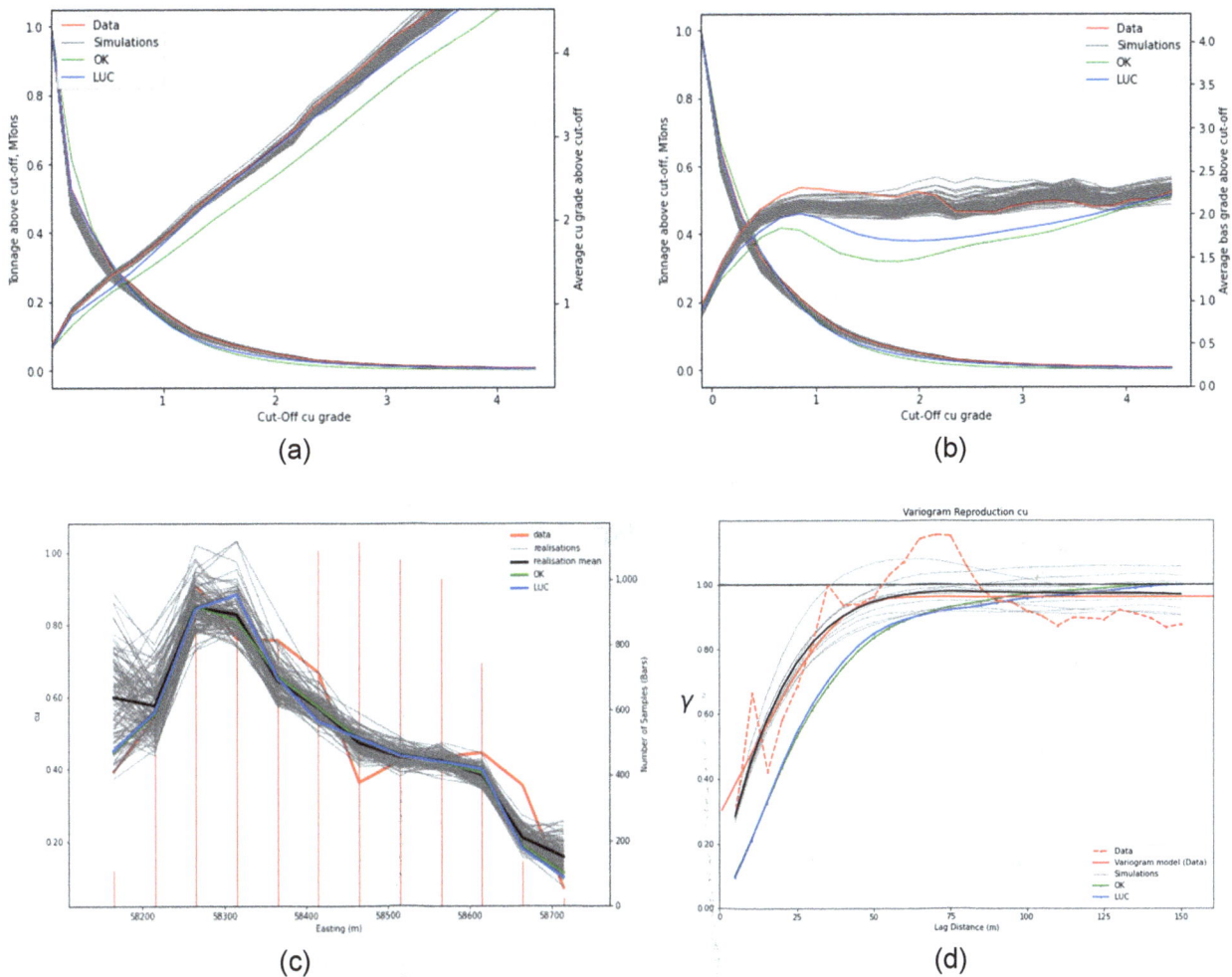

FIG 7 – Spatial simulation of grades in one spatial domain: (a) GT_cu; (b) BaS grade-tonnage curve based on copper cut-off; (c) swath-plot; (d) variogram reproduction (standardised). Red line is for data samples, grey lines are for realisations, green line is for OK, and blue line is for LUC.

According to the validation graphs, for the simulations (grey lines in Figure 7) all univariate relationships are reproduced, as evidenced by the matching grade-tonnage curves (Figure 7a). Additionally, multivariate relationships are also successfully replicated, demonstrated by honouring cross-variate grade-tonnage curve in (Figure 7b). The method is also able to address non-stationarity, as shown by reproduction of swath plots (Figure 7c) and the visual change of grades from the high-grade area to lower grades (Figure 6a to 6b). Lastly, variogram graphs in Figure 7d confirms that spatial continuity is honoured well in the simulations models.

Ordinary kriging model (green lines in Figure 7) exhibits high degree of smoothing and misrepresents amount of high-grade tonnes and average grade above non-zero cut-offs (Figure 7a). Cu:S relations are poorly reproduced at 1–2 per cent Cu cut-off range (Figure 7b) which is the cut-off used to make design decisions of stopes. Such a mismatch would affect all downstream processes and would incur risk for copper production targets, affect processing plant performance, creates additional costs for reactive actions such as material movements, potential operational delays and smelter operation adjustments.

Trends are captured well in OK model (Figure 7c), However, copper continuity is vastly overstated (variability misrepresented) due to smoothing effect of kriging. Figure 7d shows variogram reproduction in standardised by sill units to demonstrate smoothing effect of models apart from variance deflation (Table 1). OK model variogram confirms longer continuity ranges and high short-scale continuity.

LUC model (blue lines in Figure 7) provides slightly lower grade-tonnage curves (exact numbers are presented in Table 2), however, similarly to OK model, LUC model misrepresents Cu:S relations that creates downside implications highlighted above. Trends are also captured well in LUC model

(Figure 7c). Finally, the variograms from the LUC model (Figure 7d) are almost identical to the OK model. This continuity, combined with the higher contrast of grades (higher grades in the high-grade area and lower grades in the low-grade area), creates a high degree of conditional bias (McLennan and Deutsch, 2002). Highly continuous connected volumes of high-grades will be considered in the mine planning process as high revenue and low cost areas, which are not confirmed by orebody knowledge or any data.

Validation with new samples

To validate the models, data from the drilling campaign from 2020 to 2022 was used as a validation data set, while the models were developed using the July 2020 data set. The validation samples are identified based on the stope ID from the 5-year plan as of July 2020, and their locations are shown in Figure 8, where the white dots represent the data used for modelling and the coloured points represent the validation data set.

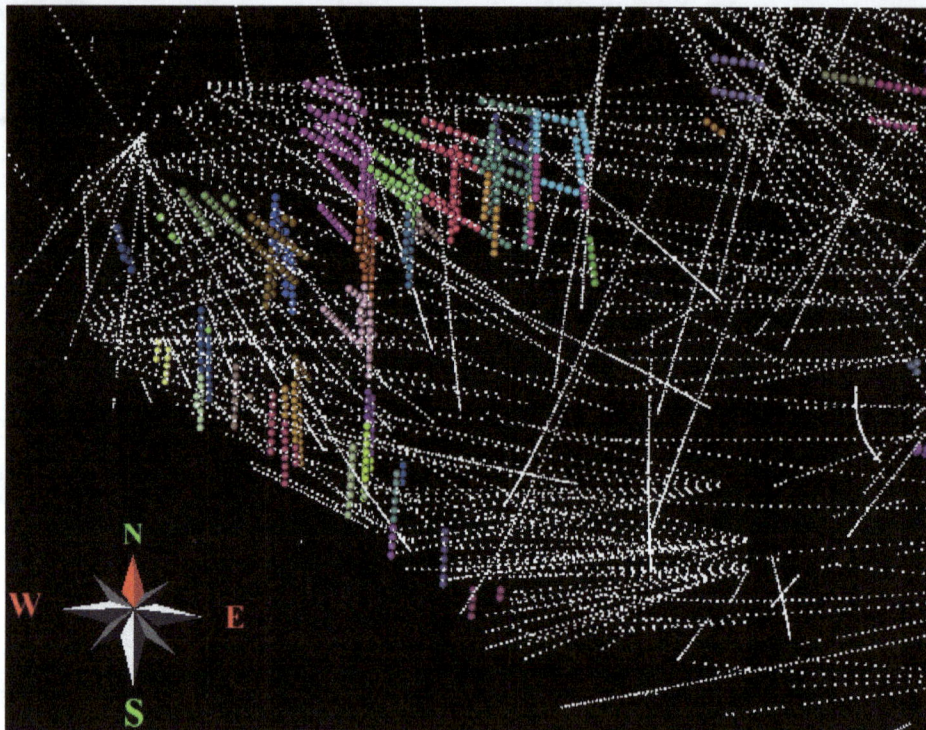

FIG 8 – Data used for modelling (white colour dots) and validation set (coloured points). Colours represent stope ID for the validation set.

To validate the models, the values of all models (and realisations) were queried at the validation set locations, and the average grade was calculated by grouping the values by stope ID. The resulting values were compared with the sample values, which were grouped by stope ID and then averaged. For instance, to compare the LUC model with the data samples, the model was queried at 890 validation sample locations, and the resulting 890 values of grades were grouped and averaged into 41 stope values, which were then compared with the 41 values from the data samples.

The univariate statistics for Cu and Cu:S are presented in Tables 3 and 4, respectively.

TABLE 3

Statistics of the validation data samples and models for Cu grades.

	Data	Simulations minimum	Simulations average	Simulations maximum	OK	LUC
proportion	41	41	41	41	41	41
mean	1.71	1.50	1.73	2.00	1.78	1.97
stdev	0.72	0.58	0.81	1.17	0.66	0.85
min	0.08	0.00	0.13	0.71	0.10	0.04
P25	1.32	0.97	1.28	1.59	1.50	1.63
P50	1.69	1.46	1.71	1.97	1.81	1.95
P75	2.05	1.85	2.10	2.49	2.17	2.45
max	3.86	2.81	4.17	6.46	3.48	4.41

TABLE 4

Statistics of the validation data samples and models for Cu:S ratio.

	Data	Simulations minimum	Simulations average	Simulations maximum	OK	LUC
proportion	41	41	41	41	41	41
mean	1.44	1.27	1.41	1.53	1.50	1.59
stdev	0.56	0.46	0.56	0.83	0.60	0.71
min	0.32	0.13	0.43	0.70	0.46	0.36
P25	1.12	0.81	1.03	1.22	1.13	1.16
P50	1.39	1.20	1.37	1.57	1.53	1.55
P75	1.74	1.54	1.71	2.10	1.81	1.90
max	2.71	2.26	2.89	5.29	3.02	3.65

Tables 3 and 4 indicate that the simulations are able to accurately capture all statistics of the actual values. The OK model also produces values consistent with the observed values, with lower variance in copper grades. However, the LUC model exhibits consistent bias towards high values of copper grades and Cu:S ratio, which are close to or higher than the maximum values observed in all simulations.

In order to further evaluate the performance of the conditional simulation model compared to the OK and LUC models, estimated-true values scatter plots, grade-tonnage curves for Cu and BaS with Cu cut-off, and an accuracy plot are presented in Figure 9 for validation with new drilling samples.

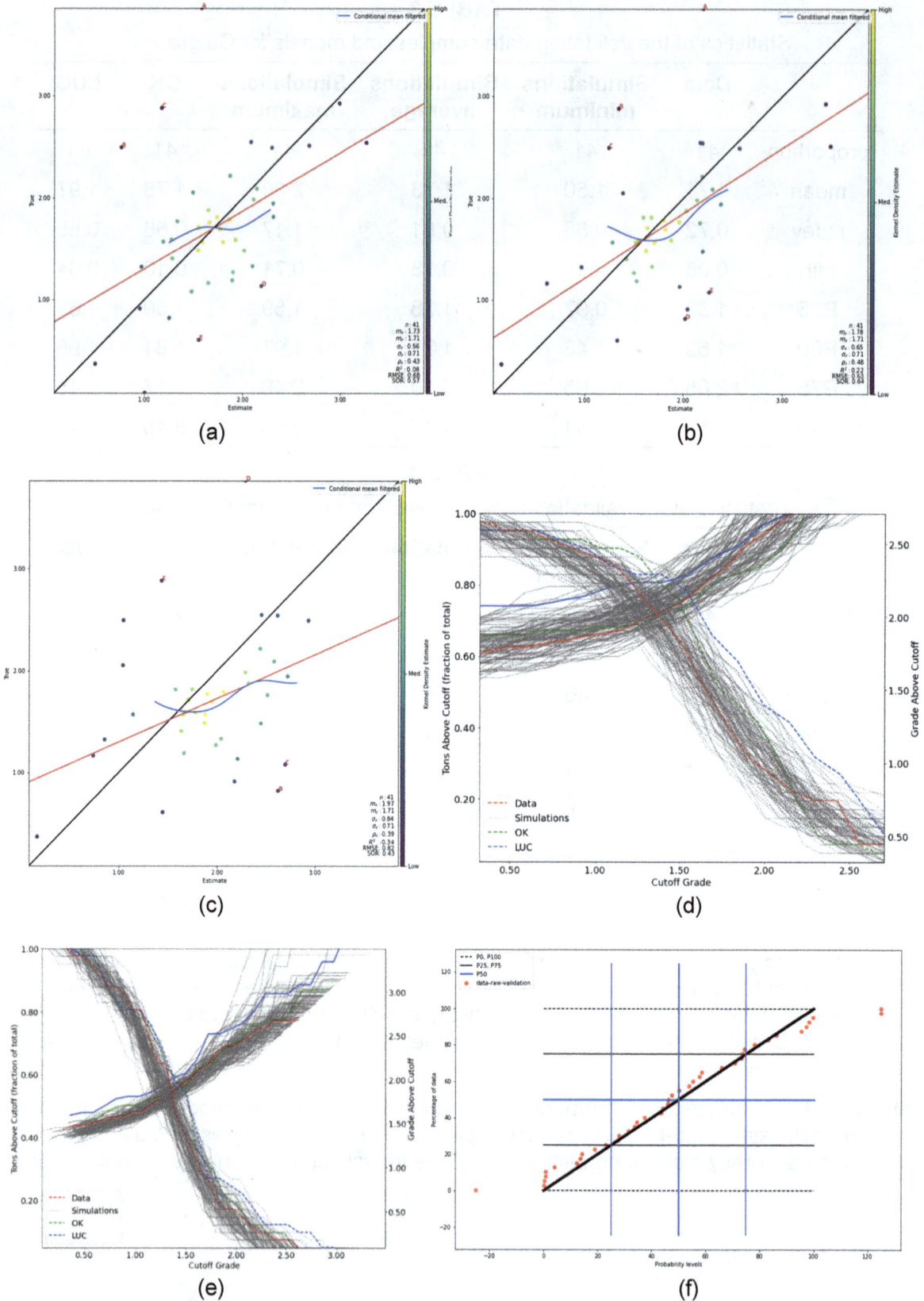

FIG 9 – Validation with new drillings: (a) The estimated-true values scatter plot for the e-type of simulations; (b) The estimated-true values scatter plot plot for OK; (c) The estimated-true values scatter plot for LUC; (d) and (e) Grade-tonnage curves for Cu and Cu:S, respectively, with the red line representing the data, the grey lines representing the simulations, the green line representing the OK model, and the blue line representing the LUC model; (f) The accuracy plot for the simulations, with points outside of the range of (0,100) indicating that not all actual values fall within the predicted confidence intervals.

Figure 9a presents a graph comparing the estimated values for the average of simulations and actual values. This graph bears a strong resemblance to the performance of the OK model, as shown in Figure 9b. Both models exhibit points that align with the 45-degree line (regression slope affected by a limited number of outliers).

The scatter plot depicting estimated-true values for the LUC model (Figure 9c) highlights a considerable conditional bias, with most of the values overstating copper grades. This bias is further substantiated in the copper grade-tonnage curves graph (Figure 9d), where both tonnages and grades (blue lines) for the LUC model are significantly overstated. At a copper cut-off of 1.5 per cent, the tonnages overstate the actual tonnages by approximately 20 per cent, and the copper grades are overestimated by 8 per cent relative to the actual average copper grade. The OK model displays a slight overcall of copper grades at a zero cut-off and overestimated tonnages at 1.5 per cent Cu (~20 per cent relative to actual tonnages).

The simulations demonstrate good concordance in both grade and tonnage for copper, as illustrated in Figure 9d, with an overall minor underestimation of average grades above the cut-off threshold (~3 per cent relative to the actuals). However, the uncertainty range of the copper grade-tonnage curves adequately encompasses the actual grade-tonnage values.

The kriging and simulations concur well with the observed Cu:S ratio in the validation data set (Figure 9e), while the LUC model exhibits an overestimation of approximately 17 per cent of Cu:S ratio compared to the actual ratio observed in the data.

Finally, the accuracy plot shown in Figure 9f represents one of the crucial measures for evaluating the performance of conditional simulations. In an ideal scenario, all data proportion versus probability intervals points should align with the 45-degree line. The OD conditional simulation model demonstrates adequate prediction of uncertainty and local distributions of grades, with only three stopes from the validation set lying outside the simulation-predicted range (red dots outside of [0,100] probabilities) compared to the data statistics and the vast majority plotting along the 1:1 line.

CONCLUSIONS

In conclusion, this paper has demonstrated the benchmarking of the simulation models against other estimation methods of OK and LUC as well as cross validation with older drilling information. The results of this benchmarking process has shown the superiority of the conditional simulation as a method in predicting recoverable resources in OD deposit avoiding the limitations of smoothing and conditional bias that are inherent in conventional deterministic approaches such as OK and LUC.

The statistical analysis and benchmarking conducted in this study have confirmed the accuracy of the method in reproducing complex non-stationary grade distributions. Additionally, the conditional simulation model provides an accurate estimation of uncertainty, which is crucial for making informed decisions in downstream mining processes.

Moving forward, the application of the conditional simulation method will be explored in the context of mine planning optimisation and decision-making under uncertainty conditions.

ACKNOWLEDGEMENTS

The authors would like to thank the management of BHP Olympic Dam for permission to publish this paper. Many geologists from across the Olympic Dam geoscience teams have contributed to material used in the paper. Specific thanks to Kathy Ehrig, Superintendent Geometallurgy, for help completing the paper.

REFERENCES

Abzalov, M, 2006. Localised uniform conditioning (LUC): A new approach for direct modelling of small blocks, *Math Geol*, 38:393–411. <https://doi.org/10.1007/s11004-005-9024-6>

Badenhorst, C, O'Connell, S and Rossi, M, 2016. New approach to recoverable resource modelling: The multivariate case at Olympic Dam, *Geostatistics Valencia*, pp 131–149. <https://doi.org/10.1007/978-3-319-46819-8_9>

Barnett, R M, Manchuk, J G and Deutsch, C V, 2014. Projection pursuit multivariate transformation, *Mathematical Geosciences*, 46:337–359. <https://doi.org/10.1007/s11004-013-9497-7>

Clark, J M and Ehrig, K J, 2019. What controls high-grade copper mineralisation at Olympic Dam?, in *Proceedings of the International Mining Geology Conference 2019*, pp 222–236 (The Australasian Institute of Mining and Metallurgy: Melbourne).

Clark, J M, Ehrig, K, Poznik, N, Cherry, A R, McPhie, J and Kamenetsky, V, 2018. Syn- to post-mineralisation structural dismemberment of the Olympic Dam Fe-oxide Cu-U-Au-Ag deposit, in *Society of Economic Geologists Annual Conference: 2018 – Metals, Mining and Society*.

Clark, J M, Passmore, M and Poznik, N, 2017. Olympic Dam rock quality designation model – an integrated approach, in *Proceedings of the Tenth International Mining Geology Conference 2017*, pp 27–36 (The Australasian Institute of Mining and Metallurgy: Melbourne).

Deutsch, C V and Journel, A G, 1998. *GSLIB: Geostatistical software library and user guide*. Oxford University Press.

Ehrig, K, Kamenetsky, V S, McPhie, J, Cook, N J and Ciobanu, C L, 2017. Olympic Dam iron oxide Cu-U-Au-Ag deposit, in *Australian Ore Deposits* (ed: G N Phillips), pp 601–610 (The Australasian Institute of Mining and Metallurgy: Melbourne).

Ehrig, K, McPhie, J and Kamenetsky, V S, 2012. Geology and Mineralogical Zonation of the Olympic Dam Iron Oxide Cu-U-Au-Ag Deposit, South Australia, *Society of Economic Geologists, Special Publication 16*, pp 237–267.

Emery, X and Arroyo, D, 2018. On a continuous spectral algorithm for simulating non-stationary Gaussian random fields, *Stoch Environ Res Risk Assess,* 32:905–919. <https://doi.org/10.1007/s00477-017-1402-3>

Goovaerts, P, 2001. Geostatistical modelling of uncertainty in soil science, *Geoderma*, 103(1–2):3–26. <https://doi.org/10.1016/S0016-7061(01)00067-2>

Isaaks, E H and Srivastava, R M, 1989. *An Introduction to Applied Geostatistics* (Oxford University Press: New York).

Journel, A G and Huijbregts, C J, 1978. *Mining Geostatistics* (Academic Press: London).

Leuangthong, O and Deutsch, C V, 2004. Transformation of residuals to avoid artifacts in geostatistical modelling with a trend, *Mathematical Geology*, 36(3):287–305.

McLennan, J and Deutsch, C, 2002. Conditional Bias of Geostatistical Simulation for Estimation of Recoverable Reserves, *CIM Bulletin*, 97.

Minniakhmetov, I and Clarke, D, 2023. SBRE framework: Application to Olympic Dam deposit, in *Proceedings of the Mineral Resource Estimation Conference 2023*, pp 285–296 (The Australasian Institute of Mining and Metallurgy: Melbourne).

Pardo-Igúzquiza, E and Chica-Olmo, M, 1993. The Fourier Integral Method: An efficient spectral method for simulation of random fields, *Mathematical Geology*, 25:77–217. <https://doi.org/10.1007/BF00893272>

Qu, J and Deutsch, C V, 2018. Geostatistical simulation with a trend using Gaussian mixture models, *Natural Resources Research*, 27(3):347–363.

Reeve, J S, Cross, K C, Smith, R N and Oreskes, N, 1990. Olympic Dam copper-uranium-gold-silver deposit, in *Geology of the Mineral Deposits of Australia and Papua New Guinea* (ed: F E Hughes), pp 1009–1035 (The Australasian Institute of Mining and Metallurgy: Melbourne).

Silva, D and Deutsch, C V, 2019. Multivariate Categorical Modelling with Hierarchical Truncated Pluri-Gaussian Simulation, *Mathematical Geosciences,* 51:527–552. <https://doi.org/10.1007/s11004-018-09782-5>

Reconciliation of resource estimates

Why I don't believe in reconciliation

S Dunham[1]

1. FAusIMM, Director, SD2 Pty Ltd, Nanango Qld 4615. Email: scott@sd2.com.au

ABSTRACT

For decades we've been focusing on reconciliation as a tool to validate mineral resource estimates. And for decades we've been misleading ourselves.

The concept of reconciliation is simple. Predict, measure, compare and assess the quality of the prediction. This four-step process reflects the ideas of statistical process control and the plan-do-check-act cycle of Shewhart and Deming fame and our intuitive understanding of using system feedback to improve performance. In an ideal world where data is robust and dense this type of feedback loop can be useful. In the mining industry however, there are some fundamental flaws with the conceptual framework.

At almost every stage of the process from pre-estimation through to final production the mining system is plagued by material challenges with data and measurement quality and quantity, system delays, and losses from a supposedly closed system. These issues can and do impact on the benefits sought from reconciliation. Instead of providing feedback leading to improved estimation, reconciliation can result in poor decisions, distracting investigations, and exacerbate the noise in an already noisy system.

Before we can rely on reconciliation as a management tool, we need to address the multiple challenges present in our everyday practices. Some of the most critical issues are addressed in this paper.

INTRODUCTION

How do you know if your mineral resource estimate (MRE) is fit-for-purpose? What makes an estimate 'good' or 'bad', 'excellent' or 'mediocre'? How do you know if your resource classification is appropriate? Is your Measured resource truly measured and is the higher confidence placed on the Measured proportion of your MRE justified? Where, how and why do you distinguish between Measured, Indicated and Inferred? These are the questions that plague a good Competent Person during estimation and reporting. These same questions underpin many components of the mining industry, ranging from finance, investor commitment and business value through to predictions of production performance and cash flow. And yet, despite our dependence on estimation and classification, there is little objective thought or research into the efficacy of the entire system.

Instead, we have reconciliation.

The idea of assessing the quality of one's own estimate is not unique to the mining industry; however, in our industry these self-assessments have come to be a fundamental tool for operators, investors and other stakeholders. This default position is flawed. We rely on reconciliation as an estimation quality metric without considering if it is appropriate or if the information from reconciliation is even meaningful. While the plan-do-check-act cycle (Shewhart, 1939; Deming, 1982) that forms the basis of reconciliation is an often-useful management tool, its success is dependent on several key assumptions: both statistical and system/procedural. In many respects common reconciliation practice does not meet these requirements. Our production data is typically incomplete, error-riddled and low precision. The system being measured is not closed meaning some input/output streams are not measured. We take measurements of the two central variables (tonnes and grade) at different points, with different precision and then assume that they are time equivalent so we can determine metal content. Taken alone any one of these issues would be sufficient to invalidate any reconciliation outcome.

RISK, RESOURCE ESTIMATION AND RECONCILIATION

Understanding estimation risk and the efficacy of prediction is (in the author's opinion) one of the most poorly understood aspects of the resource industry. There are several observations that lead to this conclusion:

- Post-completion analysis of estimates is extremely rare. In the constant flux of production, new sampling and alternative geological interpretation and multi-year operational timespans, objective analysis of the quality of any single estimate (including the risk classification scheme or confidence limits) is under-valued. Even when such an analysis is attempted (eg Elliot *et al*, 1997) the complexity of the operational processes and outcomes can distort the findings.

- We willingly accept the view of the Competent Person without any knowledge of how well they have performed in the past. Indeed, in the absence of post-completion analysis it is almost impossible to know if the Competent Person is indeed competent. Is the author a Competent Person? How do you know? Is the prediction by a resource expert any better than a prediction made at the same time with the same data by someone with no experience in the industry or are we dealing with uncertainty and unstable systems and fooling ourselves into believing we know what is happening?

- We routinely mix and match the concepts of global and local estimation accuracy and precision, assuming that an estimate that is accurate and precise at a global scale can be interpreted as being equally accurate and precise at a local scale. What does 'local' mean? How local is 'local'? Can an estimate be globally accurate and locally inaccurate and still be considered a 'good' estimate? And how do the concepts of global and local map onto the conventional resource classification schema (eg the JORC Code (2004))?

- We are plagued by unrecognised hindsight bias. This bias reveals itself in the way we construct stories to fit the perceived data. When there is some difference between an estimate and 'reality' we jump to conclusions more often fitted to our pre-existing beliefs. This is a human trait (Kahneman, 2012; Taleb, 2019) we all too willingly ignore. Our ability to create stories from observations is much stronger than our willingness to admit we cannot accurately predict the future. If it were otherwise, we would be able to predict the future as well as we rationalise the past.

- When some form of objective analysis is attempted, many of the factors that may drive the apparent difference between actual performance and the estimate are not considered. Critical estimation factors such as the precision and accuracy of the sample data, the uncertainty in the underlying statistical framework (eg the assumption of stationarity and/or the validity of the variogram model) and the impact of support are often ignored in the false assumption that the estimate has a uniform distribution of risk and uncertainty.

- Our estimation practices almost mandate the removal of 'outlier observations'. Any sample deemed too extreme to 'fit' is either capped or removed before we begin predictions. Mapping data that is too difficult to incorporate into the interpretation is discarded. These values, as rare as they may be and as difficult as they may be to manage, are real data. By ignoring them or arbitrarily reducing their influence we are *a priori* imposing limits on the quality of our estimates. Not understanding the data is not a reasonable excuse to ignore it.

- Our measurement systems are incomplete, inaccurate and imprecise. While much emphasis is placed on acquiring the best sample data and understanding the implication of data sparsity during estimation, the same level of emphasis is rarely (if ever) applied to operational production data. The measurements of tonnes and grade are separated; one system component may measure the production tonnage and not the grade of those tonnes whereas a second system component may measure the grade and not the tonnes. We may combine imprecise measures (eg truck counts) with precise measures (weighbridge) and ignore the implications of the differences in precision of the two measures. We also ignore the sample delimitation differences between tonnes and grade, assuming that a measure of weightometer tonnes can be ascribed the grade from a slurry sample as if the two were identically delimited from identical populations.

- In most operations, we assume that the cycle from estimate to production is a closed system. Reconciliation ignores the impact of estimation precision and the consequent ore-waste misallocation. Without robust sampling of the mining waste stream it is not possible to know if ore-grade mineralisation is being sent to the waste dump or being left behind in stopes. Thus, any reconciliation analysis cannot provide meaningful information and we are left to make decisions based on wishful thinking.

- Often, questions of ore quality or material type are ignored or downplayed during estimation and/or reconciliation. While there is an increased focus on these matters (the geometallurgical approach), differences between the theoretical material classification and the operational material classification (driven by a conservative approach) can and does lead to unrecognised reconciliation error.

These factors affect every resource estimate and every reconciliation system. They affect the information we provide to managers and investors and yet they are hidden from view as if to even raise the subject would be a betrayal of the discipline of resource estimation and production forecasting. Before we can rely on the plan-do-check-act cycle or other statistical process control concepts such as 'in control' and 'out of control' systems we need to address these sources of distortion.

In other words, we need to stop relying on the narrative fallacy and adopt the stance of an empirical sceptic. Recognise that we are dealing with uncertainty and therefore act accordingly.

SOME EXAMPLES

These concepts are easily demonstrated using some simple synthetic cases presented in Table 1 and Figure 1. Each case is compared to a production outcome where both the ore stream and waste stream are known exactly, herein called 'reality'. Each of the six cases reconcile at 100 per cent over a 12 month period for tonnes and grade. The differences lie in their month-to-month variation and in the grade of the waste stream.

TABLE 1

Showing 12-month reconciliation results: (a) With only knowledge of the ore stream; and (b) With knowledge of both the waste and ore streams.

A. Ore Stream Knowledge Only

Ore Stream Metal Variance vs Reality						
Month	Case 1	Case 2	Case 3	Case 4	Case 5	Case 6
January	138%	138%	83%	92%	236%	236%
February	45%	45%	27%	61%	69%	69%
March	863%	863%	518%	575%	411%	411%
April	264%	264%	158%	352%	115%	115%
May	396%	396%	237%	264%	88%	88%
June	122%	122%	73%	162%	40%	40%
July	200%	200%	120%	133%	175%	175%
August	155%	155%	93%	207%	79%	79%
Septembe	99%	99%	59%	66%	137%	137%
October	44%	44%	27%	59%	80%	80%
Novembe	69%	69%	41%	46%	82%	82%
Decembe	61%	61%	329%	81%	94%	94%
Total	100%	100%	100%	100%	100%	100%

Variance Statistics						
average	205%	205%	147%	175%	134%	134%
min	44%	44%	27%	46%	40%	40%
max	863%	863%	518%	575%	411%	411%
Range	818%	818%	491%	529%	370%	370%

B. Ore and Waste Stream Knowledge

Ore Plus Waste Metal Variance vs Reality						
Month	Case 1	Case 2	Case 3	Case 4	Case 5	Case 6
January	110%	286%	84%	88%	172%	290%
February	75%	75%	57%	90%	109%	189%
March	157%	251%	119%	126%	73%	129%
April	147%	294%	112%	177%	91%	196%
May	168%	168%	127%	134%	112%	291%
June	80%	80%	61%	96%	45%	102%
July	222%	222%	169%	177%	156%	294%
August	219%	1005%	166%	262%	184%	242%
Septembe	140%	364%	106%	112%	166%	191%
October	62%	62%	47%	74%	100%	216%
Novembe	104%	188%	79%	83%	111%	130%
Decembe	59%	83%	215%	71%	76%	86%
Total	106%	192%	106%	106%	106%	180%

Variance Statistics						
average	129%	257%	112%	124%	116%	196%
min	59%	62%	47%	71%	45%	86%
max	222%	1005%	215%	262%	184%	294%
Range	163%	943%	168%	191%	139%	208%

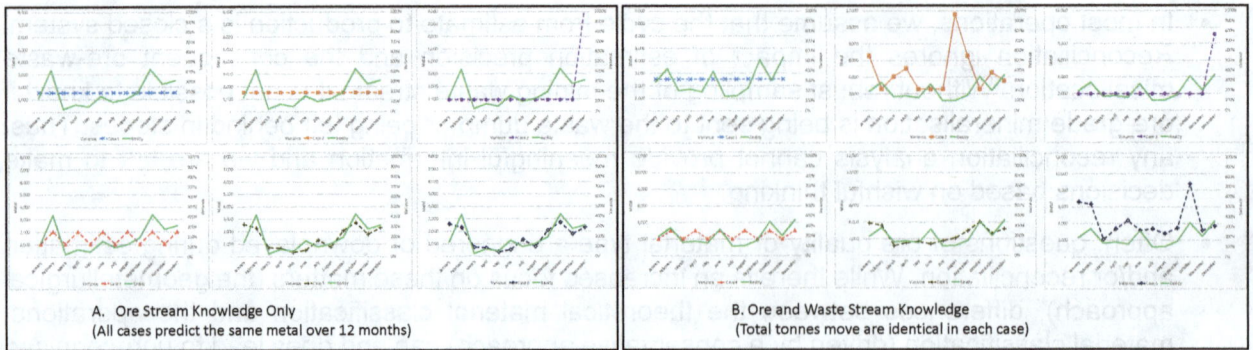

FIG 1 – Showing 12-month reconciliation results: (a) With only knowledge of the ore stream; and (b) With knowledge of both the waste and ore streams.

The cases are:

- Case 1 and Case 2 predicted a constant ore stream with no month-to-month variation. For example, an annualised volume divided into 12 equal amounts.

- Case 3 predicts a constant under-call on grade for 11 months before a spike in grade in the 12 month brings the predicted metal back to agreement with reality over the full 12 months.

- Case 4 predicts constant tonnes but has a month-to-month cyclical error in grade.

- Case 5 and Case 6 both predict the same ore production with variable tonnes and grade each month, typical of an operation with a monthly schedule or monthly reconciliation process.

The results of these six cases are informative. Recall that each case reconciles exactly to 100 per cent of both tonnes and grade over 12 months (Table 1a). Taken at that granularity each estimate could be considered fit-for-purpose. Month-by-month however, the cases show marked differences. The estimate versus reality can be extreme to the point where some form of management intervention would be expected. Cases 5 and 6 look to be 'the best' both graphically (Figure 1a) and in terms of the average difference to reality (134 per cent).

All of these estimates would be viewed as not meeting the needs of the operation and no doubt some form of root-cause analysis would ensue, followed by changes (eg the application of factors) all in the attempt to improve the quality of the estimate. But are these estimates wrong? The above results are based solely on examination of the ore stream. When the waste stream is considered and assuming both tonnes and grade are known do things look different?

The first observation from Table 1b is that each estimate predicts more metal than reality. The differences ranging from 106 per cent to 192 per cent. The second observation is that the average variation decreases and the range of month-to-month variation also decreases. Of the six cases presented, Cases 1, 3, 4 and 5 all perform well over 12 months. The best performing case on-average is Case 3 – the example that under-predicts constantly for 11 months with a large grade spike in month 12. In Cases 2 and 6 the grade of the waste stream is much higher (0.60 and 0.56) indicating a large ore-waste misallocation problem.

With knowledge of the waste stream tonnes, and especially grade, the apparent meaning of the reconciliation performance changes. Case 3 looks much more reasonable, Cases 2 and 6 looks unacceptable whereas before they were the better performing examples.

While these examples might seem trivial, they highlight a fundamental flaw in common reconciliation practice. Without accurate knowledge of the *waste* stream as well as the *ore* stream using reconciliation performance as a measure of the performance of an estimate is not appropriate and it is extremely rare for such knowledge to exist. We do not routinely sample the waste stream. We do not focus as closely on the tonnes of waste, their spatial location or variability. We do not account for the known occurrence of ore-waste misallocation *even though this concept is central to our estimation practices*.

WHAT DOES IT ALL MEAN ANYWAY?

Putting aside the flaw of incomplete knowledge, let's consider what those reconciliation results mean and how they may be interpreted. There are several questions:

- What performance metric should be used to determine an estimate is not fit-for-purpose and at what threshold should that metric trigger action?

- How is the threshold determined?

- Does the threshold reflect some capability of the system or is it a heuristic that may be unachievable for a given operation with a certain geology, production system and measurement capability?

- If there is a 'reconciliation error', how should that error be addressed?

Let's look at these individually.

In the author's experience, the most common approach to reconciliation is tracking of resource estimate versus grade control estimate versus ore treatment plant production. The variation between each is calculated as a ratio, tabulated and plotted on a time-series chart each month. Individual monthly results are then considered and when the variation is deemed beyond some reasonable expectation an explanation for the variation is reported (the narrative fallacy). The most common threshold for triggering the need for an explanation is a result outside a ±10 per cent margin.

The reliance on a ±10 per cent margin is an example of a heuristic that has no real basis. The author has not seen any logical rationale for the choice of 10 per cent versus 9 per cent, 11 per cent, 15 per cent or 5 per cent. Nor has the author seen any consideration of the difference between ±10 per cent in a higher-grade region versus a lower grade region – the proportional effect is rarely considered during reconciliation.

When queried, the common explanation for 10 per cent is that it 'feels right' or it is the degree of variation where the business's performance starts to be affected. Both statements have little to no evidence to support them. Furthermore, choosing a ±10 per cent threshold does not allow for the differences between different types of deposit or different types of operation or different management capability. Should we expect a sublevel cave operation and a selectively mined open pit to both achieve ±10 per cent? The theoretical selectivity is different, implying the ore-waste misallocation is different and therefore the reconciliation performance will be different.

The second aspect of this reconciliation process is the provision of an explanation. Some event, sequence of events or some characteristic of either the estimate or the production system is identified as causing the variation and, in some cases, some corrective action is taken. The provision of an explanation when examining past events is a well-known flaw in human thinking. We are a storytelling species. As discussed in Taleb (2019) and Kahneman, Sibony and Sunstein (2021), our ability to link causal events in hindsight is much greater than our ability to forecast the implication of events into the future. The past seems predictable when seen from the future but the reverse is not true. Considering the uncertainty in our estimates, the variability in our production processes, the errors in our measurement systems and the absence of a closed system, any explanation of poor reconciliation is unlikely to represent the true situation. Furthermore, taking action to address some assumed 'root-cause' when we have incomplete knowledge seems more likely to increase the noise in the system than to address any true shortcoming.

An example of adding noise to the system is the all-too-common application of reconciliation factors.

The global factor problem

When there is some apparent bias between an estimate and the perceived reality (often the ore treatment plant) it is common to see some sort of adjustment applied to the estimate. This usually happens between the mineral resource estimate and the ore reserve estimate or between some form of short-term estimate and the mine plan. These adjustments are typically a multiplicative factor and they are usually applied to the estimated grade. This approach is also problematic. It is an overly simplistic 'solution' to a complex question and it ignores both the available information and some fundamental mathematics. It seems strange that we are willing to invest weeks or months of effort to perfect an estimate and then override that same estimate with a naïve modification based on a flawed view of the estimation-planning-operation framework.

Adjustment factors are usually applied globally. That is, every part of the deposit is treated identically. This assumes that any under (or over) performance is uniform. It is a rare deposit where the grade is uniform or even the variance of the grade is uniform. Applying a uniform adjustment ignores this reality. In the rare cases where one particular material type, estimation domain or location is identified as causing estimation problems, it is even less appropriate to use a uniform adjustment. Significantly, the variation in geometry between the domain and the mining system is not addressed by a single factor; however, this is a central contributor to ore-waste misallocation, ore loss and dilution.

A simple adjustment factor ignores complexity and invalidates the precision implied by the original estimate. (The author notes that the assumed estimation precision is likely underestimated. Conventional estimates using weighted-averages in highly-skewed distributions will rarely exhibit the same degree of variability as the real world.)

Statistics and factors

Beyond the spatial challenges associated with applying an adjustment factor, there are some basic statistical problems. These problems are a direct consequence of our lack of knowledge of the properties of the waste stream (ie because we are not working in a closed system). Adjustment factors are conventionally applied to the estimate grade when there is some perceived problem with the estimate. But what factor and what approach?

The average grade above a cut-off can be changed in one of three ways:

1. The application of a mean shift (ie adding or subtracting a constant).
2. Changing the spread (dispersion) of the estimate.
3. Some combination of the two – this is how a multiplication factor works.

The metal above cut-off is a function of both the mean and the variance. This is easy to envisage when dealing with normal (Gaussian) distributions (Figure 2). When dealing with the positively skewed grade distributions commonly found in mineralised systems the results are less easily understood (Figure 3).

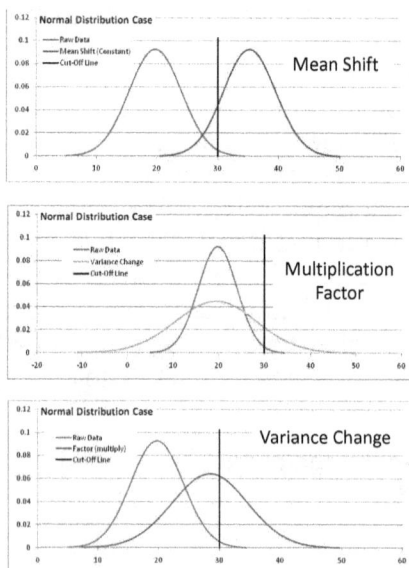

The average grade above cut-off is the same in each of these examples. Different factoring approach applied from the same starting distribution (blue curve)

Which method is correct?

FIG 2 – The impact of different factoring approaches on a Gaussian distribution.

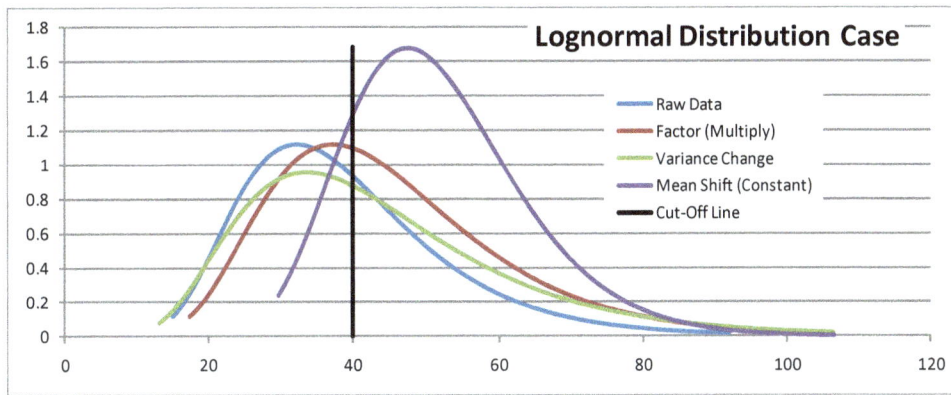

FIG 3 – Mean shift, variance change and multiplicative factors have a less intuitive impact on skewed distributions.

Applying a multiplicative factor implies that the original estimate is both biased and that the variance is incorrect. The subsequent adjustment may exacerbate ore-waste misallocation errors leading to increased reconciliation problems which in turn can lead to changes in the model and estimate. The entire system becomes unstable, estimation precision decreases resulting in a downward cycle where every adjustment or every attempt to improve the estimate simply results in poorer predictions.

IS THERE A SOLUTION?

The challenges outlined in this paper are not unknown, they are not hard to recognise and their impact on common reconciliation practice is large. And yet we still rely on reconciliation as a primary management control tool. Why? The idea of measuring performance and using that measure to improve satisfies our desire to believe we are in control. It satisfies our need to link cause and effect. This need to understand is pervasive. The 'narrative fallacy' (Taleb, 2019) is so strong it drives an entire genre across multiple industries; that of the commentator. From financial commentators ascribing market moves to some external event one day only to attribute a movement in the opposite direction to the same event on the next day, to sports commentators whose very careers are dependent on creating drama and speculation about the performance of athletes and teams, attaching weight to trivial deviations from expected performance and creating a narrative from speculation.

Is the practice of reconciliation the same? Are we no wiser than sports commentators, hiding behind our erudite explanations when any causal event is unknowable in the face of uncertainty?

This is a challenging problem. We are dealing with complex, non-linear systems with the potential for multiple aspects to cause the same apparent deviations. In other words, we are working in a field where the forward prediction and backwards explanation are asymmetrical. We have a reverse engineering problem of the kind outlined in Taleb (2019); it is easier to predict how an ice cube would melt into a puddle than, looking at a puddle, to guess the shape of the ice cube that may have caused it.

In many respects we cannot solve the problems inherent in a simplistic view of reconciliation. We can however, improve our understanding of what the limited, incomplete data is telling us, how we use that understanding and how we make decisions in the face of uncertainty. The following approaches can be useful, with caution:

- Stop thinking of reconciliation as some hard fact. It is at best fuzzy and at worse non-factual. Taking action to 'fix' a model when there is a reconciliation 'problem' can decrease system performance through out-of-control feedback loops.

- Recognise the limits of the system and make decisions accordingly. Do not take radical steps to improve where there is a perceived problem without verifying and validating. Falsify your theory, design experiments to disprove your reconciliation-based narrative.

- Look for differences that can be explained by precision instead of dogmatically focusing on accuracy.

- Consider trends carefully, remembering that an apparent bias in an estimate may be due to some aspect of the reconciliation and measurement system, it may be due to the timing of the outcome. (The author has direct knowledge of a 'reconciliation problem' that was entirely the result of not allowing for shipping adjustments in concentrate grade.)

- Get better at measuring operational performance. Fill in the gaps, measure more product streams, match the measurement of tonnes and grade/quality parcels.

- Instead of using heuristic performance limits, develop predictions of system performance using simulation tools. And remember the simulations are models – understand the implicit assumptions they are based upon.

- Beware of assuming that past performance is a good predictor of future outcomes. We recognise this issue during estimation (or should recognise) however it is often forgotten during production. A change in geology, a change in domain, a change in the average grade or a change in the *grade variability* can all invalidate any assumption that the current reconciliation results apply to next month or next year's production.

There are solutions, they lie in understanding our collective cognitive biases, thinking more and reacting less. They depend on improved measurement, less Gaussian-based reasoning and more acceptance of variability and noise. Mining is not immune to the black swan problem and the confusion of absence of evidence for evidence of absence (the round-trip fallacy).

We know the model is wrong. We have yet to recognise that our perception of reality is likewise a model, coloured by our memories, experiences, perceptions and by our mental processes.

CONCLUSION

There is a problem measuring and quantifying performance in the mining industry. The historically poor quality of production measurement systems combined with systemic deficiencies in tracking and quantifying the mine waste stream invalidate any analysis based on conventional reconciliation practice. The unacknowledged presence of ore-waste misallocation at the time of production results in ill-informed decision-making at all levels of the industry. From a resource estimation perspective incorrect interpretation of reconciliation outcomes can drive changes to future models and estimates leading to a perpetual cycle of investigations into system noise. Furthermore, relying on reconciliation to 'calibrate' or 'validate' an estimate will inevitably increase the ore-waste misallocation and cause unnecessary operational losses.

This problem is not new or novel. In the author's experience it is well recognised but ignored, with the implications underestimated and the value losses accepted as a normal part of the industry operating model. Outside of the resource estimation discipline the concept of ore-waste misallocation is poorly understood and its implications untested. Instead of addressing the data challenge directly managers, investors and other stakeholders rely on the mantra of 'the model is wrong'.

We know the model is wrong. What we do not know is how wrong. Until there is routine and rigorous tracking of both the ore and waste streams, with both tonnage and grade measurements from the same sample populations, until our reconciliation practices include investigations of selectivity, model-machine interaction and other quantifiable known variables, reconciliation will remain a futile exercise.

And beyond these known-unknowns lie the true challenges with reconciliation. The limits to predictability, the black swans, the unknown-unknowns. Evaluating our estimates without acknowledging out-of-context problems is pointless.

No, I do not believe in reconciliation.

REFERENCES

Deming, W E, 1982. *Out of the Crisis*, Center for Advanced Engineering Study, Massachusetts Institute of Technology, Cambridge, Massachusetts.

Elliot, S M, Snowden, D V, Bywater, A, Standing, C A and Ryba, A, 1997. Reconciliaton of the McKinnons Gold Deposit, Cobar, New South Wales, in *Proceedings of the Third International Mining Geology Conference*, pp 113–122 (The Australasian Institute of Mining and Metallurgy: Melbourne).

JORC, 2004. Australasian Code for Reporting of Exploration Results, Mineral Resources and Ore Reserves (The JORC Code) [online]. Available from: <http://www.jorc.org> (The Joint Ore Reserves Committee of The Australasian Institute of Mining and Metallurgy, Australian Institute of Geoscientists and Minerals Council of Australia).

Kahneman, D, Sibony, O and Sunstein, C, 2021. *Noise* (William Collins).

Kahneman, D, 2012. *Thinking, Fast and Slow* (Penguin Press: Sydney).

Shewhart, W A, 1939. *Statistical Method from the Viewpoint of Quality Control* (Graduate School, Department of Agriculture, Washington).

Taleb, N N, 2019. *The Black Swan: The Impact of the Highly Improbable*, second edition, with a new section: 'On Robustness and Fragility' (Random House Audio).

AUTHOR INDEX